10th Summer School on Theoretical Physics 2009

Symmetry and Structural Properties of Condensed Matter

Journal of Physics: Conference Series
Volume 213

Myczkowce, Poland
2-9 September 2009

ISBN: 978-1-61738-264-2

Printed from e-media with permission by:

Curran Associates, Inc.
57 Morehouse Lane
Red Hook, NY 12571

Some format issues inherent in the e-media version may also appear in this print version.

Printed by Curran Associates, Inc. (2010)

Institute of Physics
Dirac House, Temple Back
Bristol BS1 6BE UK

Phone: 44 1 17 929 7481
Fax: 44 1 17 920 0979

techtracking@iop.org

Additional copies of this publication are available from:

Curran Associates, Inc.
57 Morehouse Lane
Red Hook, NY 12571 USA
Phone: 845-758-0400
Fax: 845-758-2634
Email: curran@proceedings.com
Web: www.proceedings.com

TABLE OF CONTENTS

INFORMATION PROCESSING, ENTANGLEMENT AND TENSOR CALCULUS

INTEGRABLE MODELS AND UNITARY SYMMETRY

FINITE SYSTEMS AND NANOPHYSICS

PREFACE

This volume contains the Proceedings of the Tenth Summer School on Theoretical Physics under the banner title 'Symmetry and Structural Properties of Condensed Matter' (SSPCM 2009). The School was organized by Rzeszow University of Technology, Poland, in cooperation with AGH University of Science and Technology, Cracow, Poland, and took place on 2–9 September 2009 in Myczkowce, Poland. With this meeting we have reached the round number ten of the series of biannual SSPCM schools, which started in 1990 and were focused on some advanced mathematical methods of condensed matter physics. The first five meetings were held in Zajaczkowo near Poznan, under the auspices of The Institute of Physics of Adam Mickiewicz University, and the last five in Myczkowce near Rzeszów, in the south-eastern part of Poland. Within these two decades several young workers who started at *kindergarten* lectures at SSPCM, have now reached their PhD degrees, professorships and authority.

Proceedings of the first seven SSPCM meetings were published as separate volumes by World Scientific, and the last two as volumes 30 and 104 of *Journal of Physics: Conference Series*. The present meeting is also the third of the last schools which put the emphasis on quantum informatics.

The main topics of our jubilee SSPCM'09 are the following:

- Information processing, entanglement, and tensor calculus,
- Integrable models and unitary symmetry,
- Finite systems and nanophysics.

The Proceedings are divided into three parts accordingly. The school gathered together 55 participants from seven countries and several scientific centers in Poland, accommodating again advanced research with young collaborators and students.

Acknowledgements

The Organizing Committee would like to express its gratitude to all participants for their many activities during the School and for creating a friendly and inspiring atmosphere within our SSPCM society. Special thanks are due to all lecturers for preparing and presenting their talks and also for several valuable discussion. We would like to express our thanks to all those who prepared manuscripts, to all referees who spent their time to significantly improve our elaborations, and thus the quality of this volume, to all members of our

International Advisory Committee and to chairmen for their polite and efficient leading of sessions.

It is our pleasure to express special gratitude to The Nicolas C Metropolis Mathematics Foundation (Los Alamos, USA) for substantial financial support of our three last SSPCM schools. Indeed, it provided an essential basis for performing these meetings. We direct this gratitude to Professor James D Louck, the President of this Foundation. Also, we were happy to hear from Professor Louck that this series of schools has had an important influence on his activities on application of unitary groups in physics. It has resulted in a new monograph: James D Louck, 'Unitary Symmetry and Combinatorics', World Scientific, New Jersey 2008, which was presented at our jubilee meeting.

Tadeusz Lulek, Andrzej Wal and Barbara Lulek

All papers published in this volume of *Journal of Physics: Conference Series* have been peer reviewed through processes administered by the proceedings Editors. Reviews were conducted by expert referees to the professional and scientific standards expected of a proceedings journal published by IOP Publishing.

The growth and zeros of Bargmann functions

A Vourdas

Department of Computing, University of Bradford, Bradford BD7 1DP, UK

E-mail: a.vourdas@Bradford.ac.uk

Abstract. The Bargmann formalism is presented with emphasis on the growth and zeros of Bargmann functions. Hadamard's theorem is expressed in a physical language and it is used to construct quantum states, with given zeros in their Bargmann function. Various examples are discussed. The time evolution of Bargmann functions and the motion of their zeros, for various Hamiltonians, is studied.

1. Introduction

Various analytic representations have been used in quantum mechanics[1, 2]. The Bargmann function in the complex plane [3] has been used in conjuction with the coherent states formalism. There are deep theorems which relate the growth of entire functions and the density of their zeros[4]. In the context of Bargmann functions, they lead to results on the over-completeness or under-completeness of sequences of coherent states[5, 6, 7, 8]. Refs [9, 10, 11, 12] have studied the time evolution of quantum states for various Hamiltonians, through the paths of the zeros of their Bargmann functions (sometimes they are called 'Husimi zeros' or 'stellar representation').

In this paper we present the Bargmann formalism with emphasis on Hadamard's theorem which constructs entire functions with a given set of zeros. We express Hadamard's theorem for Bargmann functions in a physical language and construct quantum states whose Bargmann function has a a given set of zeros (with density smaller than a certain value)[8]. The time evolution of Bargmann functions and the motion of their zeros, for various Hamiltonians, is studied. During the time evolution some zeros might cease to exist or new zeros might appear (we call this 'creation and annihilation of zeros').

The aim of the paper is to present these ideas, which involve advanced topics from the theory of analytic functions, in a pedagogical and physical language. The general theory is complemented with several examples.

2. Bargmann functions

We consider a harmonic oscillator and let $a^\dagger, a$ be the creation and annihilation operators. $D(z)$ are displacement operators, $|z\rangle$ are coherent states and $|N\rangle$ are number eigenstates:

$$|z\rangle = D(z)|0\rangle = \exp\left(-\frac{1}{2}|z|^2\right)\sum_{N=0}^{\infty} z^N (N!)^{-1/2}|N\rangle; \qquad D(z) = \exp(za^\dagger - z^*a) \tag{1}$$

Let $|f\rangle$ be an arbitrary normalized ket state

$$|f\rangle = \sum_{N=0}^{\infty} f_N |N\rangle; \qquad \sum_{N=0}^{\infty} |f_N|^2 = 1 \tag{2}$$

We use the notation

$$\langle f| = \sum_{N=0}^{\infty} f_N^* \langle N|; \qquad |f^*\rangle = \sum_{N=0}^{\infty} f_N^* |N\rangle; \qquad \langle f^*| = \sum_{N=0}^{\infty} f_N \langle N| \tag{3}$$

In the Bargmann representation the state $|f\rangle$ is represented by the entire function

$$f(z) = \exp\left(\frac{1}{2}|z|^2\right)\langle f^*|z\rangle = \exp\left(\frac{1}{2}|z|^2\right)\langle z^*|f\rangle = \sum_{N=0}^{\infty} f_N z^N (N!)^{-1/2} \tag{4}$$

The scalar product is given by

$$\langle f|g\rangle = \int_{\mathbb{C}} [f(z)]^* g(z) \exp(-|z|^2) d\mu(z) = \sum_N f_N^* g_N; \qquad d\mu(z) = \frac{dz_R dz_I}{\pi} \tag{5}$$

The indices R, I in a complex number, indicate its real and imaginary parts, correspondingly. As examples, we give the Bargmann function for number eigenstates and for coherent states:

$$|N\rangle \;\rightarrow\; f(z) = \frac{z^N}{N!}; \qquad |\zeta\rangle \;\rightarrow\; f(z) = \exp\left[\zeta z - \frac{|\zeta|^2}{2}\right] \tag{6}$$

We also consider the squeezed state:

$$\begin{aligned} |w; r, \theta, \lambda\rangle_{\rm sq} &= S(r,\theta,\lambda)|w\rangle \\ S(r,\theta,\lambda) &= \exp\left[-\frac{1}{4} r e^{-i\theta}(a^\dagger)^2 + \frac{1}{4} r e^{i\theta} a^2\right] \exp[i\lambda a^\dagger a] \end{aligned} \tag{7}$$

Its Bargmann function is [8, 2]

$$\begin{aligned} f(z) &= (1-|\alpha|^2)^{1/4} \exp\left(\frac{\alpha}{2} z^2 + \beta z + \epsilon\right); \qquad \alpha = -\tanh\left(\frac{1}{2} r\right) e^{-i\theta} \\ \beta &= w e^{i\lambda}(1-|\alpha|^2)^{1/2}; \qquad \epsilon = -\frac{\alpha^*}{2} w^2 e^{2i\lambda} - \frac{1}{2}|w|^2, \end{aligned} \tag{8}$$

An operator Θ is represented with the following function

$$\Theta \;\rightarrow\; \Theta(z, \zeta^*) = \exp\left[\frac{1}{2}(|\zeta|^2 + |z|^2)\right] \langle z^*|\Theta|\zeta^*\rangle = \sum_{N,M} z^M \Theta_{MN} \zeta_N^* \tag{9}$$

where $\Theta_{MN} = \langle M|\Theta|N\rangle$. It acts on the Bargmann function $f(z)$ of the state $|f\rangle$ as follows:

$$\Theta|f\rangle \;\rightarrow\; \int_{\mathbb{C}} d\mu(z)\; \exp(-|\zeta|^2)\; \Theta(z, \zeta^*) f(z) \tag{10}$$

As examples, we consider the creation and annihilation operators for which

$$a^\dagger \;\rightarrow\; z \exp(z\zeta^*); \qquad a \;\rightarrow\; \zeta^* \exp(z\zeta^*) \tag{11}$$

This can be proven using Eq.(9). In addition to this 'integral representation', the creation and annihilation operators can also be represented with the differential operators

$$a^\dagger \;\rightarrow\; z; \qquad a \;\rightarrow\; \partial_z \tag{12}$$

This can be proved by acting with these operators on the Bargmann function $z^N/(N!)$ of the number eigenstate $|N\rangle$ and then confirm that $a^\dagger|N\rangle = (N+1)^{1/2}|N+1\rangle$ and $a|N\rangle = N^{1/2}|N-1\rangle$.

3. Growth and density of zeros of Bargmann functions

The growth of an analytic function $f(z)$ is described with two non-negative numbers, the order ρ and the type σ. We refer to the literature [4] for the formal definitions. For our purposes let

$$M(R) = \max_{R=|z|}(|f(z)|) \tag{13}$$

Then $M \approx \exp(\sigma R^{\rho})$ as $R \to \infty$.

A growth (ρ, σ) is smaller than (ρ', σ') if $\rho < \rho'$ (in which case we do not need to compare the types) or if both $\rho = \rho'$ and $\sigma < \sigma'$. Below we use the notation $\mathcal{H}(\rho, \sigma)$ for the space of functions with growth smaller than (ρ, σ).

Convergence of the scalar product of Eq.(5), implies that Bargmann functions should have growth smaller than $(\rho = 2, \sigma = 1/2)$ (i.e., the Bargmann space is $\mathcal{H}(2, 1/2)$).

The zeros of analytic functions are isolated. We consider a general analytic function $f(z)$ with an infinite number of zeros ζ_N (i.e., $f(\zeta_N) = 0$), which we label them as:

$$0 < |\zeta_1| \le |\zeta_2| \le |\zeta_3| \le \ldots; \tag{14}$$

Let $n(R)$ be the number of terms of this sequence within the circle $|z| = R$. The density of the sequence $\{\zeta_N\}$ is described with two non-negative numbers, the convergence exponent η and δ. We refer to the literature [4] for the formal definitions, but for our purposes it is sufficient to say that $n(R) \approx \delta R^{\eta}$ as $R \to \infty$. It is known that η is the infimum of positive numbers κ such that

$$\sum_{N=1}^{\infty} \frac{1}{|\zeta_N|^{\kappa}} < \infty \tag{15}$$

We say that the density (η, δ) is smaller than the density (η', δ') if $\eta < \eta'$ (in which case we do not compare δ, δ') or if both $\eta = \eta'$ and $\delta < \delta'$.

For general analytic functions it is known that [4] $\eta < \rho$ or both $\eta = \rho$ and $\delta \le \sigma\rho$. Bargmann functions have smaller growth than $(\rho = 2, \sigma = 1/2)$ and this implies that $\eta < 2$ or both $\eta = 2$ and $\delta \le 1$. Therefore the density of zeros of Bargmann functions is smaller that $(\eta = 2, \delta = 1)$. The zeros of Bargmann functions have an interesting physical meaning. Eq.(4) shows that if ζ is a zero of $f(z)$, then the coherent state $|\zeta\rangle$ is orthogonal to $|f^*\rangle$.

4. Hadamard's theorem for Bargmann functions

The Weierstrass factors are defined as

$$E(z, 0) = 1 - z; \qquad E(z, p) = (1 - z)\exp\left[z + \frac{z^2}{2} + \ldots + \frac{z^p}{p}\right] \tag{16}$$

where p is a positive integer. The exponential terms in the Weierstrass factors in Eq.(16) are important for the convergence of the infinite products below.

The Weierstrass canonical product corresponding to the sequence $\{\zeta_N\}$ is

$$\Pi_p(z) = z^m \prod_{N=1}^{\infty} E\left(\frac{z}{\zeta_N}, p\right). \tag{17}$$

It converges absolutely and it has growth with order ρ equal to the convergence exponent η of the set of zeros $\{\zeta_N\}$.

Let $f(z)$ be an entire function of order ρ with zeros $\{\zeta_N\}$. Hadamard's theorem states that an entire function of order ρ can be factorised as

$$f(z) = z^m \left[\prod_{N=1}^{\infty} E\left(\frac{z}{\zeta_N}, p\right)\right] \exp[Q_q(z)] \tag{18}$$

Here $p \leq \rho$, $Q_q(z)$ is a polynomial of degree $q \leq \rho$ and m is the multiplicity of the zero at the origin. For Bargmann functions $\rho \leq 2$ and therefore $p, q = 0, 1, 2$.

We discuss the physical meaning of Eq.(18), starting with the term $\exp[Q_q(z)]$. Using the notation $|Q_q\rangle$ for the state represented by the Bargmann function $\exp[Q_q(z)]$, we get (up to a normalisation constant)

$$\begin{aligned} \exp[Q_0(z)] &= \exp(u) &\rightarrow\quad& |Q_0\rangle = |0\rangle \\ \exp[Q_1(z)] &= \exp(Az+u) &\rightarrow\quad& |Q_1\rangle = |A\rangle_{\rm coh} \\ \exp[Q_2(z)] &= \exp\left(\frac{\alpha}{2}z^2+\beta z+\epsilon\right) &\rightarrow\quad& |Q_2\rangle = |A; r, \theta, \lambda\rangle_{\rm sq} \end{aligned} \tag{19}$$

where the relation between α, β, ϵ and A, r, θ, λ is given in Eq.(8). Therefore, the $\exp[Q_q(z)]$ with $q = 0, 1, 2$ is the Bargmann function for the vacuum state, a coherent state and a squeezed state, correspondingly.

We next discuss the physical meaning of the Weierstrass terms. Using Eq.(12) we express the Weierstrass factors in terms of creation operators as

$$E\left(\frac{a^\dagger}{\zeta}, p\right) = \left(1 - \frac{a^\dagger}{\zeta}\right) \exp\left[\frac{a^\dagger}{\zeta} + \frac{(a^\dagger)^2}{2\zeta^2} + \ldots + \frac{(a^\dagger)^p}{p\zeta^p}\right]; \qquad p = 0, 1, 2 \tag{20}$$

and the state corresponding to Eq.(18) as

$$|f\rangle = \mathcal{N}(a^\dagger)^m \left[\prod_{N=1}^{\infty} E\left(\frac{a^\dagger}{\zeta_N}, p\right)\right] |Q_q\rangle; \qquad p, q = 0, 1, 2 \tag{21}$$

where $\mathcal{N}$ is a normalization constant. Given any sequence of complex numbers $\{\zeta_N\}$ with density smaller that $(\eta = 2, \delta = 1)$ we construct with Eq.(21) a state whose Bargmann function has these zeros. Below we discuss several examples.

4.1. Example: the Mittag-Leffler function

An example of a state whose Bargmann function has growth with a given order ρ and given type σ (which may or may not be integers) is [8, 2]

$$|\rho, \sigma\rangle = \sum_{N=0}^{\infty} f_N |N\rangle; \qquad f_N = \mathcal{N} \frac{e^{iN\theta}\sigma^{N/\rho}(N!)^{1/2}}{\Gamma(\frac{N}{\rho}+1)} \tag{22}$$

where $\mathcal{N}$ is the normalisation constant

$$\mathcal{N} = \left[\sum_{N=0}^{\infty} \frac{\sigma^{2N/\rho} N!}{[\Gamma(\frac{N}{\rho}+1)]^2}\right]^{-\frac{1}{2}} \tag{23}$$

$\mathcal{N}$ is finite when $0 \leq \rho < 2$; and also when $\rho = 2$ and $\sigma < \frac{1}{2}$. The Bargmann function of this state is $\mathcal{N} E_{1/\rho}(e^{i\theta}\sigma^{1/\rho}z)$ where $E_\nu(\zeta)$ is the Mittag-Leffler function (e.g.,[13]).

4.2. Example: the Riemann $\xi(Az)$ function

As an example we consider the Riemann $\xi(z)$ function [14, 15]

$$\xi(z) = \frac{1}{2} z(z-1)\pi^{-z/2}\Gamma\left(\frac{z}{2}\right)\zeta(z). \tag{24}$$

where $\zeta(z)$ is the Riemann zeta function

$$\zeta(z) = \sum_{N=1}^{\infty} \frac{1}{N^z}. \tag{25}$$

The Riemann zeta function has been used in quantum mechanics in [16, 17].

$\xi(z)$ is an analytic function in the complex z-plane and satifies the functional relation

$$\xi(z) = \xi(1-z). \tag{26}$$

Therefore if w_n is a zero of $\xi(z)$ then $1-w_n$ is also a zero. It can be proved that all zeros are in the strip $0 \le \Re(w_n) \le 1$. The Riemann hypothesis is that all zeros are in the line $\Re(w_n) = 1/2$.

The function $\xi(z)$ can be written in the product form[14, 15]

$$\xi(z) = \xi(0) \prod_{w_n} \left(1 - \frac{z}{w_n}\right) = \xi(0) \prod_{\Im(w_n)>0} \left[1 - \frac{z(1-z)}{w_n(1-w_n)}\right] \tag{27}$$

The order of its growth is $\rho = 1$ and therefore $\xi(Az)$ qualifies as a Bargmann function. The corresponding ket state is

$$|A\rangle_{\mathrm{Rie}} = \mathcal{N}\xi(0) \prod_{w_n} \left(1 - \frac{a^\dagger}{A^{-1}w_n}\right)|0\rangle = \mathcal{N}\xi(0) \prod_{\Im(w_n)>0} \left[1 - \frac{Aa^\dagger - A^2(a^\dagger)^2}{w_n(1-w_n)}\right]|0\rangle \tag{28}$$

where $\mathcal{N}$ is a normalization constant.

4.3. *Example: the function* $\sin(\pi Az)$ *(Schrödinger cat)*

The entire function $\sin(\pi Az)$, where A is a complex number ($A \neq 0$), is factorized as [4, 18]

$$\sin(\pi Az) = \pi Az \prod_{N=1}^{\infty} \left(1 - \frac{z^2}{N^2 A^{-2}}\right) \tag{29}$$

In this example the zeros are $w_{2N} = -NA^{-1}$, $w_{2N+1} = NA^{-1}$ (where N is a natural number). The origin is also a zero (i.e., $m = 1$). Also $Q_q(z) = 0$. Its growth has order $\rho = 1$ and $\sin(\pi Az)$ qualifies as a Bargmann function. The corresponding state is the following superposition of two coherent states (Schrödinger cat):

$$\begin{aligned} |A\rangle_{\sin} &= \mathcal{N}\pi A a^\dagger \prod_{N=1}^{\infty} \left(1 - \frac{(a^\dagger)^2}{N^2 A^{-2}}\right)|0\rangle \\ &= [2 - 2\exp(-2|\pi A|^2)]^{-1/2}[|i\pi A\rangle - |-i\pi A\rangle] \end{aligned} \tag{30}$$

The index 'sin' in the notation indicates that the corresponding Bargmann function is sinusoidal. In this example, the zeros can be related to interference between the coherent states $|i\pi A\rangle$ and $|-i\pi A\rangle$.

4.4. *Example: the function* $\frac{1}{\Gamma(Az)}$

Another example which is related to the previous one but has only half of its zeros, is the $\frac{1}{\Gamma(Az)}$, where $\Gamma(z)$ is the Gamma function and A is a complex number ($A \neq 0$). The relationship is seen in the following formula:

$$\sin(\pi Az) = -\frac{\pi}{Az\Gamma(Az)\Gamma(-Az)} = \frac{\pi}{\Gamma(Az)\Gamma(1-Az)} \tag{31}$$

The entire function $\frac{1}{\Gamma(Az)}$ is factorized as follows:

$$\frac{1}{\Gamma(Az)} = Az\left[\prod_{N=1}^{\infty}\left(1+\frac{z}{NA^{-1}}\right)\exp\left(-\frac{z}{NA^{-1}}\right)\right]\exp(\gamma Az) \tag{32}$$

where $\gamma = 0.577$ is the Euler constant[4, 18]. The growth of $1/\Gamma(Az)$ has order $\rho = 1$ and therefore it qualifies as Bargmann function. The corresponding ket state, for which we use the notation $|A\rangle_{\rm Gam}$, is

$$|A\rangle_{\rm Gam} = \mathcal{N}\exp\left(\frac{\gamma^2|A|^2}{2}\right)(Aa^\dagger)\left[\prod_{N=1}^{\infty}\left(1+\frac{a^\dagger}{NA^{-1}}\right)\exp\left(-\frac{a^\dagger}{NA^{-1}}\right)\right]|\gamma A\rangle_{\rm coh} \tag{33}$$

where $\mathcal{N}$ is a normalization factor.

4.5. Example: the Weierstrass sigma function
We consider the Weierstrass sigma function,

$$\sigma(z|2^{-1}A, i2^{-1}A) = z\prod_{M,N}\left[1-\frac{z}{w_{MN}}\right]\exp\left[\frac{z}{w_{MN}}+\frac{z^2}{2w_{MN}^2}\right]; \qquad w_{MN} = A(M+iN) \tag{34}$$

where A is a positive number, M, N are integers, and the product includes all $(M, N) \neq (0, 0)$[4, 18]. The density of the zeros w_{MN} has $\eta = 2$ and $\delta = \pi/A^2$. Therefore, it can be used as a Bargmann function only if $A > \pi^{1/2}$.

The Weierstrass sigma function has growth with order $\rho = 2$ and type $\sigma = \pi/(2A^2)$. For $A > \pi^{1/2}$ we have $\sigma < 1/2$ and the Weierstrass sigma function qualifies as Bargmann function. The corresponding state, for which we use the notation $|A\rangle_{\rm Wei}$, is

$$|A\rangle_{\rm Wei} = \mathcal{N}(a^\dagger)\left[\prod_{M,N}\left(1-\frac{a^\dagger}{w_{MN}}\right)\exp\left(\frac{a^\dagger}{w_{MN}}+\frac{(a^\dagger)^2}{2w_{MN}^2}\right)\right]|0\rangle; \qquad A > \pi^{1/2} \tag{35}$$

where $\mathcal{N}$ is a normalization constant. We note that ref.[19] used the Weierstrass sigma function in the context of the completeness of coherent states.

5. Time evolution: motion of the zeros

In this section we consider a system with a Hamiltonian H, which at time $t = 0$ is in some initial state. We study the time evolution of this state, and the motion of the zeros of its Bargmann function.

We assume that at time $t = 0$ the system is in the state $|f\rangle$ and has Bargmann function $f(z)$ with zeros $\{w_N\}$. At time t the system is in the state $|f(t)\rangle = \exp(itH)|f\rangle$ and has Bargmann function $f(z;t)$ with zeros $\{w_N(t)\}$.

The number of zeros corresponding to $|f(t)\rangle$ may change as a function of time. This is obvious from the fact that any two states can be related with a unitary transformation. Below we give examples which show this explicitly.

5.1. Systems with the Hamiltonian $H_1 = \kappa a^\dagger + \kappa^* a$
In the case of systems with Hamiltonian $H_1 = \kappa a^\dagger + \kappa^* a$ the evolution operator is $\exp(itH_1) = D(it\kappa)$. It is easily seen that the Bargmann function $f(z;t)$ of the state $|f(t)\rangle$ is given by

$$f(z;t) = f(z+it\kappa^*)\exp\left(-\frac{1}{2}|t\kappa|^2 + it\kappa z\right) \tag{36}$$

From this it follows that the motion of the zeros of the Bargmann function, is given by

$$w_N(t) = w_N - it\kappa^* \tag{37}$$

In this case the relative position of the zeros does not change with time.

5.2. Systems with the Hamiltonian $H_2 = \Omega a^\dagger a$

In the case of systems with Hamiltonian $H_2 = \Omega a^\dagger a$ we prove that the Bargmann function $f(z;t)$ of the state $|f(t)\rangle$ is

$$f(z;t) = f[z \exp(i\Omega t)] \tag{38}$$

From this it follows that the motion of the zeros of the Bargmann function, is given by

$$w_N(t) = w_N \exp(-i\Omega t) \tag{39}$$

In this case also the relative position of the zeros does not change with time.

5.3. Creation and annihilation of zeros

We consider a system described by the evolution operator [20]

$$U(t) = \exp[i\gamma_N(t)]|N\rangle\langle N|; \qquad U(t)[U(t)]^\dagger = \mathbf{1} \tag{40}$$

where $\gamma_N(t)$ are real functions of time (such that $\gamma_N(0) = 0$). At $t = 0$ the system is in the coherent state $|A\rangle_{\text{coh}}$ and the corresponding Bargmann function has no zeros. At time t the system evolves into the state

$$|f\rangle = \exp\left(-\frac{|A|^2}{2}\right) \sum_{N=0}^{\infty} \frac{A^N \exp[i\gamma_N(t)]}{(N!)^{1/2}} |N\rangle \tag{41}$$

and its Bargmann function is

$$f(z) = \exp\left(-\frac{|A|^2}{2}\right) \sum_{N=0}^{\infty} \frac{(Az)^N \exp[i\gamma_N(t)]}{N!} \tag{42}$$

This function can have zeros. For example, we assume that at some time t_0

$$\gamma_{2N}(t_0) = \pi N; \qquad \gamma_{2N+1}(t_0) = \pi \tag{43}$$

Then

$$f(z) = \exp\left(-\frac{|A|^2}{2}\right) [\cos(Az) - \sinh(Az)] \tag{44}$$

It is easily seen that this function has at least one zero. Therefore the number of zeros may change as a function of time.

6. Discussion

Given any sequence of complex numbers with density less than ($\eta = 2, \delta = 1$), we have used Hadamard's theorem to construct in Eq.(21), a state whose Bargmann function has these zeros. Examples which use as Bargmann functions the Riemann $\xi(Az)$ function, the $\sin(\pi Az)$, and the $\frac{1}{\Gamma(Az)}$, were discussed.

The time evolution of Bargmann functions, has been studied. It has been shown that the Hamiltonians H_1, H_2 preserve the zeros in all Bargmann functions, and they also preserve the relative position of these zeros. We also gave an example where creation and annihilation of zeros occurs.

The work is a contribution to the study of quantum systems through the zeros of their Bargman functions.

References

[1] Perelomov A M 1986 *Generalized Coherent States and their Applications* (Berlin, Springer)
[2] Vourdas A 2006 *J. Phys.* A **39** R65
[3] Bargmann V 1961 *Commun. Pure Appl. Math.* **14** 187
[4] Boas R P 1954 *Entire Functions* (New York, Academic)
Levin B Ja 1964 *Distribution of Zeros of Entire Functions* (Rhode Island, American Math. Soc)
Levin B Ja 1996 *Lectures on Entire Functions* (Rhode Island, American Math. Soc)
[5] Bargmann V, Butera P, Girardello L and Klauder J R 1971 *Rep. Math. Phys.* **2** 211
[6] Bacry H, Grossmann A and Zak J 1975 *Phys. Rev.* B **12** 1118
Boon M and Zak J 1978 *Phys. Rev.* B **18** 6744
Boon M and Zak J 1981 *J. Math. Phys.* **22** 1090
Boon M, Zak J and Zuker J 1983 *J. Math. Phys.* **24** 316
[7] Seip K and Reine J 1992 *Angew. Math.* **429** 91
Lyubarskii Yu I 1992 *Adv. Sov. Math.* **11** 167
Seip K, Wallsten R and Reine J 1992 *Angew. Math.* **429** 107
[8] Vourdas A 1997 *J. Phys.* A **30** 4867
Vourdas A 2003 *J. Opt. B: Quantum Semiclass. Opt.* **5** S413
[9] Berry M V 1977 *Phil. Trans. R. Soc.* **287** 237
Berry M V 1987 *Proc. R. Soc.* A **413** 183
[10] Kurchan J, Leboeuf P and Saraceno M 1989 *Phys. Rev.* A **40** 6800
Voros A 1989 *Phys. Rev.* A **40** 6814
Leboeuf P and Voros A 1990 *J. Phys.* A **23** 1765
Leboeuf P 1991 *J. Phys.* A **24** 4575
Nonnenmaher S and Voros A 1997 *J. Phys.* A **30** L677
[11] Hannay J H 1996 *J. Phys.* A **29** L101
Hannay J H 1998 *J. Phys.* A **31** L755
[12] Korsch H J, Múller C and Wiescher H 1997 *J. Phys.* A **30** L677
Toscano F and de Almeida A M O 1999 *J. Phys.* A **32** 6321
[13] Erdelyi A, Magnus W, Oberhettinger F and Tricomi F 1954 *Higher Transcedental Functions* vol 3 (New York, Mc Graw Hill)
[14] Titshmarsh E C 1987 *The Theory of the Riemann Zeta Function* (New York, Clarendon Press)
[15] Edwards H M 2001 *Riemann's Zeta Function* (New York, Dover)
[16] Berry M V and Keating J P 1992 *Proc. R. Soc. London* **437)** 151
Berry M V and Keating J P 1999 *SIAM Review*, **41** 236
Voros A 2000 *J. Phys.* A **33** 7423
[17] Coffey M 2007 *Phys. Lett.* A **362** 352
[18] Remmert R 1998 *Classical Topics in Complex Function Theory* (Berlin, Springer)
[19] Perelomov A M 1971 *Theo. Math. Phys.* **6** 156
[20] Vourdas A and Bishop R F 1989 *Phys. Rev.* A **39** 214

Frames and fusion frames in quantum optics

Andrzej Jamiołkowski
Institute of Physics, Nicholas Copernicus University, 87–100 Toruń, Poland
E-mail: jam@fizyka.umk.pl

Abstract. The main purpose of this paper is to show that the notions of frames and fusion frames introduced in non-harmonic Fourier analysis are also very natural in discussion of some basic problems in theory of open quantum systems and, in particular, in quantum optics. Frames are collections of vectors in a Hilbert space which assure a natural representation of each vector in the space, but may have infinitely many different representations for any given vector. For a given quantum system represented in a Hilbert space $\mathcal{H}$, the question of minimal number η of observables $Q_1, \ldots, Q_\eta$, whose expectation values at some instants $t_1, \ldots, t_p$ determine the statistical state of the system is discussed. We assume that the time evolution of the system in question is governed by a semigroup of linear transformations with generator $\mathbb{L}$.

1. Introduction

It is a fundamental observation concerning quantum physics that the concept of vector space plays a prominent role in the formulation of basic notions of quantum systems theory. The observation that pure states of all microscopic systems satisfy the superposition principle led to postulate that the states of any quantum system must be represented by elements of complex linear space. On the other hand, all quantum observables should be represented by self-adjoint operators acting on this space, and as such also constitute a real vector space. A beautiful discussion of this issues was given by R. Feynman in his lectures on physics [1].

One of the main concepts in the study of vector spaces is the concept of basis for a given space, which allows every vector to be uniquely represented as a linear combination of the basis elements. However, the assumption of linear independence of vectors which constitute a basis is very restrictive; in many situations it is impossible to find vectors which both fulfill the basis requirements and also satisfy natural external conditions demanded by particular applications. We will see examples of such situation in some problems discussed in this paper. In such cases we need to look for more flexible types of spanning collections of vectors or, in more general situations, of spanning collections of subspaces. Frames and fusion frames provide these alternatives. It is a very important fact, that they not only have great variety of applications in many disciplines, but also have a rich theory from a purely mathematical point of view.

One of the basic consequences of the standard assumptions of quantum mechanics is the observation that determination of an unknown state can be achieved by appropriate measurements only if we have at our disposal a set of identically prepared copies of the system in question. Moreover, to set up a successful approach to the above problem of state reconstruction, one has to identify a collection of observables, *a quorum* [2], such that their expectation values contain complete information about the state of the system under consideration. The problems of state determination have gained new relevance in recent years, following the realization that quantum systems and their evolutions can perform tasks such as teleportation, secure communication or dense coding (c.f. e.g. [3]). In this new situation the development of methods to control and measure quantum states becomes an issue of practical interest. It is important to realize that if we identify the quorum of observables, then we also have a possibility to determine the expectation values of physical quantities (observables) for which *no* measuring apparatuses are available [4].

The idea of stroboscopic tomography for open quantum systems appeared for the first time in the beginning of 1980's (although expressed in different terms [5, 6, 7]). In particular, using the concept of observability, in [6] and [7] the question of the minimal number of observables $Q_1, \ldots, Q_\eta$ for which the quantum systems can be $(Q_1, \ldots, Q_\eta)$-reconstructible was discussed.

On the other hand, theory of *frames*, which are collections of vectors that provide robust and usually non-unique representations of vectors, has been the subject of research in the last two decades and has been applied in these disciplines where redundancy played a vital and useful role. However, in some applications it is natural to model and describe considered systems by collections of families of subspaces, and to split a large frame system into a set of much smaller frame systems in these subspaces. This has led to the development of a suitable theory based on *fusion frames* (families of subspaces), which provides the framework to model these more complex applications [8, 9]. In particular, a sequence of the so-called *Krylov subspaces* which appear naturally in stroboscopic tomography [7] and are defined by

$$\mathcal{K}_k(\mathbb{L}, Q) := \mathrm{Span}_{\mathbb{R}^1}\left\{Q, \mathbb{L}Q, \ldots, \mathbb{L}^{k-1}Q\right\}, \tag{1}$$

where Q is a fixed observable and $\mathbb{L}$ is a generator of time evolution of the system in question, constitutes a fusion frame in the Hilbert-Schmidt space $\mathcal{B}_h(\mathcal{H})$ (cf. below Sect 3.) if

$$\boxplus_{i=1}^{r} \mathcal{K}_\mu(\mathbb{L}, Q_i) = \mathcal{B}_h(\mathcal{H}). \tag{2}$$

In the above equality μ denotes the degree of the minimal polynomial of the superoperator $\mathbb{L}$ and $Q_1, \ldots, Q_r$ represent fixed observables. The symbol $\boxplus$ denotes Minkowski sum of subspaces (2) (cf. e.g. [10]). We recall that for two subspaces K_1 and K_2 of the vector space $\mathcal{H}$, by $K_1 \boxplus K_2$ one understands the smallest subspace of $\mathcal{H}$ which contains K_1 and K_2.

It is well known that the Krylov subspaces $\mathcal{K}_k(\mathbb{L}, Q)$ for $k = 1, 2, \ldots$ form a nested sequence of subspaces of increasing dimensions that eventually become invariant under $\mathbb{L}$. Hence for a given Q, there exists an index $\mu = \mu(Q)$, often called the *grade of Q with respect to* $\mathbb{L}$ for which

$$\mathcal{K}_1(\mathbb{L}, Q) \subsetneq \cdots \subsetneq \mathcal{K}_\mu(\mathbb{L}, Q) = \mathcal{K}_{\mu+1}(\mathbb{L}, Q) = \mathcal{K}_{\mu+2}(\mathbb{L}, Q) \cdots . \tag{3}$$

It is easy to see, that for a given operator Q, $\mu(Q)$ is equal to the degree of the minimal polynomial of $\mathbb{L}$ with respect of Q. Clearly, $\mu(Q) \leq \mu(\mathbb{L})$ where $\mu(\mathbb{L})$ denotes the degree of the minimal polynomial of superoperator $\mathbb{L}$ (cf. e.g. [11]). Now, let us observe that even if observables $Q_1, \ldots, Q_r$ are linearly independent, the Krylov subspaces $\mathcal{K}_k(\mathbb{L}, Q_i)$ for $i = 1, \ldots, r$ can have nonempty intersections. At the same time they can constitute a fusion frame for the space of all observables $\mathcal{B}_h(\mathcal{H})$.

The organisation of the paper is as follows: In Section 2 we summarize some concepts and results of the theory of frames; Section 3 presents the main ideas of stroboscopic tomography. We conclude the paper in Section 4 by discussing some applications of the notions of frames and fusion frames to problems of quantum optics.

2. Frames and fusion frames

Frames were first introduced by Duffin and Schaeffer in 1952 as a natural notion that appeared during their research in nonharmonic Fourier analysis [12]. After more than three decades

Daubechies, Grossman and Meyer [13] in 1986 initiated the use of frame theory in description of signal processing. Today, frame theory plays an important role in dozens of applied areas, cf. e.g. [14, 15].

Let us consider a Hilbert space $\mathcal{H}$ ($\dim \mathcal{H} = d < \infty$) with scalar product $\langle \cdot, \cdot \rangle$ which is linear in the second argument. A collection of vectors $\mathcal{F} = \{f_i : i \in \mathcal{I}\}$, $f_i \in \mathcal{H}$, is called *a frame* if there are two positive constants $\alpha, \beta > 0$ such that for every vector $x \in \mathcal{H}$

$$\alpha \parallel x \parallel^2 \leq \sum_{i\in\mathcal{I}} |\langle f_i, x\rangle|^2 \leq \beta \parallel x \parallel^2 . \tag{4}$$

One assumes that the number of vectors f_i is greater or equal to d. The frame is *tight* when the constants α and β are equal, $\alpha = \beta$. If $\alpha = \beta = 1$, then $\mathcal{F}$ is called a *Parseval frame*. The numbers $\langle f_i, x\rangle$ are called *frame coefficients*.

For a given frame $\mathcal{F}$ we can introduce the *analysis* Θ and *synthesis* Θ^* operators. They are defined by the equality

$$\Theta(x) = \sum_{i\in\mathcal{I}} \langle f_i, x\rangle |e_i\rangle, \tag{5}$$

where $|e_i\rangle$ stands for standard basis in $\mathbb{C}^m$ (we will consider only finite dimensional frames, so that $\mathcal{I} = \{1, \ldots, m\}$). Composing Θ with its adjoint operator Θ^*, we obtain the *frame operator*

$$F : \mathcal{H} \to \mathcal{H}, \tag{6}$$

defined by

$$Fx := \Theta^*\Theta x = \sum_{i=1}^{m} \langle f_i, x\rangle |f_i\rangle. \tag{7}$$

It is not difficult to see that any collection of vectors $\{|f_i\rangle\}_{i=1}^m$ constitutes a frame for the vector space $\mathcal{N} := \mathrm{span}\{|f_i\rangle\}_{i=1}^m$, $\mathcal{N} \subseteq \mathcal{H}$. On the other hand a family of elements $\{|f_i\rangle\}_{i=1}^m$ in $\mathcal{H}$ is a frame for $\mathcal{H}$ if and only if $\mathrm{span}\{|f_i\rangle\}_{i=1}^m = \mathcal{H}$. This means that a frame might contain more elements than necessary for it to be a basis. In particular, if $\{|f_i\rangle\}_{i=1}^m$ is a frame for $\mathcal{H}$ and $\{|g_i\rangle\}_{i=1}^n$ is an arbitrary finite collection of elements in $\mathcal{H}$, then the set $\{|f_1\rangle, \ldots, |f_m\rangle, |g_1\rangle, \ldots, |g_n\rangle\}$ is also a frame for $\mathcal{H}$.

Generally speaking, frame theory is the study of how $\{|f_i\rangle\}_{i=1}^m$ should be chosen in order to guarantee that the frame operator $\Theta^*\Theta$ is well-conditioned. In particular, $\{|f_i\rangle\}_{i=1}^m$ is a frame for $\mathcal{H}$ if there exist frame bounds α, β such that

$$\alpha\, \mathbb{I} \leq \Theta^*\Theta \leq \beta\, \mathbb{I}, \tag{8}$$

and is a tight frame iff $\Theta^*\Theta = \alpha\mathbb{I}$. It is an obvious observation that $F = \Theta^*\Theta$ is a self-adjoint and invertible operator.

Fusion frame theory (theory of frames of subspaces) is an emerging mathematical theory that provides a natural framework for performing a distributed data processing in many fields [8, 9]. In particular, one can apply these ideas in quantum state tomography. The notion of fusion frame was introduced in [8] and further developed by [9]. A fusion frame in a Hilbert space $\mathcal{H} \cong \mathbb{C}^d$ is a finite collection of subspaces $\{W_i\}_{i=1}^m$ of $\mathcal{H}$, such that there exist constants $0 < \alpha < \beta < \infty$ satisfying for any $|\varphi\rangle \in \mathcal{H}$ the two inequalities

$$\alpha \parallel |\varphi\rangle \parallel^2 \leq \sum_{i=1}^{m} \parallel \mathbb{P}_i|\varphi\rangle \parallel^2 \leq \beta \parallel |\varphi\rangle \parallel^2, \tag{9}$$

where $\mathbb{P}_i$ denotes the orthogonal projection on W_i. In other words, a collection $\{W_i\}_{i=1}^m$ is a fusion frame if and only if

$$\alpha\,\mathbb{I} \le \sum_{i=1}^{m} \mathbb{P}_i \le \beta\,\mathbb{I}. \tag{10}$$

The constants α and β are called fusion frame bounds. An important class of fusion frames is the class of tight fusion frames, for which $\alpha = \beta$. This equality leads to the operator relation $\sum_{i=1}^{m} \mathbb{P}_i = \alpha\mathbb{I}$.

A graphic representation of the various types of frames is given in Fig. 1.

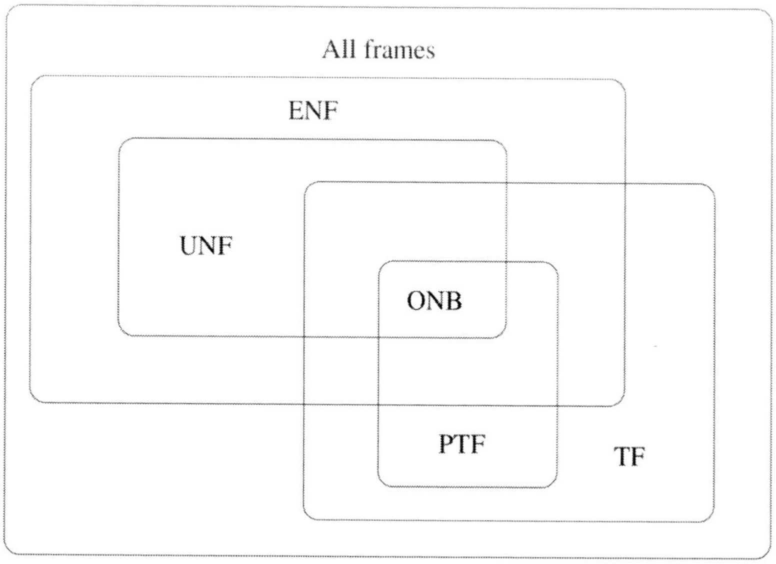

Figure 1. Types of frames; ENF:= Equal-norm frames, TF:= Tight frames, UNF:= Unit-norm frames, PTF:= Parseval tight frames, ONB:= Orthonormal bases

Now, let us recall that for a given operator $M : \mathcal{H} \to \mathcal{H}$ and a given fixed nonzero vector $|x\rangle \in \mathcal{H}$, one introduces the kth Krylov subspace of $\mathcal{H}$ by the equality

$$\mathcal{K}_k(M, x) := \operatorname{span}\{|x\rangle, M|x\rangle, \ldots, M^{k-1}|x\rangle\}. \tag{11}$$

The above definition can also be written as

$$\mathcal{K}_k(M, x) := \operatorname{span}\{p(M)|x\rangle;\ \deg(p) \le k-1\}, \tag{12}$$

where p denotes an arbitrary polynomial and $\deg(p)$ is its degree. It is an obvious observation that the size of a Krylov subspace depends on both M and $|x\rangle$. Note also that there exists such k that $\mathcal{K}_k(M, x) = \mathcal{K}_{k+1}(M, x)$ and this k is the degree of the minimal polynomial of M with respect to $|x\rangle$. If by $\mu(\lambda, M)$ we denote the minimal polynomial of the operator M, then the minimal polynomial of M with respect to any vector $|x\rangle \in \mathcal{H}$ divides $\mu(\lambda, M)$.

For a given operator $M : \mathcal{H} \to \mathcal{H}$ Krylov subspaces generated by a fixed set of vectors $|x_1\rangle, \ldots, |x_r\rangle$ constitute a fusion frame in $\mathcal{H}$ if the following equality is satisfied

$$\boxplus_{i=1}^{r} \mathcal{K}_\mu(M, x_i) = \mathcal{H}, \tag{13}$$

where $\boxplus$ denotes the so-called Minkowski sum of subsets of $\mathcal{H}$ and $\mu := \deg \mu(\lambda, M)$. It is important that we have to use the notion of Minkowski sum of subspaces instead of direct sum. This fact follows from the observation that even if the vectors $|x_1\rangle, \ldots, |x_r\rangle$ are linearly independent, then the Krylov subspaces $\mathcal{K}_\mu(M, x_i)$, $i = 1, \ldots, r$ can partially overlap. It can be easily seen that every Krylov subspace $\mathcal{K}_\mu(M, x)$ is essentially spanned on all vectors of the form $M^k|x\rangle$, where $k = 0, 1, \ldots$, and it is the smallest invariant subspace of the operator M containing $|x\rangle$.

3. Stroboscopic tomography of open quantum systems

During the two decades – 1960s and 1970s – the theory of quantum isolated systems was generalized. The so-called theory of open quantum systems (systems interacting with their environments) was established, and the main sources of inspiration for it were the quantum optics and the theory of lasers. This led to the generalization of states (now density operators are considered as natural representation of quantum states), and to generalized description of time evolution of quantum states. At that time the concepts of so-called quantum master equations – which preserve positive semi-definiteness of density operators – and the idea of quantum communication channel were born, cf. e.g. [16, 17, 18, 19, 20, 21]. On the mathematical level, this approach initiated the study of semigroups of completely positive maps and their generators. Now, we summarize the main ideas and methods of description of open quantum systems.

The time evolution of a quantum system of finitely many degrees of freedom (a qudit), coupled with an infinite quantum system, usually called reservoir, can be described, under certain limiting conditions, by a one-parameter semigroup of maps (cf. e.g. [5, 17]). Let $\mathcal{H}$ be the Hilbert space of the first system ($\dim \mathcal{H} = d$) and let

$$\Phi(t) : \mathcal{B}_h(\mathcal{H}) \to \mathcal{B}_h(\mathcal{H}), \quad t \in \mathbb{R}^1_+, \tag{14}$$

be a dynamical semigroup, where $\mathcal{B}_h(\mathcal{H})$ denotes the real vector space of all self-adjoint operators on $\mathcal{H}$. If one introduces the scalar product of operators A, B by the formula $\langle A, B\rangle = \mathrm{Tr}(A^*B)$, then $\mathcal{B}_h(\mathcal{H})$ can be considered as another inner product space, namely so-called Hilbert-Schmidt space with norm defined by $\| \rho \|^2 = \mathrm{Tr}(\rho^*\rho)$. States of the system are described by density operators $\rho \in S(\mathcal{H})$, where

$$S(\mathcal{H}) := \{\rho \in \mathcal{B}_h(\mathcal{H});\ \rho \geq 0, \mathrm{Tr}\,\rho = 1\}. \tag{15}$$

Usually one assumes that the family of linear superoperators $\Phi(t)$ satisfies

(i) $\Phi(t)$ is trace preserving, $t \in \mathbb{R}^1_+$,
(ii) $\| \Phi(t)\rho \| \leq \| \rho \|$ for all $\rho \in \mathcal{B}_h(\mathcal{H})$,
(iii) $\Phi(t_1) \circ \Phi(t_2) = \Phi(t_1 + t_2)$,

for all t_1, t_2 from $\mathbb{R}^1_+$, and if $t \to 0$, then $\lim \Phi(t) = \mathbb{I}$. Since such defined $\Phi(t)$ is a contraction, it follows from the Hille-Yosida theorem [17, 5] that there exists a linear superoperator $\mathbb{K} : \mathcal{B}_h(\mathcal{H}) \to \mathcal{B}_h(\mathcal{H})$ such that $\Phi(t) = \exp(t\mathbb{K})$ for all $t \geq 0$ and

$$\frac{\mathrm{d}\rho(t)}{\mathrm{d}t} = \mathbb{K}\rho(t), \tag{16}$$

where $\rho(t) = \Phi(t)\rho(0)$. One should stress that above conditions for semigroup $\Phi(t)$ imply preservation of positivity of density operators, $\rho(0) \geq 0 \Rightarrow \rho(t) = \Phi(t)\rho(0) \geq 0$ for all $t \in \mathbb{R}^1_+$. Now, the above equation (usually called *master equation*) defines the assignment (trajectory of $\rho(0)$)

$$\mathbb{R}^1_+ \ni t \mapsto \rho(t) \in S(\mathcal{H}), \tag{17}$$

provided that we know the initial state of the system $\rho(0) \in S(\mathcal{H})$. The fundamental question of stroboscopic tomography reads: *What can we say about the trajectories (initial state $\rho(0)$) if the only information about the system in question is given by the mean values*

$$E_i(t_j) = \mathrm{Tr}\,(Q_i\rho(t_j)), \tag{18}$$

of, say r, linearly independent self-adjoint operators $Q_1,\ldots,Q_r$ at some time instants $t_1,\ldots,t_p$, where $r < d^2-1$ and $t_j \in [0,T]$ for $j = 1,\ldots,p$, $T > 0$. In other words, the problem of the stroboscopic tomography consists in the reconstruction of the initial state $\rho(0)$, or a current state $\rho(t)$, for any $t \in \mathbb{R}^1_+$, by known expectation values (18). To be more precise we introduce the following description. Suppose that we can prepare a quantum system repeatedly in the same initial state and we make a series of experiments such that we know the expectation values $E_Q(t) = \mathrm{Tr}\,(Q\rho(t))$ for a fixed set of observables $Q_1,\ldots,Q_r$ at different time instants $t_1 < t_2 < \cdots < t_p$. The basic question is:

Can we find the expectation value of any other operator $Q \in \mathcal{B}_h(\mathcal{H})$, that is any other observable from $\mathcal{B}_h(\mathcal{H})$, knowing the set of measured outcomes of a given set $Q_1,\ldots,Q_r$ at $t_1,\ldots,t_p$, i.e.

$$\begin{array}{cccc} Q_1 & Q_2 & \cdots & Q_r \end{array}$$
$$\begin{bmatrix} E_1(t_1) & E_2(t_1) & \cdots & E_r(t_1) \\ E_1(t_2) & E_2(t_2) & \cdots & E_r(t_2) \\ \cdots & \cdots & \cdots & \cdots \\ \cdots & \cdots & \cdots & \cdots \\ E_1(t_p) & E_2(t_p) & \cdots & E_r(t_p) \end{bmatrix} \begin{array}{c} t_1 \\ t_2 \\ \vdots \\ \\ t_p \end{array} \tag{19}$$

where $0 \le t_1 < t_2 < \cdots < t_p \le T$, for an interval $[0,T]$?

If the problem under consideration is static, then the state of a d-level open quantum system (a qudit) can be uniquely determined only if $r = d^2-1$ expectation values of linearly independent observables are at our disposal. However, if we assume that we know the dynamics of our system i.e. we know the generator $\mathbb{K}$ or $\mathbb{L} := (\mathbb{K})^*$ (in Heisenberg picture) of the time evolution, then we can use the stroboscopic approach based on (19). In general, we use the term "state-tomography" to denote any kind of state-reconstruction method.

With reference to the terminology used in system theory, we introduce the following definition: A d-level open quantum system $\mathbb{S}$ is said to be $(Q_1,\ldots,Q_r)$-reconstructible on the interval $[0,T]$, if for every two trajectories defined by the equation (16) there exists at least one instant $\hat{t} \in [0,T]$ and at least one operator $Q_k \in \{Q_1,\ldots,Q_r\}$ such that

$$\mathrm{Tr}\,(Q_k\rho_1(\hat{t})) \neq \mathrm{Tr}\,(Q_k\rho_2(\hat{t})). \tag{20}$$

The above definition is equivalent to the following statement [5, 11]. A d-level open quantum system $\mathbb{S}$ is $(Q_1,\ldots,Q_r)$-reconstructible on the interval $[0,T]$, iff there exists at least one set of time instants $0 < t_1 < \cdots < t_p \le T$ such that the state trajectory can be uniquely determined by the correspondence

$$[0,T] \ni t_j \longmapsto E_i(t_j) = \mathrm{Tr}\,(Q_i\rho(t_j)), \tag{21}$$

for $i = 1,\ldots,r$ and $j = 1,\ldots,p$.

Let us observe that in the above definition of reconstructibility we discuss the problem of verifying whether the accessible information about the system in question is sufficient to determine the state uniquely and we do not insist on determining its explicitly.

The positive dynamical semigroup $\{\Phi(t),\ t \in \mathbb{R}^1_+\}$ is determined by the generator

$$\mathbb{K} : \mathcal{B}_h(\mathcal{H}) \to \mathcal{B}_h(\mathcal{H}) \tag{22}$$

(the Schrödinger picture) and it is related to the generator $\mathbb{L}$ of the semigroup in Heisenberg picture by the duality relation

$$\mathrm{Tr}[Q(\mathbb{K}\rho)] = \mathrm{Tr}[(\mathbb{L}Q)\rho]. \tag{23}$$

For a given set of observables $Q_1, \ldots, Q_r$, the subspace spanned on the operators $Q_i, \mathbb{L}Q_i, \ldots, (\mathbb{L})^{k-1}Q_i$, will be denoted by

$$\mathcal{K}_k(\mathbb{L}, Q_i) := \mathrm{Span}_{\mathbb{R}^1}\left\{Q_i, \mathbb{L}Q_i, \ldots, \mathbb{L}^{k-1}Q_i\right\}, \tag{24}$$

as the Krylov subspace in the Hilbert-Schmidt space $\mathcal{B}_h(\mathcal{H})$. If $k = \mu$, where μ is the degree of the minimal polynomial of the generator $\mathbb{L}$, then the subspace $\mathcal{K}_\mu(\mathbb{L}, Q_i)$ is an invariant subspace of the superoperator $\mathbb{L}$ with respect to Q_i. It can be easily seen that the subspace $\mathcal{K}_\mu(\mathbb{L}, Q_i)$ is essentially spanned on all operators of the form $(\mathbb{L})^k Q_i$, where $k = 0, 1, \ldots$. Furthermore, it is smallest invariant subspace of the superoperator $\mathbb{L}$ containing Q_i (i.e. the common part of all invariant subspaces of the operator $\mathbb{L}$ containing Q_i).

One can now formulate the sufficient conditions for the reconstructibility of a d-level open quantum system (c.f. [5, 11]).

Let $\mathbb{S}$ be a d-level open quantum system with the evolution governed by an equation of the form $\dot{Q}(t) = \mathbb{L}Q(t)$ (the Heisenberg picture), where $\mathbb{L}$ is the generator of the dynamical semigroup $\Phi(t) = \exp(t\mathbb{L})$. Suppose that by performing measurements, the correspondence

$$[0, T] \ni t_j \longmapsto E_i(t_j) = \mathrm{Tr}\left(\rho(0)Q_i(t_j)\right) \tag{25}$$

can be established for fixed observables $Q_1, \ldots, Q_r$ at selected time instants $t_1, \ldots, t_p$. The system $\mathbb{S}$ is $(Q_1, \ldots, Q_r)$-reconstructible if

$$\bigboxplus_{i=1}^{r} \mathcal{K}_\mu(\mathbb{L}, Q_i) = \mathcal{B}_h(\mathcal{H}). \tag{26}$$

The above condition has been obtained by using the polynomial representation of the semigroup $\Phi(t)$ [11]. Indeed, if $\mu(\lambda, \mathbb{L})$ denotes the minimal polynomial of the generator $\mathbb{L}$ and $\mu = \deg \mu(\lambda, \mathbb{L})$, then $\Phi(t) = \exp(t\mathbb{L})$ can be represented in the form

$$\Phi(t) = \sum_{k=0}^{\mu-1} \alpha_k(t)\mathbb{L}^k, \tag{27}$$

where the functions $\alpha_k(t)$ for $k = 0, \ldots, \mu - 1$ are those solutions of the scalar linear differential equation with *characteristic polynomial* $\mu(\lambda, \mathbb{L})$ for which equalities

$$\left.\frac{\mathrm{d}^l \alpha_k(t)}{\mathrm{d}t^l}\right|_{t=0} = \delta_{lk}, \tag{28}$$

are satisfied [22]. Since the functions $\alpha_k(t)$ are mutually independent, therefore for arbitrary $T > 0$ there exists at least one set of moments $t_1, \ldots, t_\mu$ ($\mu = \deg \mu(\lambda, \mathbb{L})$) such that

$$0 \le t_1 < t_2 < \cdots < t_\mu \le T \tag{29}$$

and $\det[\alpha_k(t_j)] \neq 0$. Taking into account these conditions one finds that the state $\rho(0)$ can be determined uniquely if operators of the form

$$f_{kl} := (\mathbb{L})^k Q_l \tag{30}$$

for $l = 1, \dots, r$ and $k = 0, 1, \dots$ span the space $\mathcal{B}_h(\mathcal{H})$. In other words, we can say that $\rho(0)$ can be determined if vectors (30) constitute a frame in Hilbert-Schmidt space $\mathcal{B}_h(\mathcal{H})$ or, equivalently, if Krylov subspaces $\mathcal{K}_\mu(\mathbb{L}, Q_l)$ for $l = 1, \dots, r$ constitute a fusion frame in $\mathcal{B}_h(\mathcal{H})$.

The question of an obvious physical interest is to find the minimal number of observables $Q_1, \dots, Q_\eta$ for which a d-level quantum system $\mathbb{S}$ with a fixed generator $\mathbb{L}$ can be $(Q_1, \dots, Q_\eta)$-reconstructible. It can be shown that for an d-level generator there always exists a set of observables $Q_1, \dots, Q_\eta$, where

$$\eta := \max_{\lambda \in \sigma(\mathbb{L})} \{\dim \operatorname{Ker}(\lambda \mathbb{I} - \mathbb{L})\}, \tag{31}$$

such that the system is $(Q_1, \dots, Q_\eta)$-reconstructible [11]. Moreover, if we have another set of observables $\widetilde{Q}_1, \dots, \widetilde{Q}_{\widetilde{\eta}}$ such that the system is $(\widetilde{Q}_1, \dots, \widetilde{Q}_{\widetilde{\eta}})$-reconstructible, then $\widetilde{\eta} \geq \eta$. The number η defined by (31) is called the *index of cyclicity* of the quantum open system $\mathbb{S}$ [11]. The symbol $\sigma(\mathbb{L})$ in (31) denotes the spectrum of the superoperator $\mathbb{L}$.

In particular, if we consider an isolated quantum system characterized by Hamiltonian H_0, then the minimal number of observables $Q_1, \dots, Q_\eta$ for which the system is $(Q_1, \dots, Q_\eta)$-reconstructible is given by

$$\eta = n_1^2 + n_2^2 + \dots + n_m^2, \tag{32}$$

where $n_i = \dim \operatorname{Ker}(\lambda_i I - H_0)$ for all $\lambda_i \in \sigma(H_0)$, $i = 1, \dots, m$ (for details cf. [5, 11]).

Now let us assume that the time evolution of a d-level quantum system $\mathcal{S}$ is described by the generator $\mathbb{L}$ given by

$$\mathbb{L}\rho = \frac{1}{2}\{[N\rho, N] + [N, \rho H]\} = -\frac{1}{2}[N, [N, \rho]], \tag{33}$$

that is, we consider the so-called Gaussian semigroup. The symbol N in (33) denotes a self-adjoint operator with the spectrum

$$\sigma(N) = \{\lambda_1, \dots, \lambda_m\}. \tag{34}$$

In the sequel n_i stands for the multiplicity of the eigenvalue λ_i for $i = 1, \dots, m$. One can assume that the elements of the spectrum of N are numbered in such a way that the inequalities $\lambda_1 < \lambda_2 < \dots < \lambda_m$ are fulfilled. The following theorem holds:

The index of cyclicity of the Gaussian semigroup with a generator $\mathbb{L}$ *given by* (33) *is expressed by the formula*

$$\eta = \max\{\kappa, \gamma_1, \dots, \gamma_r\}, \tag{35}$$

where $r = (m-1)/2$ *if* m *is odd or* $r = (m-2)/2$ *if* m *is even, and*

$$\kappa := n_1^2 + n_2^2 + \dots + n_m^2, \tag{36}$$

$$\gamma_k := 2\sum_{i=1}^{m-k} n_i\, n_{i+k}. \tag{37}$$

In order to prove the above theorem and to determine the value of η for the generator $\mathbb{L}$ defined by (33) we must find the number of nontrivial invariant factors of the operator $\mathbb{L}$. Let us observe that if $\sigma(N) = \{\lambda_1, \dots, \lambda_m\}$, then the spectrum of the operator $\mathbb{L}$ is given by

$$\sigma(\mathbb{L}) = \Big\{\nu_{ij} \in \mathbb{R};\ \nu_{ij} = (\lambda_i - \lambda_j)^2,\ i, j = 1, \dots, m\Big\}. \tag{38}$$

The above statement follows from the fact that the operator $\mathbb{L}$ can also be represented as

$$\mathbb{L} = N^2 \otimes I + I \otimes N^2 - 2N \otimes N\,, \tag{39}$$

where I denotes the identity in the space $\mathcal{H}$. Since N is self-adjoint therefore the algebraic multiplicity of λ_i, i.e. the multiplicity of λ_i as the root of the characteristic polynomial of N, is equal to the geometric multiplicity of λ_i, $n_i = \dim \operatorname{Ker}(\lambda_i I - N)$. Of course, we have $n_1 + \ldots + n_m = \dim \mathcal{H}$.

From (39) we can see that the multiplicities of the eigenvalues of the operator $\mathbb{L}$ are not determined uniquely by the multiplicities of $\lambda_i \in \sigma(N)$. But if we assume that $\lambda_1 < \ldots < \lambda_m$ and $\lambda_k = (k-1)c + \lambda_1$, where $k = 1, \ldots, m$, and $c = \text{const} > 0$, then the multiplicities of all eigenvalues of $\mathbb{L}$ are given by

$$\gamma_{|i-j|} = \dim \operatorname{Ker}\left[(\lambda_i - \lambda_j)^2 \mathbb{I} - \mathbb{L}\right] \tag{40}$$

for $i \neq j$ and

$$\dim \operatorname{Ker}(\mathbb{L}) = n_1^2 + \ldots + n_m^2 = \kappa \tag{41}$$

when $i = j$. Now, as we know, the minimal number of observables $Q_1, \ldots, Q_\eta$ for which the qudit $\mathcal{S}$ can be $(Q_1, \ldots, Q_\eta)$-reconstructible is given by (31), so in our case

$$\eta = \max_{i,j=1,\ldots,m} \left\{ \dim \operatorname{Ker}\left[(\lambda_i - \lambda_j)^2 \mathbb{I} - \mathbb{L}\right] \right\}, \tag{42}$$

where $\lambda_i \in \sigma(N)$. Using the above formulae and the inequality $\gamma_k < \kappa$ for $k > r$, where r is given by $(m-1)/2$ if m is odd and $(m-2)/2$ if m is even, we can observe that also without the assumption $\lambda_k = (k-1)c + \lambda_1$ one obtains

$$\eta = \max\{\kappa, \gamma_1, \ldots, \gamma_r\}\,. \tag{43}$$

This completes the proof.

Another natural question arises: what are the criteria governing the choice of time instants $t_1, \ldots, t_\mu$? The following theorem holds:

Let us assume that $0 \le t_1 < t_2 < \ldots < t_\mu \le T$. Suppose that the mutual distribution of time instants $t_1, \ldots, t_\mu$ is fixed, i.e. a set of nonnegative numbers $c_1 < \ldots < c_\mu$ is given and $t_j := c_j t$ for $j = 1, \ldots, \mu$, and $t \in \mathbb{R}_+$. Then for $T > 0$ the set

$$\tau(T) := \left\{ (t_1, \ldots, t_\mu) : \quad t_j = c_j t, \quad 0 \le t \le \frac{T}{c_\mu} \right\}$$

contains almost all sequences of time instants $t_1, \ldots, t_\mu$, i.e. all of them except a finite number.

As one can check, the expectation values $E_i(t_j)$ and the operators $(\mathbb{L}^*)^k Q_i$ are related by the equality

$$E_i(t_j) = \sum_{k=0}^{\mu-1} \alpha_k(c_j t) \Big((\mathbb{L}^*)^k Q_i, \rho_0 \Big)\,, \tag{44}$$

where we assume that $t_j = c_j t$ and the bracket $(\cdot, \cdot)$ denotes the Hilbert-Schmidt product in $\mathcal{B}_h(\mathcal{H})$. One can determine ρ_0 from (44) for all those values $t \in \mathbb{R}_+$ for which the determinant $\Omega(t)$ is different from zero, i.e.

$$\Omega(t) := \det\left[\alpha_k(c_j t)\right] \neq 0\,. \tag{45}$$

One can prove that the range of the parameter $t \in \mathbb{R}_+$ for which $\Omega(t) = 0$ consists only of isolated points on the semiaxis $\mathbb{R}_+$, i.e. does not possess any accumulation points on $\mathbb{R}_+$. To this end let us note that since the functions $t \to \alpha_k(t)$ for $k = 0, 1, \ldots, \mu - 1$, are analytic on $\mathbb{R}$, the determinant $\Omega(t)$ defined by (45) is also an analytic function of $t \in \mathbb{R}$. If $\Omega(t)$ can be proved to be nonvanishing identically on $\mathbb{R}$, then, making use of its analyticity, we shall be in position to conclude that the values of t, for which $\Omega(t) = 0$, are isolated points on the axis $\mathbb{R}$.

It is easy to check that for $k = \mu(\mu - 1)/2$

$$\left.\frac{d^k \Omega(t)}{dt^k}\right|_{t=0} = \prod_{1 \le j < i \le \mu} (c_i - c_j) . \tag{46}$$

According to the assumption $c_1 < c_2 < \ldots < c_\mu$, we have $\Omega^{(k)}(0) \neq 0$ if $k = \mu(\mu - 1)/2$. This means that the analytic function $t \to \Omega(t)$ does not vanish identically on $\mathbb{R}$ and the set of values of t for which $\Omega(t) = 0$ cannot contain accumulation points. In other words, if we limit ourselves to an arbitrary finite interval $[0, T]$, then $\Omega(t)$ can vanish only on a finite number of points belonging to $[0, T]$. This completes the proof.

4. Frames in stroboscopic tomography

As we have seen the concepts of frames and fusion frames appear in stroboscopic tomography in natural way. The conclusion is based on the discussed above polynomial representations of semigroups which describe evolutions of open systems. The possibility to represent the semigroup $\Phi(t) = \exp(t\mathbb{L})$ in the form

$$\Phi(t) = \sum_{k=0}^{\mu-1} \alpha_k(t)\mathbb{L}^k, \tag{47}$$

where μ stands for the degree of the minimal polynomial of the superoperator $\mathbb{L}$ and $\alpha_k(t)$, $k = 0 \ldots, \mu - 1$, denote some functions of the eigenvalues of $\mathbb{L}$ gives the equality (26) as a sufficient condition for stroboscopic tomography. On the other hand, this equality means that the Krylov subspaces $\mathcal{K}_\mu(\mathbb{L}, Q_i)$, $i = 1, \ldots, r$, constitute a fusion frame in the Hilbert-Schmidt space $\mathcal{B}_h(\mathcal{H})$ of all observables. Moreover, this also means that the collection of vectors

$$f_{jk} := \mathbb{L}^k Q_j, \tag{48}$$

for $j = 1, \ldots, r$ and $k = 0, 1, \ldots, \mu - 1$, constitute a frame in $\mathcal{B}_h(\mathcal{H})$ and the system in question is $(Q_1, \ldots, Q_r)$-reconstructible. In this case every element Q of the space $\mathcal{B}_h(\mathcal{H})$ can be represented as

$$Q = \sum_{j,k} \langle F^{-1} f_{jk}, Q\rangle f_{jk} = \sum_{j,k} \langle f_{jk}, Q\rangle F^{-1} f_{jk}, \tag{49}$$

where F denotes the frame operator of the collection of vectors (48). One can say even more. If $Q \in \mathcal{B}_h(\mathcal{H})$ also has another representation $Q = \sum_{j,k} c_{jk} f_{jk}$ for some scalar coefficients c_{jk}, $j = 1, \ldots, r$ and $k = 0, 1, \ldots, \mu - 1$, then

$$\sum_{jk} |c_{jk}|^2 = \sum_{jk} |\langle F^{-1} f_{jk}, Q\rangle|^2 + \sum_{j,k} |c_{jk} - \langle F^{-1} f_{jk}, Q\rangle|^2. \tag{50}$$

It is obvious that every frame in finite-dimensional space contains a subset that is a basis. As a conclusion we can say that if $\{f_{jk}\}$ is a frame but not a basis, then there exists a set of scalars $\{d_{jk}\}$ such that $\sum_{j,k} d_{jk} f_{jk} = 0$. Therefore, any fixed element Q of $\mathcal{B}_h(\mathcal{H})$ can also be represented as

$$Q = \sum_{j,k} \left(\langle F^{-1} f_{jk}, Q\rangle + d_{jk} \right) f_{jk}. \tag{51}$$

The above equality means that every $Q \in \mathcal{B}_h(\mathcal{H})$ has many representations as superpositions of elements from the set (48). But according to equality (50) among all scalar coefficients $\{c_{jk}\}$ for which

$$Q = \sum_{j,k} c_{jk} f_{jk}, \tag{52}$$

the sequence $\{\langle F^{-1} f_{jk}, Q\rangle\}$ has minimal norm. This is a general method in frame theory [23] and at the same time the main observation connected with the idea of stroboscopic tomography.

In conclusion, one can say that the Krylov subspaces $\mathcal{K}_\mu(\mathbb{L}, Q_i)$ in the space $\mathcal{B}_h(\mathcal{H})$ generated by the superoperator $\mathbb{L}$ can be used in an effective way for procedures of stroboscopic tomography if they constitute appropriate fusion frames in this space. Some details of this approach will be given elsewhere [24].

Acknowledgments

The main ideas of this paper were presented during the conference organized by Professors Barbara and Tadeusz Lulek in Myczkowce 2009. It is a great pleasure for me to thank them for their hospitality and for the effort of all organizers of this meeting.

References

[1] Feynman R 1963 *The Feynman Lectures on Physics* Vol. 3 (Addison-Wesley)
[2] Band W and Park J L 1979 *Am. J. Phys.* **47** 188
[3] Nielsen M A and Chuang I 2000 *Quantum Computation and Quantum Information* (Cambridge: Cambridge Univ. Press.)
[4] Weigert S 2000 *New Insight in Quantum Mechanics* Eds. H.-D. Doebner et al. (Singapore: World Scientific)
[5] Jamiołkowski A 1982 *On Observability of Classical and Quantum Stochastic Systems* (Toruń; N. Copernicus Univ. Press) in Polish
[6] Jamiołkowski A 1983 *Internat. J. Theoret. Phys.* **22** 369
[7] Jamiołkowski A 1986 *Frequency Domain and State Space Methods for Linear Systems* Eds. C I Byrnes and A. Lindquist (Amsterdam: Elsevier) p 347
[8] Casazza P G and Kutyniok G 2004 *Frames of subspaces* Contemp. Math. **345** 87
[9] Casazza P G et al. 2008 *Appl. Comput. Harmon. Anal.* **25** 114
[10] Farouki R. T. et al. 2001 *Geometriae Dedicata* **85** 283
[11] Jamiołkowski A 2000 *Rep. Math. Phys.* **46** 469
[12] Duffin R J and Schaeffer A C 1952 *Transactions AMS* **72** 341
[13] Daubechies I, Grossman A and Meyer Y 1986 *J. Math. Phys.* **27** 1271
[14] Kovačević J and Chebira A 2008 *An Introduction to Frames* (Boston-Delft: NOW)
[15] Heil Ch (Ed.) 2006 *Harmonic Analysis and Applications* (Boston: Birkhäuser)
[16] Kraus K 1971 *Ann. Phys.* **64** 119
[17] Kossakowski A 1972 *Rep.Math. Phys.* **3** 247
[18] Jamiołkowski A 1974 *Rep.Math. Phys.* **5** 415
[19] Gorini V et al. 1976 *J. Math. Phys.* **17** 149
[20] Lindblad G 1976 *Comm. Math. Phys.* **48** 119
[21] Ohya M 1981 *J. Math. Anal. Appl.* **84** 318
[22] Zadeh L A and Desoer C A 1968 *Linear System Theory* (New York: Mc Graw-Hill)
[23] Christensen O 2008 *Frames and Bases* (Boston: Birkhäuser)
[24] Jamiołkowski A, in preparation

Positive maps, doubly stochastic matrices and new family of spectral conditions

Dariusz Chruściński

Institute of Physics, Nicolaus Copernicus University,
Grudziądzka 5/7, 87–100 Toruń, Poland

E-mail: `darch@fizyka.umk.pl`

Abstract. We provide partial classification of positive maps in matrix algebras which is based on a family of spectral conditions. This construction recovers many well known examples of positive maps. We displayed interesting connection between positive maps and doubly stochastic matrices.

1. Introduction

One of the most important problems of quantum information theory [1] is the characterization of mixed states of composed quantum systems. In particular it is of primary importance to test whether a given quantum state exhibits quantum correlation, i.e. whether it is separable or entangled. For low-dimensional systems there exists simple necessary and sufficient condition for separability. The celebrated Peres-Horodecki criterion [2, 3] states that a state of a bipartite system living in $\mathbb{C}^2 \otimes \mathbb{C}^2$ or $\mathbb{C}^2 \otimes \mathbb{C}^3$ is separable iff its partial transpose is positive. Unfortunately, for higher-dimensional systems there is no single *universal* separability condition.

It turns out that the above problem may be reformulated in terms of positive linear maps in operator algebras: a state ρ in $\mathcal{H}_1 \otimes \mathcal{H}_2$ is separable iff $(\mathrm{id} \otimes \Phi)\rho$ is positive for any positive map Φ which sends positive operators on $\mathcal{H}_2$ into positive operators on $\mathcal{H}_1$. Therefore, a classification of positive linear maps between operator algebras $\mathcal{B}(\mathcal{H}_1)$ and $\mathcal{B}(\mathcal{H}_2)$ is of primary importance. Unfortunately, in spite of the considerable effort, the structure of positive maps is rather poorly understood [4]-[12] (for the recent review paper see [13]).

In the present paper we perform partial classification of positive linear maps which is based on spectral conditions. Actually, the presented method enables one to construct maps with a desired degree of positivity — so called k-positive maps with $k = 1, 2, \ldots, d = \min\{\dim \mathcal{H}_1, \dim \mathcal{H}_2\}$. Completely positive (CP) maps correspond to d-positive maps, i.e. maps with the highest degree of positivity. These maps are fully classified due to Stinespring's theorem [16]. Now, any positive map which is not CP can be written as $\Phi = \Phi_+ - \Phi_-$, with $\Phi_\pm$ being CP maps. However, there is no general method to recognize the positivity of Φ from $\Phi_+ - \Phi_-$. We show that suitable spectral conditions satisfied by the pair (Φ_+, Φ_-) guarantee the k-positivity of $\Phi_+ - \Phi_-$.

There is an interesting connection between positive maps in matrix algebras and doubly stochastic matrices. We provide several examples of maps and the corresponding families of doubly stochastic matrices. It was argued that such matrices may occur in the very basic description of quantum phenomena [14] (for the interesting discussion see [15]). It is hoped that the connection between positive maps and doubly stochastic matrices may shed a new light into the problem of positive maps. This part of our paper is inspired very much by the beautiful lecture by James Louck (see this volume).

2. Quantum entanglement, Schmidt number and all that

Consider a Hilbert space being the tensor product $\mathcal{H}_1 \otimes \mathcal{H}_2$, with $\dim \mathcal{H}_i = d_i < \infty$. It is clear that $\mathcal{H}_1 \otimes \mathcal{H}_2$ is isomorphic to the space of linear operators from $\mathcal{L}(\mathcal{H}_1, \mathcal{H}_2)$, i.e. each $\psi \in \mathcal{H}_1 \otimes \mathcal{H}_2$ corresponds to some linear operator $F : \mathcal{H}_1 \longrightarrow \mathcal{H}_2$. Indeed, taking orthonormal basis $\{e_i\}$ $(i = 1, \ldots, d_1)$ in $\mathcal{H}_1$ and $\{f_\alpha\}$ $(\alpha = 1, \ldots, d_2)$ in $\mathcal{H}_2$, one has

$$\psi = \sum_{i=1}^{d_1} \sum_{\alpha=1}^{d_2} \psi_{i\alpha}\, e_i \otimes f_\alpha \ , \tag{1}$$

where $\psi_{i\alpha}$ are complex coefficients. Defining

$$Fe_i := \sum_{\alpha=1}^{d_2} \psi_{i\alpha}\, f_\alpha \ ,$$

one finds

$$\psi = \sum_{i=1}^{d_1} e_i \otimes Fe_i \ , \tag{2}$$

which establishes the above-mentioned isomorphism. Let us note that the normalization condition

$$\langle \psi | \psi \rangle = \sum_{i=1}^{d_1} \sum_{\alpha=1}^{d_2} |\psi_{i\alpha}|^2 = 1 \ , \tag{3}$$

gives rise to

$$\mathrm{tr}\, FF^* = 1 \ , \tag{4}$$

for the corresponding operator F. Denote by

$$s_1 \geq s_2 \geq \ldots \geq s_d \ , \quad d := \min\{d_1, d_2\} \ ,$$

the singular values of F. One introduces the *Schmidt rank* of ψ:

$$\mathrm{SR}(\psi) := \mathrm{rank}\, F \ , \tag{5}$$

that is, the number of non-vanishing singular values of F. It is well known that ψ is separable if and only if $\mathrm{SR}(\psi) = 1$. Otherwise it is entangled.

Now, the vector space $\mathcal{L}(\mathcal{H}_1, \mathcal{H}_2)$ may be equipped with the following family of k-norms:

$$||F||_k^2 := \sum_{i=1}^{k} s_i^2 \ , \tag{6}$$

where s_i denote the singular values of F, and $1 \leq k \leq d$. Note, that for $k = 1$ one reproduces the operator norm

$$||F||_1^2 := s_1^2 = ||F||^2 \ , \tag{7}$$

whereas for $k = d$ one reproduces the Hilbert-Schmidt norm

$$||F||_d^2 := \sum_{i=1}^{d} s_i^2 = \mathrm{Tr}\, F^*F = ||F||_{\mathrm{HS}}^2 \ . \tag{8}$$

Moreover, it is clear that

$$||F||_1 \leq \ldots \leq ||F||_d . \tag{9}$$

Note, that if F corresponds to a normalized vector $\psi \in \mathcal{H}_1 \otimes \mathcal{H}_2$, then $||F||_d \equiv 1$ from the normalization condition $\langle\psi|\psi\rangle = 1$. Interestingly, if ψ is separable, then

$$||F||_1 = \ldots = ||F||_d = 1 . \tag{10}$$

If ψ is maximally entangled, i.e. $s_1^2 = \ldots = s_d^2 = 1/d$, then

$$||F||_k^2 = \frac{k}{d} , \tag{11}$$

for $k = 1, \ldots, d$.

3. Positive maps in matrix algebras

Let M_d denote a set of $d \times d$ complex matrices. A linear map

$$\Phi : M_{d_1} \longrightarrow M_{d_2} , \tag{12}$$

is called positive if it maps semi-positive definite matrix $X \in M_{d_1}$ into semi-positive definite matrix $\Phi(X) \in M_{d_2}$. It is called unital if

$$\Phi(\mathbb{I}_{d_1}) = \mathbb{I}_{d_2} , \tag{13}$$

where $\mathbb{I}_d$ is a unit matrix in M_d. It is called trace preserving if

$$\mathrm{Tr}\,\Phi(X) = \mathrm{Tr}\,X , \tag{14}$$

for all $X \in M_{d_1}$. Note that trace preserving positive maps map quantum states of d_1–level system into quantum states of d_2–level system. It is clear that a map (12) is positive if it satisfies

$$\langle\psi|\Phi(|\phi\rangle\langle\phi|)|\psi\rangle \geq 0 , \tag{15}$$

for each $\phi \in \mathbb{C}^{d_1}$ and each $\psi \in \mathbb{C}^{d_2}$. Unfortunately, the above condition is very difficult to check. Let us recall that a similar condition for a positivity of a matrix $X \in M_d$

$$\langle\psi|X|\psi\rangle \geq 0 , \tag{16}$$

for all $\psi \in \mathbb{C}^d$ reduces to the spectral problem for X: all eigenvalues of X have to be nonnegative. It is no longer true for (15). This property does not reduce to any spectral condition and it makes the problem of positive maps very nontrivial.

Let $\mathrm{id}_k : M_k \longrightarrow M_k$ denote an identity map, i.e. $\mathrm{id}_k(X) = X$ for any $X \in M_k$. A positive map (12) is called k-positive if

$$\mathrm{id}_k \otimes \Phi : M_k \otimes M_{d_1} \longrightarrow M_k \otimes M_{d_2} , \tag{17}$$

is positive. Note, that $M_k \otimes M_d$ may be considered as an algebra of $k \times k$ matrices with entries from M_d. Finally, Φ is called completely positive (CP) if it is k-positive for $k = 1, 2, \ldots$. One shows that Φ is CP iff it is d-positive, where $d = \min\{d_1, d_2\}$. Denoting by $\mathcal{P}_k$ a convex cone of k-positive maps one arrives at the following chain of cones

$$\text{positive maps} = \mathcal{P}_1 \supset \mathcal{P}_2 \supset \ldots \supset \mathcal{P}_d = \text{CP maps} . \tag{18}$$

Now, CP maps are fully characterized due the celebrated results of Stinespring, Kraus and Choi [16, 17] any CP map (12) may be represented as follows

$$\Phi(X) = \sum_\alpha V_\alpha X V_\alpha^* , \tag{19}$$

where the so-called Kraus operators V_α are linear operators $V_\alpha : \mathbb{C}^{d_1} \longrightarrow \mathbb{C}^{d_2}$. It turns out that CP property reduces to a simple spectral condition. Introducing so called Choi-Jamiołkowski matrix [6, 19]

$$\widehat{\Phi} := \sum_{i,j=1}^{d_1} |e_i\rangle\langle e_j| \otimes \Phi(|e_i\rangle\langle e_j|) , \tag{20}$$

where $\{e_1, \ldots, e_{d_1}\}$ defines an orthonormal basis in $\mathbb{C}^{d_1}$, one shows that Φ is CP if and only iff the corresponding matrix $\widehat{\Phi}$ is semi-positive definite. It is no longer true for k-positive maps with $k < d$. Now, the condition for k-positivity of Φ may be rewritten as the following condition for the corresponding matrix $\widehat{\Phi}$

$$\langle \psi | \widehat{\Phi} | \psi \rangle \geq 0 , \tag{21}$$

for all $\psi \in \mathbb{C}^{d_1} \otimes \mathbb{C}^{d_2}$ such that $\mathrm{SN}(\psi) \leq k$. For $k = 1$ it gives therefore

$$\langle \psi_1 \otimes \psi_2 | \widehat{\Phi} | \psi_1 \otimes \psi_2 \rangle \geq 0 , \tag{22}$$

for $\psi_j \in \mathbb{C}^{d_j}$. We stress that the above condition can not be reduced for any spectral condition for the matrix $\widehat{\Phi}$.

4. Positive maps: examples

Let us illustrate our definitions with a few simple examples.

Example 1 Transposition in M_d:

$$\Phi(X) = X^T . \tag{23}$$

The map is trivially positive since transposition does not change the spectrum of the Hermitian operator. Moreover, it is trace preserving and unital. For example for $d = 2$ the corresponding $\widehat{\Phi}$ matrix reads as follows

$$\widehat{\Phi} = \left(\begin{array}{cc|cc} 1 & \cdot & \cdot & \cdot \\ \cdot & \cdot & 1 & \cdot \\ \hline \cdot & 1 & \cdot & \cdot \\ \cdot & \cdot & \cdot & 1 \end{array} \right) , \tag{24}$$

where dots represent zeros. Its spectrum is given by $\{-1, 1, 1, 1\}$ and hence it is not CP. It is well known that transposition is not CP for any d (actually, it is even not 2-positive).

Example 2 A map in M_d known as the reduction map

$$\Phi(X) = \mathbb{I}_d \mathrm{Tr}\, X - X . \tag{25}$$

It is easy to check for positivity: indeed for any rank-1 projector $P = |\psi\rangle\langle\psi|$ one gets

$$\Phi(P) = \mathbb{I}_d - P = P^\perp , \tag{26}$$

where $P^\perp$ defines a projector onto a hyperplane orthogonal to ψ. Hence, $\Phi(P)$ being a projector is semi-positive definite. Again, for $d = 2$ it is easy to find the corresponding $\widehat{\Phi}$ matrix

$$\widehat{\Phi} = \left(\begin{array}{cc|cc} \cdot & \cdot & \cdot & -1 \\ \cdot & 1 & \cdot & \cdot \\ \hline \cdot & \cdot & 1 & \cdot \\ -1 & \cdot & \cdot & \cdot \end{array} \right) , \tag{27}$$

with the spectrum given by $\{-1,1,1,1\}$. It shows that reduction map is positive but not CP. Interestingly, one may show that composing reduction with transposition one obtains a CP map.

Example 3 A family of Choi maps in M_d

$$\Phi_\mu(X) = \mathbb{I}_d \mathrm{Tr}\, X - \mu X\ , \quad \mu \leq 1\ . \tag{28}$$

Note, that for $\mu = 1$ it reproduces the reduction map (25). Now, it is well known [18] that if

$$\frac{1}{k+1} < \mu \leq \frac{1}{k}\ , \tag{29}$$

then Φ_μ is k-positive but not $(k+1)$-positive. Hence, if $\mu \leq 1/d$, then Φ_μ is CP. Again, for $d = 2$ it is easy to find the corresponding $\widehat{\Phi}_\mu$ matrix

$$\widehat{\Phi}_\mu = \left(\begin{array}{cc|cc} 1-\mu & \cdot & \cdot & -\mu \\ \cdot & 1 & \cdot & \cdot \\ \hline \cdot & \cdot & 1 & \cdot \\ -\mu & \cdot & \cdot & 1-\mu \end{array} \right)\ , \tag{30}$$

with the spectrum given by $\{1-2\mu, 1, 1, 1\}$. Clearly, for $\mu \leq 1/2$ the map is CP.

5. Positive maps vs. doubly stochastic matrices

Linear positive maps generalize a notion of a nonnegative matrix, i.e. a matrix X with nonnegative entries $X_{ij} \geq 0$. Such matrices play important role in classical stochastic dynamics. Consider two stochastic systems described by discrete probability distributions $p = (p_1, \ldots, p_{d_1})$ and $q = (q_1, \ldots, q_{d_2})$. Now, a transition $p \rightarrow q$ is described by a linear map $p = Sq$, where S is a real $d_2 \times d_1$ matrix such that $S_{ij} \geq 0$, and $\sum_i S_{ij} = 1$. If $d_1 = d_2 = d$ and

$$S_{ij} \geq 0\ , \quad \sum_{i=1}^{d} S_{ij} = 1\ , \quad \sum_{j=1}^{d} S_{ij} = 1\ , \tag{31}$$

then S is called doubly stochastic (or bistochastic). A doubly stochastic matrix has one extra property – it keeps equilibrium distribution $p_0 = (1/d, \ldots, 1/d)$ invariant.

Let us observe that each trace preserving, unital positive map $\Phi : M_d \longrightarrow M_d$ gives rise to a family of doubly stochastic matrices. Indeed, take two orthonormal basis $\{e_1, \ldots, e_d\}$ and $\{f_1, \ldots, f_d\}$ in $\mathbb{C}^d$ and define

$$S_{ij} := \langle e_i | \Phi(|f_j\rangle\langle f_j|) | e_i \rangle\ . \tag{32}$$

By the very definition of positivity of Φ one has $S_{ij} \geq 0$. Now

$$\sum_{i=1}^{d} S_{ij} = \sum_{i=1}^{d} \langle e_i | \Phi(|f_j\rangle\langle f_j|) | e_i \rangle = \mathrm{Tr}\, \Phi(|f_j\rangle\langle f_j|)\ , \tag{33}$$

and using $\mathrm{Tr}\, \Phi(X) = \mathrm{Tr}\, X$ one gets $\mathrm{Tr}\, \Phi(|f_j\rangle\langle f_j|) = 1$. Moreover, one finds

$$\sum_{j=1}^{d} S_{ij} = \langle e_i | \sum_{j=1}^{d} \Phi(|f_j\rangle\langle f_j|) | e_i \rangle\ . \tag{34}$$

Taking into account that $\sum_{j=1}^{d} \Phi(|f_j\rangle\langle f_j|) = \Phi(\sum_{j=1}^{d} |f_j\rangle\langle f_j|) = \Phi(\mathbb{I}_d)$ and the fact that Φ is unital, i.e. $\Phi(\mathbb{I}_d) = \mathbb{I}_d$, one obtains $\langle e_i | \sum_{j=1}^{d} \Phi(|f_j\rangle\langle f_j|) | e_i \rangle = 1$. It shows that S_{ij} defined in (32) is doubly stochastic.

Example 4 Consider reduction map (25) in M_2. Let us take the standard orthonormal basis in $\mathbb{C}^2$ — $\{e_1, e_2\}$ — and consider

$$S_{ij} := \langle e_i|\Phi(|e_j\rangle\langle e_j|)|e_i\rangle \ . \tag{35}$$

One easily finds

$$S_{11} = S_{22} = 0 \ , \qquad S_{12} = S_{21} = 1 \ ,$$

which defines a permutation (and hence doubly stochastic) matrix.

Example 5 Consider Choi map (28) in M_2. One easily finds

$$S = (1-\mu)\begin{pmatrix} 1 & \cdot \\ \cdot & 1 \end{pmatrix} + \mu\begin{pmatrix} \cdot & 1 \\ 1 & \cdot \end{pmatrix} , \tag{36}$$

that is, Birkhoff decomposition of S into convex combination of permutation matrices.

Example 6 Consider the following family of maps in M_3:

$$\Phi_{abc}\begin{pmatrix} X_{11} & X_{12} & X_{13} \\ X_{21} & X_{22} & X_{23} \\ X_{31} & X_{32} & X_{33} \end{pmatrix} = \tag{37}$$

$$= \frac{1}{a+b+c}\begin{pmatrix} aX_{11}+bX_{22}+cX_{33} & -X_{12} & -X_{13} \\ -X_{21} & aX_{22}+bX_{33}+cX_{11} & -X_{23} \\ -X_{31} & -X_{32} & aX_{33}+bX_{11}+cX_{22} \end{pmatrix} ,$$

with $a, b, c \geq 0$. Now, Φ_{abc} is positive if and only if the following conditions are satisfied

(i) $a \geq 0$,

(ii) $a+b+c \geq 2$,

(iii) if $a \leq 1$, then $bc \geq (1-a)^2$.

Moreover, Φ_{abc} is CP iff $a \geq 2$. Note, that Φ_{abc} is unital and trace preserving. The corresponding doubly stochastic matrix S reads as follows

$$S = \frac{1}{a+b+c}\begin{pmatrix} a & b & c \\ c & a & b \\ b & c & a \end{pmatrix} , \tag{38}$$

that is

$$S = a'\begin{pmatrix} 1 & \cdot & \cdot \\ \cdot & 1 & \cdot \\ \cdot & \cdot & 1 \end{pmatrix} + b'\begin{pmatrix} \cdot & 1 & \cdot \\ \cdot & \cdot & 1 \\ 1 & \cdot & \cdot \end{pmatrix} + c'\begin{pmatrix} \cdot & \cdot & 1 \\ 1 & \cdot & \cdot \\ \cdot & 1 & \cdot \end{pmatrix} . \tag{39}$$

Hence, one finds again the Birkhoff decomposition of S into convex combination of permutation matrices with

$$a' = \frac{a}{a+b+c} \ , \quad b' = \frac{b}{a+b+c} \ , \quad c' = \frac{c}{a+b+c} \ . \tag{40}$$

Interestingly, S is not only doubly stochastic but circulant as well.

Let us recall that a class of $d \times d$ doubly stochastic matrices defines a convex polytope in R^{d^2} known as the Birkhoff polytope B_d. The principal fact about doubly stochastic matrices is the Birkhoff theorem [22] which states that the set B_d of doubly stochastic matrices of order d is the convex hull of the set of permutation matrices (of order d), and furthermore that the vertices (extreme points) of B_d are precisely the permutation matrices. It means that any $S \in B_d$ may be represented as

$$S = \sum_\pi a_\pi P_\pi \ , \tag{41}$$

where π enumerates permutations of order d and P_π is a $d \times d$ permutation matrix, i.e.

$$(P_\pi)_{ij} = \delta_{i,\pi(j)} \ . \tag{42}$$

Coefficients a_π satisfy: $a_\pi \geq 0$ and $\sum_\pi a_\pi = 1$. It should be stressed that not all permutation matrices are linearly independent. It turns out [22] that among $d!$ elements there are exactly $b_d = (d-1)^2 + 1$ linearly independent.

Interestingly, any $S \in B_d$ gives rise to the unital, trace preserving CP map in M_d. Indeed, using (41) let us define $\Phi : M_d \longrightarrow M_d$ by

$$\Phi(X) = \sum_\pi a_\pi P_\pi X P_\pi^* \ , \tag{43}$$

which defines Kraus representation of Φ with the corresponding Kraus operators $V_\pi = \sqrt{a_\pi}\, P_\pi$. Note that (43) is unital and trace preserving. Indeed, one has

$$\Phi(\mathbb{I}_d) = \sum_\pi a_\pi P_\pi P_\pi^* = \mathbb{I}_d \ , \tag{44}$$

where we used the fact that P_π defines an orthogonal matrix. Moreover,

$$\mathrm{Tr}\, \Phi(X) = \sum_\pi a_\pi \mathrm{Tr}\, (P_\pi X P_\pi^*) = \sum_\pi a_\pi \mathrm{Tr}\, X = \mathrm{Tr}\, X \ . \tag{45}$$

The above properties are equivalent to

$$\sum_\pi V_\pi V_\pi^* = \sum_\pi V_\pi^* V_\pi = \mathbb{I}_d \ .$$

It shows that any classical stochastic operation defined by the Birkhoff decomposition (41) may be generalized to the corresponding quantum operation defined by a CP map (43).

6. Spectral class of positive maps

Consider a vector space of linear operators $\mathcal{L}(\mathbb{C}^{d_1}, \mathbb{C}^{d_2})$. It is D-dimensional ($D = d_1 d_2$) Hilbert space equipped with the Hilbert-Schmidt scalar product

$$(F, G)_{\mathrm{HS}} := \mathrm{Tr}\, F^* G \ . \tag{46}$$

Let $\{V_1, \ldots, V_D\}$ denotes an orthonormal (with respect to (46)) basis in $\mathcal{L}(\mathbb{C}^{d_1}, \mathbb{C}^{d_2})$. Let us define two CP maps

$$\Phi_-(X) := \sum_{\alpha=1}^{L} \lambda_\alpha V_\alpha X V_\alpha^* \ , \tag{47}$$

and

$$\Phi_+(X) := \sum_{\alpha=L+1}^{D} \lambda_\alpha V_\alpha X V_\alpha^* \ , \tag{48}$$

with $1 \leq L < D$, such that

$$\lambda_\alpha \geq 0 \ , \quad \alpha = 1, \ldots, L \ , \tag{49}$$

and

$$\lambda_\alpha > 0 \ , \quad \alpha = L+1, \ldots, D \ , \tag{50}$$

It implies that the corresponding matrices $\widehat{\Phi}_\pm$ satisfy

$$\widehat{\Phi}_- \geq 0 \ , \quad \widehat{\Phi}_+ > 0 \ . \tag{51}$$

Finally, let us define

$$\Phi := \Phi_+ - \Phi_- \ . \tag{52}$$

Our main result consists in the following

Theorem 1 *Let* $\sum_{\alpha=1}^{L} ||V_\alpha||_k^2 < 1$. *If the following spectral conditions are satisfied*

$$\lambda_\alpha \geq \mu_k \ , \quad \alpha = L+1, \ldots, D \ , \tag{53}$$

where

$$\mu_\ell := \frac{\sum_{\alpha=1}^{L} \lambda_\alpha ||V_\alpha||_\ell^2}{1 - \sum_{\alpha=1}^{L} ||V_\alpha||_\ell^2} \ , \tag{54}$$

then Φ *is k-positive. If moreover* $\sum_{\alpha=1}^{L} ||V_\alpha||_{k+1}^2 < 1$ *and*

$$\mu_{k+1} > \lambda_\alpha \ , \quad \alpha = L+1, \ldots, D \ , \tag{55}$$

then Φ *being k-positive is not* $(k+1)$*-positive.*

For the proof we refer to [20].

7. Spectral class – examples

Surprisingly this simple construction recovers many well known examples of positive maps.

Example 7 Transposition in M_2. One finds

$$\Phi_-(X) = \lambda_1 V_1 X V_1^* \ , \tag{56}$$

and

$$\Phi_+(X) = \sum_{\alpha=2}^{4} \lambda_\alpha V_\alpha X V_\alpha^* \ , \tag{57}$$

where $\lambda_1 = \ldots = \lambda_4 = 1$, and

$$V_1 = \frac{1}{\sqrt{2}} \left(\begin{array}{cc} 0 & 1 \\ -1 & 0 \end{array} \right) \ , \ V_2 = \left(\begin{array}{cc} 1 & 0 \\ 0 & 0 \end{array} \right) \ , \ V_3 = \left(\begin{array}{cc} 0 & 0 \\ 0 & 1 \end{array} \right) \ , \ V_4 = \frac{1}{\sqrt{2}} \left(\begin{array}{cc} 0 & 1 \\ 1 & 0 \end{array} \right) \ . \tag{58}$$

It is easy to check that

$$(V_\alpha, V_\beta)_{\rm HS} = \delta_{\alpha\beta} \ . \tag{59}$$

Moreover

$$||V_1||_1^2 = \frac{1}{2} \ , \quad ||V_1||_2^2 = 1 \ . \tag{60}$$

One easily finds $\mu_1 = 1$ and hence condition (53) is trivially satisfied $\lambda_\alpha \geq \mu_1$ for $\alpha = 2, 3, 4$.

Example 8 Reduction map in M_d

$$\Phi(X) = \mathbb{I}_d \mathrm{Tr}\, X - X \ . \tag{61}$$

One finds

$$\Phi_-(X) = \lambda_1 V_1 X V_1^* \ , \tag{62}$$

and

$$\Phi_+(X) = \sum_{\alpha=2}^{D} \lambda_\alpha V_\alpha X V_\alpha^* \ , \tag{63}$$

where $\lambda_1 = d - 1$ and $\lambda_2 = \ldots = \lambda_D = 1$. Moreover

$$V_1 = \frac{1}{\sqrt{d}} \mathbb{I}_d \ , \tag{64}$$

and the remaining V_α are arbitrary $d \times d$ matrices such that

$$(V_\alpha, V_\beta)_{\mathrm{HS}} = \delta_{\alpha\beta} \ , \tag{65}$$

and $\mathrm{Tr}\, V_\alpha = 0$ (for $\alpha = 2, \ldots, D$). Again, one finds $\mu_1 = 1$ and hence condition (53) is trivially satisfied $\lambda_\alpha \geq \mu_1$ for $\alpha = 2, \ldots, D = d^2$. Now, since V_1 corresponds to the maximally entangled state one has $1 - ||V_1||_2^2 = (d-2)/d < 1$. Hence, condition (55)

$$\mu_2 = 2\frac{d-2}{d-1} > \lambda_\alpha \ , \quad \alpha = 2, \ldots, D \ , \tag{66}$$

implies that Φ is not a 2-positive.

For other examples of positive maps belonging to our spectral class see [20, 21].

8. Conclusions

We provide partial classification of positive linear maps based on family of spectral conditions. It is shown that our class does contain well known examples of positive maps. We displayed interesting connection between positive (unital and trace preserving) maps and doubly stochastic matrices. It turns out [20] that our scheme may be easily generalized to the multipartite setting. The subject deserves further studies in order to investigate further properties of positive maps from the spectral class.

Acknowledgements

It is a pleasure to thank James Louck for interesting discussions during the series of conferences in Myczkowce. I thank Barbara and Tadeusz Lulek for creating very friendly atmosphere and for kind hospitality during the conference in Myczkowce. This work was partially supported by the Polish Ministry of Science and Higher Education Grant No 3004/B/H03/2007/33 and by the Polish Research Network *Laboratory of Physical Foundations of Information Processing*.

References

[1] Nielsen M A and Chuang I L 2000 *Quantum computation and quantum information* (Cambridge:Cambridge University Press) 2000.
[2] Peres A 1996 *Phys. Rev. Lett.* **77** 1413
[3] Horodecki P 1997 *Phys. Lett. A* **232** 333
[4] Størmer E 1963 *Acta Math.* **110** 233
[5] Arveson W 1969 *Acta Math.* **123** 141
[6] Choi M - D 1975 *Lin. Alg. Appl.* **10** 285; *ibid* **12** 95

[7] Woronowicz S L 1976 *Rep. Math. Phys.* **10** 165
[8] Woronowicz S L 1976 *Comm. Math. Phys.* **51** 243
[9] Robertson A G 1983 *Quart. J. Math. Oxford* (2) **34** 87
[10] Robertson A G 1983 *Proc. Roy. Soc. Edinburh Sect.* A **94** 71; **34** 87
[11] Chruściński D and Kossakowski A. 2008 *J. Phys. A: Math. Theor.* **41** 215201; *ibid* **41** 145301
[12] Størmer E 2008 *J. Funct. Anal.* **254** 2303
[13] Chruściński and Kossakowski A 2007 *Open Systems and Inf. Dynamics* **14** 275
[14] Landé A 1965 *New Foundations of Quantum Mechanics* (Cambridge: Cambridge University Press)
[15] Louck J D 1997 *Found. Phys.* **27** 1085
[16] Paulsen V 2003 *Completely Bounded Maps and Operator Algebras* (Cambridge: Cambridge University Press)
[17] Kraus K 1983 *States, Effects and Operations: Fundamental Notions of Quantum Theory*, (Springer Verlag)
[18] Terhal B and Horodecki P 2000 *Phys. Rev.* A **61** 040301
[19] Jamiołkowski A 1972 *Rep. Math. Phys.* **3** 275
[20] Chruściński D and Kossakowski A 2009 *Comm. Math. Phys.* **290** 1051
[21] Chruściński D, Kossakowski A and Sarbicki G 2009 *Phys. Rev.* A **80** (2009) 042314.
[22] Louck J D 2008 *Unitary Symmetry and Combinatorics* (Singapore: World Scientific)

Correlation and entanglement measures in trapped few-particle systems

Anna Okopińska

Institute of Physics, Jan Kochanowski University, Świętokrzyska 15, 25-406 Kielce, Poland

E-mail: okopin@fuw.edu.pl

Abstract. Quantum correlations in few-particle systems are discussed in dependence on the shape of the external confining potential. The energy spectra, as well as the natural orbitals and their occupation numbers, are determined by using the optimized configuration interaction method. Considering the inter-particle interactions of the contact type, the comparison of various entanglement measures is given for the full range of interaction strength from strongly attractive to repulsive values. Entropic correlation measures are also discussed for the case of long-range Coulomb interactions between the particles.

1. Introduction

Interest in theoretical studies of few-particle quantum systems has grown enormously in early 1980s since their experimental realization in the form of semiconductor quantum dots (QDs) [1] or few-ion systems in electromagnetic traps. Other realizations soon became possible due to the rapid development of trapping techniques for ultracold gases [2] and the application of optical lattices to isolate systems containing a few atoms or molecules [3]. The possibility of studying single few-body objects in these experiments represents a remarkable advantage over systems existing in nature, such as atoms or nuclei, that are not directly observable but inferred from experiments on macroscopic bodies containing enormous numbers of individuals. Another advantage of the artificial structures is that their parameters, such as the trapping potential, the number of constituents N, and interaction strength, may be controlled at will [2, 3]. The possibility of fabricating low-dimensional artificial few-body structures gave impetus for their detailed theoretical studies. Two-dimensional systems interacting via Coulomb force are the subject of interest in connection to semiconducting QDs, the perpendicular dimension of which is usually negligible [1]. Studies of one-dimensional systems became increasingly popular after their experimental realization in ultracold gases [4]. The interatomic interaction at low temperatures may be considered as a contact one, $U(\mathbf{r}_1, \mathbf{r}_2) = g_{3D}\delta(\mathbf{r}_1 - \mathbf{r}_2)$ with $g_{3D} = \frac{4\pi\hbar^2 a_{3D}}{m}$ and a_{3D} being the scattering length. Quasi-one dimensional systems are created by using trapping potential with the perpendicular confinement length $a_\perp$ much smaller than the longitudinal one. In this case the effective interaction potential is well approximated by $U(x_1, x_2) = g_{1D}\delta(x_1 - x_2)$, where $g_{1D} \approx \frac{4\hbar^2 a_{3D}}{m a_\perp (a_\perp - 1.46 a_{3D})}$ [3], which makes the theoretical description relatively simple.

Artificial few-body systems are under intense research for possible applications in quantum information processing [5]. The parallel theoretical and experimental studies give also hope of better understanding the strongly-coupled many-body physics. In particular, characterizing states of many-body systems using tools and measures developed in quantum information theory appears useful and inspiring. One of the major conceptual tasks is the quantification of entanglement and analysis of its behavior in many-body states. This work aims to discuss quantum correlations and entanglement measures in the case of two-particle systems confined by different forms of external potentials. For one-dimensional two-particle systems with contact interaction we compare the behavior of various entanglement characteristics in the full range of positive and negative values of the effective interaction strength g_{1D}. For two-dimensional quantum systems with Coulombic interaction the influence of anisotropy of the harmonic confinement on entanglement properties is studied. The numerical approach is based on the Rayleigh-Ritz method in the basis of harmonic oscillator eigenfunctions with nonlinear parameters. We show that the strategy of spectral trace minimization appears effective in determining the energy spectrum and entanglement properties of the two-particle systems.

The plan of our work is as follows. In section 2 various correlation and entanglement measures are presented. A brief description of the numerical approach is given in section 3. In section 4 the efficiency of the spectral trace optimization in solving various two-particle problems is demonstrated. Section 5 is devoted to conclusion.

2. Quantum state characteristics

A quantum state of many-body system is described by N-particle wave function $|\Psi>=: \Psi(\mathbf{r},\sigma)$ that fulfils the Schrödinger equation

$$\widehat{H}\Psi(\mathbf{r},\sigma) = E\Psi(\mathbf{r},\sigma), \qquad \mathbf{r} = \{\vec{r_1},...,\vec{r_N}\},\ \sigma = \{\sigma_1,...,\sigma_N\}, \tag{1}$$

where σ_i denotes the spin of a particle. We consider the case of spin independent Hamiltonian

$$\widehat{H} = \sum_{i=1}^{N} -\frac{\hbar^2}{2m}\Delta_i + V(\vec{r_i}) + \sum_{i<j}^{N} U(\vec{r_i},\vec{r_j}), \tag{2}$$

when the $N-$particle wave function factorizes into $\Psi(\mathbf{r},\sigma) = \chi(\sigma)\phi(\mathbf{r})$. The spin part $\chi(\sigma)$ is a linear combination of the products of single-particle spin states $\chi_{i_1}(\sigma_1)...\chi_{i_N}(\sigma_N)$, whereas the spatial wave function depends on the interaction $U(\vec{r_i},\vec{r_j})$, and the trapping potential $V(\vec{r})$ through the partial differential equation

$$\sum_{i=1}^{N}[-\frac{\hbar^2}{2m}\Delta_i + V(\vec{r_i}) + \sum_{i<j}^{N} U(\vec{r_i},\vec{r_j})]\phi(\mathbf{r}) = E\phi(\mathbf{r}) \tag{3}$$

with an appropriate boundary condition. In the case of bound states, which is studied in the present work, the condition is of the form $\phi(\mathbf{r}) = 0,\ \nabla\phi(\mathbf{r}) = 0, |\mathbf{r}| \to \infty$, which means that $\phi(\mathbf{r}) \in \mathcal{L}^2$. The potentials $U(\vec{r_i},\vec{r_j})$ and $V(\vec{r})$ are assumed to be real, the solution of the Schrödinger equation $\phi(\mathbf{r})$ may be thus chosen as a real function, which is, however, unseparable if $U(\vec{r_i},\vec{r_j}) \neq 0$. In the case of identical particles, the total wave function must be antisymmetric under exchange of any pair of particles if they are fermions (half-odd-integer spin) and symmetric if they are bosons (integer spin).

2.1. Correlation and entanglement measures

The concept of entanglement is closely related to quantum correlations. It is commonly accepted that for identical particles the correlations arising purely from their indistinguishable nature should be excluded from the definition of quantum correlations [6, 7]. Therefore, for bosonic (fermionic) particles, the symmetrized (anti-symmetrized) products of one-particle functions are considered as non-entangled and deviations from such states are used to measure the amount of entanglement. Many ways of measuring the amount of correlation have been developed in quantum chemistry and quantum information science. The most used energetic, statistical and entropic measures will be described in subsections 2.2, 2.3, 2.4, respectively.

2.2. Correlation energy

The energetic correlation measure is usually defined with respect to the mean-field approximation that amounts to including the correlations due to particle statistics, and neglecting those that are due to the inter-particle interaction. The correlation energy is defined as the difference between the approximate and the exact energy of the system. For systems composed of fermionic particles, the mean-field coincides with the Hartree-Fock approximation and so defined measure is used in quantum chemistry since long time [8]. For bosonic systems the mean-field approximation is obtained by representing the many-body state by the product of identical spatial orbitals $\phi(\mathbf{r}) = \varphi(\vec{r}_1)....\varphi(\vec{r}_N)$ [2]. The energy functional, defined as the mean-field value of the Hamiltionian (1), reads

$$E[\varphi,\varphi^*] = \int \varphi^*(\vec{r})\left[-\frac{\hbar^2}{2m}\Delta + V(\vec{r})\right]\varphi(\vec{r})d^3r + \frac{(N-1)}{2}\int\int U(\vec{r},\vec{r}')|\varphi(\vec{r})\varphi(\vec{r}')|^2 d^3r d^3r' \tag{4}$$

and its minimization under the condition of constancy of the particle number provides the nonlinear Gross-Pitaevskii (GP) equation

$$\left[-\frac{\hbar^2}{2m}\Delta + V(\vec{r}) + \int U(\vec{r},\vec{r}')|\varphi(\vec{r}')|^2 d^3r'\right]\varphi(\vec{r}) = \mu\varphi(\vec{r}), \tag{5}$$

where the chemical potential μ has to be determined from the normalization condition $\int |\varphi_{GP}(\vec{r})|^2 d^3r = 1$. The solution of the GP equation yields the mean-field approximation to the energy $E_{GP} = E[\varphi_{GP},\varphi^*_{GP}]$, which determines the correlation energy by

$$E_c = E_{GP} - E_{exact}. \tag{6}$$

2.3. Kutzelnigg coefficient

Among various statistical correlation coefficients discussed by Kutzelnigg et al. [9] the simplest correlation measure is a scalar coefficient that for identical particles is defined as

$$\tau_{ij} = \frac{\langle\vec{r}_1\cdot\vec{r}_2\rangle_{ij} - \langle\vec{r}\rangle_i^2}{\langle\vec{r}^2\rangle_i - \langle\vec{r}\rangle_i^2}, \tag{7}$$

where the expectation values are weighted with the probability density $|\phi(\vec{r}_1,...,\vec{r}_N)|^2$ and integrated over the remaining variables. The Kutzelnigg coefficient is bounded by $|\tau_{ij}| \leq 1$.

2.4. Entropic entanglement measures

Entropic correlation measures are based on the Schmidt decomposition [5]. Here we consider the spatial correlations for a pure $N-$particle state $|\phi>$. Upon dividing the full system on two parts 1 and 2, the Schmidt decomposition takes a form

$$|\phi\rangle = \sum_l k_l|v_l\rangle|u_l\rangle, \tag{8}$$

where $\{v_l\}$ and $\{u_l\}$ are the orthonormal basis sets in the two Hilbert spaces. The spatial wave function $|\phi>$ defines the density matrix

$$\rho = |\phi><\phi| \tag{9}$$

that also contains full information on the state and may be conveniently used to define various quantities describing some of the system properties. For a chosen partition into two subsystems, the reduced density matrices are obtained by taking the trace over the degrees of freedom of the remaining part, namely

$$\rho_1 = Tr_2\rho, \ \rho_2 = Tr_1\rho. \tag{10}$$

The resulting decompositions of the reduced density matrices are given by

$$\rho_1 = \sum_l \lambda_l |v_l\rangle\langle v_l|, \ \rho_2 = \sum_l \lambda_l |u_l\rangle\langle u_l|, \tag{11}$$

where the number of nonzero terms in both sums is the same and is called the Schmidt rank. The common eigenvalues of both matrices, $\lambda_l = k_l^2$, are termed occupancies because $\sum_l \lambda_l = 1$. For distinguishable particles the state is non-entangled if its Schmidt rank equals to 1, otherwise the whole set of occupancies $\{\lambda_l\}$ characterizes the amount of entanglement in the considered state. In order to describe the entanglement by the one quantity, various measures have been constructed from the coefficients k_l, such as

- *von Neumann entropy* [10] $S_{vN} = -Tr\rho Log_2\rho - \sum k_l^2 log_2 k_l^2 = -\sum \lambda_l log_2 \lambda_l$
- *effective number of Schmidt modes* [11] $K = 1/Tr[\rho^2] = 1/\sum \lambda_l^2$
- *linear entropy* [12] $L = 1 - Tr[\rho^2] = 1 - 1/K$
- *robustness* [13] $R = (Tr[\phi])^2 - 1 = (\sum k_l)^2 - 1$

For identical particles the issue is more complicated then for distinguishable ones, since the correlations arising from the particle statistics have to be excluded from the definition of entanglement measures [6, 7]. This will be discussed later when we consider the two-boson and two-fermion systems.

3. Optimized Configuration Interaction Method

3.1. Configuration Interaction Method

One of most efficient ways to solve differential equations is the spectral approach which in quantum mechanics is known under the name of the *Rayleigh-Ritz* (RR) method [14]. The approach consists in approximating the solution of the Schrödinger equation as a finite linear combination

$$\phi(\mathbf{r}) \approx \phi^{(M)}(\mathbf{r}) = \sum_{\alpha=0}^{M-1} c_\alpha \psi_\alpha(\mathbf{r}), \tag{12}$$

where the trial functions $\psi_\alpha(\overrightarrow{r})$ are taken from some Hilbert space basis that may be assumed orthonormal for convenience. Treating the coefficients c_α as variational parameters, one obtains a set of linear equations

$$\sum_{\beta=0}^{M-1} (H_{\alpha\beta} - \varepsilon\delta_{\alpha\beta})c_\beta = 0, \ \alpha = 0, 1, ..., M-1, \tag{13}$$

where $H_{\alpha\beta} = \langle\alpha|\widehat{H}|\beta\rangle$ denotes the matrix elements of the Hamiltonian between the states of the basis, $|\alpha\rangle = \psi_\alpha$. The eigenvectors of the above equations $\{c_\beta\}_\alpha$ determine Mth order approximations to wave functions $\phi_\alpha^{(M)}(\overrightarrow{r})$. The corresponding eigenvalues $\varepsilon_\alpha^{(M)}$ are the variational upper bounds to the exact bound-state energies of $\alpha = 0, ..., M-1$ states. The accuracy of the method can be systematically improved by increasing the number M of included basis functions, which yields successive approximations to the larger and larger number of states. In this way a desired part of the spectrum may be determined with the approximate eigenvalues monotonically converging to the exact bound-state energies [14]. Considering the few-particle case it is convenient to use basis sets of a proper symmetry under exchange of particle, i.e., taking the basis functions in the bosonic (fermionic) case as permanents (determinants) constructed from the one-particle basis functions. In quantum chemistry such an approach is termed configuration interaction (CI) method. In condensed matter physics the method is known under the name of "exact diagonalization", although the results become exact only in the limit $M \to \infty$.

3.2. Nonlinear Parameters

The convergence properties of the RR method depend strongly on the choice of the expansion basis. The more similar are the trial functions to the exact eigenfunctions of the considered problem, the quicker is the convergence. Fortunately, flexibility of the method can be increased by introducing a set of adjustable parameters $\Omega = \{\Omega_1, ..., \Omega_K\}$ into the functions of the basis $|\alpha\rangle_\Omega$ and exploiting the freedom in choosing their values so as to improve the convergence of the scheme [15]. Optimization strategies may be divided into two classes: those where nonlinear parameters are fixed so as to obtain the quickest convergence to the minimum energy value, and those where they are fixed before performing RR diagonalization. The former approach applies to a particular level, the approximate energy of which is determined only after diagonalization of the RR matrix. Establishing the quickest convergence, requires thus repeating diagonalization many times for different values of nonlinear parameters, which leads to computationally expensive algorithms of iterative optimization [16]. Another problem is that the energy as a function of nonlinear parameters exhibits numerous local minima and deterministic optimization algorithms do not guarantee achieving a global minimum. This may be overcome by the stochastic variational approach where the search of the best values of nonlinear parameters is performed starting from randomly chosen sets of initial values. In this approach good convergence for small atoms, molecules and few-nucleon systems is obtained [17], but the approximate wave functions are not orthogonal, since calculation are performed for each of energy levels independently. Therefore, the strategy of fixing nonlinear parameters by optimization of the convergence rate is mainly used when only the ground state of the system is of interest and high accuracy is required.

Whenever a larger part of the spectrum is to be determined, fixing the values of nonlinear parameters before diagonalization of the RR matrix is more advantageous, since numerical calculations are much simpler and approximate eigenstates are automatically orthonormal. Various strategies of fixing nonlinear parameters are possible and their convergence properties have to be checked a posteriori for a given system. Fixing nonlinear parameters so as to ensure the proper asymptotic behavior of trial functions, has proven effective for simple systems [18], but an extension of this approach to other systems and other basis sets is problematic, since asymptotic expansions which are uniformly valid in the nonlinear parameters are difficult to construct. Therefore, ad hoc methods are more popular, e.g. with nonlinear parameters fixed so as to minimize the expectation value of the Hamiltonian in a chosen state of the trial basis, or so as to vanish some nondiagonal elements of the RR matrix. Depending on the considered system, one or the other gives a better convergence rate, but none of the method can be regarded as justified by any principle.

3.3. Spectral Trace Optimization

We observed [19] that the choice of nonlinear parameters might be based on the principle of minimal sensitivity [20] which states that nonphysical parameters introduced into calculation should be fixed so as to make the given order approximation to the physical quantity as less sensitive as possible to the variations of them. The only physical quantity which may be calculated before diagonalization of the RR matrix is its trace $Tr_M H = \sum_{\alpha=1}^{M} {}_\Omega < \alpha|\widehat{H}|\alpha >_\Omega$, which represents the Mth order approximation to the sum of energies of the M lowest bound-states. Therefore, in the optimized RR method [19] we require the values of nonlinear parameters in Mth order calculation to be chosen so as to render the trace of the truncated matrix $Tr_M H$ stationary

$$\frac{\delta}{\delta\Omega_i} Tr_M H = 0, \quad i = 1, ..., K. \tag{14}$$

The effectiveness of the optimized RR scheme may be demonstrated for one-dimensional oscillators with various powers of anharmonicity k, described by the Hamiltonians

$$\widehat{H} = -\frac{1}{2}\frac{d^2}{dx^2} + \frac{\omega^2 x^2}{2} + \lambda x^k, \tag{15}$$

using the basis of the eigenfunctions of the harmonic oscillator $\widehat{H} = -\frac{1}{2}\frac{d^2}{dx^2} + \frac{\Omega^2 x^2}{2}$ of an arbitrary frequency Ω, that are given by

$$\langle x|n\rangle_\Omega = \psi_n^\Omega(x) = \left(\frac{\Omega}{\sqrt{\pi}2^n n!}\right)^{\frac{1}{2}} H_n(\Omega x) e^{-\frac{\Omega^2 x^2}{2}}. \tag{16}$$

We have shown [19] that with Ω fixed in Mth order calculation so as to fulfil the stationarity condition (14), well converged results for the spectrum of the quartic oscillator ($k = 4$) in both the single- and double-well cases are obtained automatically. A particularly good test is provided by quasi-solvable example of the sextic oscillator $\widehat{H} = -\frac{1}{2}\frac{d^2}{dx^2} - \frac{35}{\sqrt{2}}x^2 + x^6$, for which nine states of even parity ($n = 0, 2, ..., 16$) are known exactly. The exponential convergence of the optimized RR scheme is evidenced in Fig.1(a) adapted from Ref. [21], where the relative errors $\delta E_n^{(N)}(\Omega_{opt})$ for the ground-state ($n = 0$) and the highest of the analytically known excited states ($n = 16$), are plotted in function of M. The quick convergence of various moments of the position operator $x_n^{k(M)}(\Omega_{opt}) = \langle n|x^k|n\rangle$, shown in Fig.1 (b) and (c) for $n = 0$ and $n = 16$ states, indicates that approximations to wave functions are also accurate. The optimized RR method works also well for oscillators with higher anharmonicities which present a more stringent test for effectiveness of nonlinear parameters optimization. Even in the difficult cases of multi-well oscillators with $k = 10$, the 50-digits accuracy is easily achieved with RR matrices of order $M < 100$ [21].

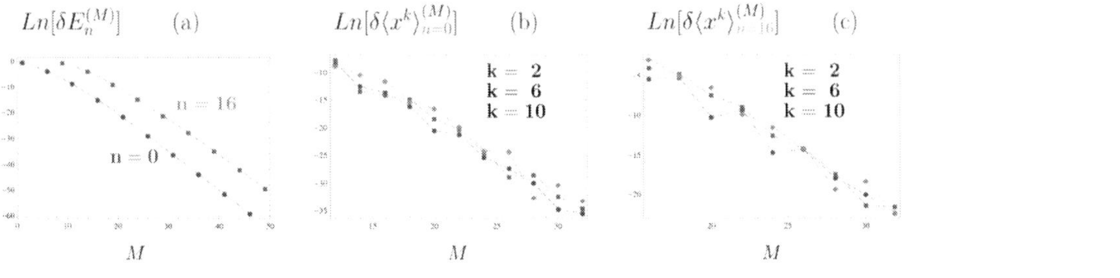

Figure 1. Relative errors, for energy $\delta E_n^{(M)}(\Omega_{opt})$ and various moments for n=0 and n=16 state of the sextic oscillator in function of the dimension M of the optimized RR matrix.

In the case of a central potential $V(r)$ at fixed angular momentum l, the Schrödinger equation for the radial function $\frac{u(r)}{r}$ reduces to the one-variable eigenequation for the radial Hamiltonian

$$\widehat{H}^{(l)} = -\frac{1}{2}\frac{d^2}{dr^2} + \frac{l(l+1)}{2r^2} + V(r). \tag{17}$$

An appropriate two-parameter basis for the RR method may be constructed from the set of eigenfunctions of the pseudo-harmonic oscillator Hamiltonian $\widehat{H} = -\frac{1}{2}\frac{d^2}{dr^2} + \frac{\Omega^2 r^2}{2} + \frac{(\gamma^2-2\gamma+3/4)}{2r^2}$ that are given by [22]

$$\langle r|n\rangle_{\gamma,\Omega} = u_n^{\gamma,\Omega}(r) = (-1)^n \frac{1}{\Gamma(\gamma)}\sqrt{\frac{2\Omega^\gamma\Gamma(\gamma+n)}{n!}}\, r^{\gamma-\frac{1}{2}} e^{-\frac{1}{2}r^2\Omega}\, {}_1F_1(-n;\gamma;\Omega r^2). \tag{18}$$

The strategy of fixing the parameters Ω and γ through the stationarity of the trace condition (14) has been tested [21] for the anharmonic oscillator Hamiltonian

$$\widehat{H}^{(l)} = -\frac{1}{2r}\frac{d^2}{dr^2}r + \frac{l(l+1)}{2r^2} + \frac{\omega^2 r^2}{2} + \lambda r^s. \tag{19}$$

The RR calculation are well defined if $\gamma > -s/2$, since only in this case the matrix elements ${}_{\Omega,\gamma}\langle m|r^k|n\rangle_{\Omega,\gamma}$ appear finite for $k = 1, 2, \ldots$ [23]. The method proved effective in determining the lower part of the spectrum even in the case of spiked oscillators with $s < -2$ which are most difficult to tackle numerically. The energy estimates in function of the truncation order M are shown in Fig.2 (adapted from Ref. [21]) on two quasi-solvable examples of the "antisextic" oscillator ($s = -6$) with $l = 0$, where ground-states energies are analytically known. The convergence is exponential, albeit its rate is generally slower than in the case of positive power oscillators. In contradistinction to the positive power case, where optimization of the Ω parameter was only important, for spiked oscillators the value of the parameter γ plays a crucial role. The optimum value of γ depends strongly on λ and grows with M. The condition $\gamma_{opt} > 3$ is fulfilled, although the smaller is λ, the closer γ_{opt} is to the value 3, where the matrix elements become singular, which explains the worsening of convergence with decreasing λ.

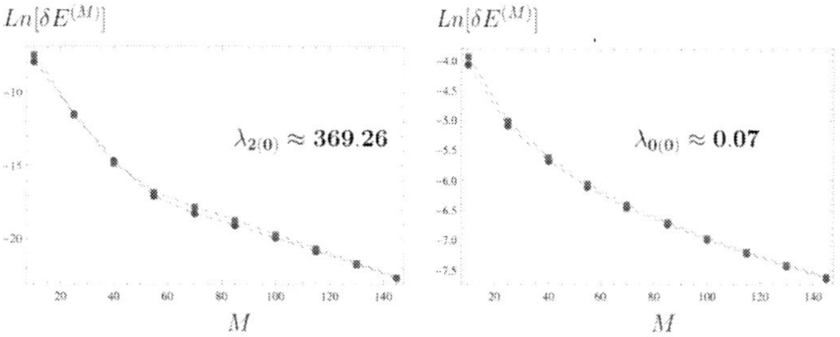

Figure 2. Relative energy errors for the ground-state of the radial "antisextic" oscillator in function of the dimension M of the optimized RR matrix for various interaction strength λ.

In Fig. 3 (adapted from Ref. [21]) the accuracy of our method of fixing nonlinear parameters is compared with that obtained with nonlinear parameters being optimum for convergence of the ground state energy of the "antisextic" oscillator [24] for the RR matrix of dimension $M = 80$. The later strategy gives very good approximation for the ground state, but the accuracy rapidly decreases for higher levels, while the trace optimization strategy provides a uniformly good approximation for all the lowest levels. This makes our strategy particularly appropriate when determination of a larger part of the quantum spectrum is required. This part of spectrum might be successfully utilized to obtain accurate results for the free energy [19], or the time evolution in anharmonic potentials [25]. With a modern software environment the optimized RR scheme may be implemented in a way allowing an arbitrary precision calculation by taking advantage of the present computer algebra abilities to deal with exact numbers. This is especially important for determining the tiny energy splitting in the case of multi-well potentials with nearly degenerate levels [21].

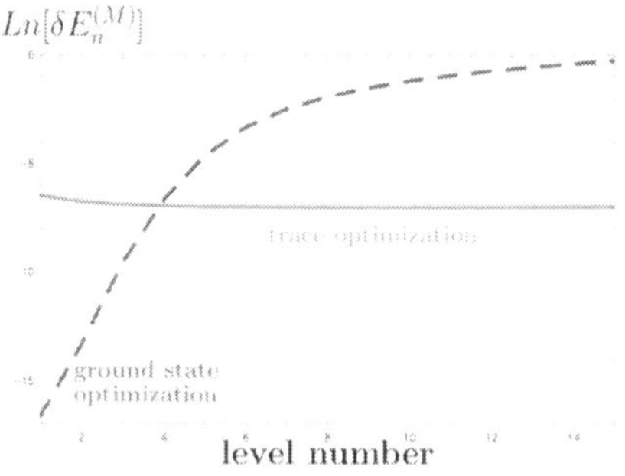

Figure 3. Relative errors of bound state energies of the spiked oscillator of $\lambda = 0.1$ in function of the level number, determined with nonlinear parameters chosen so as to optimize the ground-state energy (blue) and the trace of the RR matrix (red).

4. Two-particle systems

In the following we will study two-particle confined systems with the Hamiltonian given by

$$\widehat{H} = -\frac{\hbar^2}{2m}\Delta_1 + V(\overrightarrow{r_1}) - \frac{\hbar^2}{2m}\Delta_2 + V(\overrightarrow{r_2}) + U(\overrightarrow{r_1}, \overrightarrow{r_2}). \tag{20}$$

In this case, the only possible partition is into two one-particle systems with identical reduced density matrices of a simple one-particle form $\rho(\overrightarrow{r}, \overrightarrow{r}') = \int \phi(\overrightarrow{r}, \overrightarrow{r}'')\phi(\overrightarrow{r}'', \overrightarrow{r}')d^3r''$, the eigenfunctions of which are just the natural orbitals (NOs), well known in quantum chemistry. The Schmidt decomposition of the two-particle function is unambiguous

$$\phi(\overrightarrow{r_1}, \overrightarrow{r_2}) = \sum k_l v_l(\overrightarrow{r_1}) v_l(\overrightarrow{r_2}), \tag{21}$$

with the coefficients k_l being real but not necessarily positive, which enables determination of occupancies $\lambda_l = k_l^2$ without necessity of calculating the density matrix. The coefficients k_l may be directly calculated using the optimized CI approach by representing the spatial wave function as

$$\phi(\overrightarrow{r_1}, \overrightarrow{r_2}) = \sum_{j \geq i} c_{ij}\psi_{ij}(\overrightarrow{r_1}, \overrightarrow{r_2}), \tag{22}$$

where $\psi_{ij}(\overrightarrow{r_1}, \overrightarrow{r_2}) = b_{ij}[\varphi_i^{\Omega}(\overrightarrow{r_1})\varphi_j^{\Omega}(\overrightarrow{r_2}) \pm \varphi_j^{\Omega}(\overrightarrow{r_1})\varphi_i^{\Omega}(\overrightarrow{r_2})]$ with $b_{ii} = 1/2$ and $b_{ij} = 1/\sqrt{2}$ for $i \neq j$, and the upper (lower) sign corresponds to bosons (fermions). Using the same basis of an arbitrary frequency Ω (16) to expand the natural orbitals as

$$v(\overrightarrow{r}) = \sum p_n \varphi_n^{\Omega}(\overrightarrow{r}), \tag{23}$$

turns the eigenproblem (21) into an algebraic equation

$$\sum (A_{mn} - k_n \delta_{mn}) p_n = 0, \tag{24}$$

where $A_{nn} = c_{nn}$ and $A_{mn} = 2^{-1/2} c_{mn}$ for $m \neq n$ with c_{nm} determined from diagonalization of $[H]_{M \times M}$. By diagonalization of the matrix $[A]_{M \times M}$, the approximate coefficients k_n and the corresponding occupancies $\lambda_n = k_n^2$ may be determined. Due to the fact that $\sum A_{nm}^2 = 1$, their numerical values satisfy $\sum k_n^2 = \sum \lambda_n = 1$. The application of the method for determining two-particle wave functions has been tested in the doctor's thesis of P.Kościk [26].

We will show the efficacy of optimization of the CI method through the stationarity of the trace condition in determining spectra and NOs of trapped two-particle systems on two examples. First is a one-dimensional system in anharmonic confinement of single- and double-well shape with the interparticle interactions of the contact type. The second is a two-dimensional system of Coulombically interacting particles confined in an anisotropic harmonic potential.

4.1. Two-boson system

The state of two identical bosons is non-entangled only if its Schmidt rank equals 1 or 2 [6, 7]. In the first case, $\lambda_0 = 1$ and the state is the product of the same natural orbital

$$\phi(\mathbf{r}_1, \mathbf{r}_2) = v_0(\mathbf{r}_1)v_0(\mathbf{r}_2). \tag{25}$$

We will call such a state "condensed", as representing a two-particle analogue of the Bose-Einstein condensate, according to the Penrose-Onsager criterion [27]. In the second case, $\lambda_0 = \lambda_1 = \frac{1}{2}$, and the Schmidt decomposition of wave function comprises two cases

$$\phi(\mathbf{r}_1, \mathbf{r}_2) = \frac{1}{\sqrt{2}}\left(v_0(\mathbf{r}_1)v_0(\mathbf{r}_2) \pm v_1(\mathbf{r}_1)v_1(\mathbf{r}_2)\right). \tag{26}$$

In both cases the two natural orbitals are equally populated and wave function can be written as a permanent of two orthogonal orbitals, $\phi(\mathbf{r}_1, \mathbf{r}_2) = \frac{1}{\sqrt{2}}\left(\eta(\mathbf{r}_1)\chi(\mathbf{r}_2) + \chi(\mathbf{r}_1)\eta(\mathbf{r}_2)\right)$, where $\eta = \frac{1}{\sqrt{2}}(v_0 - iv_1)$, $\chi = \frac{1}{\sqrt{2}}(v_0 + iv_1)$ for the plus sign, and $\eta = \frac{1}{\sqrt{2}}(v_0 - v_1)$, $\chi = \frac{1}{\sqrt{2}}(v_0 + v_1)$ for the minus sign in (26). We will call the states of this kind "fragmented", as being two-particle examples of the many-body fragmentation phenomenon [28].

Here we consider a one-dimensional system of two bosons in an anharmonic trapping potential

$$V(x) = wx^2 + x^4. \tag{27}$$

For $w < 0$ the potential has two wells, the depth of which decreases with increasing w, undergoing a transition to single-well behavior at $w = 0$. For $w \geq 0$ the potential shape changes from purely quartic to a harmonic one in the limit $w \to \infty$. A pleasant feature of one-dimensional bosonic systems is their solvability for arbitrary trapping potential in the strong repulsion limit. At $g_{1D} \to \infty$, the $N-$boson wave-function may be represented via Bose-Fermi mapping as the modulus of the wave function of N noninteracting fermions [29], which is known as the Tonks-Girardeau gas and provides useful asymptotic results for verifying numerical calculations performed at large g_{1D}. In the case of a two-particle system, the $g_{1D} \to \infty$ limit of the ground state wave function is given by $\psi^{TG}_{2,Bose}(x_1, x_2) = \frac{1}{\sqrt{2}}|\psi_0(x_1)\psi_1(x_2) - \psi_1(x_1)\psi_0(x_2)|$, where ψ_0 and ψ_1 are the two lowest one-particle states.

The harmonic limit of the trapping potential (27) is the easiest to discuss, since in this case the two-particle eigenfunctions $\phi(\mathbf{r}_1, \mathbf{r}_2)$ are known exactly in an implicit form [30] and the NO's occupancies may be determined by performing their diagonalization numerically. The three largest occupancies, $\lambda_0, \lambda_1, \lambda_2$ for the ground state are plotted in Fig.4 in function of the interaction strength. At $g_{1D} = 0$, the value of λ_0 equals 1, and only in this case all other occupancies do vanish, which means that the state of two harmonically confined bosons is non-entangled only if the bosons do not interact. The weakly entangled "condensed state", where only one NO is significantly occupied, is realized only in the case of very weak interaction, $|g_{1D}| < 0.1$. With increasing $|g_{1D}|$ the entanglement grows, which shows up in the smooth increase of λ_1 and λ_2 at the cost of λ_0, for both repulsive and attractive interactions. In the repulsive case, the occupancies monotonically approach their Tonks-Girardeau limits $\lambda_0^{TG} \approx 0.7745$, $\lambda_1^{TG} \approx 0.1765$ and $\lambda_2^{TG} \approx 0.0414$. In this limit all the remaining NOs are also occupied, however the sum of their occupancies is only about 0.0076. In the case of strongly attractive interaction, the situation is different, since the ground state does not tend to the fermion-like state. In fact, the fermionized state exists but there are infinitely many states of energy lower than this. As can be seen in Fig.4, at $g_{1D} \to -\infty$ the occupancies of NO's become uniformly small, which means that in this limit the ground state is maximally entangled.

Having obtained a sufficient amount of the coefficients k_l, we determined the accurate approximations to the correlation measures discussed in Sect.2.1. The results are presented

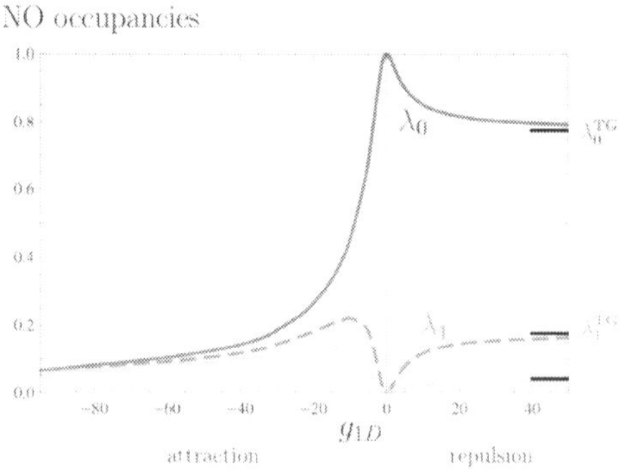

Figure 4. The largest occupancies λ_0, λ_1 and λ_2 for the harmonically confined two-boson system in function of g_{1D}, their Tonks-Girardeau limits λ_0^{TG}, λ_1^{TG} and λ_2^{TG} are marked by horizontal lines.

in Fig. 5 in comparison with the correlation energy and the Kutzelnigg coefficient. All the correlation measures grow monotonically with increasing interaction strength $|g_{1D}|$, and all but the correlation energy remain bounded in the limit $g_{1D} \to \infty$.

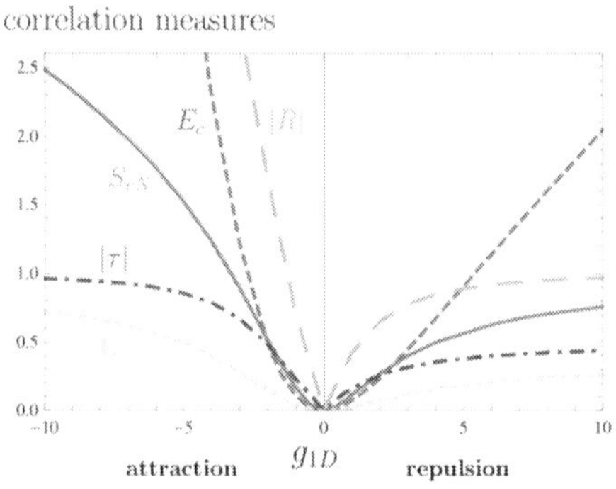

Figure 5. Comparison of the correlation energy E_c, robustness R, von Neumann entropy S_{vN}, linear entropy L, and the Kutzelning coefficient τ, in function of the interaction strength g_{1D}, for the two-boson system confined in the harmonic oscillator potential.

The role played by correlations can be examined further by testing the number of NOs that are needed for accurate determination of the two-particle wave function by the expansion (21) truncated at the nth term, $\psi^{nNO}(x_1, x_2)$. The successive approximation to the ground state energy $E^{nNO} = \langle\psi^{nNO}|H|\psi^{nNO}\rangle/||\psi^{nNO}||^2$ obtained by including up to four most occupied orbitals are shown in Fig.6 in function of the interaction strength and compared with the exact result and the GP approximation. As $|g_{1D}|$ increases, more and more orbitals in the sum (21) are necessary to obtain a satisfactory approximation to the energy. The one-orbital approximation and the GP energy E_{GP} are very close up to $|g_{1D}| \approx 5$, only for larger values of $|g_{1D}|$ the GP approximation performs slightly better than E^{1NO}. However, the difference between the exact energy and the GP approximation (correlation energy) starts to be noticeable already at $|g_{1D}| > 0.1$ and grows monotonically for increasing interaction strength. This shows that the correlation effects may be neglected in determining the ground state energy only in the case of very weak interactions.

The correlations effects are best visible in the two-boson probability density. In Fig.7 the approximate GP two-particle density is compared with the approximations obtained by taking into account one, two, and three NOs, and the exact density for strongly correlated case ($g_{1D} = 5$). In this case the bosons are located with great probability on diametrical positions with respect to the center of the trap, which shows up in the exact two-particle density. The GP distribution does not capture this behavior. The one-orbital approximation differs from the GP approximation but also fails. Only after taking three NOs into account the approximate two-particle distribution properly describes the exact behavior.

Journal of Physics: Conference Series **213** (2010) 012004 doi:10.1088/1742-6596/213/1/012004

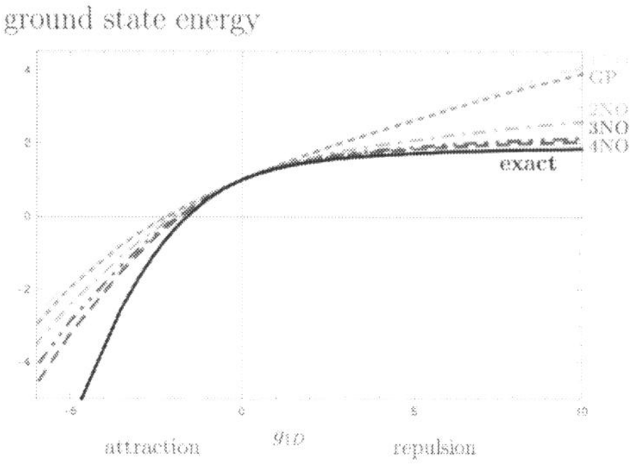

Figure 6. Ground-state energy of the harmonically confined two-boson system in the approximation including n natural orbitals (nNO) for $n = 1, 2, 3, 4$ compared with the GP approximation and the exact result.

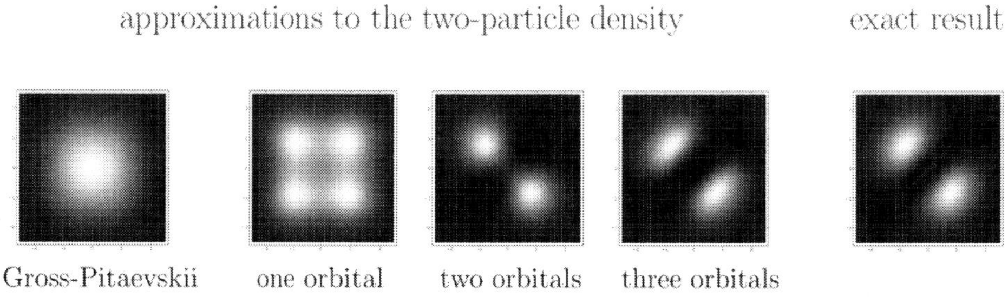

Figure 7. The GP density $|\psi_{GP}(x_1, x_2)|^2$ (left column) compared with approximations $|\psi^{nNO}(x_1, x_2)|^2$ for $n = 1, 2$ and 3 orbitals (three middle columns) and the exact result $|\psi(x_1, x_2)|^2$ (right column) for two harmonically confined boson interacting with the strength $g_{1D} = 5$.

In the case of anharmonic confinement (27) we applied the optimized CI method with the harmonic oscillator one-particle basis (16) to solve the two-boson problem. With Ω fixed through the stationarity of the trace condition (14) a sufficiently large set of NOs may be obtained so as to determine the von Neumann entropy to a desired accuracy. In Fig.8 the results are plotted in function of $g1D$ for several values of the parameter w. For $w > 0$, when the trapping potential is single-well shaped, the dependence of von Neumann on g_{1D} is similar as in the harmonic confinement case. For decreasing w, the increase of the von Neumann entropy becomes more rapid and around $w \approx -3$ the character of the dependence changes markedly. The deeper is the double-well potential the smaller is the value of g_{1D} for which the Tonks-Girardeau behavior $S_{vN}^{TG} = 1$ is approximately achieved. This tendency has been observed before for the δ-split [31] and double-well potential [32].

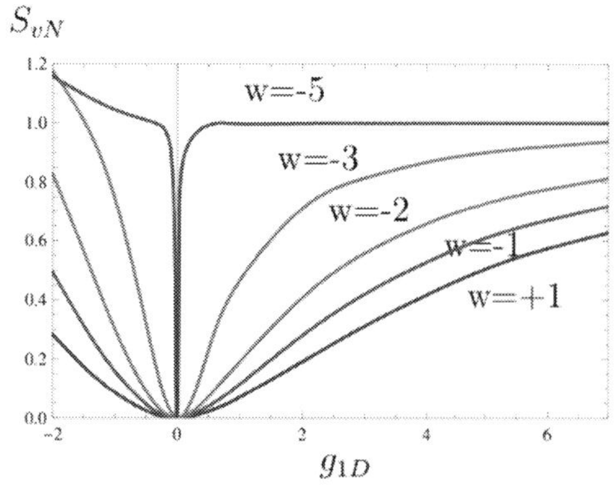

Figure 8. The von Neumann entropy for anharmonic confining potential $V(x) = wx^2 + x^4$ for $w = 1, -1, -2, -3, -5$ as a function of $g1D$.

The analysis of the Tonks-Girardeau limit, $g_{1D} \to \infty$, is presented in function of the parameter w in Fig.9. The two largest occupancies, shown on the left side of the figure, approach the Tonks-Girardeau values of the harmonically confined system in the limit of $w \to \infty$. For finite values of w, the anharmonic term in the potential (27) plays a role, being the more important the smaller is the value of w. Below $w \approx -3$, when the confinement potential becomes deeply double-well shaped, the occupancy of the orbital v_1 increases more rapidly and for $w < -5$ becomes practically equal to the occupancy of the lowest orbital ($\lambda_0 \approx \lambda_1 \approx 0.5$), so the state is almost perfectly fragmented and non-entangled. The Tonks-Girardeau limit of the von Neumann entropy in function of w is shown in the right part of Fig.9. In the limit of $w \to -\infty$ the state becomes non-entangled as $S_{vN} \to 1$ and the Schmidt rank tends to 2 with $\lambda_0 \approx \lambda_1 \to 1/2$. However, the dependence on w is not monotonous and S_{vN} takes also the value 1 at a finite w, but the corresponding ground state is entangled, since the values of two lowest occupancies (marked by crosses on the figure) do not sum up to one and the remaining Schmidt eigenvalues do not vanish. For identical bosons the value of entropy alone is not sufficient to assert whether the state is entangled or not, which was also observed in the case of delta-split potential [32].

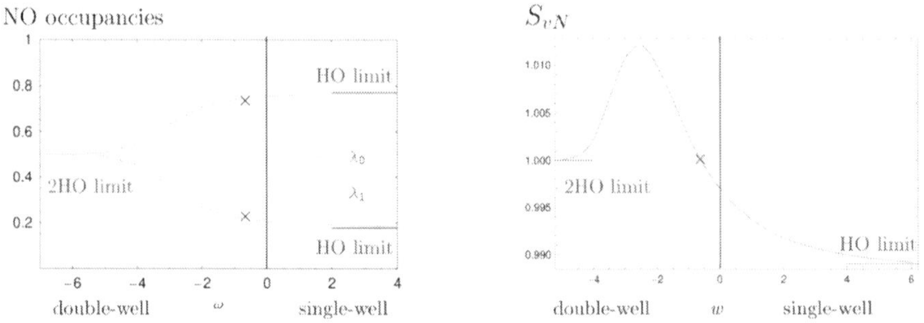

Figure 9. The two largest NO occupancies (left) and the von Neumann entropy (right) in the anharmonic confining potential $V(x) = wx^2 + x^4$ in the Tonks-Girardeau limit. The limits of a harmonic confinement $w \to \infty$ (HO limit) and of two independent harmonic wells $w \to -\infty$ (2HO limit) are marked.

4.2. Two fermions

A two-dimensional system of charged particles in an anisotropic harmonic potential is commonly used as a model of a semiconductor QD [1]. The Hamiltonian of a two-electron QD is given by

$$\hat{H} = \sum_{i=1}^{2}[\frac{\mathbf{p}_i^2}{2} + \frac{1}{2}(\omega_x^2 x_i^2 + \omega_y^2 y_i^2)] + \frac{e^2}{|\mathbf{r}_2 - \mathbf{r}_1|}. \tag{28}$$

After rescaling $E \to e^4 E$ and $\omega_{x,y} \to 2e^4\omega_{x,y}$, the Schrödinger equation takes the form

$$\left[\sum_{i=1}^{2}[-\frac{1}{2}\nabla_{r_i}^2 + 2\omega_x^2 x_i^2 + 2\omega_y^2 y_i^2] + \frac{1}{|\mathbf{r}_2 - \mathbf{r}_1|}\right]\phi(\mathbf{r}_1, \mathbf{r}_2) = E\phi(\mathbf{r}_1, \mathbf{r}_2), \tag{29}$$

where the dimensionless parameters $\omega_{x,y}$ denote the ratio of confinement to Coulombic interaction energies. The center of mass motion may be separated out and solved exactly. The relative motion equation

$$[-\frac{\partial^2}{\partial x^2} - \frac{\partial^2}{\partial y^2} + \omega_x^2 x^2 + \omega_y^2 y^2 + \frac{1}{\sqrt{x^2+y^2}}]\phi^r = E^{\mathbf{r}}\phi^r \tag{30}$$

is separable in only two particular cases: $\omega_x = \omega_y = \omega_\rho$ (in polar coordinates) and $\omega_y : \omega_x = 2$ (in parabolic coordinates). Both cases are quasi-solvable. The closed-form solutions exist at particular values ω_ρ of the isotropic confinement [33] and at particular values $\omega_y = 2\omega_x$ of the anisotropic confinement [34]. The values of frequencies at which the exact solutions are known correspond to the weak confinement case ($\omega_{x,y} << 1$), where the correlations between the electrons are the strongest and tend to create the so-called Wigner molecules. This regime is especially interesting, as the numerical calculations are rather demanding. The exact solutions will thus serve us to check the results of calculation performed by the optimized RR method.

We aim to discuss how the anisotropy of confinement influences the entanglement properties of the system. Considering the two-fermion ground-state with symmetric spatial wave function, which corresponds to the singlet spin structure, we determined the coefficients k_l of the Schmidt decomposition of the two-particle function at different aspect ratios of the confinement frequencies. The resulting von Neumann entropy is presented in Fig.10 in function of the dimensionless parameter $\sqrt{\omega_x^{-1}}$ that represents the ratio of interaction to confinement energies. It can be seen that for weak interactions the entanglement properties do not depend on the shape of the trap. For stronger interactions the dependence on the aspect ratio of harmonic frequencies grows, the von Neumann entropy being smaller at stronger anisotropies.

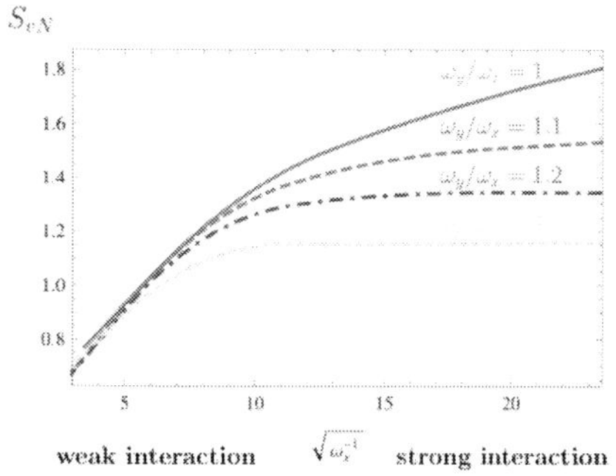

Figure 10. Von Neumann spatial entropy as a function of $\sqrt{\omega_x^{-1}}$ for anisotropy ratios ω_x/ω_y = 1, 1.1, 1.2, and 2.

5. Conclusion

We applied the optimized CI method to calculate spectra of two-particle confined systems. Optimization of the scheme has been done by introducing nonlinear parameters into the functions of the one-particle basis and fixing their values by minimization of the trace of the truncated Hamiltonian matrix. Such optimization strategy proved efficient in solving one-particle problems to arbitrary precision with the computational cost much lower than that of iterative optimization of nonlinear parameters. In our scheme the desired part of energy spectrum is determined to a desired accuracy in one run and the approximate eigenvectors are mutually orthogonal, which enables effective determination of various system properties.

In the two-particle case the optimized CI method enables a direct derivation of the natural orbital occupancies and determination of entropic entanglement measures. Comparison of various entanglement measures has been performed for the ground state of two bosons in a one-dimensional trap. The dependence of the von Neumann entropy on the interaction strength has been discussed on the example of anharmonic confinement potential when its shape changes from a harmonic to a deep double-well. Influence of anisotropy on spatial entanglement properties has been discussed for a singlet ground-state of the two-dimensional QD consisting of two Coulombically interacting electrons confined in an anisotropic harmonic potential. The increase of correlations with increasing anisotropy has been demonstrated.

Acknowledgments
Thanks go to Dr. Przemysław Kościk for valuable computational assistance and stimulating discussions of the results.

References
[1] Jacak L, Hawrylak P, Wójs A 1997 *Quantum Dots* (Berlin: Springer)
[2] Pethick C J and Smith H 2002 *Bose-Einstein Condensation in Dilute Gases* (Cambridge: Cambridge University Press)
[3] Bloch I, Dalibard J and Zwerger W 2008 Rev. Mod. Phys. **80** 885
[4] Kinoshita T, Wenger T, and Weiss D S 2004 Science **305** 1125
[5] Nielsen M A and Chuang I L 2000 *Quantum Computation and Quantum Information* (Cambridge: Cambridge University Press)
[6] Eckert K, Schliemann J, Bruß D and Lewenstein M (2002) Ann.Phys.(N.Y.) **299** 88
[7] Ghirardi G, Marinatto L 2004 Phys. Rev. **A 70** 012109
[8] Lövdin P-O 1959 *J.Mol.Spectr.* **3** 46
[9] Kutzelnigg W, del Re G, Bertier G 1968 Phys. Rev. **172** 49
[10] Bennett C H, Bernstein H J, Popescu S, and Schumacher B 1996 Phys.Rev.**A 53** 2046
[11] Grobe R, Rzążewski K and Eberly J H 1994 J.Phys. **B 27** L503
[12] Wu L-A, Sarandy M S, Lidar D A, and Sham L J 2006 Phys. Rev.**A 74** 052335
[13] Vidal G and Tarrach R 1999 Phys.Rev. **A 59** 141
[14] Reed M and Simmons B 1978 *Methods of Modern Mathematical Physics, Vol.4* (New York: Academic) p.82
[15] Hylleraas E A 1929 *Zeit.Phys.* **54** 347
[16] Rychlewski J and Komasa J 2003 in *Explicitly Correlated Gaussian Functions in Chemistry and Physics* (Dordrecht: Kluwer) p.91
[17] Suzuki Y and Varga K 1998 *Stochastic Variational Approach to Quantum-Mechanical Few-Body Problems* (Berlin: Springer)
[18] Hill R N 1995 Phys.Rev.**A 51** 4433
[19] Okopińska A 1987 Phys.Rev.**D36** 1273
[20] Stevenson P M 1981 Phys.Rev.**D23** 2961
[21] Kościk P and Okopińska A 2007 J. Phys. A: Math. Theor. **40** 10851
[22] Davidson P M 1932 Proc.R.Soc.London **130** 459
[23] Hall R L and Saad N 2000 J. Phys.**A33** 569
[24] Saad N Hall R L and Katatbeh Q D 2005 J.Math.Phys. **A46** 022104
[25] Amore P, Aranda A, Fernandez F M and Jones H, Phys. Lett. **A340** (2005) 87
[26] Kościk P 2009 Doctor's thesis, Jan Kochanowski University, Kielce
[27] Penrose O and Onsager L 1956 Phys. Rev. **104** 576
[28] Nozières P and Saint James D 1982 J. Phys. (Paris) **43** 1133
[29] Girardeau M 1960 *J.Math.Phys.* **1** 516
[30] Busch T, Englert B, Rzążewski K and Wilkens M 1998 Found. Phys. **28** 549
[31] Murphy D S, McCann J F, Goold J and Busch T 2007 Phys.Rev. **A76** 053616
[32] Murphy D S and McCann J F 2008 Phys.Rev. **A77** 063413
[33] Taut M 1994 J. Phys. A: Math. Gen.**27** 1045
[34] Kościk P and Okopińska A 2007 J. Phys. A: Math. Theor. **40** 1045

Journal of Physics: Conference Series **213** (2010) 012005 doi:10.1088/1742-6596/213/1/012005

Quantum information and phase transitions: fidelity and state distinguishability

V R Vieira

Centro de Física das Interacções Fundamentais
and Departamento de Física
Instituto Superior Técnico
Universidade Técnica de Lisboa
Lisboa, Portugal

E-mail: vrv@cfif.ist.utl.pt

Abstract. We review the definition and properties of fidelity as a measure of distinguishability of quantum states, presenting some of its geometric aspects.

We apply then fidelity to the study of phase transitions in condensed matter systems, both at zero and finite temperatures. It does not rely on any already known order parameter, being useful not only in the usual Landau type phase transitions but also in phase transitions displaying topological order.

The so called fidelity susceptibility and its relation to the usual thermodynamic susceptibilities and response functions are discussed.

Finally, we consider the fidelity between different reduced density matrices for a subsystem of a larger system.

1. Introduction

In the last years, there have been very important developments in an interdisciplinary approach involving different fields of physics and quantum information and computation [1].

Probably the most well known of these developments involves entanglement. Entanglement was considered by Schrödinger, already in 1935, at the time of the Einstein-Podolsky-Rosen paradox, not one, but rather the characteristic trait of quantum mechanics, the one that enforces its entire departure from classical lines of thought. Physicists became used to it, in particular in the study of the correlation functions of physical systems, and of their subsystems, but only recently entanglement was taken seriously as an important resource for quantum information science, in particular in quantum computation, communication and cryptography [2].

Important steps have also been given in quantum computation, both theoretically, with the development of quantum algorithms, and experimentally, with the development of new physical systems and tools, regarding the applications.

Quantum mechanics is a probabilistic theory, allowing us to make predictions on the possible outcomes of experiments performed on physical systems, which have been previously prepared. It is then natural to expect that the techniques of estimation and the methods of statistical inference used in information theory would be very valuable in addressing the questions and problems posed by quantum mechanics, not only at a fundamental level, but also in its applications to different areas in physics, namely condensed matter physics.

We will consider the application of fidelity, a measure of distinguishability between probability distributions, to the distinguishability between quantum states of a system and to the study of phase transitions in condensed matter physics, stressing some of its geometrical features.

2. Classical Fidelity

Classically, the fidelity between two probability distributions p_i and q_i is defined by

$$F(p,q) = \sum_i \sqrt{p_i q_i} \tag{1}$$

(sometimes, in the literature, fidelity is defined as the square of this quantity). It can be understood as the inner product of the two unit vectors $\sqrt{p}$ and $\sqrt{q}$, of components $\sqrt{p_i}$ and $\sqrt{q_i}$ respectively. One has $0 \leq F(p,q) \leq 1$, with $F(p,q) = 1$, only when the two probability distributions are the same.

Fidelity itself is not a distance, but can be used to define distances. The Bures angle is defined by

$$D_A(p,q) = \arccos F(p,q) = \theta. \tag{2}$$

Geometrically, it represents the distance between the two vectors, measured along the sphere of unit length. The Bures distance is defined by

$$D_B(p,q) = \sqrt{2(1 - F(p,q))} = 2\sin\frac{\theta}{2} \tag{3}$$

and represents the length of the chord between the two points in the sphere. Finally, the so called sine distance is defined by

$$S_B(p,q) = \sqrt{1 - F^2(p,q)} = \sin\theta. \tag{4}$$

If the two probability distributions are infinitesimally close, these distances become the same

$$ds^2 = \sum_i (d\sqrt{p_i})^2 = \frac{1}{4}\sum_i \frac{dp_i^2}{p_i}, \tag{5}$$

which is the Fisher metric (in the case of several parameters λ^μ one has $dp_i = \sum_\lambda d\lambda^\mu \frac{\partial p_i}{\partial \lambda^\mu}$). This parameter dependence induces a metric in parameter space from the metric in distribution space which we are considering.

Sometimes, it is convenient to consider non-normalized distributions $\vec{r}$ and then normalize them, according to $\sqrt{p} = \frac{\vec{r}}{r}$. From $dr = \frac{\vec{r}}{r} \cdot d\vec{r}$, one has

$$d\sqrt{p} = \frac{d\vec{r}_\perp}{r} = \frac{d\vec{r}}{r} - \frac{\vec{r}}{r}\left(\frac{\vec{r}}{r} \cdot \frac{d\vec{r}}{r}\right), \tag{6}$$

removing the component of $d\vec{r}$ along $\vec{r}$, such that $\vec{r} \cdot d\vec{r}_\perp = 0$. The inner product between two deviations $d\sqrt{p}_1$ and $d\sqrt{p}_2$, from a probability distribution $\sqrt{p}$, is then given by

$$d\sqrt{p}_2 \cdot d\sqrt{p}_1 = \frac{d\vec{r}_{2\perp}}{r} \cdot \frac{d\vec{r}_{1\perp}}{r} = \frac{d\vec{r}_2 \cdot d\vec{r}_1}{r^2} - \frac{(d\vec{r}_2 \cdot \vec{r})(\vec{r} \cdot d\vec{r}_1)}{r^4}, \tag{7}$$

which is the Fubini-Study metric. For a single variation, it reduces to

$$D^2_{FS}(d\vec{r}) = d\sqrt{p}^2 = \left(\frac{d\vec{r}_\perp}{r}\right)^2 = \left(\frac{d\vec{r}}{r}\right)^2 - \left(\frac{\vec{r} \cdot d\vec{r}}{r^2}\right)^2. \tag{8}$$

One sees then the Fisher metric as the metric of the unit sphere, also described by the Fubini-Study metric.

The Fisher metric is associated to the Fisher information, i.e. the variance of $\ln p$, given by

$$\langle(\delta \ln p)^2\rangle = \sum_i p_i(\delta \ln p_i)^2 = \sum_i \frac{(\delta p_i)^2}{p_i} \tag{9}$$

$$= \sum_i \delta p_i \delta \ln p_i = -\langle\delta^2 \ln p\rangle, \tag{10}$$

since its expectation $\langle\delta \ln p\rangle = \sum_i p_i \delta \ln p_i = \delta \sum_i p_i = 0$.

In information theory, the Fisher information is used in connection to the Cramér-Rao inequality to set a lower bound on the variance of estimators [3]. Using the Cauchy-Schwarz inequality $\sqrt{\text{var}(T)\text{var}(V)} \geq |\text{cov}(V,T)|$, with $V = \frac{d\ln p}{d\lambda}$, $\text{var}(V)$ is the Fisher information and $\text{cov}(V,T) =< \frac{d\ln p}{d\lambda}T >= \sum_i \frac{dp_i}{d\lambda}T$.

The fidelity between two probability distributions described by $\vec{r}_1 = \vec{r}+\lambda d\vec{r}_1$ and $\vec{r}_2 = \vec{r}+\lambda d\vec{r}_2$ is given by

$$F = 1 - \frac{1}{2}D^2_{FS}(d\vec{r}_1 - d\vec{r}_2)\lambda^2 + O(\lambda^3), \tag{11}$$

i.e. the deviation of the fidelity from 1 is, in lowest order, the square of the Fubini-Study metric of the difference of the variations. One can vary either of them or both. If the variations are the same, the fidelity stays equal to 1.

The derivation of eq. 11 is the same as for the quantum case, treated in the next section.

3. Quantum Fidelity

We will define the fidelity between two states of a quantum system considering first the case of pure states, next the case where one of the states is a mixed state but the other is still a pure state, and finally, the case of two mixed states.

3.1. Pure States

If the two states of the quantum system are pure, being specified by the normalized wave functions $|\varphi_1\rangle$ and $|\varphi_2\rangle$ (with the states given by the respective projectors), the fidelity between them is defined by

$$F(|\varphi_1\rangle\langle\varphi_1|, |\varphi_2\rangle\langle\varphi_2|) = |\langle\varphi_2|\varphi_1\rangle|, \tag{12}$$

i.e. is given by the modulus of the overlap, in agreement to its physical interpretation as the probability of the transition between those two states.

In general, we prefer to work with a non normalized wave function $|\psi\rangle$. The normalized wave function is then given by $|\varphi\rangle = \frac{|\psi\rangle}{\sqrt{\langle\psi|\psi\rangle}}$. Its phase can be changed by a gauge transformation, one of the basic invariances of quantum mechanics, transforming a wave $|\psi\rangle$ into $|\psi'\rangle = e^{i\chi}|\psi\rangle$, giving $|d\psi'\rangle = e^{i\chi}(|d\psi\rangle + id\chi|\psi\rangle)$. Projectors are phase independent and therefore gauge invariant. Only relative phases are relevant and the physical results should be invariant under gauge transformations.

The variation of $|\varphi\rangle$ is given by

$$|d\varphi\rangle = \frac{|d\psi\rangle}{\sqrt{\langle\psi|\psi\rangle}} - \frac{1}{2}\frac{|\psi\rangle}{\sqrt{\langle\psi|\psi\rangle}}\frac{\langle d\psi|\psi\rangle + \langle\psi|d\psi\rangle}{\langle\psi|\psi\rangle}. \tag{13}$$

One does not have $\langle\psi|d\varphi\rangle = 0$, since the normalization condition of the wave function implies $\langle d\varphi|\psi\rangle + \langle\psi|d\varphi\rangle = 0$, instead. However, one can make the decomposition

$$|d\varphi\rangle = |d\varphi\rangle_\perp + |d\varphi\rangle_\parallel \tag{14}$$

with

$$|d\varphi\rangle_\perp = \frac{|d\psi\rangle}{\sqrt{\langle\psi|\psi\rangle}} - \frac{|\psi\rangle}{\sqrt{\langle\psi|\psi\rangle}}\frac{\langle\psi|d\psi\rangle}{\langle\psi|\psi\rangle} = \left(I - \frac{|\psi\rangle\langle\psi|}{\langle\psi|\psi\rangle}\right)\frac{|d\psi\rangle}{\sqrt{\langle\psi|\psi\rangle}} \tag{15}$$

$$|d\varphi\rangle_{||} = -\frac{1}{2}\frac{\langle d\psi|\psi\rangle - \langle\psi|d\psi\rangle}{\langle\psi|\psi\rangle}\frac{|\psi\rangle}{\sqrt{\langle\psi|\psi\rangle}} = i\alpha|\varphi\rangle, \tag{16}$$

with $\langle\psi|d\varphi\rangle_\perp = 0$ and $\langle\varphi|d\varphi\rangle_{||} = i\alpha$, where $\alpha = \frac{i}{2}\frac{\langle d\psi|\psi\rangle - \langle\psi|d\psi\rangle}{\langle\psi|\psi\rangle}$ is real, in agreement with the normalization condition $\langle d\varphi|\varphi\rangle + \langle\varphi|d\varphi\rangle = 0$, for $|\varphi\rangle$.

Due to the elimination of the component along $|\psi\rangle$, the variation $|d\varphi\rangle_\perp$ is gauge invariant.

The condition $\langle\psi|\varphi\rangle = 0$ defines the horizontal tangent space, in the language of fiber bundles, and also parallel transportation, in differential geometry and in the study of geometric phases, e.g. the Berry phase [4, 5]. Outside the unit sphere one should use the more general condition $\alpha = 0$, instead.

The overlap between two deviations $|d\varphi_1\rangle$ and $|d\varphi_2\rangle$ from a normalized wave function $|\varphi\rangle$ is given by

$$\langle d\varphi_2|d\varphi_1\rangle = {}_\perp\langle d\varphi_2|d\varphi_1\rangle_\perp + {}_{||}\langle d\varphi_2|d\varphi_1\rangle_{||} \tag{17}$$

$${}_\perp\langle d\varphi_2|d\varphi_1\rangle_\perp = \frac{\langle d\psi_2|d\psi_1\rangle}{\langle\psi|\psi\rangle} - \frac{\langle d\psi_2|\psi\rangle}{\langle\psi|\psi\rangle}\frac{\langle\psi|d\psi_1\rangle}{\langle\psi|\psi\rangle} \tag{18}$$

$${}_{||}\langle d\varphi_2|d\varphi_1\rangle_{||} = \frac{1}{4}\frac{\langle\psi|d\psi_2\rangle - \langle d\psi_2|\psi\rangle}{\langle\psi|\psi\rangle}\frac{\langle d\psi_1|\psi\rangle - \langle\psi|d\psi_1\rangle}{\langle\psi|\psi\rangle}, \tag{19}$$

with the second term real and symmetric in $|d\varphi_1\rangle$ and $|d\varphi_2\rangle$.

Separating the real and imaginary parts we have $\langle d\varphi_2|d\varphi_1\rangle = g(|d\varphi_1\rangle, |d\varphi_2\rangle) + ih(|d\varphi_1\rangle, |d\varphi_2\rangle)$. The real part is symmetric in its arguments and the imaginary part is antisymmetric. They are given by

$$g(|d\varphi_1\rangle, |d\varphi_2\rangle) = \frac{1}{2}\frac{\langle d\psi_2|d\psi_1\rangle + \langle d\psi_1|d\psi_2\rangle}{\langle\psi|\psi\rangle} - \frac{1}{4}\frac{\langle d\psi_1|\psi\rangle + \langle\psi|d\psi_1\rangle}{\langle\psi|\psi\rangle}\frac{\langle d\psi_2|\psi\rangle + \langle\psi|d\psi_2\rangle}{\langle\psi|\psi\rangle} \tag{20}$$

$$h(|d\varphi_1\rangle, |d\varphi_2\rangle) = \frac{1}{2}\frac{\langle d\psi_2|d\psi_1\rangle - \langle d\psi_1|d\psi_2\rangle}{\langle\psi|\psi\rangle} - \frac{1}{2}\frac{\langle d\psi_2|\psi\rangle\langle\psi|d\psi_1\rangle - \langle d\psi_1|\psi\rangle\langle\psi|\psi_2\rangle}{\langle\psi|\psi\rangle^2}. \tag{21}$$

In differential geometry, $g(|d\varphi_1\rangle, |d\varphi_2\rangle)$ defines the Riemann metric and $h(|d\varphi_1\rangle, |d\varphi_2\rangle)$ the connection, leading to the Berry phase.

For a single variation $|d\varphi\rangle$, the overlap $\langle d\varphi|d\varphi\rangle$ reduces to

$$\langle d\varphi|d\varphi\rangle = \frac{\langle d\psi|d\psi\rangle}{\langle\psi|\psi\rangle} - \frac{\langle d\psi|\psi\rangle\langle\psi|d\psi\rangle}{\langle\psi|\psi\rangle^2} + \left(\frac{i}{2}\frac{\langle d\psi|\psi\rangle - \langle\psi|d\psi\rangle}{\langle\psi|\psi\rangle}\right)^2. \tag{22}$$

The first term is the Fubini-Study metric

$$D^2_{FS}(|d\psi\rangle) = {}_\perp\langle d\varphi|d\varphi\rangle_\perp = \frac{\langle d\psi|d\psi\rangle}{\langle\psi|\psi\rangle} - \frac{\langle d\psi|\psi\rangle\langle\psi|d\psi\rangle}{\langle\psi|\psi\rangle^2} \tag{23}$$

and the second term is real and positive. The condition of parallel transportation is therefore the one which minimizes $\langle d\varphi|d\varphi\rangle$, rendering it also gauge invariant and leading to the Fubini-Study metric.

Some of the peculiarities and subtleties of quantum mechanics, namely those of the so called wave function renormalization, come from the fact that its fundamental metric is, in fact, this metric.

The overlap $\langle d\varphi|d\varphi\rangle$ can be also rewritten as

$$\langle d\varphi|d\varphi\rangle = \frac{\langle d\psi|d\psi\rangle}{\langle\psi|\psi\rangle} - \left(\frac{1}{2}\frac{\langle d\psi|\psi\rangle + \langle\psi|d\psi\rangle}{\langle\psi|\psi\rangle}\right)^2. \tag{24}$$

It is interesting to take $|\psi\rangle$ itself already normalized, i.e. that $\langle\psi|\psi\rangle = 1$, and compare what happens to eqs. 22 and 24.

Given that normalized wave functions $|\varphi\rangle$ and projectors $P_\varphi = |\varphi\rangle\langle\varphi|$, satisfying the condition $Tr\ P_\varphi^2 = Tr\ P_\varphi = 1$, are directly related, it is instructive to look also at the variation of a projector. Either directly or using eqs. 14 and 16 we find

$$\begin{aligned} dP_\varphi &= (|d\varphi\rangle_\perp + i\alpha|\varphi\rangle)\langle\varphi| + c.c. \\ &= |d\varphi\rangle_\perp\langle\varphi| + |\varphi\rangle_\perp\langle d\varphi|, \end{aligned} \tag{25}$$

involving only the perpendicular term, the one which is gauge invariant.

We establish now the relation between the fidelity and the Fubini-Study metric.

Taking $|\psi_1\rangle = |\psi\rangle + |d\psi_1\rangle\lambda$ and $|\psi_2\rangle = |\psi\rangle + |d\psi_2\rangle\lambda$, with λ real, and defining $\alpha_i = \frac{\langle\psi|d\psi_i\rangle}{\langle\psi|\psi\rangle}$ and $\beta_{ij} = \frac{\langle d\psi_i|d\psi_j\rangle}{\langle\psi|\psi\rangle}$ one has

$$\begin{aligned} F^2 &= \frac{\langle\psi_2|\psi_1\rangle\langle\psi_1|\psi_2\rangle}{\langle\psi_2|\psi_2\rangle\langle\psi_1|\psi_1\rangle} \\ &= \frac{(1+(\alpha_1+\alpha_2^*)\lambda+\beta_{21}\lambda^2)(1+(\alpha_2+\alpha_1^*)\lambda+\beta_{12}\lambda^2)}{(1+(\alpha_2+\alpha_2^*)\lambda+\beta_{22}\lambda^2)(1+(\alpha_1+\alpha_1^*)\lambda+\beta_{11}\lambda^2)} \\ &= 1-((\beta_{11}-\beta_{12}-\beta_{21}+\beta_{22})+(\alpha_1^*-\alpha_2^*)(\alpha_1-\alpha_2))\,\lambda^2+O(\lambda^3) \\ &= 1-D_{FS}^2\lambda^2+O(\lambda^3) \end{aligned} \tag{26}$$

or, equivalently, $F \simeq 1 - \frac{1}{2}D_{FS}^2\lambda^2$, where

$$D_{FS}^2 = \frac{(\langle d\psi_1| - \langle d\psi_2|)(|d\psi_1\rangle - |d\psi_2\rangle)}{\langle\psi|\psi\rangle} - \frac{(\langle d\psi_1| - \langle d\psi_2|)|\psi\rangle}{\langle\psi|\psi\rangle}\frac{\langle\psi|(|d\psi_1\rangle - |d\psi_2\rangle)}{\langle\psi|\psi\rangle} \tag{27}$$

is the square of the Fubini-Study metric for $|d\psi_1\rangle - |d\psi_2\rangle$, i.e. $D_{FS}^2(|d\psi_1\rangle - |d\psi_2\rangle)$.

The deviation from 1 of the fidelity between those two pure states is, in lowest order, the square of the Fubini-Study metric of the difference of the variations.

The proof of eq. 11 is similar, but with all quantities real.

3.2. One Mixed State

In this section we motivate physically the definition of fidelity between a pure and a mixed state. In the next section, we will give the general definition, retrieving the cases where at least one of the states is pure, as particular cases.

Since the square of the fidelity between pure states is $F^2(|\varphi_1\rangle\langle\varphi_1|, |\varphi_2\rangle\langle\varphi_2|) = |\langle\varphi_2|\varphi_1\rangle|^2 = \langle\varphi_1|\varphi_2\rangle\langle\varphi_2|\varphi_1\rangle$ and recalling that the state of a system in a statistical mixture of pure states

$|\varphi_{2,i}\rangle$, with probabilities p_i, is $\rho_2 = \sum_i |\varphi_{2,i}\rangle p_i \langle\varphi_{2,i}|$, it is natural to expect that, averaging with the probabilities p_i, one should have $F^2(|\varphi\rangle\langle\varphi|, \rho) = \langle\varphi|\rho|\varphi\rangle$, leading to

$$F(|\varphi\rangle\langle\varphi|, \rho) = \sqrt{\langle\varphi|\rho|\varphi\rangle}. \tag{28}$$

The fidelity between a pure and a mixed state is then the square root of the matrix element of the density matrix on the pure state.

3.3. Mixed States

The case of two mixed states can be treated using the concept of purification of a mixed state.

Every mixed state ρ in a Hilbert space H can be viewed as the reduced state of some pure state in a larger Hilbert space. It is always possible to find a second Hilbert space H' and a wave function $|\varphi\rangle$ in $H \otimes H'$ such that $\rho = Tr_{H'}|\varphi\rangle\langle\varphi|$.

The fidelity between two mixed states ρ_1 and ρ_2 is defined using the fidelity between their purifications $|\varphi_1\rangle$ and $|\varphi_2\rangle$. Since the process of purification is not unique, due to arbitrariness in the choice of the phases, one looks for the maximum over all possible purifications. The fidelity is then defined as [6]

$$F(\rho_1, \rho_2) = \max_{|\varphi_1\rangle, |\varphi_2\rangle} |\langle\varphi_2|\varphi_1\rangle|. \tag{29}$$

In general, it is sufficient to use for H' a copy of the Hilbert space itself. Since ρ is Hermitian the set of its eigenvectors $|e_i\rangle$, with eigenvalues p_i, is a basis of H. One chooses then a basis of vectors $|f_i\rangle$ of H', not necessarily the same. The state $|\varphi\rangle = \sum_i \sqrt{p_i}|e_i\rangle \otimes |f_i\rangle = (\sqrt{\rho} \otimes I)|\Psi\rangle$, where $|\Psi\rangle = \sum_i |e_i\rangle \otimes |f_i\rangle$, is a purification of the state ρ. It is already in the Schmidt decomposition form, defined in the Appendix.

This approach, with its characteristic duplication of the degrees of freedom, has been applied also in other areas of physics [7, 8]. In particular, it is the basis of the so called Thermo Field Dynamics [8, 9] (where a "thermal state" is defined, to treat equilibrium quantum statistical mechanical systems by field theoretical methods using Wick's theorem as in the zero temperature formalism), similar to the Keldish formalism.

When one or both states are pure, the general definition leads to those already given. From its definition, several properties of the fidelity between mixed states can be derived: it is symmetric in its arguments, $0 \leq F(\rho_1, \rho_2) \leq 1$, reaching its maximum value only when the two states are the same, etc.

In general [6, 10], given two mixed states ρ_1 and ρ_2, we can have the purifications $|\varphi_1\rangle = \sum_i \sqrt{p_{1,i}}|e_i\rangle \otimes |f_i\rangle$ and $|\varphi_2\rangle = \sum_i \sqrt{p_{2,i}}|\tilde{e}_i\rangle \otimes |g_i\rangle$. The different basis are related by arbitrary unitary transformations $|\tilde{e}_i\rangle = V|e_i\rangle$, $|f_i\rangle = U_1|e_i\rangle$ and $|g_i\rangle = U_2|\tilde{e}_i\rangle$. Using $\sum_i A|e_i\rangle \otimes |e_i\rangle = \sum_i |e_i\rangle \otimes A^T|e_i\rangle$, which can be verified evaluating the matrix elements, one has

$$\begin{aligned} |\varphi_1\rangle &= \sum_i \sqrt{\rho_1}|e_i\rangle \otimes U_1|e_i\rangle = \sum_i \sqrt{\rho_1}U_1^T|e_i\rangle \otimes |e_i\rangle \\ |\varphi_2\rangle &= \sum_i \sqrt{\rho_2}U_2^T|\tilde{e}_i\rangle \otimes |\tilde{e}_i\rangle = \sum_i \sqrt{\rho_2}U_2^T V V^T|e_i\rangle \otimes |e_i\rangle. \end{aligned}$$

The inner product of the two purifications is

$$\langle\varphi_2|\varphi_1\rangle = \sum_i \langle e_i| \left(U_2^T V V^T\right)^\dagger \sqrt{\rho_2}\sqrt{\rho_1}U_1^T|e_i\rangle. \tag{30}$$

Using the left polar decomposition (defined in the Appendix) of an operator to write $\sqrt{\rho_2}\sqrt{\rho_1} = U|\sqrt{\rho_2}\sqrt{\rho_1}|$, where the modulus of the operator is given by $|\sqrt{\rho_2}\sqrt{\rho_1}|^2 =$

$\left(\sqrt{\rho_2}\sqrt{\rho_1}\right)^\dagger \sqrt{\rho_2}\sqrt{\rho_1} = \sqrt{\rho_1}\rho_2\sqrt{\rho_1}$ and U is unitary, we find

$$\begin{aligned}\langle\varphi_2|\varphi_1\rangle &= Tr\ \left(U_2^T V V^T\right)^\dagger U|\sqrt{\rho_2}\sqrt{\rho_1}|U_1^T \\ &= Tr\ |\sqrt{\rho_2}\sqrt{\rho_1}|U_1^T\left(U_2^T V V^T\right)^\dagger U. \qquad (31)\end{aligned}$$

Since $|Tr\ |A|U| \leq Tr\ |A|$, for U unitary, with equality only when $U = I$, as can be easily seen using the basis in which $|A|$ is diagonal, we finally arrive at the general expression for the fidelity

$$F(\rho_1, \rho_2) = Tr\ |\sqrt{\rho_2}\sqrt{\rho_1}| = Tr\ \sqrt{\sqrt{\rho_1}\rho_2\sqrt{\rho_1}}. \qquad (32)$$

As already stated, it reduces to eq. 12, in the case of two pure states, and to eq. 28, in the case of a pure and a mixed state.

Two different situations may occur: the commutative and the non commutative cases. In the commutative case, the two states commute, having therefore a common basis of eigenvectors $|\varphi_i\rangle$. Both states can be diagonalized simultaneously, being described by the probabilities $p_{1,i}$ and $p_{2,i}$. In this case, the fidelity reduces to the classical fidelity expression

$$F(\rho_1, \rho_2) = \sum_i \sqrt{p_{1,i}p_{2,i}} \qquad (33)$$

for these probability distributions.

In the non commutative case not only the eingenvalues change but also the eigenvectors themselves change, rotating in the Hilbert space. In general, using eqs. 6 and 25, we have

$$\begin{aligned}d\sqrt{\rho} &= \sum_i |\varphi_i\rangle \frac{dr_{\perp i}}{r}\langle\varphi_i| + \sum_i \frac{r_i}{r}\left(|d\varphi_i\rangle_\perp\langle\varphi_i| + |\varphi_i\rangle_\perp\langle d\varphi_i|\right) \qquad (34)\\ d\rho &= \sum_i |\varphi_i\rangle dp_i\langle\varphi_i| + \sum_i p_i\left(|d\varphi_i\rangle_\perp\langle\varphi_i| + |\varphi_i\rangle_\perp\langle d\varphi_i|\right) \qquad (35)\end{aligned}$$

for the variation of a mixed state. The first terms are purely classical, whereas the second terms are a quantum feature.

In the case of mixed states it is possible to define different metrics, with different properties and drawbacks [11]. We will define the metric directly from the fidelity as $F = 1 - \frac{1}{2}ds^2 + \cdots$, finding the so called Bures metric [12, 6]. The maximization condition in the definition of the fidelity is comparable to the transversality condition in the case of pure states.

For simplicity, we take $\rho_1 = \rho$ fixed and $\rho_2 = \rho + d\rho + \frac{1}{2}d^2\rho + \cdots$, with $Tr\ d\rho = Tr\ d^2\rho = 0$. We must solve $\sqrt{\sqrt{\rho_1}\rho_2\sqrt{\rho_1}} = \rho + X + \frac{1}{2}Y + \cdots$, with X first order and Y second order, for X and Y. One finds $\rho X + X\rho = \sqrt{\rho}d\rho\sqrt{\rho}$ and $\rho Y + 2(d\rho)^2 + Y\rho = \sqrt{\rho}(d^2\rho)\sqrt{\rho}$, leading, on the basis in which ρ is diagonal, to $X_{ij} = \frac{\sqrt{p_i}d\rho_{ij}\sqrt{p_j}}{p_i+p_j}$ and $Y_{ii} = -\frac{1}{p_i}(d\rho)^2_{ii} + d^2\rho_{ii}$. Finally, one has $Tr\ X = \frac{1}{2}Tr\ d\rho = 0$ and

$$ds^2 = -Tr\ Y = \frac{1}{2}\sum_{ij}\frac{|d\rho_{ij}|^2}{p_i + p_j}, \qquad (36)$$

which is the Bures metric [13].

In the discussion of the Bures metric and of its parallel transport condition it is convenient [5] to rewrite ds^2 in terms of G, defined by

$$d\rho = \rho G + G\rho. \qquad (37)$$

as

$$ds^2 = \frac{1}{2}Tr\ d\rho G. \qquad (38)$$

4. Fidelity and Phase Transitions

Phase Transitions is one of the most challenging areas in Condensed Matter Physics.

Traditionally, the phase transitions considered were the so called classical or thermal phase transitions, i.e. at finite temperatures, driven by thermal fluctuations [14]. Typically, in the thermodynamic limit, the partition function or some of its derivatives become non analytic, displaying singularities at some points. Phase transitions and critical phenomena where one of the topics driving the development of the renormalization group [15].

More recently, there has been a great interest in the so called quantum phase transitions, i.e. at zero temperatures, driven by quantum fluctuations [16]. Typically, the ground state (or some low lying excited state) of the system changes qualitatively with the variation of some parameter in which the Hamiltonian of the system depends.

Most phase transitions are associated with a symmetry breaking order parameter and some type of Landau-Ginzburg-Wilson free energy is normally used. However, there are some phase transitions which do not fit into the Landau-Ginzburg-Wilson paradigm of phase transitions, as the Beresinskii-Kosterlitz-Thouless phase transition [17, 18, 19] and the so called topological phase transitions [20], with applications in topological quantum computation.

The physical properties of the system change drastically across the boundary between two different phases [21, 22]. The fidelity between states should then decrease noticeably when the states approach that boundary and in particular if those states belong to different sites of it. Fidelity simply compares different states of a system, without being necessary any knowledge about some predefined order parameter. It is then a very natural tool to use and can be considered as an order parameter by itself.

In the next sections we see how the ideas presented above can be immediately applied to the study of phase transitions and have been applied to the study of several models.

4.1. Zero Temperature

At zero temperature, the system is usually in its ground state, which depends on the parameters defining the system Hamiltonian, like interaction strength, magnetic or electric fields, etc, defining the phase space.

Since we are dealing with pure states, the fidelity is simply the modulus of the overlap between their wave functions.

The behavior of the ground state and of the lowest energy levels are fundamental for the understanding of the phase transitions and of the fidelity.

4.2. Finite Temperatures

In quantum statistical mechanics two ensembles are usually considered: the canonical and the grand canonical ensembles, where the system is coupled to an environment exchanging only energy or energy and particles with the system (in this context the zero temperature case, just considered, is called the micro canonical ensemble). In the Appendix we summarize the derivation of these ensembles from information theory, using the Boltzmann-Gibbs-Shannon-von Neumann entropy.

The density matrix for the canonical ensemble is

$$\rho = \frac{e^{-\beta\mathcal{H}}}{Z} \tag{39}$$

where $\beta = \frac{1}{k_B T}$ is the inverse of the absolute temperature T (k_B being the Boltzmann constant) and $Z = Tr\, e^{-\beta\mathcal{H}}$ is the partition function. The density matrix for the grand canonical ensemble is

$$\rho = \frac{e^{-\beta(\mathcal{H}-\mu N)}}{Z} \tag{40}$$

with the partition function given by $Z = Tr\ e^{-\beta(\mathcal{H}-\mu N)}$.

In these cases, the general formula for the fidelity must be used. Temperature is an additional parameter to be considered in the definition of the phase space.

In the commutative case, the fidelity, expressed in terms of partition functions, is

$$F(\rho_1, \rho_2) = \frac{\overline{Z}}{\sqrt{Z_1 Z_2}} \tag{41}$$

where Z_1 and Z_2 are the partition functions for the two states and $\overline{Z}$ is the partition function obtained replacing $\beta\mathcal{H}$ by the average $\frac{1}{2}\left(\beta_1(\mathcal{H}_1 - \mu_1 N) + \beta_2(\mathcal{H}_2 - \mu_2 N)\right)$, reducing to $\overline{\beta}(\mathcal{H} - \mu N)$ with $\overline{\beta} = \frac{1}{2}(\beta_1 + \beta_2)$, for a variation of the temperature, or $\beta\overline{\mathcal{H}}$, with $\overline{\mathcal{H}} = \mathcal{H}_0 + \overline{\lambda}\mathcal{H}'$, with $\overline{\lambda} = \frac{1}{2}(\lambda_1 + \lambda_2)$, if the Hamiltonian can be divided into to commuting parts $[\mathcal{H}_0, \mathcal{H}'] = 0$.

As in the zero temperature case level crossings are very important for the behavior of phase transitions and of the fidelity. These level crossings can occur already for the finite systems or only in the thermodynamic limit.

In some applications, it is interesting to consider a small part of a larger system. One is then led to consider the reduced density matrix of that subsystem and the fidelity between two such reduced density matrices [23]. Even if the large system is in the ground state in general the small system will be in a mixed state.

5. Fidelity Susceptibility

Fidelity depends on two density matrices, i.e. on two points of phase space. In practice, one fixes, at a small value, the difference $\Delta\lambda = \lambda_2 - \lambda_1$ of the parameters defining them, and varies their center of gravity $\overline{\lambda} = \frac{1}{2}(\lambda_2 + \lambda_1)$, exploring the phase space, in particular the boundaries between phases.

It is then natural to consider the so called fidelity susceptibility [24, 25] defined by $F = 1 - \frac{1}{2}\chi_F(\Delta\lambda)^2 + \cdots$, for zero temperature, or by $\ln F = -\frac{1}{2}\chi_F(\Delta\lambda)^2 + \cdots$, for finite temperatures, which depends on a single point in phase space only.

The derivation we have done for eq. 26, using independent variations of the non normalized wave functions cannot be used, and must be redone, using the normalized wave functions instead. In second order of the (variation) of the driving parameter we have $|\varphi(\lambda)\rangle = |\varphi\rangle + |\varphi'\rangle\lambda + \frac{1}{2}|\varphi''\rangle\lambda^2 + \cdots$ These derivatives must satisfy constraints required by the normalization condition $\langle\varphi|\varphi\rangle = 1$ given by $\langle\varphi'|\varphi\rangle + \langle\varphi|\varphi'\rangle = 0$ and $\langle\varphi''|\varphi\rangle + 2\langle\varphi'|\varphi'\rangle + \langle\varphi|\varphi''\rangle = 0$. We have then $\langle\varphi_2(\lambda)|\varphi_1(\lambda)\rangle = 1 + \left(\langle\varphi_2'|\varphi\rangle + \langle\varphi|\varphi_1'\rangle\right)\lambda + \frac{1}{2}\left(\langle\varphi_2''|\varphi\rangle + 2\langle\varphi_2'|\varphi_1'\rangle + \langle\varphi|\varphi_1''\rangle\right)\lambda^2 + \cdots$ Using those constraints, the square of the fidelity is

$$\begin{aligned} F^2 &= \langle\varphi_2(\lambda)|\varphi_1(\lambda)\rangle\langle\varphi_1(\lambda)|\varphi_2(\lambda)\rangle \\ &= 1 - \left(\langle\varphi_1'| - \langle\varphi_2'|\right)\left(1 - |\varphi\rangle\langle\varphi|\right)\left(|\varphi_1'\rangle - |\varphi_2'\rangle\right)\lambda^2 + \cdots \\ &\simeq 1 - {}_\perp\!\left(\langle\varphi_1'| - \langle\varphi_2'|\right)\left(|\varphi_1'\rangle - |\varphi_2'\rangle\right)_\perp \lambda^2, \end{aligned}$$

with $|\varphi\rangle_\perp$ given by eq. 15. We find again the Fubini-Study metric, due to the presence of the perpendicular projector $1 - |\varphi\rangle\langle\varphi|$. This result can be more simply understood noticing that $F^2 = Tr\ P_{\psi_2}P_{\psi_1}$ and using eq. 25 and the constraints resulting from $Tr\ P_\varphi^2 = 1$ by differentiation, i.e. $Tr\ P_\varphi dP_\varphi = 0$ $Tr\ P_\varphi d^2 P_\varphi = -Tr\ (dP_\varphi)^2$.

From the Schrödinger equation $(\mathcal{H}(\lambda) - E(\lambda))|\varphi(\lambda)\rangle = 0$, it follows immediately that $(d\mathcal{H}(\lambda) - dE(\lambda))|\varphi(\lambda)\rangle + (\mathcal{H}(\lambda) - E(\lambda))\,d|\varphi(\lambda)\rangle = 0$. One finds then the Hellman-Feynman theorem $dE = \langle\varphi|d\mathcal{H}|\varphi\rangle$ and that $d|\varphi\rangle = \frac{1}{E-\mathcal{H}}(d\mathcal{H} - dE)|\varphi\rangle$, where the resolvent is the pseudo-inverse (obtained inverting the non-zero eigenvalues and doing nothing do those which are zero), or, equivalently introducing explicitly the orthogonal projector, as usually done in the time independent perturbation theory. Applying this result to the ground state, the fidelity susceptibility is

$$\chi_F = {}_\perp\langle d\varphi|d\varphi\rangle_\perp = \langle 0|\left(d\mathcal{H} - dE\right)\frac{1}{(E_0 - \mathcal{H})^2}\left(d\mathcal{H} - dE\right)|0\rangle. \tag{42}$$

Inserting a decomposition of the identity (with the ground state excluded) one finally has

$$\chi_F = \sum_{n\neq 0}\frac{d\mathcal{H}_{0n}d\mathcal{H}_{n0}}{(E_0 - E_n)^2}, \tag{43}$$

where $d\mathcal{H}_{mn} = \langle m|d\mathcal{H}|n\rangle$.

The matrix element structure resembles that of the second order perturbation theory expression for the energy, but with the energy difference denominator squared, suggesting that the fidelity should be a more sensitive quantity for the detection of phase transitions, diverging faster [26].

For finite temperatures, in the commutative case, the fidelity susceptibility coincides with the usual thermodynamic susceptibilities or response functions [24, 25]. Writing $F = e^{-\frac{1}{2}\chi_F(\Delta\lambda)^2}$ the divergence of the susceptibility is clearly associated to a drop of the fidelity at the phase transitions, in agreement with our intuition.

In the Appendix we have showed that $d\ln Z = -Ed\beta + Nd(\beta\mu) - \beta\langle\frac{\partial\mathcal{H}}{\partial\lambda}\rangle d\lambda$. From $\frac{\partial\ln Z}{\partial\beta} = -E$, we have $\chi_F = \frac{\partial^2\ln Z}{\partial\beta^2} = \langle(\mathcal{H} - \langle\mathcal{H}\rangle)^2\rangle = k_BT^2\frac{\partial E}{\partial T} = k_BT^2C_v$, proportional to the specific heat (in the canonical ensemble), for a variation in temperature. Similarly, we have $\chi_F = \frac{\partial^2\ln Z}{\partial\lambda^2} = \beta^2\langle(\mathcal{H}' - \langle\mathcal{H}'\rangle)^2\rangle = \beta^2\chi_F$, proportional to the appropriate susceptibility, for a variation of some other parameter, in the commutative case. If λ is a magnetic field and $\mathcal{H}'$ is a spin component, χ_F is the usual longitudinal magnetic susceptibility.

From eq. 35 the matrix elements of $d\rho$, in the basis where ρ is diagonal, are $\langle\varphi_i|d\rho|\varphi_j\rangle = dp_i\delta_{ij} + p_i\langle d\varphi_i|\varphi_j\rangle + p_j\langle\varphi_i|d\varphi_j\rangle = dp_i\delta_{ij} + (p_i - p_j)\langle d\varphi_i|\varphi_j\rangle$. The Bures metric becomes then

$$ds^2 = \frac{1}{4}\sum_i\frac{(dp_i)^2}{p_i} + \frac{1}{2}\sum_{i\neq j}|\langle d\varphi_i|\varphi_j\rangle|^2\frac{(p_i - p_j)^2}{p_i + p_j}, \tag{44}$$

called, in the literature [27], the classical and the non classical terms, respectively. The non classical terms lead to the so called Uhlmann phase [28], as also numerically found [25].

In the commutative case, the Bures metric reduces to the Fisher metric, given by the first term. Since the second term is always non negative, the Fisher metric is a lower bound of the Bures metric, in general.

Alternatively to this fidelity susceptibility approach, one may also look for the singularities of the Riemann metric tensor [29].

6. Application to Specific Models

Fidelity has been applied with considerable success to the different types of thermal, quantum and topological phase transitions.

Applications have involved spin, electronic or bosonic systems, both for finite systems and in the thermodynamic limit, using the solutions of exactly solvable models (typically in one dimension models), exact diagonalization of finite systems, mean field approximations, renormalization group analysis, finite scaling, density matrix renormalization group, etc.

For a recent, complete and exhaustive review, see [30], and references therein.

7. Conclusions

We have reviewed the notion of fidelity as a measure of distinguishability of probability distributions in information theory and its extension to the quantum states, stressing some of its geometric features. We have also showed how to apply it to the study of phase transitions in condensed matter systems.

We have tried to keep the notation as simple as possible, but pointing out the connection to differential and information geometry, at the same time.

We have tried to organize the presentation in a pedagogical manner and we hope that this review will be useful as an introduction to more advanced topics such as the study of the geometric Berry and Uhlmann phases.

Acknowledgments

Valuable discussions with Nikola Paunković and Pedro Ribeiro are acknowledged. This work was supported by FCT Project PTDC/FIS/64926/2006, Portugal.

Some mathematical results

Matrix singular value decomposition

Given a matrix A (of dimension $(m \times n)$, not necessarily square) the matrices $A^\dagger A$ and $AA^\dagger$ are Hermitian and therefore each of them have a complete set of orthogonal eigenvectors, with real non negative eigenvalues. Having obtained the eigenvectors $|\alpha_i\rangle$ with eigenvalues λ_i, of $A^\dagger A$, the vectors $|\beta_i\rangle$, defined by $\sqrt{\lambda_i}|\beta_i\rangle = A|\alpha_i\rangle$, for $\lambda_i \neq 0$, are eigenvectors of $AA^\dagger$, also with eigenvalue λ_i. If necessary, the other eigenvectors of $AA^\dagger$ can be found, completing the orthogonal basis. Some attention must be paid to the zero eigenvalues, in particular when the matrix is not square. Acting with A on the decomposition of the identity, written as $I = \sum_i |\alpha_i\rangle\langle\alpha_i|$, we obtain $A = \sum_i |\beta_i\rangle\sqrt{\lambda_i}\langle\alpha_i|$. Its matrix elements can be written in the form $\langle b|A|a\rangle = \sum_i \langle b|\beta_i\rangle\sqrt{\lambda_i}\langle\alpha_i|a\rangle$, i.e. we can write A in the form $V^\dagger \Lambda U$, with U, V unitary and Λ, of dimension $(m \times n)$, with diagonal elements only.

Matrix polar decompositions

Defining $J = \sum_i |\alpha_i\rangle\sqrt{\lambda_i}\langle\alpha_i| = \sqrt{A^\dagger A}$, $K = \sum_i |\beta_i\rangle\sqrt{\lambda_i}\langle\beta_i| = \sqrt{AA^\dagger}$ and $U = \sum_i |\beta_i\rangle\langle\alpha_i|$ we have $A = UJ = KU$. Note that $Tr\ \sqrt{A^\dagger A} = Tr\ \sqrt{AA^\dagger}$.

When A is square, U is unitary and $A = UJ$ and $A = KU$ are called the left and right polar decompositions of A, respectively. This is a generalization of the usual polar decomposition of a complex number into modulus and phase.

Schmidt decomposition

A state of a composite system $|\psi\rangle$ can be written using the basis states $|e_i\rangle$ $|f_j\rangle$ of the Hilbert spaces H_A and H_B, as $|\psi\rangle = \sum_{ij} c_{ij}|e_i\rangle \otimes |f_j\rangle$. Using the singular value decomposition we can write $c_{ij} = \sum_k \alpha_{ik} l_k \beta_{kj}$, with l_k positive. Defining $|a_k\rangle = \sum_i |e_i\rangle\alpha_{ik}$ and $|b_k\rangle = \sum_j \beta_{kj}|f_j\rangle$, we find $|\psi\rangle = \sum_k l_k|a_k\rangle|b_k\rangle$.

Information theory and thermodynamics

We derive now the Boltzmann factor for the canonical and grand canonical ensembles and obtain the combined expression of the first and second laws of thermodynamics from the Boltzmann-Gibbs-Shannon-von Neumann entropy.

Starting with a general treatment, we consider the problem of determining the least biased density matrix ρ, knowing the expectation values of the observables F^α, given that $Tr\ \rho F^\alpha = f^\alpha$, $\alpha = 1, \cdots, N$, besides the probability normalization condition $Tr\ \rho = 1$.

The answer is to maximize the Boltzmann-Gibbs-Shannon-von Neumann entropy $S = -k_B Tr\ \rho \ln \rho$, where k_B is the Boltzmann's constant, subject to those constraints, i.e. maximize $\frac{S}{k_B} - \lambda_0(Tr\ \rho - 1) - \sum_{\alpha=1}^{N} \lambda_\alpha (Tr\ \rho F^\alpha - f^\alpha)$, where λ_α are Lagrange multipliers. This approach is also known as the Jaynes principle [31, 32].

Using the cyclic property of the trace to write $d\ Tr\ f(\rho) = Tr\ f'(\rho) d\rho$, the solution is given by

$$\rho = \frac{e^{-\sum_\alpha \lambda_\alpha F^\alpha}}{Z}, \tag{B.1}$$

where $Z = e^{\lambda_0 + 1} = Tr\ \ e^{-\sum_\alpha \lambda_\alpha F^\alpha}$ is the partition function.

The problem under consideration can change, either because the expectation values f^α change or because the observables F^α themselves change with some external parameter θ, as for example the eigen-energies of a system when its volume V or magnetic field changes. Using again the cyclic property of the trace, one has then

$$d \ln Z = -\sum_\alpha d\lambda_\alpha f^\alpha - \sum_\alpha \lambda_\alpha \langle \frac{dF^\alpha}{d\theta} \rangle d\theta, \tag{B.2}$$

where $\langle \frac{dF^\alpha}{d\theta} \rangle$ denotes the expectation values, defined by ρ, of the variation of the observables. Using eq. B.1 the entropy becomes

$$\frac{S}{k_B} = \ln Z + \sum_\alpha \lambda_\alpha f^\alpha. \tag{B.3}$$

Its variation is given by

$$d\frac{S}{k_B} = \sum_\alpha \lambda_\alpha \left(df^\alpha - \langle \frac{dF^\alpha}{d\theta} \rangle d\theta \right) \tag{B.4}$$

having used eq. B.2. Since one has $\frac{\partial \frac{S}{k_B}}{\partial f^\alpha} = \lambda_\alpha$, i.e. the Lagrange multipliers are the partial derivatives of the entropy with respect to the averages f_α, and $\frac{\partial \frac{S}{k_B}}{\partial \theta} = -\sum_\alpha \lambda_\alpha \langle \frac{dF^\alpha}{d\theta} \rangle$, it follows that

$$\ln Z = \frac{S}{k_B} - \sum_\alpha \frac{\partial \frac{S}{k_B}}{\partial f^\alpha} f^\alpha, \tag{B.5}$$

is the Légendre transform of the entropy with respect to the expectation values of the observables.

If the entropy is extensive, i.e. a homogeneous function of degree 1 of its variables, including the external parameter, then we will have $\frac{S}{k_B} = \sum_\alpha \lambda_\alpha \left(f^\alpha - \langle \frac{dF^\alpha}{d\theta} \rangle \theta \right)$ and $\ln Z = -\sum_\alpha \lambda_\alpha \langle \frac{dF^\alpha}{d\theta} \rangle \theta$.

Turning now to thermodynamics, the observable to be considered in the canonical ensemble is the energy $E = \langle \mathcal{H} \rangle$ and in the grand canonical ensemble are the energy and the particle number. The Lagrange multiplier associated to the energy is $\beta = \frac{1}{k_B T}$, the inverse of the absolute temperature T. The Lagrange multiplier associated to the particle number N is $-\beta\mu$, where μ is the chemical potential. The density matrices for these ensembles are then given by eqs. 39 and 40.

We find that $d\frac{S}{k_B} = \beta (dE - \langle d\mathcal{H} \rangle) - \beta\mu dN$. Defining $\langle \frac{d\mathcal{H}}{dV} \rangle = -p$, where V is the volume, we have $dS = \frac{dE}{T} + \frac{p}{T} dV - \frac{\mu}{T} dN$, or equivalently,

$$dE = -pdV + TdS + \mu dN, \tag{B.6}$$

which is the combined expression of the first and second laws of thermodynamics.

We also have $\ln Z = \frac{S}{k_B} - \beta E + \beta\mu N = -\beta(E - TS - \mu N)$ and $d \ln Z = -E d\beta + N d(\beta\mu) + \beta p dV$.

Finally, we have $\ln Z = \beta p V$, for an extensive system.

References

[1] Nielsen M A and Chuang I L 2000 *Quantum Computation and Quantum Information* (Cambridge, England: Cambridge University Press)
[2] Amico L, Fazio R, Osterloh A and Vedral V 2008 *Rev. Mod. Phys.* **80** 517
[3] Amari S and Nagaoka H 2000 *Methods of information geometry* (Transactions of mathematical monographs; v. 191, American Mathematical Society, Oxford University Press)
[4] Nakahara M 1990 *Geometry, Topology and Physics* (Bristol: Institute of Physics Publishing)
[5] Uhlmann A 1993 *Rep. Math. Phys.* **33** 253
[6] Uhlmann A 1976 *Rep. Math. Phys.* **9** 273
[7] Louisell W H 1973 *Quantum Statistical Properties of Radiation* (John Wiley & Sons)
[8] Takahashi Y and Umezawa H 1975 *Collective Phenomena* **2** 55–80 [reprinted in Int. J. Mod. Physics B 10, 1755 (1996)]
[9] Umezawa H, Matsumoto H and Tachiki M 1982 *Thermofield Dynamics and Condensed States* (North-Holland)
[10] Jozsa R 1994 *J. Mod. Opt.* **41** 2315
[11] Sommers H J and Życzkowski K 2003 *J. Phys. A* **36** 10083
[12] Bures D J C 1969 *Trans. Amer. Math. Soc.* **135** 199
[13] Hübner M 1993 *Phys. Lett. A* **179** 226
[14] Domb C 1996 *The Critical Point: A Historical Introduction to the Modern Theory of Critical Phenomena* (CRC Press)
[15] Goldenfeld N 1992 *Lectures on Phase Transitions and the Renormalization Group* (Frontiers in Physics 85, Westview Press)
[16] Sachdev S 2000 *Quantum Phase Transitions* (Cambridge, England: Cambridge University Press)
[17] Beresinskii V L 1971 *Sov. Phys. JETP* **32** 493
[18] Kosterlitz J M and Thouless D J 1973 *J. Phys. C* **6** 1181
[19] Kosterlitz J M 1974 *J. Phys. C* **7** 1046
[20] Wen X G 2004 *Quantum Field Theory of Many-Body Systems* (New York: Oxford University)
[21] Zanardi P and Paunkovic N 2006 *Phys. Rev. E* **74** 031123
[22] Zhou H Q and Barjaktarevic J P 2008 *J. Phys. A: Math. Theor.* **41** 412001
[23] Paunkovic N, Sacramento P D, Nogueira P, Vieira V R and Dugaev V K 2008 *Phys. Rev. A* **77** 052302
[24] You W L, Li Y W and Gu S J 2007 *Phys. Rev. E* **76** 022101
[25] Paunkovic N and Vieira V R 2008 *Phys. Rev. E* **77** 011129
[26] Yu W C, Kwok H M, Cao J and Gu S J 2009 *Phys. Rev. E* **80** 021108
[27] Zanardi P, Venuti L C and Giorda P 2007 *Phys. Rev. A* **76** 062318
[28] Uhlmann A 1986 *Rep. Math. Phys.* **24** 229
[29] Zanardi P, Giorda P and Cozzini M 2007 *Phys. Rev. Lett.* **99** 100603
[30] Gu S J arXiv:0811.3127v1
[31] Jaynes E T 1957 *Physical Review* **106** 620
[32] Jaynes E T 1957 *Physical Review* **108** 171

Flux qubits on semiconducting quantum ring

M Szopa and E Zipper
Instytut Fizyki, Uniwersytet Slaski w Katowicach, Poland
E-mail: szopa@us.edu.pl

Abstract. The ability to control the quantum state of a single electrons in a quantum ring made of a semiconductor is at the heart of recent developments towards a scalable quantum computer. A peculiar dispersion relation of quantum rings allows to steer the ground state properties by the magnetic flux and offers spin and orbital degrees of freedom for quantum manipulations. We show that such ring can be effectively reduced to the two-state system forming a qubit on orbital or spin degrees of freedom.

1. Introduction

Quantum information and computation are one of the fastest expanding areas of modern physics. The basic units of quantum computation are the two state systems, usually called qubits, which are the superpositions of quantum states [1]. These include microscopic objects such as ions in the electromagnetic traps, atoms in cavities, nuclear spins in molecules [2] and charge and flux states in mesoscopic solid state systems [3, 4, 5]. The latter are easier to integrate using standard circuit technology but the large number of degrees of freedom makes it more difficult to maintain the coherence.

To built the qubit a necessary condition which has to be fulfilled is that the system under consideration has to be treated in a good approximation as a two state quantum system. Thus the energy of next lower and higher lying states has to be at least five qubit splittings away to avoid leakage of quantum information [4, 6]. An attractive candidate for a solid state qubit is based on semiconducting quantum dot, which allow controlled coupling of one or more electrons using rapidly switchable voltages applied to electrostatic gates. Recently a flux qubit built on semiconducting quantum ring has been proposed [7].

The size of a flux qubit (similar to a big molecule) places it on the border line between the two kinds of quantum bits discussed above. The small number of degrees of freedom together with the small size should help to decouple it from the environment but simultaneously it can be easily coupled to other qubits. The system is an arbitrary doubly connected conductor or semiconductor. The phase coherence length of an electron in the system must be longer then its circumference. The two opposite persistent current (PC) states with a well defined orbital magnetic moments are the basis states. Quantum tunneling via the controllable potential barrier close to one of the degeneracy point leads to a formation of a qubit states beeing the quantum superposition of two persistent current states [7]. In that model we considered spinless electrons which was a reasonable assumption for metallic rings but an oversimplification for semiconductor ones.

In the present paper we include spin into considerations. In such a system apart from orbital, there is a spin degree of freedom resulting in a forthfold deeneracy in full symmetry systems.

To make a qubit i.e. a two state system we have to break this symmetry. It can be done by the application of the magnetic field B_e parallel to the axis of the ring and/or by breaking spatial symmetry of the system. Because we consider orbital as well as spin magnetic moments of an electron, the appropriate qubits will be called orbital and spin qubits respectively. The general state is of the form

$$|Q_{orb,spin}\rangle = |n_l\rangle \otimes |s\rangle. \tag{1}$$

where n is the orbital angular momentum quantum number yielding at a given B_e persistent currents corresponding to orbital momentum parallel $l \equiv\uparrow$ and antiparallel $l \equiv\downarrow$ to external field and $s = \pm\frac{1}{2}$ is the spin quantum number inducing the spin magnetic moment $\mu_s = g_s\mu_B s$, where g_s is the Lande factor. In principle it is feasible to make o two state system, well separated from the other states, based on the general form (1). It would require however, the mixing potential acting simultaneuosly on the spin and orbit degrees of freedom. To make it easier to fabricate, we consider in the present paper two kinds of systems, having fixed one of the degrees of freedom. The first is the system which is in a superposition of states with fixed s and well defined orbital angular quantum numbers n and m, degenerate at B_e, corresponding to opposite PC $l \equiv\uparrow,\downarrow$; we call it an *orbital* qubit

$$|Q^{\pm}_{orb}\rangle = \alpha|n_\uparrow, s\rangle \pm \beta|m_\downarrow, s\rangle, \tag{2}$$

where $|\alpha|^2 + |\beta|^2 = 1$. The second type of qubit is a spin qubit, which is formed by the states of the same orbital number n but opposite, well defined, spin quantum numbers

$$|Q^{\pm}_{spin}\rangle = \left|n, +\frac{1}{2}\right\rangle, \quad \left|n, -\frac{1}{2}\right\rangle. \tag{3}$$

In Chapter 2 we discuss the formation of orbital and spin qubits. In Chapter 3 we present coherent manipulations on qubits, the conclusions are given in Chapter 4.

2. Orbital and spin qubits on semiconductor quantum rings.

Let us consider a defect-free mesoscopic semiconducting quasi $1D$ ring of radius R ($2\pi R < L_\phi$) in the presence of static magnetic flux ϕ, $\phi = B_e \cdot \pi R^2$, B_e is the applied magnetic field perpendicular to the plane of the ring. The quantized energy spectrum of electrons is numbered by the orbital quantum number $n = 0, \pm 1, ...$ and spin quantum number $s = \pm\frac{1}{2}$

$$E_{n,s}(\phi') = \frac{\hbar^2}{2m^* R^2}\left[\left(n - \phi'\right)^2 + 2g_s\frac{m^*}{m_e}s\phi'\right], \tag{4}$$

where $\phi' = \frac{\phi}{\phi_0}$, $\phi_0 = \frac{h}{e}$, and m^*/m_e is the ratio of the effective mass to the normal mass of an electron. We assume $T \ll \Delta_{min}$ where Δ_{min} is the smallest energy gap in the quantized energy spectrum, in order to neglect thermal excitations.

At $\phi = 0$ the energy levels with $n \neq 0$ are fourfold degenerate due to orbital (clockwise and anticlockwise persistent current) and spin degrees of freedom. In order to built a qubit one has to remove the fourfold degeneracy. This can be done by the application of B_e which couples independently to spin and orbital moments, changing the energy of the states so that they are no longer equivalent. Let us take as an example GaAs quantum rings of the radius $R = 40nm$ where $g_s = -0.4$ and $m^*/m_e = 0.067$ [8]. The energy spectrum (4) is shown in Fig.1.

At first we develop a qubit on the spin degrees of freedom. For the formation of qubit, to make it relatively insensitive to the fluctuations of the flux, one should choose a nondegenerate orbital state in a flat part of the orbital energy spectrum at integer ϕ'. The energy gap between the ground n and the first excited $n' = n \pm 1$ states of the same spin is independent on n and

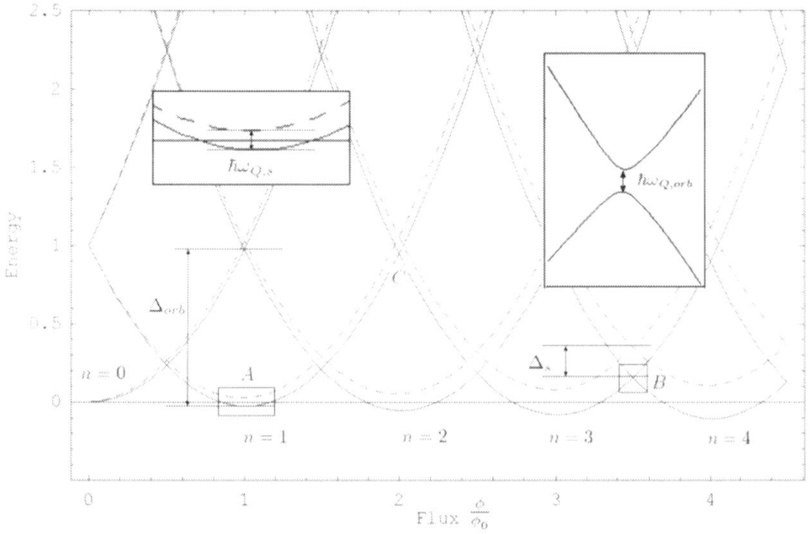

Figure 1. The energy spectrum (4) of an electron in a 1D quantum ring of the radius $R = 40nm$ made of GaAs with spin $s = -\frac{1}{2}$ (solid line) and $s = \frac{1}{2}$ (dashed line). The spin qubit A has spin splitting gap $\hbar\omega_{Q,s}$ much smaller than the gap to the first excited state Δ_{orb}. The orbital qubit B has energy splitting $\hbar\omega_{Q,orb}$ 5 times smaller than the spin splitting Δ_s. The orbital qubits can be also build at integral $\frac{\phi}{\phi_0}$ (e.g. point C). The energy unit is $\frac{\hbar^2}{2m^*R^2} \simeq 3.4K$.

equal to $\Delta_{orb} = \frac{\hbar^2}{2m^*R^2} \simeq 3.4K$. On the other hand the spin splitting $\hbar\omega_{Q,s}$ is linear in ϕ' but in small fields e.g. at $\phi' = 1$, where the ground state correspond to $n = 1$ (see point A in Fig.1) it is

$$\hbar\omega_{Q,s} = E_{1,\frac{1}{2}} - E_{1,-\frac{1}{2}} = 0.18K, \tag{5}$$

i.e. much smaller than the orbital gap. If the ring is occupied by a single electron ($N_e = 1$) the spin qubit is formed at A. Such a two level system formed by electron spin-up and down states is one of the most promising candidates for a qubit. It is separated from the higher excited states by the energy gap yielding the ratio $\Delta_{orb}/\hbar\omega_{Q,s} \gg 5$ and therefore is robust against fluctuations of the electric and magnetic field [4, 6].

For the formation of an orbital qubit one should choose a pair of degenerate orbital states well separated from the higher states. As the energy spectrum of the orbital motion is periodic and spin splitting linear in ϕ', to obtain a considerable spin splitting would require the application of stronger magnetic field B_e. If we introduce to the ring a controllable potential barrier [7], the tunneling of electron close to the degeneracy points will lead to the superposition of states with opposite persistent current but with the *same* spin direction leading to a formation of the orbital qubit. For example let us assume $\phi' = 3.5$ and the number of electrons $N_e = 1$. Then $E_{3,s} = E_{4,s}$ and the point B in Fig.1 accuires a spin splitting

$$\Delta_s = E^0_{3(4),\frac{1}{2}} - E^0_{3(4),-\frac{1}{2}} = 0.63K \tag{6}$$

In the presence of the energy barrier [7] of finite lenght a and height V_0 quantum tunneling (which conserves the spin direction) should thus lead to an orbital qubit i.e. a quantum superposition of the two opposed current states with fixed spin $s = -\frac{1}{2}$. Assuming resonable lenght a and height V_0 of the barrier qubit energy splitting $\hbar\omega_{Q,orb} \sim 0.12K$ is five times smaller than Δ_s and the unwanted excitations to higher states will be strongly reduced [4, 6]. To increase the $\Delta_s/\hbar\omega_{Q,orb}$ ratio even further one can take higher fields, e.g. half integral $\phi' \simeq 10$, which are still feasible to obtain. In a similar way an orbital qubit with fixed opposite spin $s = \frac{1}{2}$ can be built for a ring with $N_e = 3$. With this number of electrons one can also build orbital qubits at integral ϕ' (e.g. point C in Fig.1).

The above discussion was based on the theoretical spectrum of the single channel ring. To support our idea one should consider a more realistic model which takes into account a few radial channels. The high-quality quantum rings have been recently fabricated on AlGaAs-GaAs [9] heterostructures and InGaAs [10] and the energy spectrum has been investigated both theoretically and experimentally.

The idea to implement a qubit in that type of structures is the construction of quantum superposition states of two degenerate (or nearly degenerate) basis states. As we have shown the basis states could be of different origin and depending on them the coherent coupling would be obtained by the tunneling via the potential barrier or by the transverse magnetic field.

3. Coherent manipulations of qubits

In order to use quantum rings in quantum computation it is necessary to establish a way to perform single qubit operations and to implement efficient quantum logic gates on pairs of qubits.

In a pseudospin notation [7, 11] the qubit hamiltonian can be written as

$$H_Q = -\frac{1}{2}B_z\widehat{\sigma}_z - \frac{1}{2}B_x\widehat{\sigma}_x. \tag{7}$$

where $\widehat{\sigma}_z, \widehat{\sigma}_x$ denote Pauli spin matrices. For the orbital qubit the term B_z beeing an effective magnetic field can be tuned by varying the magnetic flux threading the ring

$$B_z = E_\downarrow - E_\uparrow = \begin{cases} \Delta\left[1 - 2\frac{\phi}{\phi_0}\right] & \text{for odd } N_e \\ -\Delta \cdot 2\frac{\phi}{\phi_0} & \text{for even } N_e \end{cases} \tag{8}$$

The effective magnetic field B_x describes the tunneling amplitude $\hbar\omega_{\downarrow\uparrow}$ between the two potential wells and can be tuned by changing the barrier parameters. With these two external control parameters the elementary single-bit operations i.e. z and x rotations [11] can be performed. The measurement of the orbital qubit states can be done by SQUID magnetometer inductively coupled to it [7]. The two orbital qubits can be coupled magnetically by the flux the circulating currents generate in a similar way as the superconducting flux qubits [4, 11, 12]. Such direct inductive coupling is always switched on. A controllable interqubit coupling which can be switched on and off can be achieved when it is mediated by an LC cirquit [13]. Two qubit operations and entanglement can be also mediated by the non-classical electromagnetic field of the high-Q microcavity [14, 15, 16].

For spin qubit B_z and B_x are real magnetic fields which can be used to perform all single qubit operations on a Bloch sphere. Cirquits of nano rings are scalable and they can be initialized to the many qubit ground state by relaxation at low temperature. Reading out the spin states can be done in a similar way as in QD using spin dependent tunneling or by spin-charge conversion [19]. The single qubit rotations can be achieved if the magnetic field could be pulsed exclusively onto one ring e.g. by a scanning probe tip.

By connecting such rings in a series one can achieve all qubit gates. The two spins can be coupled to each other by an effective exchange interaction via tunneling barriers with the strengh which can be controlled by gate voltages [17]. Both types of qubits can be driven by microwave pulses resonant with the level separation.

The performance of qubits is limited by the interaction with noisy environment. Relaxation and decoherence of orbital qubits has been discussed in [7, 18]. It was shown there that the relaxation and decoherence times are of the order of one to several microsec and that the quality factor giving the number of quantum logic operations is of the order of 10^4.
At present the general mechanisms of relaxation and decoherence of spin qubits in QDs are known, for a review see e.g. [19]. For QRs, that is the ring-shaped QDs, these mechanisms should be similar (although some caution is necessary because of the additional orbital degree of freedom) giving the relaxation and decoherence times of the order of milisec.

4. Conclusions

Quantum computing is based on the controlled evolution of quantum mechanical state. Contrary to classical bits qubits can be in a quantum superposition of different logical values and coherently coupled into entangled states. It is unclear wheather atomic physics implementations could ever be scaled up to large-scale quantum computation. Interest in solid state nanosystems in quantum information processing is motivated by the prospect of scalable device fabrication. In this paper we discussed the formation of spin and orbital qubits on semiconductor quantum rings and coherent manipulations performed on them. Because the basic states are the magnetic moment states the qubits will be sensitive to the flux noise, but relatively insensitive to the charge noise. As semiconductor quantum rings are now experimentally accesible these results demonstrate the feasibility of operating single or few electrons in quantum rings as quantum bits. Thus nanometre-scale devices in the form of rings are, besides quantum dots, promising physical realizations of systems to be used for quantum state engineering.

References

[1] Nielsen M A Chuang I L 2000 *Quantum Computation and Quantum Information* (England: Cambridge University Press)
[2] Raimond J M, Brune M, Haroche S 2001 *Rev. Mod. Phys.* **73** 565
[3] Nakamura Y, Pashkin Yu A, Tsai J. S 1999 *Nature* **398** 786
[4] Mooij J E, Orlando T P, Levitov L S, Tian L, van der Wal C H 1999 *S Lloyd Science* **285** 1036; Mooij J E, Harmans C J P M 2005 *New J. Phys.* **7** 219
[5] Migliore R, Messina A 2003 *Phys. Rev. B* **67** 134505, Migliore R, Messina A 2005 *Phys. Rev. B* **72** 214508
[6] Mooij J E, private communication
[7] Zipper E, Kurpas M, Szelag M, Dajka J, Szopa M 2006 *Phys. Rev. B* **74** 125426
[8] Voskoboynikov O, Yiming Li, Hsiao-Mei Lu, Cheng-Feng Shih and P. Lee 2002 *Phys. Rev. B* **66** 155306
[9] Fuhrer A, Luscher S, Ihn T, Henzel T, Ensslin K, Wegscheider W, Bichler M 2001 *Nature* **413** 822; Ihn T, Fuhrer A, Ensslin K, Wegscheider W, Bichler M 2005 *Physica E* **26** 225
[10] Lorke A, Luyken R J, Govorov A O, Kotthaus J P 2000 *Phys. Rev. Lett.* **84**, 2223
[11] Makhalin Y, Schoen G, Schnirman A 2001 *Rev. Mod. Phys.* **73** 357
[12] Plantenberg J H, de Groot P C, Harmans C J P M, Mooij J E 2007 *Nature* **447** 836
[13] Makhalin Y, Schoen G, Schnirman A 2000 *J. Low. Temp. Phys.* **118** 751
[14] Zipper E, Kurpas M, Dajka J, Kuś M, 2008 *J. Phys. Condens. Matter.* **20** 275219
[15] Kurpas M, Zipper E 2008 *Eur. Phys. J.* **D50** 201
[16] Kurpas M, Dajka J, Zipper E 2009 *J. Phys. Condens. Matter.* **21**, 235602
[17] Zipper E, Szopa M, Marganska-Lyzniak M, to be published
[18] Kurpas M, Dajka J, Zipper E 2007 *phys. stat. sol. b* **244** 2470
[19] Hanson R, Kouwenhoven L P, Petta J R, Tarucha S, Vandersypen L M K 2007 *Rev. Mod. Phys.* **79** 1217

Symmetry Analysis in Parametrisation of Complex Systems

W Sikora and J Malinowski

Faculty of Physics and Applied Computer Science, AGH - University of Science and Technology, Al. Mickiewicza 30, 30-059 Kraków, Poland

E-mail: sikora@novell.ftj.agh.edu.pl

Abstract. The symmetry analysis method based on the theory of group representations is used for description of complex systems and their behavior in this work. The first trial of using the symmetry analysis in modeling of behavior of complex social system is presented. The evacuation of large building scenarios are discussed as transition from chaotic to ordered states, described as movements of individuals according to fields of displacements, calculated correspondingly to given scenario. The symmetry of the evacuation space is taken into account in calculation of displacements field – the displacements related to every point of this space are presented in the coordinate frame in the best way adapted to given symmetry space group, which is the set of basic vectors of irreducible representation of given symmetry group. The results got with using the symmetry consideration are compared with corresponding results calculated under assumption of shortest way to exits (Voronoi assumption).

1. Introduction

In presented work behavior of systems of many individual elements organized by different type of connections and interactions are regarded. The example of such complex systems (CS-s), investigated by physicists are solid states, build with huge number of atoms. The special type of them are crystals, in which atoms fill discreet space characterized by strictly defined symmetry, described by one of 230 crystallographic space groups. Quite different example of complex systems, investigated by sociologist are social networks, build with huge number of people (regarded as some individual objects). Both types of systems require big number of parameters for description of their properties and behavior under the action of external factors. One of important subject of the interest of scientists is prediction of new states of CS-s, which may be realized after transformation, when the initial state of the system is known, and description of these states by minimal number of relevant parameters. It is known, that presentation of model structures in the frame of coordinate axis chosen in the best matching to the symmetry of the problem, provides the simplest (requiring the lowest number of independent parameters) description of the system properties. As usually, the time evolution of the systems – also social systems – takes place in strictly given spaces, with well known symmetry properties. Space symmetry conditions impose strong restrictions on the evolution of the systems. Not all types of behavior are allowed. The symmetry properties should be taken into account in creation of models for extracting these possibilities of system evolutions, which are permitted by symmetry of the space and for choosing the best coordinate axis for their description, related to the symmetry of the system. The symmetry analysis based on the theory of group representations gives the opportunity to realize this task.

2. Symmetry analysis in solid state physics

At the case of crystals, under assumption, that the group-subgroup relation is preserved in the transformation procedure, (what means, that the structure of the system is not drastically changed), the symmetry analysis (SA) [1] based on the theory of space groups and their representations gives opportunity to propose the set of all possible types of final structures. It uses all restrictions following from strictly given symmetry of the space, where the atoms are localized in the initial phase. Because the space has limitations (must fulfill the symmetry conditions) not all evolutions in it are able. The SA explicitly indicates which evolutions are permitted. These new types of orderings – presented in the basic vectors of irreducible representations of initial phase symmetry group, as the coordinate frame best related to the symmetry – is described by as small as possible number of free parameters. Such presentation allows also discussion the connections between different types of transformations. For realization of the mentioned problems the computer program MODY based on the representations of symmetry space groups [2], and another associated program – SPLIT [3], had been elaborated. They offer to find all possibilities of new scalar, vector and tensor type crystal properties allowed by the initial symmetry of the crystal. The SA successfully had been applied in the investigations of phase transitions in crystals. The examples of published results may be found for discussion of structural phase transitions (both displacements and order-disorder types) [4], [5], orderings of magnetic moments [6], quadrupols [7], and also orderings of molecules and clusters [8].

3. Symmetry analysis in description of social systems

The description of behavior of social systems depending on many different factors is not simple. Construction of models approximating their real behavior, with significant restriction of number of used parameters is needed. Because the SA method is general, and includes the type of objects localized in the given space only by their transformation properties under the action of symmetry operations, it was putt to the test, is possible to use SA to prediction of the evolutions of the social system. The first trial had been made for modeling the evacuation of high and large office building scenario. In this model the transformation of social system takes place in the building, which is symmetrical and span the space for this system. In order to application of SA the assumptions concerning evacuation models had been made as given below:

1. The long, high building constructed with identical fractions repeated in three perpendicular directions presents the space, where the behavior of social system is regarded. The architecture of the building presented at Fig.1 allows with approximation (under the periodic conditions, similar as for the crystals) the description of its symmetry by Pmm2 space group

Figure 1. The architecture of the building.

2. In the "elementary cell" the existing exits and statistical distribution of individuals over points of the floor- the (x; y; 0) positions - satisfy the initial symmetry assumed as Pmm2 (probability of occupation of each site at the floor is the same). Such "elementary cell" may be visualized as at Fig.2.

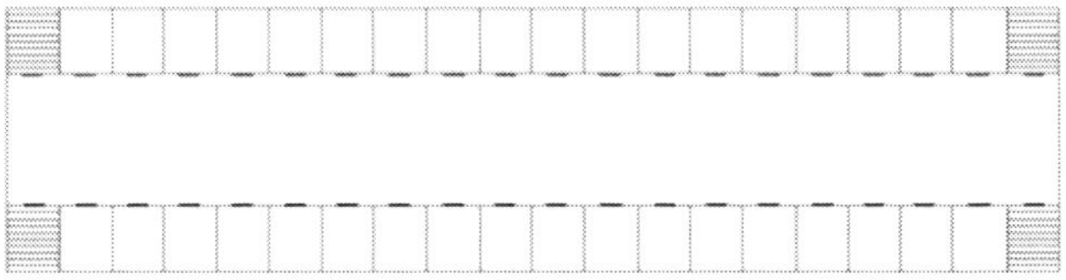

Figure 2. The projection of one building fraction (the elementary cell).

3. Some accidence appears, which leads to the movement of individuals:
   a) chaotic movement, because the knowledge about the possibility of leaving the building through each of exits is the same. Than the probability of displacements each of individuals in each direction is the same, what leads to the symmetry of "superstructure", (space and the function defined on this space, describing probability of displacements of each point) the same as the symmetry of the space only. The probability of speeds of movements of all individuals are also the same. It is the initial phase of the system.
   b) ordered movement, the state attained in phase transition under the action of external information realizing the good evacuation, where displacements of all individuals are going in strictly given directions with strictly given speeds. At each situation the symmetry of function describing the displacements related to each point of the cell floor should be the same, as the symmetry of allowed space.
   b1) the information indicates that all exits are allowed, the symmetry of the space das not change
   b2) the information indicates that some of the exits are forbidden, the symmetry of the allowed space is lower than the symmetry of the initial one
   b3) another situations may appear which lead to lowering the symmetry of allowed space (different accidents scenario).

The symmetry analysis of possible system transformations for Pmm2 space group, and $k = (0; 0; 0)$ (it means, that the situation in each cell is the same), the 4i Wyckof positions ($r_{1n} = (x_n; y_n; 0)$) have at the case of polar functions four irreducible representations (IR-s) allowed for realization of displacement phase transition – tau1, tau2, tau3 and tau4 (the labeling of IR-s is given according to Kovalev Tables of IR-s [9]). Each of them is one dimensional and each of them is allowed three times, what means, that for each representation three different functions may be defined as IR basic vectors. Because these functions are one-dimensional and real, they directly, with accuracy to one multiplication factor for each of them – one free parameter, describe the displacements related to given set of points. As may be seen from the Table 1 the displacements of positions belonging to one set of equivalent positions are strongly correlated. Because only the displacements in the *xy* plane are taken into account in the models, only versions I and II should be included to calculations (all Ci are zero).

Table 1. Displacements vectors (x,y,z) at positions in the elementary cell

Representation and destination group	Version	Displacements vectors (x,y,z) at positions in the elementary cell, which form the basic vector function				
		Free parameter	Pos. 4i			
			1: $(x, y, 0)$	2: $(1-x, 1-y, 0)$	3: $(1-x, y, 0)$	4: $(x, 1-y, 0)$
Tau1 destination group Pmm2	I	A_1	(1,0,0)	(-1,0,0)	(-1,0,0)	(1,0,0)
	II	B_1	(0,1,0)	(0,-1,0)	(0,1,0)	(0,-1,0)
	III	C_1	(0,0,1)	(0,0,1)	(0,0,1)	(0,0,1)
Tau2 destination group P112	I	A_2	(1,0,0)	(-1,0,0)	(1,0,0)	(-1,0,0)
	II	B_2	(0,1,0)	(0,-1,0)	(0,-1,0)	(0,1,0)
	III	C_2	(0,0,1)	(0,0,1)	(0,0,-1)	(0,0,-1)
Tau3 destination group Pm11	I	A_3	(1,0,0)	(1,0,0)	(-1,0,0)	(-1,0,0)
	II	B_3	(0,1,0)	(0,1,0)	(0,1,0)	(0,1,0)
	III	C_3	(0,0,1)	(0,0,-1)	(0,0,1)	(0,0,-1)
Tau4 destination group P1m1	I	A_4	(1,0,0)	(1,0,0)	(1,0,0)	(1,0,0)
	II	B_4	(0,1,0)	(0,1,0)	(0,-1,0)	(0,-1,0)
	III	C_4	(0,0,1)	(0,0,-1)	(0,0,-1)	(0,0,1)

As are shown in the Table 1, the representation Tau1 describe the displacements distribution with symmetry the same as have the initial phase, so it may be applied to construction the model corresponding to 3.b1 assumption. The last three representations follow to displacements distribution, which symmetry is lower than Pmm2 group of initial phase. They lead to evacuation scenarios corresponding to 3.b2 assumption, with two forbidden exits. (Each representation is related to different choice of forbidden exits, realizing corresponding symmetry.)

As the example the fields of displacements following from SA getting for Tau1 and Tau2 representations are shown correspondingly at the Fig 3a and Fig 4a. For comparison the fields of displacements calculated for both situations under the assumption of the "shortest way to exit" (Voronoi criterion) are presented at the Fig 3b and Fig 4b.

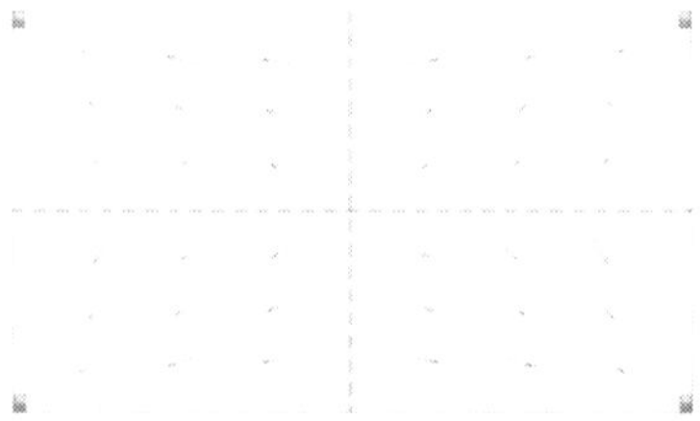

Figure 3a. Symmetric Tau1 field of displacements with Pmm2 symmetry.

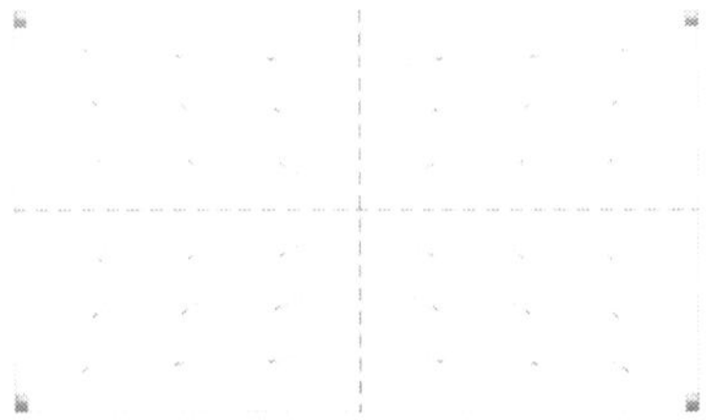

Figure 3b. Voronoi field of displacements with Pmm2 symmetry.

Figure 4a. Symmetric Tau2 field of displacements with P112 symmetry.

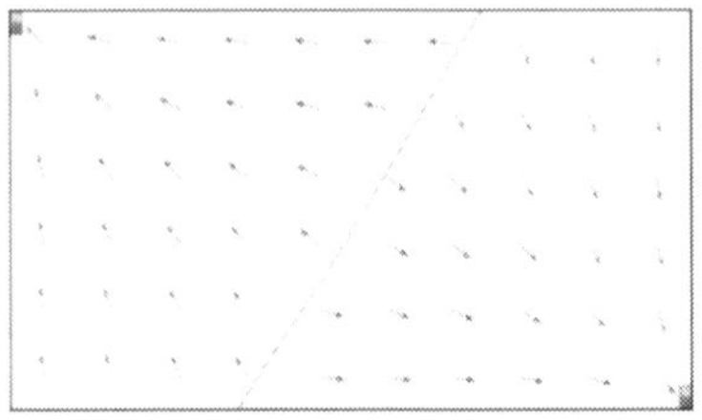

Figure 4b. Voronoi field of displacements with P112 symmetry.

As may be seen Tau 1 presents the full agreement with Voronoi, whereas Tau2 and Voronoi lead to different displacement distributions.

The results of presented symmetry analysis may be used in computer simulations of evacuation, realizing different scenarios, where possible are, for example, investigations of time of evacuation and number of collisions. It will give the possibility to make the best choice between different models.

Acknowledgement
This work is partially supported by EU program SOCIONICAL (FP7 No 231288).

References

[1] Izyumov Yu A, Syromyatnikov V N 1990 *Phase Transitions and Crystal Symmetry* (Dordrecht, Kluwer Academic Publishers) chapter 2

[2] Sikora W *et. al.* 2004 *J. Appl. Cryst.* **37** 1015-1019 (http://novell.ftj.agh.edu.pl/sikora/mody.htm)

[3] Malinowski J http://www.ftj.agh.edu.pl/~malinowski/split.html

[4] Sikora W, Malinowski J and Figiel H 2007 *J. Alloys Compd.* **446-447** 423-428 doi:10.1016/j.jallcom.2006.12.092

[5] Sikora W, Kuna A 2008 *Acta Physica Polonica* A **113** 1211-1224

[6] Fischer P, Pomjakushin V, Keller L, Daoud-Aladine A, Sikora W, Domman A and Hulliger F 2005 *Phys. Rev.* B **72** 134413

[7] Sikora W, Pytlik L, Białas F and Malinowski J 2007 *J. Alloys Compd.* **442** 61-69 doi: 10.1016/j.jallcom.2006.08.34

[8] Sikora W, Malinowski J, Kuna A, Pytlik L 2008 *Journal of Physics: Conference Series* **104** 012023; doi: 10.1088/1742-6596/104/012023

[9] Kovalev O. V. 1993. *Representations of the Crystallographic Space Groups,* ed Stokes H. T. & Hatch D. M. (London, Gordon & Breach)

Particle-hole symmetry in algebraic Bethe Ansatz for the XXX model

R Stagraczynski and T Lulek
Rzeszow University of Technology, The Faculty of Mathematics and Applied Physics, Powstancow Warszawy 6, 35-959 Rzeszow, Poland
E-mail: rstag@prz.edu.pl

Abstract. It is well known that the space of all quantum states of the XXX model for a magnetic ring of N nodes, each with the spin 1/2, decomposes into sectors with r spin deviations, $r = 0, 1, 2, ..., N$ [1, 2, 3, 4]. The sectors r and $N - r$ are related mutually by the particle-hole transformation which exchanges the signs + and − on each node. We discuss here effects of this transformation on the formalism of algebraic Bethe Ansatz, in particular on the form of the monodromy matrix, the main tool of this formalism. We derive explicitly appropriate transformation rules for C_N- orbits of magnetic configurations and the corresponding Fourier transformations within the bases of wavelets. In particular, we point out some important phase relations between orbits on both sides of the equator.

1. Introduction

In this paper we discuss the particle-hole symmetry within the algebraic Bethe Ansatz [5], which consists in interchange of + and − signs at each node. We are therefore not restricted to the sectors $r = 1, ...$ up to the equator, but consider the total space of quantum states of the magnet.

We use here a notation in which $\widetilde{N} = \{j = 1, 2, ..., N\}$ denotes the set of nodes, labelled by consecutive integers j, $1 \leqslant j \leqslant N$, and each magnetic configuration is described by the mapping: $f : \widetilde{N} \mapsto \widetilde{2}$ where $\widetilde{2} := \{0, 1\}$, thus any magnetic configuration is given by the form:

$$f = |i_1, i_2, ..., i_N >, \; i_j \in \widetilde{2} \text{ for } j \in \widetilde{N}, \tag{1}$$

where "1" and "0" stands for " − " and " + ", respectively.

2. The Heisenberg magnetic ring

Let the mapping J assign a unique integer to each magnetic configuration

$$J : \widetilde{2}^{\widetilde{N}} \mapsto \mathbb{N}, J(f) = \sum_{j \in \widetilde{N}} i_j 2^{N-j} + 1. \tag{2}$$

Thus the lexical order of magnetic configurations is defined in terms of J as: $f < f'$. if $J(f) < J(f')$ The mapping J is in fact a slightly modified "binary → decimal" conversion, and "+1" shift arises from the presence of ferromagnetic vacuum.

In terms of J, the action of the transitional symmetry group of the ring (C_N group) on any magnetic configuration can be defined by:

$$C_N(J(f)) := (J(f)-1)/2 + 2^{N-1}((J(f)-1)\,mod\ 2)+1 \tag{3}$$

It is well known that each configuration can be also displayed by its number of spin deviations $\mathbf{r} = \{0, 2, ..., r\}$, the relative position vector $\mathbf{t} = (t_1, t_2, ..., t_r)$where t_α is the distance between the α'th and $(\alpha+1)$mod r spin deviation on the chain, and j is the consecutive number of configuration within the C_N-orbit labelled by $\mathbf{t}$.
Relative position vector $\mathbf{t}$ specifies the distribution of Bethe pseudoparticles within the C_N-orbit [6]. The bases of C_N- orbits is constructed with respect to:

- the lexical order of the parameter $r = 0, 1, 2, ..., N$
- the first element of each C_N- orbit has the property $J(f_1) = max\{J(f)\}$ for f-s within this orbit.
- for a fixed r, the orbits are ordered according to descending values of $J(f_1)$ of the corresponding initial element
- consecutive elements in each orbit are generated by the action of C_N

According to this convention the transform matrix G from the lexical bases to that of C_N- orbits for magnetic ring of $N = 4$ nodes has the form:

G=

$J(i_1,...,i_4)$	1	2	3	4	5	6	7	8	9	10	11	12	13	14	15	16
$J(r,\mathbf{t},j)$	1	9	5	3	2	13	7	4	10	11	6	15	8	12	14	16

Since the bases of C_N- orbits is just a reordered lexical one, the G operator shows a bijection between those two bases.

3. Particle-hole symmetry

The particle-hole symmetry Q exchanges the signs of spin "+" and "-" on each node. Operation Q can be formally defined as

$$Q : \tilde{2}^{\tilde{N}} \mapsto \tilde{2}^{\tilde{N}},\ Q|f> = |(i_1+1)mod2, ..., (i_N+1)mod2> \tag{4}$$

The operation Q has the following properties:

- no fixed point: $Qf \neq f$
- $Q^2 = id_{\tilde{2}^{\tilde{N}}}$
- Q is a bijection

In the lexical bases, the matrix of Q has the antidiagonal form $Q_{ij} = \delta_{i,2^N-j+1}$ To investigate the action of particle-hole symmetry on C_N- orbits it is convenient to write matrix of operator Q in the bases of orbits.

The action of Q on the C_N- orbits of Heissenberg magnet with $N = 4$ nodes can be presented as:

		r	0	1				2						3				4
		t	()	(4)				(1,3)				(2,2)		(1,1,2)				(1,1,1,1)
r	**t**	j	1	1	2	3	4	1	2	3	4	1	2	1	2	3	4	1
0	()	1																1
1	(4)	1												0	1	0	0	
		2												0	0	1	0	
		3												0	0	0	1	
		4												1	0	0	0	
2	(1,3)	1						0	0	1	0							
		2						0	0	0	1							
		3						1	0	0	0							
		4						0	1	0	0							
	(2,2)	1										0	1					
		2										1	0					
3	(1,1,2)	1		0	0	0	1											
		2		1	0	0	0											
		3		0	1	0	0											
		4		0	0	1	0											
4	(1,1,1,1)	1	1															

The particle-hole symmetry usually exchanges the orbit. Case of $N = 4$ is exceptional since the orbits on "the equator" i.e. $r = N/2$ are preserved. Unimportant 0's inside the matrix are omitted.

In order to provide the momentum representation of the space $\mathcal{H} = lc_{\mathbb{C}}\widetilde{2}^{\widetilde{N}}$ of all quantum states of the magnetic ring e.g. in the bases of wavelets, the set $\widetilde{N}$ should be replaced by the Brillouin zone B [7].

$$|B,k> = \frac{1}{\sqrt{N}} \sum_{j\in\widetilde{N}} \exp[\frac{-2i\pi jk}{N}] |\widetilde{N}, j> \tag{5}$$

where k is the quasimomentum. Transformation to the bases of wavelets can be written as:

$$\hat{F}|N,j> = |B, \beta(j)>,$$

where the Fourier operator $\widehat{F}$ for a regular orbit of C_N is defined by the formula

$$<\widetilde{N}, j_1|\hat{F}|\widetilde{N}, j_2> = \frac{1}{\sqrt{N}}\omega^{-j_1 j_2}, \ \omega = \exp[\frac{2i\pi}{N}] \tag{6}$$

and β is a bijection defined by:

$$\beta : \widetilde{N} \mapsto B, \beta(j) = j \mod B$$

the bijection β for the $N = 4$

j	1	2	3	4
$\beta(j)$	1	2	-1	0

An example of the Q operator matrix in the wavelets bases for N=4

		r	0	1				2						3				4
		t	()	(4)				(1,3)				(2,2)		(1,1,2)				(1,1,1,1)
r	**t**	k	0	1	2	-1	0	1	2	-1	0	2	0	1	2	-1	0	0
0	()	0																1
1	(4)	1												0	0	i	0	
		2												0	-1	0	0	
		-1												-i	0	0	0	
		0												0	0	0	1	
2	(1,3)	1						0	0	-1	0							
		2						0	1	0	0							
		-1						-1	0	0	0							
		0						0	0	0	1							
	(2,2)	2										-1	0					
		0										0	1					
3	(1,1,2)	1		0	0	-i	0											
		2		0	-1	0	0											
		-1		i	0	0	0											
		0		0	0	0	1											
4	(1,1,1,1)	0	1															

The action of Q operation preserves the minimal and maximal pseudomomentum k, just adding the phase ± 1, pseudomomenta of other states usually changes and may have imaginary phases.

To achieve the exact solutions of Bethe Ansatz (rigged string configurations) let us choose the Hamilton operator in the form required by the Monodromy matrix formalism [5]

$$H_N = \frac{1}{4}\sum_{j\in\tilde{N}} (\sigma_j^x\sigma_{j+1}^x + \sigma_j^y\sigma_{j+1}^y + \sigma_j^z\sigma_{j+1}^z - I_N) \tag{7}$$

where

$$\sigma_j^a = I\otimes ...I\otimes \underbrace{\sigma^a}_{j} \otimes I...\otimes I, \ \text{a=x,y,z} \tag{8}$$

$$I_N = I\otimes ...\otimes I$$

and $\sigma^x, \sigma^y, \sigma^z$ are standard Pauli matrices and I is the 2×2 unit matrix.
It is convenient to transform that Hamiltonian into the bases of orbits or wavelets. The block form of H_N in these bases allows us to find solutions for each block separately. Thus allows to use rigged configurations bases, and the corresponding transformation matrix for $N=4$ has the form:

6G=

		r	0	1				2						3				4
		S	2	2	1			2	1			0		2	1			2
		$\nu\mathcal{L}$	-	g	[0] 2	[1] 2	[2] 2	gg	[0] 2g	[1] 2g	[2] 2g	[0/0] 0	[0][ ] 0	ggg	[0] 2gg	[1] 2g	[2] 2gg	gggg
		k	0	0	1	2	-1	0	1	2	-1	0	2	0	1	2	-1	0
r	**t**	j																
0	()	1	6															
1	(4)	1		3	-3i	-3	3i											
		2		3	-3	3	-3											
		3		3	3i	-3	-3i											
		4		3	3	3	3											
2	(1,3)	1						$\sqrt{6}$	-3i	0	3i	$-\sqrt{3}$	-3					
		2						$\sqrt{6}$	-3	0	-3	$-\sqrt{3}$	3					
		3						$\sqrt{6}$	3i	0	-3i	$-\sqrt{3}$	-3					
		4						$\sqrt{6}$	3	0	3	$-\sqrt{3}$	3					
	(2,2)	1						$\sqrt{6}$	0	$-3\sqrt{2}$	0	$-2\sqrt{3}$	0					
		2						$\sqrt{6}$	0	$3\sqrt{2}$	0	$-2\sqrt{3}$	0					
3	(1,1,2)	1												3	-3i	-3	3i	
		2												3	-3	3	-3	
		3												3	3i	-3	-3i	
		4												3	3	3	3	
4	(1,1,1,1)	1																6

The transformation matrix consists of eigenvectors, here for convenience multiplied by 6.

Once the eigenproblem is solved the action of Q on the exact solutions of XXX Heisenbeg magnet can be easily explored.

For ring consisting of 4 nodes the corresponding Q operator can be written in form:

			r	0	1				2						3				4
			S	2	2	1			2	1			0		2	1			2
			$\nu\mathcal{L}$	-	g	[0] 2	[1] 2	[2] 2	gg	[0] 2g	[1] 2g	[2] 2g	[0/] 0	[0] 0	ggg	[0] 2gg	[1] 2gg	[2] 2gg	gggg
r	S	$\nu\mathcal{L}$	k	0	0	1	2	-1	0	1	2	-1	0	2	0	1	2	-1	0
0	2	-	0																1
1	2	g	0												1	0	0	0	
	1	[0] 2	1												0	-i	0	0	
		[1] 2	2												0	0	-1	0	
		[2] 2	-1												0	0	0	i	
2	2	gg	0						1	0	0	0	0	0					
	1	[0] 2g	1						0	-1	0	0	0	0					
		[1] 2g	2						0	0	-1	0	0	0					
		[2] 2g	-1						0	0	0	-1	0	0					
		[0/] 0	0						0	0	0	0	1	0					
		[0] 0	2						0	0	0	0	0	1					
3	2	ggg	0		1	0	0	0											
	1	[0] 2ggg	1		0	i	0	0											
		[1] 2ggg	2		0	0	-1	0											
		[2] 2ggg	-1		0	0	0	-i											
4	2	gggg	0	1															

where g dentotes the Goldstone boson. Particle-hole operator in rigged string bases consists of diagonal submatices placed on antidiagonal. Operation Q replaces the state with the one of the same type. Allowed phases on "the equator" are ± 1. When acting on the state (below/beyond) the equator Q (adds/substracts) some goldston bosons, and the allowed phases are $\{\pm 1, \pm i\}$.

4. Conclusions

Despite its simplicity, the particle-hole symmetry operator has a non trivial form in various useful bases of Bethe Ansatz. We have discussed here its action on: the bases of magnetic configuration in the lexical order, C_N-orbits, wavelets and the exact Bethe Ansatz solutions.

References

[1] Bethe H 1931 *Z. Physik* **71** 205 (in German; English translation in: D.C.Mattis, The Many-Body Problem, World
[2] Voynarovich F 1982 *J. Phys. C: Solid State Phys.* **15** 6397
[3] Ceulemans A Cojocaru SP and Chibotaru 2001 *Eur. Phys. J.* **B 21** 511
[4] Caspers WJ, Labuz M, Wal A, Kuzma M and Lulek T 2003 *J. Phys. A: Math. Gen.* **36** 5369
[5] Faddeev LD and Takhtajan LA 1981 *LOMI* **109** 134 (in Russian; English translation: 1984 *J. Sov. Math.* **24** 241)
[6] Jakubczyk D, Lulek T, Jakubczyk P and Lulek B 2006 *Journal of Physics: Conference Series* **30** 188
[7] Lulek T, Lulek B, Jakubczyk P and Jakubczyk D 2006 *Journal of Physics: Conference Series* **30** 209

Magnonic qudit and algebraic Bethe Ansatz

B Lulek and T Lulek

Rzeszow University of Technology, The Faculty of Mathematics and Applied Physics, Powstancow Warszawy 6, 35-959 Rzeszow, Poland

E-mail: `tadlulek@prz.edu.pl`

Abstract. A magnonic qudit is proposed as the memory unit of a register of a quantum computer. It is the N-dimensional space, extracted from the 2^N-dimensional space of all quantum states of the magnetic Heisenberg ring of N spins 1/2, as the space of all states of a single magnon. Three bases: positional, momentum, and that of Weyl duality are described, together with appropriate Fourier and Kostka transforms. It is demonstrated how exact Bethe Ansatz (BA) eigenfunctions, classified in terms of rigged string configurations, can be coded using a collection of magnonic qudits. To this aim, the algebraic BA is invoked, such that a single magnonic qudit is prepared in a state corresponding to a magnon in one of the states provided by spectral parameters emerging from the corresponding BA equations.

1. Introduction

It is known the Bethe Ansatz (BA) [1, 2, 3], which provides the exact solution to the eigenproblem of the Heisenberg Hamiltonian for a magnetic ring of N nodes with the spin 1/2 and isotropic nearest neighbour interactions, has an adequate formulation in terms of quantum computations [4, 5]. We recall here briefly that the set $\tilde{2} = \{+, -\} = \{0, 1\}$ of single-node spin projections can be recognized as a *bit*, i.e. the memory unit of a (classical) computer, the set $\tilde{N} = \{1, 2, \ldots, N\}$ of magnetic nodes describes the size of computer memory, and the set $\tilde{2}^{\tilde{N}} = \{f : \tilde{N} \to \tilde{2}\}$ of all magnetic configurations is the register of a classical computer. Moreover, the linear span of the bit $\tilde{2}$ over the field $\mathbb{C}$ of complex numbers becomes a *qubit*, and the space $\mathcal{H}$ of all quantum states of the magnetic ring, i.e. the span of the set $\tilde{2}^{\tilde{N}}$, can serve a a register of a quantum computer. In this way, the results of BA can be adapted within the midst of quantum informatics.

The adaptation mentioned above, with the spin 1/2 in the role of a qubit, is quite natural but not the only way of use of BA. Here we propose another comparison of BA and quantum informatics, based on the notion of a magnonic qudit [5]. Our approach also uses the magnetic ring of N spins 1/2 interacting via the Heisenberg Hamiltonian as a prototypic model, but proposes another extraction of the single-particle space, namely the subspace consisting of all states with a single spin deviation from the ferromagnetic saturation state. Clearly, this is the space isomorphic to $\mathbb{C}^N$, with the states corresponding to localization of this deviation on a single node as an orthonormal basis. We refer to this space as to the magnonic qudit $h = \mathbb{C}^N$, and to this basis, labeled by nodes, as to the computational basis. Our aim is to demonstrate how an arbitrary exact Bethe Ansatz eigenfunction for the sector of r overturned spins, $r \leq N/2$, can be encoded within at most r magnonic qudits. Our demonstration uses the formalism of algebraic Bethe Ansatz [3].

2. The magnonic qudit

Let $|\tilde{N}j\rangle = |+\ldots-+\ldots++\rangle$, $j\in\tilde{N}$, be the magnetic configuration with the single overturned spin at the j-th node. We treat the set $\{|\tilde{N}j\rangle\,|\,j\in\tilde{N}\}$ as the computational orthonormal basis, and the linear closure of this set over the field $\mathbb{C}$ of complex numbers,

$$h = lc_{\mathbb{C}}\{|\tilde{N}j\rangle\,|\,j\in\tilde{N}\} \tag{1}$$

is a linear unitary space referred to as the magnonic qudit. Within BA, h is the space of all quantum states of the system, consisting of a *single* spin deviation, or a single Bethe pseudoparticle on the magnetic ring $\tilde{N}$. We assume it to be a single memory unit of a (hypothetical) quantum computer, so that the algebra

$$\operatorname{End} h = lc_{\mathbb{C}}U(N) \tag{2}$$

of all operators in h, which can be seen as the linear closure of the unitary group $U(N)$, encompasses all single-qudit quantum gates.

Within the Schwinger description [6] of finite-dimensional Hilbert spaces, the computational basis vectors $|\tilde{N}j\rangle$ acquires the meaning of the *position* of a Bethe pseudoparticle [7, 8, 9], whereas the (finite) Fourier transform

$$|Bk\rangle = \frac{1}{\sqrt{N}}\sum_{j\in\tilde{N}} \mathrm{e}^{-2\pi ikj/N}|\tilde{N}j\rangle \tag{3}$$

yields the basis of *quasimomenta* in the magnonic qudit h. Here $k\in B$ is the quasimomentum of a pseudoparticle, and

$$B = \left\{k = 0,\pm1,\pm2,\ldots,\begin{array}{ll}\pm(N/2-1), N/2 & \text{for } N \text{ even}\\ \pm(N-1)/2 & \text{for } N \text{ odd}\end{array}\right\} \tag{4}$$

is the Brillouin zone of the chain $\tilde{N}$, that is, the dual group of the cyclic group C_N on the ring $\tilde{N}$ - the translational symmetry group of the magnetic ring. The two bases in h, positions $|\tilde{N}j\rangle$ and quasimomenta $|Bk\rangle$, are related mutually by the Fourier operator $\hat{F}\in\operatorname{End}h$ as

$$\hat{F}|\tilde{N}j\rangle = |B\beta(j)\rangle,\quad j\in\tilde{N}, \tag{5}$$

where $\beta:\tilde{N}\to B$ is a bijection between the positions and quasimomenta of the system, given by

$$\beta(j) = j \bmod B,\quad j\in\tilde{N}, \tag{6}$$

where $j\,\mathrm{mod}B$ denotes reduction of an integer $j \bmod N$ to the range provided by Eq. (4) (quasimomenta $j > N/2$ become negative). Within the basis of positions, matrix elements of the Fourier operator are

$$\langle\tilde{N}j_1\,|\,\hat{F}|\tilde{N}j_2\rangle = \frac{1}{\sqrt{N}}\,\mathrm{e}^{-2\pi ij_1j_2/N},\quad j_1,j_2\in\tilde{N}, \tag{7}$$

in accordance with Eq. (3). It is worth to observe that the operator $\hat{P}=\hat{F}^2$ satisfies

$$\hat{P}|\tilde{N}j\rangle = |\tilde{N},-j \bmod N\rangle,\quad j\in\tilde{N}, \tag{8}$$

$$\hat{P}|Bk\rangle = |B,-k\rangle,\quad k\in B, \tag{9}$$

and thus can identified as the parity on the ring $\tilde{N}$.

Schwinger operators $\hat{U} \in \mathrm{End}\, h$ and $\hat{V} \in \mathrm{End}\, h$, defined as shifts in the position and momentum bases, that is by

$$\hat{U}|\tilde{N}j\rangle = |\tilde{N}, (j+1) \bmod N\rangle, \quad j \in \tilde{N}, \tag{10}$$

and

$$\hat{V}|Bk\rangle = |B, (k+1) \bmod B\rangle, \quad k \in B, \tag{11}$$

respectively, provide a natural basis in the algebra $\mathrm{End}\, h$ of observables for this system, namely

$$\mathrm{End}\, h = lc_{\mathbb{C}}\{U^j V^j \mid j \in \tilde{N}, k \in B\} \cong lc_{\mathbb{C}}(\tilde{N} \times B). \tag{12}$$

In this way, the set $\tilde{N} \times B$ acquires the meaning of the "classical" phase space for the magnonic qudit.

The magnonic qudit h can be also seen as the carrier space of the transitive representation $R^{\{N-1,1\}}$ of the symmetric group Σ_N acting on the set $\tilde{N}$, with Young subgroup $\Sigma^{\{N-1,1\}} = \Sigma_{N-1} \times \Sigma_1 \subset \Sigma_N$ as the stabilizer. The decomposition

$$R^{\{N-1,1\}} = \Delta^{\{N\}} + \Delta^{\{N-1,1\}} \tag{13}$$

into irreps Δ^λ of Σ_N provides a third basis in h, in a form

$$|\mu\lambda y\rangle = \sum_{j \in \tilde{N}} K_{j,\lambda y} |\tilde{N}j\rangle, \tag{14}$$

with $K_{j,\mu y}$ being Kostka matrices at the level of bases [10] for the transitive representation R^μ with $\mu = \{N-1,1\}$ denoting the weight, λ being the shape of the Young diagram corresponding to rhs of Eq. (13), and $y \in SYT(\lambda)$ denoting the standard Young tableau of the shape λ. We refer to the basis (14) of magnonic qudit to as the irreducible basis of the Schur-Weyl duality, since it is related to mutually dual actions $A : \Sigma_N \times \mathcal{H} \to \mathcal{H}$ and $B : U(2) \times \mathcal{H} \to \mathcal{H}$ of the symmetric group Σ_N and the unitary group $U(2)$ (the latter realized by its action on $lc_{\mathbb{C}}\tilde{2}$) on the space $\mathcal{H}$ of all quantum states of the magnetic ring.

3. The algebraic form of Bethe Ansatz

The algebraic BA, proposed by Faddeev and Takhtajan [3] (cf. also [11]), expresses the famous exact eigenstates of Bethe [1] in terms of some operators which can be readily interpreted in the language of quantum gates [12]. It is rooted in the identity of Dirac [13], which involves a pair of single-node Pauli matrices. For the case of two qubits, that is for the space $\mathbb{C}^2 \otimes \mathbb{C}^2$, Dirac identity reads

$$\vec{\sigma}_1 \cdot \vec{\sigma}_2 + I \otimes I = u_{12} = \begin{pmatrix} 1 & 0 & 0 & 0 \\ 0 & 0 & 1 & 0 \\ 0 & 1 & 0 & 0 \\ 0 & 0 & 0 & 1 \end{pmatrix}, \tag{15}$$

where $\sigma^x, \sigma^y, \sigma^z$ are standard Pauli matrices, which form, together with the two-dim unit matrix I, a basis of the algebra $\mathrm{End}\,\mathbb{C}^2$ - the algebra which encompasses all single-qubit gates, $\vec{\sigma}_j$ is the vector matrix for the qubit labeled by j (at this moment, $j \in \{1, 2\}$, and thus scalar product $\vec{\sigma}_1 \cdot \vec{\sigma}_2 \in \mathrm{End}\,(\mathbb{C}^2 \otimes \mathbb{C}^2)$ reads

$$\vec{\sigma}_1 \cdot \vec{\sigma}_2 = \sigma^x \otimes \sigma^x + \sigma^y \otimes \sigma^y + \sigma^z \otimes \sigma^z, \tag{16}$$

$u_{12} \in \mathrm{End}\,(\mathbb{C}^2 \otimes \mathbb{C}^2)$ is the transposition operator for the two qubits, presented in the computational basis by the rhs matrix of Eq. (15). We recall here that the formal identification of $\vec{s}_j = \frac{1}{2}\vec{\sigma}_j$ as the vector operator of the electron spin localized at the j-th node was the key ingredient of the Heisenberg theory of magnetism at the early stages of development of quantum mechanics.

The space $\mathcal{H}$ of all quantum states of the magnet is the N-th tensor power $(\mathbb{C}^2)^{\otimes N}$ of a single qubit $\mathbb{C}^2$, so that its computational basis is determined by the set

$$\tilde{2}^{\tilde{N}} = \{f : \tilde{N} \to \tilde{2}\} \tag{17}$$

of all magnetic configurations, i.e. mappings $f : \tilde{N} \to \tilde{2}$. Each magnetic configuration $f \in \tilde{2}^{\tilde{N}}$ corresponds to a separable state

$$|f\rangle = |f(1)\rangle \otimes \ldots \otimes |f(N)\rangle \in \mathcal{H}. \tag{18}$$

Accordingly, the algebra $\mathrm{End}\,\mathcal{H}$ is generated from single-node operators by

$$\hat{s}_j^\alpha = I \otimes \ldots \otimes \hat{s}^\alpha \otimes \ldots \otimes I \in \mathrm{End}\,\mathcal{H}, \quad j \in \tilde{N}, \quad \alpha \in \{x, y, z\}, \tag{19}$$

with s^α corresponding to the j-th factor of the tensor power $(\mathbb{C}^2)^{\otimes N}$and unit I to all the other ones.

The algebraic BA is formulated within an extended system consisting of $N+1$ qubits $\mathbb{C}^2$, N of which correspond to nodes of the magnetic ring, and the last qubit, denoted by $V = \mathbb{C}^2$, is referred to as an auxiliary one. The space of the extended system is $\mathcal{H} \otimes V$, so that the auxiliary qubit can be seen as an "environment" of the magnetic ring. Now, one introduces the Lax operator, or, more specifically, a family of operators in $\mathcal{H} \otimes V$, labeled by a continuous variable $\lambda \in \mathbb{C}$, called the spectral parameter, and defined for each pair of qubits which constitute the extended system. It is convenient to define the Lax operator $L_{ja}(\lambda) \in \mathrm{End}\,(\mathcal{H} \otimes V)$ for the pair (j, a) of qubits, the former of which corresponds to the magnetic node $j \in \tilde{N}$, and the latter is the auxiliary one, by the formula

$$L_{ja}(\lambda) = \lambda\,\mathrm{id}_{\mathcal{H}} \otimes \mathrm{id}_V + i\vec{s}_j \cdot \vec{\sigma}_a, \tag{20}$$

where $\mathrm{id}_{\mathcal{H}} = I^{\otimes N}$ and $\mathrm{id}_V = I$ are unit operators in the corresponding spaces. Two remarks related to this definition can be inserted now. Firstly, the Lax operator is defined for the whole extended system, but it acts effectively only in two factors, the qubit $(\mathbb{C}^2)_j$ at the node j and the auxiliary qubit V. It thus acts as a "mouse", attached to the register $\mathcal{H}$ of a quantum computer. Secondly, the relation in Eq. (20) distinguishes the auxiliary qubit by use of Pauli operators for it, and spin operators for qubits of magnetic nodes, but this distinction is just conventional. In particular, it allows us to use the Dirac identity (15), which yields

$$L_{ja}(\lambda) = \left(\lambda - \frac{i}{2}\right) \mathrm{id}_{\mathcal{H}} \otimes \mathrm{id}_V + iu_{ja}, \tag{21}$$

where the transposition operator u_{ja} for qubits $(\mathbb{C}^2)_j$ and V is embedded into $\mathrm{End}\,(\mathcal{H} \otimes V)$, i.e. $u_{ja} \in \mathrm{End}\,(\mathcal{H} \otimes V)$, in such a way that it acts trivially on all other $(N-1)$ qubits of the space $\mathcal{H} \otimes V \cong (\mathbb{C}^2)^{\otimes N+1}$. The Lax operator in the computational basis of the auxiliary qubit V, that is, in the basis of the environment of the magnetic ring, has the form

$$L_{ja}(\lambda) = \begin{pmatrix} \lambda + i\hat{s}_j^z & i\hat{s}_j^- \\ i\hat{s}_j^+ & \lambda - i\hat{s}_j^z \end{pmatrix}, \tag{22}$$

where $\hat{s}^\pm = \hat{s}_j^x \pm \hat{s}_j^y \in \mathrm{End}\,\mathcal{H}$. It is thus a 2×2 matrix, with entries being operators in $\mathcal{H}$.

The Lax operator provides a kind of connection between qubits of the extended system. The product of Lax operators over consecutive nodes of the ring $\tilde{N}$ defines the monodromy matrix

$$\mathcal{M}(\lambda) = L_{1a}(\lambda)L_{2a}(\lambda)\dots L_{Na}(\lambda) \in \mathrm{End}\,(\mathcal{H}\otimes V), \tag{23}$$

presented in the computational basis of the auxiliary qubit V as

$$\mathcal{M}(\lambda) = \begin{pmatrix} A(\lambda) & B(\lambda) \\ C(\lambda) & D(\lambda)\,, \end{pmatrix} \tag{24}$$

with $A(\lambda)$, $B(\lambda)$, $C(\lambda)$, $D(\lambda) \in \mathrm{End}\,\mathcal{H}$ being some operators in $\mathcal{H}$. We sketch here some important features of these operators.

All these operators, A, B, C and D, are polynoms on the spectral parameter λ, of degree not larger than N. It justifies the adjective "algebraic" BA. The trace

$$F(\lambda) = A(\lambda) + B(\lambda), \tag{25}$$

i.e. the transfer matrix of Baxter [14], yields the family of mutually commuting operators,

$$[F(\lambda), F(\lambda')] = 0, \quad \lambda, \lambda' \in \mathbb{C}, \tag{26}$$

for arbitrary values of the spectral parameter λ, and thus yields the subalgebra of integrals of motion in $\mathrm{End}\,\mathcal{H}$ – the key for integrability of the XXX model.

We explore in this note the fact, that the operators $B(\lambda)$, when acting on the ferromagnetic vacuum, $|\mathrm{vac}\rangle = |++\dots+\rangle$, create the magnon with the spectral parameter λ. In our notation it can be written as

$$B(\lambda)|\mathrm{vac}\rangle = |\lambda\rangle \in h, \tag{27}$$

with $|\lambda\rangle$ being a state of the magnonic qudit h, introduced in Section 2. The state $|\lambda\rangle$ is fully determined by the creation operator $B(\lambda)$, for a given value of the spectral parameter λ. We show in the next section how to encode an arbitrary exact BA eigenstate as a sequence of magnonic qudits, each in a state represented by an appropriate spectral parameter, emerging from BA equations.

4. Encoding of rigged string configurations in the magnonic qudit

In order to encode exact BA eigenstates using the magnonic qudit, we first recall that (i) the space $\mathcal{H}$ of all quantum states of the magnetic ring decomposes as

$$\mathcal{H} = \sum_{r=0}^{N} \oplus \mathcal{H}^{(r)} \tag{28}$$

into subspaces $\mathcal{H}^{(r)}$ with r overturned spins, or, equivalently, with the z-projection $M = N/2 - r$ of the total spin of the magnet; (ii) each subspace $\mathcal{H}^{(r)}$ decomposes further as

$$\mathcal{H}^{(r)} = \sum_{r'=0}^{\min(r,N-r)} \oplus \mathcal{H}^{(r,r')} \tag{29}$$

into subspaces $\mathcal{H}^{(r,r')}$ with the total spin $S = N/2 - r'$; (iii) each subspace $\mathcal{H}^{(r,r')}$ is spanned on the orthonormal basis $|\nu\mathcal{L}\rangle$ of exact BA eigenstates, called rigged string configurations. Here $\nu \vdash r'$ is a partition of the integer r' called a configuration of strings, each each part of ν, or equivalently, each row of the Young diagram of ν, is a string with the length l equal to the number of boxes in the row, and $\mathcal{L}$ is the collection of riggings, i.e. integers (quantum numbers), one for each string, which characterize its state of motion, and vary within a prescribed range, dependent on the length l. More specificly, let m_l be the number of l-strings in the string configuration $\nu \vdash r'$, so that

$$\sum_l l m_l = r'. \tag{30}$$

Then

$$\mathcal{L} = \{L_{lv} \mid l = 1, 2, \ldots, v = 1, 2, \ldots, m_l\}, \tag{31}$$

and the riggings L_{lv} satisfy selection rules

$$0 \le L_{lv} \le L_{lv'} \le P_l \quad \text{for } 1 \le v \le v' \le m_l, \tag{32}$$

where

$$P_l = N - 2Q_l, \tag{33}$$

and Q_l is the number of boxes in the first l columns of the Young diagram of ν (c.f., e.g. [4] for more details).

In this way, each BA eigenstate $|\nu\mathcal{L}\rangle$ is determined by a rigging L_{lv}, associated with each *row* (l, v) of the Young diagram ν. The algebraic BA provides an equivalent determination of this eigenstate, in terms of a collection of r' spectral parameters λ_{lv}^m, such that each *box* (l, v, m) of the Young diagram ν corresponds to a distinct spectral parameter. These parameters are labeled, for each string (l, v), by the integer or half-integer m, varying by step 1 within the range

$$-\frac{l-1}{2} \le m \le \frac{l-1}{2}, \tag{34}$$

so that the quantity $(l-1)/2$ resembles "the spin of an l-string". It is worth to add that the string hypothesis of Bethe implies that for $N \to \infty$ the spectral parameters tend to the form

$$\lambda_{lv}^m = \lambda_{lv} + im \tag{35}$$

with real λ_{lv} and m, and that in the most of the known cases for a finite N the actual results for spectral parameters are quite close to the asymptotic form (35) (but there are severe exceptions in several, but asymptoticly remote cases - c.f. e.g. [15]).

We arrive at a conclusion that each exact eigenstate $|\nu\mathcal{L}\rangle$ cane be written within algebraic BA in a form

$$|\nu\mathcal{L}\rangle = \prod_{(l,v,m)\subset\nu} \hat{B}\left(\lambda_{lv}^m\right)|\text{vac}\rangle, \tag{36}$$

where the collection $\{\lambda_{lv}^m\}$ of spectral parameters satisfies the algebraic system of r' equations (Bethe equations), of the standard form

$$\left(\frac{\lambda_{lv}^m - \frac{i}{2}}{\lambda_{lv}^m + \frac{i}{2}}\right)^N = \prod_{(l',v',m')\neq(lmv)} \frac{\lambda_{lv}^m - \lambda_{l'v'}^{m'} - i}{\lambda_{lv}^m + \lambda_{l'v'}^{m'} + i} \tag{37}$$

and the product in the rhs of Eq. (36) runs over all boxes of the Young diagram ν of the configuration of strings. Thus r' copies of the magnonic qudit h described in Section 2, one copy for each state $\hat{B}(\lambda_{lv}^m)|\text{vac}\rangle \in h$, are sufficient for construction of the state $|\nu\mathcal{L}\rangle$. In other words, the collection of r' qudits forms a register in a quantum computer composed of qudits, for the rigged string configuration $|\nu\mathcal{L}\rangle$.

5. Discussion

We have described a hypothetical unit of memory of a quantum computer, extracted from the space of quantum states of the one-dimensional Heisenberg ring, as the space $h = \mathbb{C}^N$ of all states with a *single* overturned spin, referred to as the magnonic qudit. To keep a close connection with the prototypic Heisenberg model of magnetism, we pointed out three natural orthonormal bases in the qudit space h: (i) the computational basis, referred to as the positional one, (ii) the basis of momenta, introduced in terms of the finite Fourier transform along the Schwinger unitary geometry, and (iii) the irreducible basis of the Schur-Weyl duality, realized by Kostka transformation at the level of bases.

We pointed out the way of encoding exact BA eigenfunctions, classified by rigged string configurations $|\nu\mathcal{L}\rangle$, in a register of a quantum of magnonic qudits. To this aim, we have reviewed in short the algebraic form of BA, based on the monodromy matrices and the associated operators A, B, C, D, dependent on the spectral parameter λ. We have pointed out the fact that a single magnonic qudit can be prepared in a state $B(\lambda)|\mathrm{vac}\rangle \in h$, which represents a single Bethe pseudoparticle with the spectral parameter λ. It follows that a collection of $r' \leq r$ magnonic qudits carries the full information on the rigged string configuration $|\nu\mathcal{L}\rangle$ when each qudit is in the state $\hat{B}(\lambda^m_{l\nu})|\mathrm{vac}\rangle$, corresponding to on of the collection of spectral parameters corresponding to this state.

References

[1] Bethe H 1931 *Z. Physik* **71** 205 (in German; English translation in: Mattis DC 1993 *The Many-Body Problem* (World Sci.: Singapore) 689)
[2] Gaudin M 1983 *La fonction d'onde de Bethe* (Masson:Paris)
[3] Faddeev LD and Takhtajan LA 1981 *LOMI* **109** 134 (in Russian; English translation: 1984 *J. Sov. Math.* **24** 241)
[4] Lulek T 2008 *J. Phys. Conf. Ser.* **104** 012043
[5] Jakubczyk P, Topolewicz S, Wal A and Lulek T 2009 *OSID* **16** 221
[6] Schwinger J 1970 *Quantum Kinematics and Dynamics* (Benjamin: New York)
[7] Lulek B, Lulek T, Wal A and Jakubczyk P 2003 *Physica* **B 337** 375
[8] Milewski J and Ambrozkożko E 2008 *J. Phys. Conf. Ser.* **104** 010040
[9] Milewski J, Banaszak G and Lulek T 2010 *OSID* **15** to appear
[10] Lulek B, Lulek T, Jakubczyk P and Jakubczyk D 2009 *Physica* **B 404** 115
[11] Faddeev LD and Takhtajan LA 1981 *Phys. Lett.* **85 A**, 375 1981
[12] Nielsen MA and Chuang IL 2000 *Quantum Computation and Quantum Information* (Cambridge University Press:Cambridge)
[13] Dirac PAM 1958 *The principles of quantum mechanics, 4th Edition* (Oxford University Press:Oxford)
[14] Baxter RJ 1982 *Exactly Solvable Models in Statistical Mechanics* (Academic Press:New York)
[15] Essler FHL, Korepin VE and Schoutens K 1992 *J. Phys. A* **25** 4115
[16] Caspers WJ, Labuz M, Wal A, Kuźma M and Lulek T 2003 *J. Phys. A: Math. Gen.* **36** 5369

Matrix Schur Functions, Permutation Matrices, and Young Operators as Inner Product Spaces

James D. Louck

Los Alamos National Laboratory Fellow, 54 Wildflower Way, Santa Fe, NM, 87506 USA

E-mail: jimlouck@aol.com

Abstract. An inner product is defined on the space of permutation matrices and the space of matrices dual to the permutation matrices is given. The relationship of permutation matrices to the expansion of matrices of fixed line-sum is discussed. This inner product carries over to the space of linear Young operators, as does the notion of the dual space. Motivation from physics for considering such algebraic structures is also given; in particular, the real, orthogonal, irreducible representations of the symmetric group originating from the matrix Schur functions are given.

1. Background, Review and Motivation

There is, perhaps, no group that has more universal applications in quantum physics than the symmetric group S_n because of the extraordinary restrictions put on the quantum states of composite systems by the Pauli principle (see Hamermesh [1]) and Wybourne [2] for such applications). The classical work of Young [3] establishing the foundations of this subject, is well-known. We begin by reviewing the definition of the symmetric group, using the one-line notation for elements of the group, which may not be familiar to all.

The symmetric group S_n may be defined to be the set of all $n!$ sequences $\{\pi\}$ in the integers $1, 2, \ldots, n$, together with a multiplication rule for multiplying each pair $\rho \in S_n, \tau \in S_n$ of elements in the set S_n, as follows:

$$\begin{aligned} S_n = \{\pi = (\pi_1, \pi_2, \ldots, \pi_n) \mid & \text{ the } n \text{ parts } \pi_i, i = 1, 2, \ldots, n \\ & \text{of } \pi \text{ are any arrangement of the integers } 1, 2, \ldots, n\}, \qquad (1) \\ \rho \times \tau = \rho\tau = (\rho_{\tau_1}, \rho_{\tau_2}, & \ldots, \rho_{\tau_n}), \text{ all } \rho, \tau \in S_n. \end{aligned}$$

Technically, the symmetric group is the pair of objects $(S_n, \times)$, since one and the same set can have different rules for multiplying its elements. It is customary, however, to denote the group simply by S_n once the rule for multiplying elements has been given, and write $\rho \times \tau$ as $\rho\tau$ for each pair of elements $\rho, \tau \in S_n$, as already anticipated in (1).

We use the one-line symbol $\pi = (\pi_1, \pi_2, \ldots, \pi_n)$ for the elements of S_n, where the meaning of this symbol is: replace 1 by π_1, replace 2 by $\pi_2, \ldots,$ replace n by π_n, which is written

$1 \mapsto \pi_1, 2 \mapsto \pi_2, \ldots, n \mapsto \pi_n$. This statement has, of course, no meaning on its own — it has a contextual meaning in that it must be applied in the context of some set of objects $X = \{X_1, X_2, \ldots, X_n\}$, where the elements of X (which themselves may be sets) are enumerated by $1, 2, \ldots, n$. For example, the permutation $\pi = (3, 4, 2, 1) \in S_4$ can be applied to $x = (x_1, x_2, x_3, x_4)$, with the result that $\pi : (x_1, x_2, x_3, x_4) \mapsto (x_3, x_4, x_2, x_1)$; or π can be applied to $y = (x_2, x_1, x_4, x_3)$ with the result that $\pi : (x_2, x_1, x_4, x_3) \mapsto (x_4, x_3, x_1, x_2)$. The instruction carried by the notation $\pi = (3, 4, 2, 1)$ is expressed verbally by: replace 1 by 3, replace 2 by 4, replace 3 by 2, replace 4 by 1, in any collection of entities labeled by $1, 2, \ldots, n$ on which the permutation is defined to act. In particular, the general permutation $\rho \in S_n$ can act on another permutation $\tau \in S_n$, in which case the rule gives $\rho \times \tau = \rho\tau = (\rho_{\tau_1}, \rho_{\tau_2}, \ldots, \rho_{\tau_n})$, all $\rho, \tau \in S_n$, as given in (1). Thus, property is called the closure property.

The elements of the set S_n satisfy the rules imposed on every abstract group: (i) the product of two elements in the set is again an element; (ii) the associative product rule $(\pi\rho)\tau = \pi(\rho\tau)$ holds for all elements; (iii) there is an identity element $e = (1, 2, \ldots, n)$ that commutes with each element; (iv) there is an inverse π^{-1} for each element. (The (unique) inverse, denoted π^{-1}, of the sequence $\pi = (\pi_1, \pi_2, \ldots, \pi_n) \in S_n$ is obtained from π by writing out the sequence $(\pi^{-1}_{\pi_1} = 1, \pi^{-1}_{\pi_2} = 2, \ldots, \pi^{-1}_{\pi_n} = n)$, and rearranging the entries in this sequence such that the **subscripts** are $1, 2, \ldots, n$ as read left-to-right.) This presentation of permutations as one-line symbols of sequences with a multiplication rule is quite useful, especially from the viewpoint of concise presentation.

The action of a permutation $\pi = (\pi_1, \pi_2, \ldots, \pi_n) \in S_n$ on an arbitrary sequence of n indeterminates $x = (x_1, x_2, \ldots, x_n)$ is defined by

$$\pi(x) = (x_{\pi_1}, x_{\pi_2}, \ldots, x_{\pi_n}). \tag{2}$$

The action of the product $\rho\tau$ of two elements of the group is defined by the rule: first apply τ to the sequence x followed by applying ρ to the sequence $\tau(x) == (x_{\tau_1}, x_{\tau_2}, \ldots, x_{\tau_n})$. Thus, we have that

$$(\rho\tau)(x) = \rho(\tau(x)) = (x_{(\rho\tau)_1}, x_{(\rho\tau)_2}, \ldots, x_{(\rho\tau)_n}). \tag{3}$$

The set of permutation matrices is a realization of the multiplication rules for the symmetric group S_n by $0-1$ matrices of order n. The permutation matrix $P_\pi, \pi \in S_n$, is defined as the following matrix:

$$P_\pi = (e_{\pi_1}\, e_{\pi_2}\, \ldots\, e_{\pi_n}), \tag{4}$$

where e_i is a the *unit column matrix* with 1 in row i and 0 in all other rows, each $i = 1, 2, \ldots, n$. Thus, the element in row i and column j of P_π is given by

$$(P_\pi)_{ij} = \delta_{i,\pi_j}. \tag{5}$$

It may be verified directly from these matrix elements that the set of $n!$ matrices of order n defined by

$$\mathbb{P}_{S_n} = \{P_\pi \,|\, \pi \in S_n\} \tag{6}$$

satisfies all the group rules for the symmetric group S_n in which now the rule of multiplication is ordinary matrix multiplication. Thus, $P_\rho P_\tau = P_{\rho\tau}$, each pair $\rho, \tau \in S_n$. In particular, $P_\pi^{(-1)} = P_\pi^T$, where T denotes matrix transposition. Thus, we have a *representation of* S_n by a set of real, orthogonal matrices of order n. A unit row matrix presentation can also be used, but we will always use the column matrix form.

The action $\pi(x)$ of a permutation on a sequence $x = (x_1, x_2, \ldots, x_n)$ can presented in matrix form as ordinary row-on-column matrix multiplication:

$$P_\pi x = x_\pi, \tag{7}$$

where now x is the column matrix $x = \mathrm{col}(x_1, x_2, \ldots, x_n)$ and x_π is the column matrix $x_\pi = \mathrm{col}(x_{\pi_1}, x_{\pi_2}, \ldots, x_{\pi_n})$. In particular, we have that

$$P_\pi e_j = e_{\pi_j} = \sum_{i=1}^{n} \delta_{i,\pi_j} e_i. \tag{8}$$

Permutation matrices are of interest on their own. This is because there is a class of matrices $\{A\}$ over the real numbers $\mathbb{R}$, or the complex numbers $\mathbb{C}$, that have an expansion in terms of permutation matrices:

$$A = \sum_{\pi \in S_n} a_\pi P_\pi, \text{each } a_\pi \in \mathbb{C}, \tag{9}$$

Before developing the properties of this class of matrices further in Sect. 2, we illustrate by three principal examples their occurrence in problems in physics and chemistry, hence, their intrinsic interest.

Each of the following sets of matrices has an expansion in terms of permutation matrices: doubly stochastic, magic square, and alternating sign, as we next briefly discuss:

(i) A doubly stochastic matrix of order n has line-sum $r = 1$ (see (18) below) with elements $a_{ij} \in \mathbb{N}$, the nonnegative real numbers. We denote the set of all doubly stochastic matrices of order n by $\mathbb{A}_n$. The existence of the expansion (9) for each $A \in \mathbb{A}_n$ is known as Birkhoff's theorem. A simple proof of property (9) (over the positive integers) is found in Brualdi and Ryser [4]. Our interest stems from the role of doubly stochastic matrices in the quantum theory of angular momentum of composite systems (see Ref. [5,6]). This theory and its relation to combinatorics has been developed extensively in Ref. [5]. It suffices here to remark that all $3n - j$ coefficients can be realized as the elements of a real orthogonal matrix, called a recoupling matrix, *each of which is a doubly stochastic matrix.*

(ii) A magic square matrix of order n has line-sum equal to arbitrary $r \in \mathbb{R}_+$, the positive integers, with elements $a_{ij} \in \mathbb{N}$. We denote the set of all magic square matrices of order n by $\mathbb{M}_n(r)$. Magic square matrices are the simplest example of the more general class of matrices of order n, denoted by $\mathbb{M}^p_{n\times n}(\alpha, \beta)$ in Ref. [5] and defined below in Sect. 4, for which the row and column sums r_i and c_j are not all equal, but are equal to the parts $r_i = \alpha_i$ and $c_j = \alpha'_j$ of the *weights* (also called *contents*) $\alpha = (\alpha_1, \alpha_2, \ldots, \alpha_n) \vdash p$ and $\alpha' = (\alpha'_1, \alpha'_2, \ldots, \alpha'_n) \vdash p$ of the pair of Gelfand-Tsetlin patterns that enumerate the irreducible integer representations of the general linear group $GL(n, \mathbb{C})$ and the unitary group $U(n)$. (The notation $\alpha \vdash p$, etc. denotes that the parts of α sum to a prescribed positive integer p.) The sets $\mathbb{M}^p_{n\times n}(\alpha, \beta)$ occur in many contexts; significantly, they are basic to the definition of what are called matrix Schur functions (see Sect. 4), all of which is discussed in great detail in Ref. [5]. Little is known about the cardinality of the set $\mathbb{M}^p_{n\times n}(\alpha, \beta)$. A magic square is the special case: $\mathbb{M}_n(r) = \mathbb{M}^{nr}_{n\times n}(\alpha, \beta)$, for which $\alpha_i = \beta_i = r$, each $i = 1, 2, \ldots, n$. Closed formulas for the cardinality of the set $\mathbb{M}_n(r)$ are known for $n = 1, 2, 3, 4, 5$, for which $|M_1(r)| = 1, |M_2(r)| = r + 1, N_3(r) = \binom{r+5}{5} - \binom{r+2}{5}$, and the explicit formulas for $n = 4, 5$ have been noted recently in Ref. [7].

(iii) An alternating sign matrix of order n has line-sum $r = 1$ with elements $a_{ij} \in \mathcal{A} = \{-1, 0, 1\}$, such that when a given row (column) is read from left (top)-to-right (bottom) and all zeros in that row (column) are ignored, the row (column) contains either a single 1, or is an alternating series $1, -1, 1, -1, \ldots, 1, -1, 1$. We denote the set of all alternating sign matrices of order n by $\mathbb{A}S_n$. The discovery and proof of the formula for the number of alternating sign matrices of order n is a fascinating story of the interrelations between topics in mathematics, as reviewed in the essay by Bressoud and Propp [8], the book by Bressoud [9], and the article by Propp [10]. The formula is

$$|\mathbb{A}S_n| = d_n = \prod_{j=0}^{n-1} \frac{(3j+1)!}{(n+j)!}. \tag{10}$$

This relation was first proved by Zeilberger [11] and augmented by a closely related, and, perhaps, more significant formula in Zeilberger [12], since (10) follows quite easily from the second result (see Ref.[5]). The integer d_n also gives the cardinality of several other sets of "combinatorial" entities. A physical interpretation of the number d_n is in counting the number of states of a two-dimensional "6-vertex " model of ice, known as square ice. Kupenberg [13] uses this 6-vertex model of statistical mechanics to give another derivation of the relation (10).

It is interesting to note that the collection of $n!$ permutation matrices of order n is a subset of each of the classes of matrices: doubly stochastic, magic square, and alternating sign, and the subset of permutation matrices is, at the same time, a basis for expanding all such matrices of order n.

We next turn to the mathematical framework for this develop of the properties of permutation matrices.

2. Algebra of Permutation Matrices

The set of permutation matrices $\mathbb{P}_{S_n}$ defined by (6) can be taken as the basis elements in a real or complex linear vector space $\mathcal{P}_n$ with an inner product: Thus, if we define the matrix A by

$$A = \sum_{\pi \in S_n} a_\pi P_\pi, \text{each } a_\pi \in \mathbb{C}, \tag{11}$$

then it is well-known that the set of all such matrices, obtained by letting the expansion coefficients take on all real (resp., complex) values, is a real (resp., complex) linear vector space. Indeed, if we take a second such matrix,

$$B = \sum_{\pi \in S_n} b_\pi P_\pi, \text{each } b_\pi \in \mathbb{C}, \tag{12}$$

then the number defined by

$$(A \,|\, B) = \mathrm{Tr}(A^\dagger B) \tag{13}$$

is an *inner product* on the vector space $\mathcal{P}_n$, where $\dagger$ denotes the complex conjugated transpose of A, and Tr denotes the trace of a matrix. Then:

$$(P_\pi \,|\, P_{\pi'}) = \sum_{i=1}^{n} \delta_{\pi_i, \pi'_i} \text{ and } (A \,|\, B) = \sum_{i=1}^{n} \sum_{\pi, \pi' \in S_n} \delta_{\pi_i, \pi'_i} a^*_\pi b_{\pi'}. \tag{14}$$

where $*$ denotes complex conjugation.

The element a_{ij} in row i and column j of a matrix $A = (a_{ij})_{1 \leq i,j \leq n}$ of the form (11) is given in terms of the expansion coefficients a_π by

$$a_{ij} = \sum_{\pi \in S_n^{i,j}} a_\pi, \tag{15}$$

where $S_n^{i,j}$ is the subset of permutations in S_n defined by

$$S_n^{i,j} = \{\pi \in S_n \mid \pi_j = i\}. \tag{16}$$

These subsets of S_n then satisfy the following union relations:

$$\bigcup_{k=1}^{n} S_n^{i,k} = S_n, \ \bigcup_{k=1}^{n} S_n^{k,j} = S_n, \ i,j = 1,2,\ldots,n. \tag{17}$$

It follows, in turn, from (15) that each row-sum r_i and each column-sum c_j of the elements of A is given, respectively, by

$$r_i = \sum_{k=1}^{n} a_{ik} = c_j = \sum_{k=1}^{n} a_{kj} = \sum_{\pi \in S_n} a_\pi. \tag{18}$$

Thus, all row and column sums are equal. We refer to this common sum as the *line-sum* of the matrix A and denote it by l_A :

$$l_A = \sum_{\pi \in S_n} a_\pi. \tag{19}$$

Each matrix $A \in \mathcal{P}_n$ is allowed to have its own line-sum l_A; that is, we do not require all such matrices to have the same line-sum.

Next, we take the product of two matrices $A, B \in \mathcal{P}_n$ and obtain:

$$AB = C = \sum_{\pi \in S_n} c_\pi P_\pi \in \ \mathcal{P}_n, \tag{20}$$

where the group multiplication rule for permutation matrices gives the expansion coefficients as

$$c_\pi = \sum_{\rho \in S_n} a_\rho b_{\rho^{-1}\pi}. \tag{21}$$

Thus, we find that the line-sum of the product matrix AB satisfies the product rule:

$$l_{AB} = l_A l_B. \tag{22}$$

A linear vector space that also has a product defined for all pairs of vectors such that the product of each pair is again a vector in the space is called an algebra — here, the *algebra of permutation matrices,* which we denote by $\mathcal{AP}_n$.

Thus far we have given no hint as to how to determine the expansion coefficients a_π for a matrix $A \in \mathcal{AP}_n$. This, in turn, raises the question of the linear independence of the set $\mathbb{P}_{S_n}$ of permutation matrices; that is, the dimensionality of the vector space we have denoted by $\mathcal{P}_n$.

As is the case for any linear vector space, a set of permutation matrices $\{P_\pi \,|\, \pi \in \Sigma_n \subset S_n\}$ is linearly independent over the real (complex) numbers if and only if the relation

$$\sum_{\pi \in \Sigma_n} a_\pi P_\pi = 0_n \text{ (zero matrix of order } n) \tag{23}$$

implies that $a_\pi = 0$, each $\pi \in \Sigma_n$. To my knowledge, my postdoctoral student, John H. Carter [14] was the first to prove: *The number b_n of linearly independent permutation matrices of order n, hence, also the dimensionality of the space $\mathcal{P}_n$ is given by*

$$\dim \mathcal{P}_n = b_n = (n-1)^2 + 1. \tag{24}$$

Curiously, we could not find this elementary result in the published literature, although probably known to many. The derivation of this number is repeated in Ref. [7], together with the proof of several sets Σ_n of permutations giving a basis set of linearly independent permutations matrices:

$$\mathbb{P}_{\Sigma_n} = \{P_\pi \,|\, \pi \in \Sigma_n \subset S_n\}. \tag{25}$$

Then, each matrix A given by (11) can now be rewritten in the form

$$A = \sum_{\pi \in \Sigma_n} a_\pi P_\pi, \text{ each } a_\pi \in \mathbb{C}. \tag{26}$$

The expansion coefficients, or coordinates a_π relative to the basis $\mathbb{P}_{\Sigma_n}$ are now unique.

The unique coordinates a_π in the expansion (26) relative to a basis Σ_n can be determined by using the dual basis to $\mathbb{P}_{\Sigma_n}$, which we denote by $\overline{\mathbb{P}}_{\Sigma_n}$. This is carried out as follows: The dual matrices are defined by the orthogonality requirements:

$$(\overline{P}_\pi \,|\, P_{\pi'}) = \delta_{\pi,\pi'}, \text{ each pair } \pi, \pi' \in \Sigma_n. \tag{27}$$

Thus, the coordinates in relation (26) are given uniquely by

$$a_\pi = (\overline{P}_\pi | A). \tag{28}$$

It is, however, difficult to determine the basis set of matrices in the dual basis:

$$\overline{\mathbb{P}}_{\Sigma_n} = \{\overline{P}_{\Sigma_n} \,|\, \pi \in \Sigma_n\}. \tag{29}$$

We have determined the general set $\overline{\mathbb{P}}_{\Sigma_n}$ in Ref. [7] for a specific basis set Σ_n. Here, we just note the example for $n = 3$, where the orthogonality property (27) can be verified explicitly.

Example. The dual permutation matrices $\overline{P}_\pi$ for $\pi \in \Sigma_3 = \{(1,2,3),$ $(1,3,2),(2,3,1),(2,1,3),(3,2,1)\}$ (the permutation $(3,1,2)$ is missing):

$$\begin{aligned}
\overline{P}_{1,2,3} &= (2\,P_{1,2,3} - P_{2,1,3} + P_{2,3,1} - P_{1,3,2} - P_{3,2,1})/3, \\
\overline{P}_{2,1,3} &= (-3\,P_{1,2,3} + 5\,P_{2,1,3} - 3\,P_{2,3,1} + 2\,P_{1,3,2} + 2\,P_{3,2,1})/9, \\
\overline{P}_{2,3,1} &= (P_{1,2,3} - P_{2,1,3} + 2\,P_{2,3,1} - P_{1,3,2} - P_{3,2,1})/3, \\
\overline{P}_{1,3,2} &= (-3\,P_{1,2,3} + 2\,P_{2,1,3} - 3\,P_{2,3,1} + 5\,P_{1,3,2} + 2\,P_{3,2,1})/9, \\
\overline{P}_{3,2,1} &= (-3\,P_{1,2,3} + 2\,P_{2,1,3} - 3\,P_{2,3,1} + 2\,P_{1,3,2} + 5\,P_{3,2,1})/9. \ \square
\end{aligned} \tag{30}$$

We point out that it is never the case that the subset $\Sigma_n \subset S_n$ of $b_n = (n-1)^2 + 1$ linearly independent permutation matrices is a subgroup of S_n. Nonetheless, it is still true that

$$AB = \sum_{\rho,\tau \in \Sigma_n} a_\rho b_\tau P_{\rho\tau} = \sum_{\pi \in \Sigma_n} d_\pi P_\pi, \tag{31}$$

for two matrices A and B written in the form (11) and (12) with coordinates a_π, b_π. But now the coordinates of the product AB are not given by the rule (21), since Σ_n is not a group. The first sum in (31) splits into a sum of two terms: those for which $\rho\tau \in \Sigma_n$ and those for which $\rho\tau \in S_n$, but $\rho\tau \notin \Sigma_n$. The permutation matrices $P_{\rho\tau}$ for $\rho\tau$ in the second split sum must be re-expressed in terms of the $P_\pi, \pi \in \Sigma_n$ to obtain the formula for the coordinates d_π of the product AB. However, the product rule $l_{AB} = l_A l_B$ for line-sums given by (22) still holds — it is a basis independent relation.

3. An Algebra of Young Operators

We recall that a linear operator is a rule for mapping vectors in a linear vector space $\mathcal{V}$ into new vectors in the same space. Let $\mathcal{V}_n$ denote a linear vector space of finite-dimension $\dim\mathcal{V}_n = n$ over the complex numbers $\mathbb{C}$. Also, let $\mathcal{V}_n$ be an inner product space with an orthonormal basis set of vectors $\mathbf{B}_n = \{|1\rangle, |2\rangle, \ldots, |n\rangle\}$ (Dirac bra-ket notation). Then, a linear operator Y_π can be defined with the property

$$Y_\pi|i\rangle = |\pi_i\rangle, \text{ each } i = 1, 2, \ldots, n. \tag{32}$$

The linear property then means that the action of $Y_\pi, \pi \in S_n$, on each vector $|\alpha\rangle = \sum_{i=1}^n \alpha_i|i\rangle \in \mathcal{V}_n$, where $\alpha = (\alpha_1, \alpha_2, \ldots, \alpha_n) \in \mathbb{C}^n$, is given by

$$Y_\pi \sum_{i=1}^n \alpha_i|i\rangle = \sum_{i=1}^n \alpha_i Y_\pi|i\rangle = \sum_{i=1}^n \alpha_i|\pi_i\rangle = \sum_{i=1}^n \alpha_{\pi_i^{(-1)}}|i\rangle = |\alpha_{\pi^{(-1)}}\rangle, \tag{33}$$

where $\alpha_{\pi^{(-1)}} = (\alpha_{\pi_1^{(-1)}}, \alpha_{\pi_2^{(-1)}}, \ldots, \alpha_{\pi_n^{(-1)}})$. The linear operator Y_π is the simplest example of a *Young operator* associated with the symmetric group S_n. The product $Y_{\pi'}Y_\pi$ of two such Young operators is then defined by $(Y_{\pi'}Y_\pi)|\alpha\rangle = Y_{\pi'}(Y_\pi|\alpha\rangle)$, which must hold for each vector $|\alpha\rangle \in \mathcal{V}_n$. This relation gives the operator identity on $\mathcal{V}_n$ expressed by $Y_{\pi'}Y_\pi = Y_{\pi'\pi}$, as required of a group action.

We can now extend the definition of a Young operator to arbitrary sums (over $\mathbf{C}$) of the *elementary* Young operators Y_π, as given by

$$Y(x) = \sum_{\pi \in S_n} x_\pi Y_\pi, \text{ each } x_\pi \in \mathbb{C}. \tag{34}$$

It is a straightforward exercise to prove the well-known result: *The set of all linear operators* $\{Y(x)\,|x \in \mathbb{C}^n\}$ *is itself a linear vector space* $\mathcal{V}_n^{op}$ *over the complex numbers.* We also refer to $Y(x)$ as a Young operator. But now we have the additional property that, for each pair of Young operators $Y(x) \in \mathcal{V}_n^{\text{op}}$ and $Y(y) \in \mathcal{V}_n^{\text{op}}$, it also the case that the product $Y(x)Y(y)$ is a Young operator; that is, we have

$$Y(x)Y(y) = Y(z) \in \mathcal{V}_n^{\text{op}};\ z_\pi = \sum_{\rho \in S_n} x_\rho y_{\rho^{(-1)}\pi}. \tag{35}$$

Thus, we have an *algebra of Young operators*. See Hamermesh [1, pp. 239-246] for a more general discussion of this algebra.)

The matrix elements of the elementary Young operator Y_π on the orthonormal basis $\mathbf{B}_n$ of the vector space $\mathcal{V}_n$ are given by

$$\langle i\,|Y_\pi|\,j\rangle = \langle i\,|\pi_j\rangle = \delta_{i,\pi_j} = (P_\pi)_{ij}. \tag{36}$$

Thus, the matrix representation of the elementary Young operator Y_π on the vector space $\mathcal{V}_n$ is by the permutation matrix P_π. This result implies:

Every relation in Sect. 2 that holds between permutation matrices P_π and their linear combinations $A, B, \ldots$ over $\mathbb{C}$ also holds between the elementary Young operators Y_π and their linear combinations $Y(a), Y(b), \ldots$ over $\mathbb{C}$.

We repeat the significant relations: (i) dimension of the vector space $\mathcal{V}_n^{\mathrm{op}}$ of Young operators: $\dim\mathcal{V}_n^{\mathrm{op}} = b_n = (n-1)^2+2$; (ii) inner product of elementary Young operators: $\langle Y_\pi\,|\,Y_{\pi'}\rangle = (P_\pi\,|\,P_{\pi'}) = \sum_{i=1}^n \delta_{\pi_i,\pi'_i}$; (iii) inner product of Young operators $Y(a)$ and $Y(b)$: $\langle Y_a\,|\,Y_b\rangle = \sum_{i=1}^n \sum_{\pi,\pi'\in\Sigma_n} \delta_{\pi_i,\pi'_i} a_\pi^* b_\pi$; (iv) elementary dual Young operators defined by the orthogonality requirements: $\langle \overline{Y}_\pi\,|\,Y_{\pi'}\rangle = (\overline{P}_\pi\,|\,P_{\pi'}) = \delta_{\pi,\pi'}$, each pair $\pi,\pi'\in\Sigma_n$; (v) coordinates in a basis: $Y(a) = \sum_{\pi\in\Sigma_n} a_\pi Y_\pi$, each $a_\pi\in\mathbb{C}$, where $a_\pi = \langle\overline{Y}_\pi|Y(a)\rangle = (\overline{P}_\pi|P(a))$.

Example 1.

$$\begin{aligned} &P_{1,2,3}+P_{2,3,1}+P_{3,1,2}-P_{1,3,2}-P_{3,2,1}-P_{2,1,3}=0 \text{ (zero matrix)}\\ &\qquad\text{if and only if}\\ &Y_{1,2,3}+Y_{2,3,1}+Y_{3,1,2}-Y_{1,3,2}-Y_{3,2,1}-Y_{2,1,3}=0 \text{ (zero operator)}. \end{aligned} \tag{37}$$

Example 2. The five relations (30) for permutation matrices and their duals carry over exactly to elementary Young operators and their duals simply by making the substitutions $P_\pi \mapsto V_\pi$ and $\overline{P}_\pi \mapsto \overline{V}_\pi$.

We hasten to point out that Young operators can be defined on other vectors spaces that give rise to a more general theory of their algebra (see Hamermesh [1]). It is nonetheless the case that the inner product $\langle Y_\pi\,|\,Y_{\pi'}\rangle = \sum_{i=1}^n \delta_{\pi_i,\pi'_i}$ can always be defined, since it depends only on the properties of the permutations $\pi,\pi'\in S_n$.

4. Matrix Schur Functions

A real, orthogonal, irreducible representation of the symmetric group is obtained directly from the specialization of a class of many-variable polynomials known as *matrix Schur functions.* Whereas an ordinary Schur function is based on a single partition, a matrix Schur function is based on a number of stacked partitions that satisfy certain "betweenness" relations. The definition and properties of the matrix Schur functions is the subject of considerable length in Ref. [5]. Here, we list in Items 1-15 some of the properties of matrix Schur functions for the present application to the irreducible representations of S_n. The matrix Schur functions are completely defined by various of these relations, as discussed and proved in Ref. 5.

(i) Notation for matrix Schur functions: $D\begin{pmatrix} m' \\ \lambda \\ m \end{pmatrix}(Z)$ (also called $D^\lambda-$ polynomials and Gelfand-Tsetlin polynomials in Ref. [5])

(ii) Matrix Schur functions are homogeneous polynomials in the indeterminates (variables) $Z^A = \prod_{1\leq i,j\leq n} z_{ij}^{a_{ij}}$, each $a_{ij} \in \mathbb{N}$:

$$D\begin{pmatrix} m' \\ \lambda \\ m \end{pmatrix}(Z) = \sum_{A\in \mathbb{M}^p_{n\times n}(\alpha,\alpha')} C\begin{pmatrix} m' \\ \lambda \\ m \end{pmatrix}(A)\frac{Z^A}{A!} \tag{38}$$

(iii) Definitions of symbols:

(a) λ is a partition with n parts: $\lambda = (\lambda_1, \lambda_2, \ldots, \lambda_n)$, each $\lambda_i \in \mathbb{N}$, with $\lambda_1 \geq \lambda_2 \geq \cdots \geq \lambda_n \geq 0$.

(b) $\binom{\lambda}{m}$ is a stacked array of partitions:

$$\binom{\lambda}{m} = \left(\begin{array}{ccccccccc} \lambda_1 & & \lambda_2 & \cdots & \lambda_j & & \cdots & \lambda_{n-1} & \lambda_n \\ & m_{1,n-1} & m_{2,n-1} & \cdots & m_{j,n-1} & \cdots & & m_{n-1,n-1} & \\ & & & & \vdots & & & & \\ & & m_{1,j} & & m_{2,j} & \cdots & m_{j,j} & & \\ & & & & \vdots & & & & \\ & & & m_{1,2} & & m_{2,2} & & & \\ & & & & m_{1,1} & & & & \end{array}\right). \tag{39}$$

in which row j is a partition with j parts, and adjacent rows $(\mathbf{m}_{j-1}, \mathbf{m}_j), j = 2, 3, \ldots, n$, satisfy the "betweenness conditions:"

$$\begin{aligned} \mathbf{m}_j &= (m_{1,j}, m_{2,j}, \ldots, m_{j,j}), j = 1, 2, \ldots, n, \\ m_{i,j-1} &\in [m_{i,j}, m_{i+1,j}], \text{ each } i = 1, 2, \ldots, j-1, \\ [a,b], a \leq b, & \text{ denotes the closed interval of} \\ & \text{nonnegative integers: } [a,b] = \{a, a+1, \ldots, b\}. \end{aligned} \tag{40}$$

(c) Lexical patterns (patterns they satisfy betweenness rule):

$$\mathbb{G}_\lambda = \left\{\binom{\lambda}{m} \middle| m \text{ is a lexical pattern}\right\}. \tag{41}$$

(iv) The notation $\begin{pmatrix} m' \\ \lambda \\ m \end{pmatrix}$ denotes a pair of lexical GT patterns that share the same partition λ, where the pattern $\binom{m'}{\lambda}$ denotes the inversion of the pattern $\binom{\lambda}{m'}$, and the shared partition λ is written only once. Thus, it is these *double GT patterns* that enumerate the matrix Schur functions $D\begin{pmatrix} m' \\ \lambda \\ m \end{pmatrix}(Z)$. in the n^2 indeterminates $Z = (z_{ij})_{1\leq i,j\leq n}$.

(v) The C-coefficients in the right-hand side in the definition of the matrix Schur functions that multiply the polynomial factor $Z^A/A!$ are, up to multiplying factors, the elements of a real orthogonal matrix. The details are not important here (see Ref, [5] for details).

(vi) The set $\mathbb{M}^p_{n\times n}(\alpha, \alpha')$:

(a) Weights of the GT patterns:

$$\alpha = (\alpha_1, \alpha_2, \ldots, \alpha_n) \text{ of } \begin{pmatrix} \lambda \\ m \end{pmatrix}; \; \alpha' = (\alpha'_1, \alpha'_2, \ldots, \alpha'_n) \text{ of } \begin{pmatrix} \lambda \\ m' \end{pmatrix}. \tag{42}$$

(b) Homogeneity properties of the polynomials:

$$\begin{aligned} &\text{degree } \alpha_i \text{ in row } z_i = (z_{i1}, z_{i2}, \ldots, z_{in}) \text{ of } Z, \\ &\text{degree } \alpha'_j \text{ in column } z^j = (z_{i1}, z_{i2}, \ldots, z_{in} \text{ of } Z, \\ &p = \textstyle\sum_i \lambda_i = \sum_i \alpha_i = \sum_i \alpha'_i. \end{aligned} \tag{43}$$

(c) $\mathbb{M}^p_{n\times n}(\alpha, \alpha')$ denotes the set of all matrix arrays A of order n such that the sum of all elements in row i of A is α_i and the sum of all elements in column j of A is α'_j, where $\lambda \vdash p, \alpha \vdash p, \alpha' \vdash p$.

(vii) The summation is over all $A \in \mathbb{M}^p_{n\times n}(\alpha, \alpha')$: *The matrix Schur functions are homogeneous polynomials of total degree p in the indeterminates Z.*

(viii) Orthogonality relations in an inner product denoted (,) (isomorphic to boson inner product):

$$\begin{aligned} &\left(Z^{A'}, Z^A\right) = A!\delta_{A',A}, \\ &\left(D\begin{pmatrix} m' \\ \lambda \\ m \end{pmatrix}(Z), D\begin{pmatrix} m''' \\ \lambda' \\ m'' \end{pmatrix}(Z)\right) = \delta_{m,m''}\delta_{m',m'''}\delta_{\lambda,\lambda'}M(\lambda), \\ &M(\lambda) = \prod_{i=1}^{n}(\lambda_i + n - i)! \Big/ 1!2!\cdots(n-1)!\text{Dim}\lambda, \\ &\text{Dim}\lambda = \prod_{1\leq i<j\leq n}(\lambda_i - \lambda_j + j - i) \Big/ 1!2!\cdots(n-1)! \text{ (Weyl)}, \\ &D^\lambda(I_n) = I_{\text{Dim}\lambda}. \end{aligned} \tag{44}$$

(ix) Invertibility:

$$\frac{Z^A}{A!} = \sum_{\lambda\vdash p} \sum_{m,m'\in\mathbb{G}_\lambda(\alpha,\alpha')} \frac{1}{M(\lambda)A!} C\begin{pmatrix} m' \\ \lambda \\ m \end{pmatrix}(A) D\begin{pmatrix} m' \\ \lambda \\ m \end{pmatrix}(Z), \tag{45}$$

each $A \in \mathbb{M}^p_{n\times n}(\alpha, \alpha')$, and $\mathbb{G}_\lambda(\alpha, \alpha')$ denotes the set of double GT patterns of weight (α, α') and shape λ.

(x) Robinson, Schensted, Knuth (RSK) identity (expresses invertibility):

$$\sum_{\lambda\vdash p} K(\lambda, \alpha)K(\lambda, \alpha') = |\mathbb{M}^p_{n\times n}(\alpha, \alpha')|. \tag{46}$$

(xi) Multiplication property: Fundamental multiplication rule:

$$D^\lambda(X)D^\lambda(Y) = D^\lambda(XY), \text{ for arbitrary } X, Y. \tag{47}$$

(xii) Transpositional symmetry: Z replaced by the transpose Z^T :

$$D^\lambda(Z^T) = (D^\lambda(Z))^T. \tag{48}$$

(xiii) Group property: If Z is specialized to be a member of any matrix group, continuous or finite, representations of that group are obtained. Indeed, this applies as well to any multiplicative matrix algebra. In particular, irreducible representations of the general linear group $GL(n,\mathbb{C})$ and the unitary group $U(n)$ are obtained:

(xiv) Diagonal property: Z restricted to the diagonal matrix denoted $\mathrm{diag}Z = \mathrm{diag}(z_1, z_2, \ldots, z_n)$:

$$D\begin{pmatrix} m' \\ \lambda \\ m \end{pmatrix}(\mathrm{diag}Z) = \delta_{m,m'} z_1^{\alpha_1} z_2^{\alpha_2} \cdots z_n^{\alpha}, \tag{49}$$

where $\alpha = (\alpha_1, \alpha_2, \cdots, \alpha_n)$ is the weight of the GT pattern $\binom{\lambda}{m}$.

(xv) Schur functions. The trace of the above relation gives the ordinary Schur functions:

$$s_\lambda(z) = \mathrm{Tr}D^\lambda(\mathrm{diag}Z) = \sum_\alpha K(\lambda, \alpha)x^\alpha, \tag{50}$$

where the sum is over the distinct weights α of $\binom{\lambda}{m}$.

(xvi) Generalized MacMahon's master theorem:

$$\frac{1}{\det(I_{n^2} - tX \otimes Y)} = \sum_{k\geq 0} t^k \sum_{\lambda\vdash k} \mathrm{Tr}D^\lambda(X)\,\mathrm{Tr}D^\lambda(Y), \tag{51}$$

in which X and Y are arbitrary matrices of order n of commuting indeterminates, and $X \otimes Y$ is their Kronecker product (see Méndez [18] and Ref. [5]).

5. Real Orthogonal Irreducible Representations of S_n

5.1. Matrix Schur Function Real Orthogonal Irreducible Representations of S_n

The irreducible repesentations of the symmetric group are obtained from the matrix Schur functions defined by (38) in two steps:

(i). Impose conditions $\lambda \vdash n; \alpha, \alpha' \in \mathbb{G}_\lambda(1^n)$:

$$D\begin{pmatrix} m' \\ \lambda \\ m \end{pmatrix}(Z) = \sum_{\pi\in S_n} C\begin{pmatrix} m' \\ \lambda \\ m \end{pmatrix}(P_\pi)\, z_{\pi_1,1} z_{\pi_2,2} \cdots z_{\pi_n,n}, \tag{52}$$

in consequence of $\mathbb{M}^p_{n\times n}(\alpha, \alpha') = \{P_\pi \,|\, \pi \in S_n\}$ and $(P_\pi)_{ij} = \delta_{i,\pi_j}$. This result is of interest on its own because it gives a whole class of matrices $D^\lambda(Z), \lambda \vdash n$, such that the determinantal multiplication rule holds:

$$D^\lambda(X)D^\lambda(Y) = D^\lambda(XY). \tag{53}$$

(ii). Set $Z = P_{\pi'}$ in (52) (and then rename π' to be π):

$$D\begin{pmatrix} m' \\ \lambda \\ m \end{pmatrix}(P_\pi) = \begin{cases} C\begin{pmatrix} m' \\ \lambda \\ m \end{pmatrix}(P_\pi), \text{ for } \alpha, \alpha' \in \mathbb{G}_\lambda(1^n), \\ 0, \text{ all other weights} \end{cases} \tag{54}$$

The corresponding matrices then have the multiplication property:

$$D^\lambda(P_\pi)D^\lambda(P_{\pi'}) = D^\lambda(P_\pi P_{\pi'}). \tag{55}$$

The set of matrices $\{D^\lambda(P_\pi) \mid \pi \in S_n\}$ constitute a real, orthogonal, irreducible representation of S_n for each partition $\lambda = (\lambda_1, \lambda_2, \ldots, \lambda_n) \vdash n$, hence, all real, orthogonal, irreducible representations, up to orthogonal similarity. The order of the matrices $D^\lambda(P_\pi)$ is now denoted by dimλ; it is equal to the number of GT patterns $\lambda \vdash n$ such that the weight of $\binom{\lambda}{m}$ is $\alpha = (1^n)$. The RSK relation (46) reduces now to the sum-of-squares identity:

$$n! = \sum_{\lambda \vdash n} K(\lambda, 1^n) K(\lambda, 1^n). \tag{56}$$

5.2. *Jucys-Murphy Real Orthogonal Representations of S_n*

The permutation $(i,j) \in S_n$ is defined to be $\pi = (\pi_1, \pi_2, \ldots, \pi_n)$, in which all parts $\pi_k = k$, except for parts i and j, for which $\pi_i = j$ and $\pi_j = i$ for $i \neq j$. The permutation (i,j) is called a *transposition*, since it interchanges i and j in its action, and leaves all other integers unchanged. It has the properties $(i,j) = (j,i) = (i,j)^{-1}$. Thus, there are in all $n(n-1)/2$ distinct transpositions in S_n, which can be chosen to be $(i,j), 1 \leq i < j \leq n$.

It is well-known at this conference from the lectures by Lulek [19], and others (see, for example, Lulek *et al.* [20], Jakubczyk *et al.* [21]), that the $n-1$ Jucys-Murphy operators (Young operators) defined by $M_2 = Y_{(1,2)}, M_3 = Y_{(1,3)} + Y_{(2,3)}, M_4 = Y_{(1,4)} + Y_{(2,4)} + Y_{(3,4)}, \ldots, M_n = Y_{(1,n)} + Y_{(2,n)} + \cdots + Y_{(n-1,n)}$ mutually commute. In our presentation, we choose to replace the Jucys-Murphy (JM) operators by the modified set $M_2, M_2 + M_3, M_2 + M_3 + M_4, \cdots$; that is, we use the modified Jucys-Murphy operators $M_k^{(n)}$ defined by

$$M_k^{(n)} = \sum_{1 \leq i < j \leq k} Y_{(i,j)}, \; k = 2, 3, \ldots, n. \tag{57}$$

This set is, of course, still an independent set of $n-1$ mutually commuting Young operators belonging to the group algebra of S_n.

We have defined in Sect. 3 the action of the elementary Young operators on a very special vector space $\mathcal{V}_n$ of dimension dim$\mathcal{V}_n = n$ with basis $\mathbf{B}_n$, and with the action of each elementary Young operator $Y_{(i,j)}$ given explicitly by relation (32). This action determines completely the properties of the set of JM operators defined by relations (57), without further assumptions. But on this space $\mathcal{V}_n$ each elementary Young operator $Y_{(i,j)}$ is represented by the permutation matrix $P_{(i,j)}$. Thus, it must be the case that the spectrum of the JM operators $M_k^{(n)}$ is exactly that obtained by diagonalizing the associated matrices, which are real symmetric; that is,

$$P_k^{(n)} = \sum_{1 \leq i < j \leq n} P_{(i,j)} = \begin{pmatrix} J_k + \frac{1}{2}k(k-3)I_k & 0_{n-k} \\ & \\ 0_{n-k} & \binom{k}{2} I_{n-k} \end{pmatrix}, \tag{58}$$

where J_k denotes the matrix of order k with elements all equal to 1, and 0_{n-k} denotes the zero matrix of order $n-k$. For $k = n$, the last row and column are to be omitted in this relation; that is, $P_n^{(n)} = J_n + \frac{1}{2}k(k-3)I_n$.

The characteristic equation of the matrix $P_k^{(n)}$ is given by

$$\left(\lambda - \binom{k-1}{2} + 1\right)^{k-1} \left(\lambda - \binom{k}{2}\right)^{n-k+1} = 0, \tag{59}$$

where $\binom{a}{m}$ denotes a binomial coefficient. Thus, we find that the real symmetric matrix $P_k^{(n)}$ of order n has two distinct eigenvalues:

$$\begin{aligned} \lambda_k^{(n)}(1) &= \binom{k-1}{2} - 1 \text{ with multiplicity } k-1, \\ \lambda_k^{(n)}(2) &= \binom{k}{2} \text{ with multiplicity } n-k+1, \end{aligned} \tag{60}$$

where this result holds for each of the matrices for $k = 2, 3, \ldots, n$, and the binomial coefficient $\binom{1}{2}$ is defined to be 0.

There always exists a real orthogonal matrix R of order n that simultaneously diagonalizes a set of commuting symmetric matrices. Thus, there exists an R such that

$$R^T\, P_k^{(n)}\, R = D_k^{(n)} = \begin{pmatrix} \left(\binom{k-1}{2} - 1\right) I_{k-1} & 0_{n-k+1} \\ & \\ 0_{n-k+1} & \binom{k}{2} I_{n-k+1} \end{pmatrix}, \tag{61}$$

where R is the same for all $k = 2, 3, \ldots, n$. For the case at hand, the matrix $R = (R_1\, R_2\, \cdots\, R_n)$ is unique up to $\pm$ sign of its columns $R_i, i = 1, 2, \ldots, n$, a result that can be proved as follows: Relation (61) implies $DR = RD$, where $D = \sum_{k=2}^{n} D_k^{(n)}$ is a diagonal matrix with **distinct** elements. But then it follows trivially that R is unique up to choice of signs of its columns. This is just the expression in matrix form that the set of $n-1$ mutually commuting symmetric matrices $P_k^{(n)}, k = 2, 3, \ldots, n$ is a complete set. Equivalently, this is the expression of the well-known fact (Lulek [19]) that the set of $n-1$ PM operators (57) is a complete set of mutually commuting Hermitian operators on the space $\mathcal{V}_n$. (The Hermitian property is expressed by $\langle a\,|M_k^{(n)}|\,b\rangle = \left(\langle b\,|M_k^{(n)}|\,a\rangle\right)^* = \sum_{i,j=1}^{n} a_i b_j^* (P_k^{(n)})_{ij}$ for each pair of vectors $|a\rangle, |b\rangle \in \mathcal{V}_n$.)

We thus obtain the JM **reducible representation** J_π given by orthogonal similarity to a permutation matrix:

$$J_\pi = R^T P_\pi R,\ \pi \in S_n, \tag{62}$$

where this representation is characterized as the unique diagonalization of the complete set of real symmetric matrices in relation (58) by the real, orthogonal matrix R, which is unique up to $\pm$ signs of its columns.

The irreducible real orthogonal JM representations can now be obtained from the matrix Schur functions in exactly the same manner as the GT representation $D^\lambda(P_\pi), \pi \in S_n$, given by (55). We simply replace P_π in relation (55) by J_π given by (62) and use the multiplication property (47) to obtain:

$$D^\lambda(J_\pi) = D^\lambda(R^T) D^\lambda\left(P_\pi\right) D^\lambda(R), \text{ each } \pi \in S_n, \tag{63}$$

where the matrix $D^\lambda(R)$ of order dimλ is a real orthogonal matrix obtained from the matrix Schur function (38) by setting $Z = R$, choosing $\lambda \vdash n$, and restricting the weights of the patterns $\binom{\lambda}{m}$ and $\binom{\lambda}{m'}$ to be $\alpha = \alpha' = (1^n)$. *As expected, each real orthogonal irreducible JM representation of S_n is orthogonally similar to the real orthogonal irreducible GT representation corresponding to the same partition $\lambda \vdash n$.*

Example. It is useful to illustrate the above theory for $n = 3$:

(i) Matrices of the commuting JM Young operators $M_k^{(3)}, k = 2, 3$:

$$\begin{aligned} P_2^{(3)} &= P_{2,1,3} = \begin{pmatrix} 0 & 1 & 0 \\ 1 & 0 & 0 \\ 0 & 0 & 1 \end{pmatrix}, \\ P_3^{(3)} &= P_{2,1,3} + P_{3,2,1} + P_{1,3,2} = \begin{pmatrix} 1 & 1 & 1 \\ 1 & 1 & 1 \\ 1 & 1 & 1 \end{pmatrix}. \end{aligned} \tag{64}$$

(ii) Simultaneous eigenvalue-eigenvector relations:
columns of $R = (R_1\, R_2\, R_3)$:

$$R_1 = \frac{1}{\sqrt{6}}\begin{pmatrix} 1 \\ -1 \\ 0 \end{pmatrix}, R_2 = \frac{1}{\sqrt{2}}\begin{pmatrix} 1 \\ 1 \\ -2 \end{pmatrix}, R_3 = \frac{1}{\sqrt{3}}\begin{pmatrix} 1 \\ 1 \\ 1 \end{pmatrix}. \tag{65}$$

simultaneous eigenvectors:

$$\begin{aligned} &P_2^{(3)} R_1 = -R_1,\ P_2^{(3)} R_2 = R_2,\ P_2^{(3)} R_3 = R_3; \\ &P_3^{(3)} R_1 = 0\, R_1,\ P_2^{(3)} R_2 = 0\, R_2,\ P_2^{(3)} R_3 = 3\, R_3. \end{aligned} \tag{66}$$

Thus, we have the simultaneous diagonalization:

$$R^T P_2^{(3)} R = \mathrm{diag}(-1, 1, 1), \quad R^T P_3^{(3)} R = \mathrm{diag}(0, 0, 3), \tag{67}$$

where R is unique up to signs of its columns because the sum of the two diagonal matrices in (67) has unique eigenvalues diag$(-1, 1, 4)$. This gives the reducible JM real orthogonal representation $J_\pi = R^T P_\pi R$ of S_3 by real orthogonal matrices of order 3. The three irreducible JM real orthogonal representations of S_3 are then given by

$$D^\lambda(J_\pi) = D^\lambda(R^T) D^\lambda(P_\pi)\, D^\lambda(R), \tag{68}$$

where $\lambda = (1, 1, 1), (2, 1, 0), (3, 0, 0)$. These are the identity representation, the real orthogonal representation of order 2, and the alternating representation.

6. Conclusions

The basic role of permutations matrices has been illustrated through their occurrence in the expansion of three types of matrices of fixed line-sum: doubly stochastic, magic squares, and alternating sign, all of which are of interest in physical applications. It has been shown that the algebra of permutation matrices stands on its own and has implications for the algebra of Young operators. The occurrence of a natural inner product for permutations allows for the unique expansion of arbitrary fixed line-sum matrices and arbitrary Young operators in terms of any basis set $\Sigma_n \subset S_n$ of $(n-1)^2 + 1$ linearly independent matrices (Young operators) by use of the dual space. Finally, the elemental role of permutation matrices has been outlined in the determination of the real, orthogonal, irreducible representations of the symmetric group by application of the matrix Schur functions, both in the Gelfand-Tsetlin basis and in the Jucys-Murphy basis.

References

[1] Hamermesh M Group Theory and Its Application to Physical Problems 1962 Reading: Addison-Wesley.
[2] Wybourne B G Proc. Conf. Symmetry and Structural Properties of Condensed Matter ed. T Lulek, W Florek and B Lulek 1996 Singapore: World Scientific p. 253.
[3] Young A The Collected Papers of Alfred Young 1871-1940, 1977 Toronto: University of Toronto Press.
[4] Brualdi R A and Ryser H J *Combinatorial Matrix Theory* **39** Cambridge: Cambridge University Press.
[5] Louck J D Unitary Symmetry and Combinatorics 2009 Singapore: World Scientific.
[6] Louck J D 1997 *Foundations of Physics* **27** 1085.
[7] Louck J D Unitary Symmetry and Combinatorics: Addendum (In preparation)
[8] Bressoud D and Propp J 1999 *Notices of the AMS* **46** 637.
[9] Bressoud D M Proofs and Confirmations: The Story of the Alternating Sign Matrix Conjecture 1999 Cambridge: Cambridge University Press.
[10] Propp J 2001 *Discrete mathematics and Theoretical Computer Proceedings* **AA (DM-CCG)** 43.
[11] Zeilberger D 1996 *Electronic J. of Combinatorics* **3** R13.
[12] Zeilberger D 1996 *J. Math.* **2** 59.
[13] Kupenberg G 1996 *Inter. Math. Res. Notes* **1996** 139.
[14] Carter J H Permutation Matrices and the Representatioin of Matrices with Fixed Line-Sum 2004 (LAUR 04-5399, Los Alamos National Laboratory, Los Alamos, NM 87545,unpublished).
[15] Macdonald I Symmetric Functions and Hall Polynomials 1979 (second edition 1995) Oxford: Oxford University Press.
[16] Knuth D E 1970 *Pac. J. Math.* **34 709**.
[17] Stanley R P Enumerative Combinatorics (vols. 1 and 2) Cambridge: Cambridge University Press.
[18] Méndez M Proc. Conf. Symmetry and Structural Properties of Condensed Matter ed T Lulek, A Wal, and B Lulek 2003 Singapore: World Scientific p 265.
[19] Lulek T Proc. Conf. Symmetry and Structural Properties of Condensed Matter ed T Lulek, B Lulek, and A. Wal 2002 Singapore: World Scientific p. 279.
[20] Lulek T Lulek B Jakubczyk D and Jakubczyk P 2006 *Physica B: Condensed Matter* **382** 162.
[21] Jakubczyk P Topolewicz S Wal A and Lulek T 2009 *Open Systems & Information Dynamics* **19** 221.

Symmetry and Structural Properties of Condensed Matter (SSPCM 2009) IOP Publishing
Journal of Physics: Conference Series **213** (2010) 012011 doi:10.1088/1742-6596/213/1/012011

Hopf algebras: motivations and examples

G. H. E. Duchamp♠, P. Blasiak◇, A. Horzela◇, K. A. Penson♡ and A. I. Solomon♣♡

♠ Université Paris 13
Laboratoire d'Informatique Paris Nord, CNRS UMR 7030
99 Av. J-B. Clément, F 93430 Villetaneuse, France

◇ H. Niewodniczański Institute of Nuclear Physics, Polish Academy of Sciences
ul. Eliasza-Radzikowskiego 152, PL 31342 Kraków, Poland
♡ Laboratoire de Physique Théorique de la Matière Condensée
Université Pierre et Marie Curie, CNRS UMR 7600
Tour 24 - 2e ét., 4 pl. Jussieu, F 75252 Paris Cedex 05, France
♣The Open University, Physics and Astronomy Department, Milton Keynes MK7 6AA, United Kingdom

E-mail:
ghed@lipn-univ.paris13.fr, penson@lptl.jussieu.fr

Abstract. This paper provides motivation as well as a method of construction for Hopf algebras, starting from an associative algebra. The dualization technique involved relies heavily on the use of Sweedler's dual.

1. Introduction

The formalism of quantum theory represents the physical concepts of states, observables and their transformations as objects in some Hilbert space $\mathcal{H}$ (as we use a minimum of structure and instead of scalar products use duality, we will denote the space below by V). Briefly, vectors in the Hilbert space describe states of a system, and linear transformations on V represent basic observables. Sets of transformations usually possess some structure such as that of a group, semi-group, Lie algebra, etc, and in general can be handled within the concept of an algebra $\mathcal{A}$. The action of the algebra on the vector space of states V and observables $V^\vee$ is carried by its representation. Hence if an algebra is to describe physical transformations it has to have representations in all physically relevant systems. This requirement leads directly to the Hopf algebra structure in physics. From the mathematical viewpoint the structure of the theory, modulo details, seems to be clear. Physicists, however, need to have some additional properties and constructions to move freely in this arena.
Hopf algebras have become ubiquitous in physics; we cite here only a few examples:

- Feynman graphs are associated with an algebra related to the Hopf algebra of rooted trees [4, 5].
- In quantum statistical mechanics, the partition function description can be shown to give rise to a Hopf algebra structure essentially a zero-dimensional analogue of the Feynman algebra cited above [17].

- Quantum groups, a special type of Hopf algebra, can provide useful models in quantum physics, for example giving a more accurate photon distribution for the laser than the usual Poisson one arising from the Glauber coherent states [13].

Here we will show how the structure of Hopf algebras enters in the context of representations.

The first issue at point is the construction of the tensor product of vector spaces which is needed for the description of composite systems. Suppose we know how a transformation acts on individual systems, *i.e.* we know its representation in each of the vector spaces V_1 and V_2, respectively. Hence the need naturally arises for a canonical construction of an induced representation of this algebra in $V_1 \otimes V_2$ which would describe its action on the composite system. Such a scheme exists and is provided by the co-product in the algebra, *i.e.* a morphism $\Delta : \mathcal{A} \to \mathcal{A} \otimes \mathcal{A}$. The physical plausibility of this construction requires the equivalence of representations built on $(V_1 \otimes V_2) \otimes V_3$ and $V_1 \otimes (V_2 \otimes V_3)$, since the composition of three systems cannot depend on the order in which it is done. This requirement forces the co-product to be co-associative. Another point is connected with the fact that from the physical point of view the vector space $\mathbb{C}$ represents a trivial system having only one property "being itself", which cannot change. Hence, one should have a canonical representation of the algebra on a trivial system, denoted by $\epsilon : \mathcal{A} \to \mathbb{C}$. Next, since the composition of any system with a trivial one cannot introduce new representations, those on V, $V \otimes \mathbb{C}$ and $\mathbb{C} \otimes V$ should be equivalent. This requirement imposes the condition on ϵ to be a co-unit in the (co-) algebra. In this way we motivate the need for a bi-algebra structure in physics. The concept of an antipode enters in the context of measurement. Measurement in a system is described in terms of $V^\vee \otimes V$ and measurement predictions are given through the canonical pairing $c : V^\vee \otimes V \to C$. Observables, described in the dual space $V^\vee$, can also be transformed and representations of appropriate algebras are given with the help of an anti-morphism $S : \mathcal{A} \to \mathcal{A}$. Physics requires that a transformation performed on the system and apparatus simultaneously should not change the measurement results; hence the pairing should trivially transform under the action of the bi-algebra. We thus obtain the condition on S to be an antipode, which is the last ingredient of a Hopf Algebra.

2. How to never forget the requirements of Hopf algebras and give a meaning to them

Let $\mathcal{A}$ be a k-associative algebra with unit (in short, a k-AAU; here k will be a (commutative) field). We call a *representation* of $\mathcal{A}$ any mapping

$$\rho : \mathcal{A} \to End(V) \tag{1}$$

which is compatible with the AAU structure of both $\mathcal{A}$ and $End(V)$ (that is, ρ is linear, $\rho(xy) = \rho(x) \circ \rho(y)$ identically[1] and $\rho(1_\mathcal{A}) = Id_V$).

Usually, one considers the datum of ρ as an "action" (on the left) of it over the k-vector space V by

$$g.v := \rho(g)(v)\ . \tag{2}$$

This way of considering the action (given by Eq. (2)) is just a change of notation and we will use either at times in order to give more flexibility or avoid confusion. Vector spaces with an action (as (2)) of $\mathcal{A}$ on them are called (left) $\mathcal{A}$-modules.

One requires nice computation rules on $\mathcal{A}$-modules; that is,

- (direct) sums: given two $\mathcal{A}$-modules, how do we construct a "natural" representation on $V_1 \oplus V_2$

[1] This is in order that V be seen as a left module, see the "additional discussion" at the end of the manuscript.

- products: same problem for $V_1 \otimes V_2$, compatible with the associativity (*i.e.* "natural" identification $V_1 \otimes (V_2 \otimes V_3) \simeq (V_1 \otimes V_2) \otimes V_3$)
- unit: same problem for the "natural" identifications $V \otimes k \simeq k \otimes V \simeq V$
- a procedure to compare left and right modules; in particular, $V^\vee$ (the algebraic dual of V) possibly being compatible with the the natural pairings $V^\vee \otimes V \to k$ and $V \otimes V^\vee \to k$

(a) **Sums**: If ρ_i; $i = 1,2$ denote the representation morphisms on V_i; $i = 1,2$. Then, one constructs

$$\rho_3(u) = \begin{pmatrix} \rho_1(u) & 0_{Hom(V_1,V_2)} \\ 0_{Hom(V_2,V_1)} & \rho_2(u) \end{pmatrix} \tag{3}$$

where

$$a = \begin{pmatrix} a_{11} & a_{12} \\ a_{21} & a_{22} \end{pmatrix} \tag{4}$$

is the conventional representation of $a \in End(V_1 \oplus V_2)$ subject to $a_{i,j} \in Hom(V_i, V_j)$; $i,j = 1,2$.

(b) **Tensor products**: If one thinks about groups, with the notation as above, one has

$$g.(x_1 \otimes x_2) = g.x_1 \otimes g.x_2 \tag{5}$$

(g acts as an automorphisms) and for Lie algebras g acts as follows

$$g.(x_1 \otimes x_2) = g.x_1 \otimes x_2 + x_1 \otimes g.x_2 \tag{6}$$

i.e. by derivations. So, one can abstract the spaces and write that the scheme of action is $g \otimes g$ in case of groups and $g \otimes I + I \otimes g$ in the case of Lie algebras.
In either case (groups and Lie algebras) the representation can be rewritten as the action of an AAU $\mathcal{A}$ (algebra of the group in the first case and enveloping algebra in the second) and the "scheme of action" of $g \in \mathcal{A}$ can be written as a tensor of order two

$$\sum g_{(1)} \otimes g_{(2)} \in \mathcal{A} \otimes \mathcal{A}. \tag{7}$$

In both these classical cases, one has

$$g.(x_1 \otimes x_2) = \sum g_{(1)}.x_1 \otimes g_{(2)}.x_2 \ . \tag{8}$$

So, one needs a mapping $\Delta : \mathcal{A} \to \mathcal{A} \otimes \mathcal{A}$, in Sweedler's notation [1, 18]

$$\Delta(a) = \sum_{(1),(2)} a_{(1)} \otimes a_{(2)} \tag{9}$$

and one defines the action of $\mathcal{A}$ on a tensor product by

$$a.(x_1 \otimes x_2) = \sum_{(1),(2)} a_{(1)}.x_1 \otimes a_{(2)}.x_2 \ . \tag{10}$$

Now one can prove [8] that Eq. (10) defines a representation of $\mathcal{A}$ for any choice of ρ_1, ρ_2 iff $\Delta : \mathcal{A} \to \mathcal{A} \otimes \mathcal{A}$ is a morphism of algebras *i.e.* if, for all $u, v \in \mathcal{A}$ and with

$$\Delta(u) = \sum_{(1),(2)} u_{(1)} \otimes u_{(2)}; \ \Delta(v) = \sum_{(3),(4)} v_{(3)} \otimes v_{(4)} \ , \tag{11}$$

one has

$$\Delta(uv) = \sum_{(1),(2),(3),(4)} u_{(1)}v_{(3)} \otimes u_{(2)}v_{(4)}. \tag{12}$$

Moreover, one can prove that this procedure is compatible with the usual identification

$$V_1 \otimes (V_2 \otimes V_3) \simeq (V_1 \otimes V_2) \otimes V_3 \tag{13}$$

for any choice of representations ρ_1, ρ_2, ρ_3 iff Δ is coassociative which means that

$$(\Delta \otimes Id) \circ \Delta(u) = (Id \otimes \Delta) \circ \Delta \ . \tag{14}$$

The intuitive idea of the LHS of (14) applied to u is obtained by splitting u in two parts (as in eq (11)) and then resplitting the first component

$$\Delta(u_{(1)}) = \sum_{(11),(12)} u_{(11)} \otimes u_{(12)}$$

then

$$(\Delta \otimes Id) \circ \Delta(u) = \sum_{(11),(12),(2)} u_{(11)} \otimes u_{(12)} \otimes u_{(2)} \tag{15}$$

and the RHS of (14) applied to u is obtained by splitting u in two parts and then resplitting the second component

$$\Delta(u_{(2)}) = \sum_{(21),(22)} u_{(21)} \otimes u_{(22)} \ .$$

then

$$(Id \otimes \Delta) \circ \Delta(u) = \sum_{(1),(21),(22)} u_{(1)} \otimes u_{(21)} \otimes u_{(22)} \tag{16}$$

so, for all $u \in \mathcal{A}$, the result of the two computations (15), (16) must coincide.

Another way to see co-asociativity which explains the name, and is also useful to produce examples and counterexamples, is through duality. For every $\phi, \psi \in \mathcal{A}^\vee$, one defines a linear form $\phi * \psi$ on $\mathcal{A}$ by

$$\langle \phi *_\Delta \psi | x\rangle = \langle \phi \otimes \psi | \Delta(x)\rangle \ . \tag{17}$$

One can prove [8] that Δ is co-associative iff $*_\Delta$ is associative.

(c) **Unit**: Here, one has to cope with the identifications $V \otimes k \simeq k \otimes V \simeq V$. One first needs a (one-dimensional) representation of $\mathcal{A}$ on k. This is exactly provided by a mapping $\epsilon : \mathcal{A} \to k$ which is a morphism of AAU *i.e.* a character. In order to have $V \otimes k \simeq V$ (resp. $k \otimes V \simeq V$), one needs to have identically (*i.e.* for all $x \in \mathcal{A}$)

$$\sum_{(1)(2)} x_{(1)}\epsilon(x_{(2)}) = x; \ resp. \sum_{(1)(2)} \epsilon(x_{(1)})x_{(2)} = x \ . \tag{18}$$

If ϵ fulfils these conditions one says that it is a counit. Again, one can invoke duality and transpose the character ${}^t\epsilon : k \to \mathcal{A}^\vee$. Setting $e_\epsilon = {}^t\epsilon(1_k)$, one has that [8] ϵ is a counit iff e_ϵ is a unity in $(\mathcal{A}^\vee, *_\Delta)$, as defined in Eq. (17).

We summarize this discussion:

- an AAU $\mathcal{A}$ is given
- there is no problem to compute direct sums of (same sided) representations of $\mathcal{A}$

- to construct tensor products one needs a mapping $\Delta : \mathcal{A} \to \mathcal{A} \otimes \mathcal{A}$
- if one wants this construction to give representations of $\mathcal{A}$, one needs that Δ be a morphism of algebras
- if one requires that the identification $V_1 \otimes (V_2 \otimes V_3) \simeq (V_1 \otimes V_2) \otimes V_3$ be compatible with the construction of representations by Δ, one needs that Δ be coassociative (*i.e.* its dual be associative)
- for the identification $V \otimes k \simeq k \otimes V \simeq V$ to be compatible with the construction of representations by Δ, one first needs a reprsentation of $\mathcal{A}$ on k, which is a character $\epsilon : \mathcal{A} \to k$ and that ϵ be a counit (*i.e.* its dual be a unit of the dual law $*_\Delta$).

At this point, if $(\mathcal{A}, *, 1_\mathcal{A}, \Delta, \epsilon)$ fulfils the preceding requirements, $(\mathcal{A}, *, 1_\mathcal{A}, \Delta, \epsilon)$ is called a bi-algebra.

The next paragraph will be devoted to two aspects of dualization: namely the dualization of the modules and of the algebra itself.

3. Dualization, antipode and Sweedler's dual.

The question of dualization is twofold.
Q1) Let $(\mathcal{B}, *, 1_\mathcal{A}, \Delta, \epsilon)$ be a bialgebra. How does one compare left and right $\mathcal{B}$-modules ? in particular, for a given module V, how to compare V and $V^\vee$?
Q2) Can one endow $\mathcal{B}^\vee$, or at least a subspace of it, with the structure of a bialgebra by dualizing the product, coproduct, unit and counit ?

Throughout this paragraph $\mathcal{A} = (\mathcal{A}, *, 1_\mathcal{A}, \Delta, \epsilon)$ is a bialgebra.

3.1. Dualization of modules

Let V be a left $\mathcal{A}$-module, then $V^\vee$ is endowed with the structure of a right $\mathcal{A}$-module by

$$\langle \psi.g | x\rangle = \langle \psi | g.x\rangle \tag{19}$$

for $\phi \in V^\vee$, $g \in \mathcal{A}$ and $x \in V$. Then, how does one compare $V^\vee$ and any left module W ?
The solution is to reverse the multiplication of operators by use of an anti-endomorphism $S : \mathcal{A} \to \mathcal{A}$, that is a linear mapping which is compatible with the identity $S(1_\mathcal{A}) = 1_\mathcal{A}$ and reverses the products $S(xy) = S(y)S(x)$ (such an antimorphism should be familiar to the reader as the adjoint or transpose of a matrix). We then form a new action on the left of $V^\vee$ by

$$\langle g *_S \psi | x\rangle = \langle \psi | S(g).x\rangle \ . \tag{20}$$

It is left to the reader to check that this is a true action on the left.
If, moreover, one wants this action to be compatible with the natural pairings $V^\vee \otimes V \to k$ and $V \otimes V^\vee \to k$, one should have, for every $\psi \in V^\vee$, $x \in V$, $g \in \mathcal{A}$ and $\Delta(g) = \sum_{(1)(2)} g_{(1)} \otimes g_{(2)}$

$$\sum_{(1)(2)} \langle g_{(1)} *_S \psi | g_{(2)} x\rangle = \sum_{(1)(2)} \langle g_{(2)} *_S \psi | g_{(1)} x\rangle = \epsilon(g)\langle \psi | x\rangle \tag{21}$$

which is possible (in full generality) iff, for all $g \in \mathcal{A}$

$$\sum_{(1)(2)} S(g_{(1)})g_{(2)} = \sum_{(1)(2)} g_{(1)}S(g_{(2)}) = \epsilon(g)1_\mathcal{A} \ . \tag{22}$$

One can prove [1, 8] that if there is a solution S of (22), it is unique and that it is an antimorphism $\mathcal{A} \to \mathcal{A}$. Such a solution, if it exists, is called an antipode for the bialgebra $\mathcal{A}$. Then, we arrive at the definition:

Journal of Physics: Conference Series **213** (2010) 012011 doi:10.1088/1742-6596/213/1/012011

Definition 3.1 *An algebra $\mathcal{A} = (\mathcal{A}, *, 1_\mathcal{A}, \Delta, \epsilon, S)$ is called a Hopf algebra iff $(\mathcal{A}, *, 1_\mathcal{A}, \Delta, \epsilon)$ is a bialgebra and S is an antipode for it.*

Example 3.2 *Let $\Sigma = \{a, b, c, \cdots\}$ be an alphabet (*i.e. *a set of variables) and Σ^* be the free monoid i.e. the set of all words (or strings) in the letters of Σ. The set Σ is endowed with the law of concatenation (which will be denoted "conc" [14] in this example) which is just the juxtaposition of the strings, namely with $u = a_1a_2\cdots a_p$ and $v = b_1b_2\cdots b_q$, one has*

$$conc(u, v) = uv = a_1a_2\cdots a_pb_1b_2\cdots b_q \ . \tag{23}$$

This is an associative law with unit (the empty string 1_{Σ^*}). The k algebra of Σ^* ($k[\Sigma^*]$) is usually denoted $k\langle\Sigma\rangle$ and known under the name of algebra of non-commutative polynomials in the variables Σ (or, due to its universal properties, the free algebra [2, 3] and below). It is the set of linear combinations

$$P = \sum_{i=1}^{n} \alpha_i w_i \tag{24}$$

and, as in the commutative case, the $w_i \in \Sigma^*$ are called the monomials of P and the $\alpha_i \in k$ its coefficients. The product of two polynomials P, Q as in (24) is defined by the concatenation of their monomials (the strings w_i). It is easily checked that $(k\langle\Sigma\rangle, conc, 1_{\Sigma^*})$ is an AAU [2, 3] which is free in the following sense. For each set-theoretical mapping $\phi : \Sigma \to \mathcal{A}$ where $\mathcal{A}$ is an AAU, there exists a unique morphism of AAUs $\bar{\phi} : k\langle\Sigma\rangle \to \mathcal{A}$ such that $\phi = \bar{\phi} \circ nat$

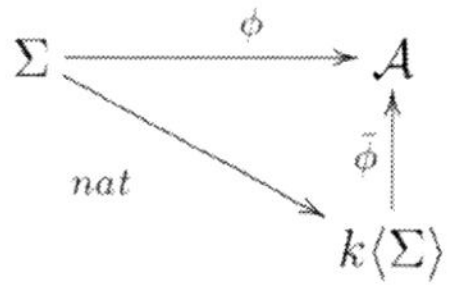

(25)

where nat is the natural mapping which sends the letter $a \in \Sigma$ to itself considered as a monomial (of degree 1). Given a partition of the alphabet, $\Sigma = G \sqcup L$ (Σ is the union of the two disjoint subsets G and L) and, for $x \in G$ (resp. $x \in L$), define

$$\Delta(x) = x \otimes x \ ; \ (\text{resp. } \Delta(x) = x \otimes 1 + 1 \otimes x) \ . \tag{26}$$

One uses (25) to extend Δ as a mapping $\Delta : k\langle\Sigma\rangle \to k\langle\Sigma\rangle \otimes k\langle\Sigma\rangle$ which can be checked to admit the following nice combinatorial description. For a word $w = a_1\cdots a_n$ let us define I_G (resp. I_L) as the set of places of letters in G (resp. in L)

$$I_L = \{i \in [1..n]_\mathbb{N} \mid a_i \in L\} \ ; \ I_G = \{i \in [1..n]_\mathbb{N} \mid a_i \in G\} \tag{27}$$

then

$$\Delta(w) = \sum_{I+J=I_L} w[I_G \cup I] \otimes w[I_G \cup J] \tag{28}$$

where for $R = \{r_1, r_2\cdots r_k\} \subset [1..n]_\mathbb{N}$ (in increasing order), $w[R]$ is the subword $a_{r_1}a_{r_2}\cdots a_{r_k}$. One can check that $(k\langle\Sigma\rangle, conc, 1_{\Sigma^*}, \Delta_{G,P}, \epsilon)$ with $\epsilon(x) = 1$ if $x \in G$ and 0 if $x \in L$ (ϵ is extended to $k\langle\Sigma\rangle$ using (25)).
The bialgebra $k\langle\Sigma\rangle$ constructed above admits an antipode iff $G = \emptyset$.

3.2. Dualization of a bialgebra

Given a bialgebra $(\mathcal{A}, *, 1_\mathcal{A}, \Delta, \epsilon)$, one wants to endow the dual space $\mathcal{A}^\vee$ with the structure of a bialgebra. Let us see, step by step what it is possible to do.

Coproduct and counit: One can always transpose the comultiplication and the counit by the formulas given in eq. (17), (18) and endow $\mathcal{A}^\vee$ with the structure of an AAU.

Dualization of the multiplication: This is exactly the point where one has to restrict the domain because it is not always possible to dualize the multiplication operation on all elements of $\mathcal{A}^\vee$, since the transpose of the multiplication (denoted by Δ_*) is a mapping $\mathcal{A}^\vee \to (\mathcal{A} \otimes \mathcal{A})^\vee$ and, in general, its image is not included in $\mathcal{A}^\vee \otimes \mathcal{A}^\vee$; the inclusion $(\mathcal{A}^\vee \otimes \mathcal{A}^\vee) \subset (\mathcal{A} \otimes \mathcal{A})^\vee$ is strict in infinite dimension. Let us call A^0 the set of all elements $\psi \in A^\vee$ such that $\Delta_*(\psi) \in \mathcal{A}^\vee \otimes \mathcal{A}^\vee$. One may characterise this set by use of the following finite orbit properties.

Proposition 3.3 *(see also [1] paragraph 2.2, [18] and [8]) Let k be a field, $\mathcal{A}$ a k-algebra and $f \in \mathcal{A}^\vee$. For all $s \in \mathcal{A}$, we define the (shifted) linear forms ${}_sf, f_s$ by ${}_sf(x) = f(xs)$ and $f_s(x) = f(sx)$.*
The following are equivalent :
i) The family $(f_s)_{s\in\mathcal{A}}$ is of finite rank in $\mathcal{A}^\vee$
ii) The family $({}_sf)_{s\in\mathcal{A}}$ is of finite rank in $\mathcal{A}^\vee$
iii) There exists a double family $(g_i, h_i)_{1\le i\le n}$ of functions in $\mathcal{A}^\vee$ such that

$$\Big(\forall x, y \in S\Big)\Big(f(xy) = \sum_{i=1}^{n} g_i(x)h_i(y)\Big) \tag{29}$$

iv) There exists $n \in \mathbb{N}$, $\lambda \in k^{1\times n}, \gamma \in k^{n\times 1}$ and $\mu : \mathcal{A} \to k^{n\times n}$ a morphism of algebras such that $(\forall s \in \mathcal{A})(f(s) = \lambda\mu(s)\gamma)$.

For proofs, see [1, 9, 10, 11, 12, 18]. The the set of elements which fulfil the equivalent conditions of proposition (3.3) is a subspace of $\mathcal{A}^\vee$ denoted $\mathcal{A}^0$, closed by all operations and co-operations, that is

- $\mathcal{A}^0$ is closed under $*_\Delta$, $\epsilon \in \mathcal{A}^0$ so that $(\mathcal{A}^0, *_\Delta, \epsilon)$ is a AAU
- $\Delta_*(\mathcal{A}^0) \subset \mathcal{A}^0 \otimes \mathcal{A}^0$ and $\delta_{1_\mathcal{A}} : f \mapsto f(1_\mathcal{A})$ is a counit for Δ_*
- $(\mathcal{A}^0, *_\Delta, \epsilon, \Delta_*, \delta_{1_\mathcal{A}})$ is a bialgebra
- ${}^tS(\mathcal{A}^0) \subset \mathcal{A}^0$ and ${}^tS$ is an antipode for the bialgebra $(\mathcal{A}^0, *_\Delta, \epsilon, \Delta_*, \delta_{1_\mathcal{A}})$

A link with the theory of automata [15, 16] is that the elements of Sweedler's dual of $k\langle\Sigma\rangle$ are precisely the series that can be recognized by finite automata [10].

4. Conclusion

In this note we introduced step by step the axioms of a Hopf algebra. The motivation for this is the following: An associative algebra with unit (AAU), which is a commonly occuring structure in physics, will inevitably lead to the necessity of dealing with representation modules, tensor products, duals, etc. The Hopf algebra structure provides us with just the machinery to manage these constructs. Further, the appropriate tools, such as product and co-product, occur in a symmetric fashion. In order to implement the passage to the symmetric, Hopf structure, a correct mathematical dualization requires the use of Sweedler's dual.

Acknowlegements The authors wish to acknowledge support from the Agence Nationale de la Recherche (Paris, France) under Program No. ANR-08-BLAN-0243-2 and from PAN/CNRS Project PICS No.4339(2008-2010). Two of us (P.B. and A.H.) wish to acknowledge support from the Polish Ministry of Science and Higher Education under Grants Nos. N202 061434 and N202 107 32/2832. Another (G.H.E.D.) would like to acknowledge support from "Projet interne au LIPN 2009" "Polyzêta functions".

References

[1] E. Abe, *Hopf Algebras*, Cambridge University Press (2004).
[2] J. Berstel, C. Reutenauer, *Rational series and their languages*. EATCS Monographs on Theoretical Computer Science, Springer, (1988).
[3] N. Bourbaki, *Algebra III*, Springer (2004)
[4] A. Connes, D. Kreimer, *From Local Perturbation Theory to Hopf and Lie Algebras of Feynman Graphs*, Letters in Mathematical Physics 56, 3-15, 2001.
[5] A. Connes and D. Kreimer, *Hopf algebras, Renormalization and Noncommutative geometry*, Comm. Math. Phys 199 (1998), no. 1, 203242, arXiv:hep-th/9808042.
[6] G. H. E. Duchamp, P. Blasiak, A. Horzela, K. A. Penson, A. I. Solomon, *Feynman graphs and related Hopf algebras*, Journal of Physics: Conference Series **30**, 107 (2006), SSPCM'05, Myczkowce, Poland. arXiv:cs.SC/0510041
[7] G. H. E. Duchamp, P. Blasiak, A. Horzela, K. A. Penson, A. I. Solomon, *A Three Parameter Hopf Deformation of the Algebra of Feynman-like Diagrams*, arXiv:0704.2522, (math-ph).
[8] G. H. E. Duchamp, P. Blasiak, A. Horzela, K. A. Penson, A. I Solomon, *Hopf Algebras in General and in Combinatorial Physics: a practical introduction*, arXiv:0802.0249 (to be published).
[9] G. H. E. Duchamp, C. Reutenauer, *Un critère de rationalité provenant de la géométrie noncommutative*, Inventiones Mathematicae, **128** (1997) 613-622.
[10] G. H. E. Duchamp, C. Tollu, *Sweedler's duals and Schützenberger's calculus*, arXiv:0712.0125 (to be published).
[11] G. P. Hochschild, *Basic theory of algebraic groups and Lie algebras*, Springer (1981).
[12] G. Jacob, *Représentations et substitutions matricielles dans la théorie matricielle des semigroupes*, Thèse, Univ. de Paris (1975).
[13] J. Katriel and A. I. Solomon, *Nonideal Lasers, Nonclassical Light and Deformed Photon States*. Phys.Rev.A 49 5149 (1994)
[14] C. Reutenauer, *Free Lie algebras*, Oxford University Press, (1993).
[15] M. P. Schützenberger, *On the definition of a family of automata*, Information and Control, **4** (1961) 245-270.
[16] M. P. Schützenberger, *On a theorem of R. Jungen*, Proc. Amer. Math. Soc. **13** (1962) 885-889.
[17] A.I. Solomon, G. Duchamp, P. Blasiak, A. Horzela and K. A. Penson, *Partition functions and graphs: A combinatorial approach*, Proceedings of the XI International Conference on Symmetry Methods in Physics (SYMPHYS-11) (Prague, Czech Republic, June 2004), in press (JINR Publishers, Dubna) arXiv:quant-ph/0409082
[18] M. E. Sweedler, *Hopf algebras*, W.A. Benjamin, New York, (1969).

Journal of Physics: Conference Series **213** (2010) 012012 doi:10.1088/1742-6596/213/1/012012

Probing the Big Bang at the Relativistic Heavy Ion Collider (RHIC) (or Probing the Big Bang 13.7 billion years later)

David M Lee

H846, P-25, Los Alamos National Laboratory,
Los Alamos, New Mexico, USA 87545

Email: dlee@lanl.gov

Abstract. The Relativistic Heavy Ion Collider (RHIC) at the Brookhaven National Laboratory in the USA is a variable energy proton-proton and ion-ion collider that is the first accelerator capable of colliding heavy ions. RHIC was designed to do experiments that provide important information about the Standard Model of particle physics, Quantum Chromodynamics (QCD). QCD predicts that in the early part of the Universe just after the Big Bang the world consisted of a Quark Gluon Plasma, a weakly interacting collection of quarks and gluons. At RHIC we can recreate the conditions of the early Universe by colliding heavy ions at 200 GeV. This paper will give a general overview of the physics motivation for studying the QGP, how our experiments are designed to study the QGP, what we have learned over the last 9 years, and what the future holds.

Introduction

The goal of heavy ion physics is to study QCD under extreme temperatures and energy densities with a great emphasis on the discovery of the quark-gluon plasma (QGP). The QGP is the nuclear medium where the neutrons or protons (baryons) are no longer bound states of quarks but instead the quark and gluon constituents of the neutrons and protons are in a deconfined state. This new state of matter has been predicted for many years but never observed [1,2,3]. The QGP is thought to be the primary form of matter that occurred within 10 μs after The Big Bang, the event that caused the formation of our universe. As such the study of the QGP is important in cosmology, astrophysics, and nuclear physics. Very quickly after the formation of the QGP, the universe cooled and the quarks and gluons bound together to remain that way for the next 13.7 billion years. The bound quarks and gluons formed hadrons and anti-hadrons but they quickly annihilate so by about 1 sec after the Big Bang only the hadrons remain. Later in time, light atoms were formed, then came the formation of stars and heavy nuclei, and finally the universe of today with galaxies and solar systems began about 8-9 billion years after the Big Bang. One of the mysteries of the universe is why we have a predominance of matter over anti-matter.

Ordinary nuclear matter is a system composed of interacting neutrons and protons that form the nucleus in the center of a much larger atomic system that includes the nucleus and electrons. The neutrons and protons, also collectively called hadrons, are composed of three quarks and gluons, the force carrier that binds the quarks together. Physicists have learned that there are six types of quarks and anti quarks. Only two are found in the universe today and they are called the up and the down quark. Protons are made of two up quarks and one down quark and the neutrons are made of two down quarks and one up quark. The other heavier quarks, the strange, charm, bottom, and the top, are not found naturally but are produced in energetic collisions of high energy particles. Hadron is a term that identifies particles that are made of quarks and are held together by the strong force. There are two types of hadrons, baryons and mesons. Mesons are not naturally occurring but are formed by the collisions of energetic particles. Mesons are composed of a quark and an anti-quark and baryons are composed of three quarks. In highly energetic collisions of particles many additional hadrons and mesons are produced some composed of the heavier quarks.

The Relativistic Heavy Ion collider at the Brookhaven National Laboratory in the USA was constructed with one of its goals to study QCD under extreme conditions and search for evidence of the QGP. Beginning in 2000, the RHIC collider has run successfully with a variety of beam conditions and energies and shed light on the sought after QGP but also discovered many puzzles.

The Relativistic Heavy Ion Collider (RHIC)

RHIC is a variable energy heavy ion collider facility that can collide 100 GeV/nucleon Au ion beams together with the goal of producing a new state of matter called the quark gluon plasma (QGP). The accelerator complex is shown in figure 1. Gold ions are produced in an ion source and accelerated to ~1 MeV/nucleon in the Tandom Van de Graaf accelerator with a charge state of +32 and then transported to the Booster accelerator where they are accelerated to 95 MeV/nucleon with a charge state of +77. The AGS (Alternating Gradient Synchroton) takes the output of the Booster and accelerates the Au ions to 8.9 GeV/nucleon with a charge state of +79 and then the ions are transported to the RHIC ring where they are accelerated to 100 GeV/nucleon. In the RHIC accelerator there are two independent counter rotating beams that are focused to collide at six intersecting positions around the ring. The ring is 3.8 km in diameter, contains 1740 superconducting magnets operating at 3.5 Tesla, and has a luminosity of 2 x $10^{26}cm^{-2}s^{-1}$.

Figure 1. The RHIC accelerator complex.

Four detectors are positioned at the intersecting regions, PHENIX, STAR, BRAHMS, and PHOBOS. The two detectors, PHENIX and STAR, are designed to cover large areas of rapidity and angle with STAR focusing on hadron detection and PHENIX focusing on rare events and leptons. The other two smaller detectors are more specific with PHOBOS measuring large amounts of untriggered data and BRAHMS covering a large range of rapidity but with limited angular range. All of the experiments are complimentary to each other and still provide overlap in key areas. The two smaller detectors finished taking data in 2006.

The STAR detectors' main feature is a large Time Projection Chamber (TPC) for measuring the tracks of the charged particles emanating from the collision vertex. The detector is shown in figure 2, idealized in the left panel and as built in the right panel. The TPC is a large cylindrical gas volume where the charged particles leave ionization tracks that drift to a wire grid array at either end that detects the ionization track. Surrounding the TPC are other types of detectors that are used to help identify the charged particles. A typical heavy ion collision event in the STAR detector is shown in figure 3.

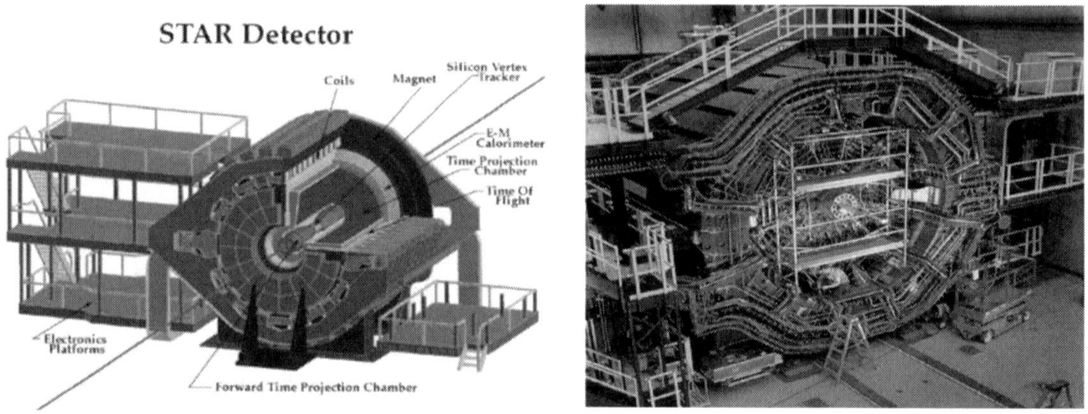

Figure 2. The STAR detector shown idealized in the left panel and as built in the right panel. The large TPC is the heart of the detector

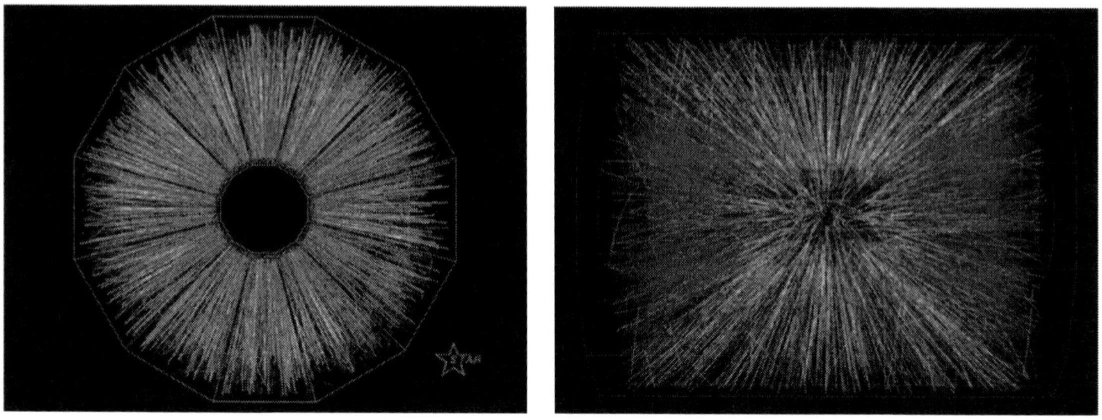

Figure 3. A typical heavy ion event from STAR. The view in the beam direction is in the left panel and the view transverse to the beam is in the right panel.

The PHENIX detector consists of four large analyzing spectrometers that are designed to measure the properties of the particles coming out of the collisions of heavy ions. The detector is shown in figure 4, idealized in the left panel and as built in the right panel. PHENIX is designed specifically to measure direct probes of the collisions such as electrons, muons, and photons. The two beams enter the detector via a beam pipe shown in the left panel running from the upper left to the lower right and collide at the center of the detector in the center of the central magnet (depicted in the left panel with a darker blue color). The two conical shaped magnets that surround the upstream and downstream portion of the beam pipe are designed to measure the momentum of the muons coming from the collisions and the series of panels in the back portion of the conical magnets identify the muons. Surrounding the central portion of the detector is the central magnet that is used for momentum analysis of the charged particles and to provide a lower momentum threshold for charged particle detection. Centered at 90 degrees to the beam line are two large detector carriages with a number of subdetectors for measuring particles trajectories, drift chambers and pad chambers, electromagnetic calorimeters to measure the electron and photon energies, cherenkov detectors to identify electrons and time-of-flight detectors to identify charged hadrons.

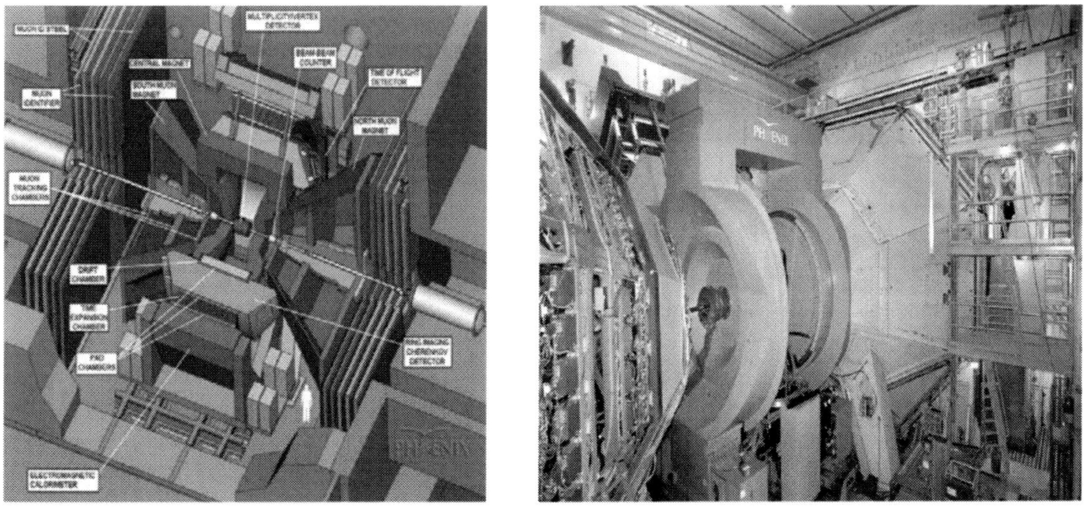

Figure 4. The left panel is an artist view of the PHENIX detector and the right panel is a photo of the actual detector. In the right panel one of the detector carriages has been moved out of view.

Both the PHENIX and STAR detectors have hundreds of thousands of signals coming from all of the subdetectors for each event and all of that information must be processed electronically and recorded for further analysis. Various electronic triggers are used to filter the raw data to reduce the extremely high data volume before it is finally archived on computer disks. The raw data volume from the PHENIX detector exceeds 700 Mbits/s, but only a small fraction of the data is important for each collision event. Before data analysis files can be produced the raw data files are preprocessed to produce condensed files for analysis and this process can take up to a year following the data run. The subsequent analysis of the data involves determining which particles in the ~1000 particle event are important (perhaps 1-3 particles). This can take a few minutes per event so many months are needed to analyze the stored data. Many different analyses are done to look at different physics processes so this time consuming process is repeated many times and in fact sometimes physics analysis are still being done 3-5 years after the yearly data run.

A Brief Review of the Quark Model

The quark model was independently proposed by physicists Murray Gell-Mann[4] and George Zweig[5] in 1964. The quark model is a part of the Standard Model of Physics. Their idea was that protons and neutrons as well as other particles that were discovered were not elementary but made up of smaller constituents which they called quarks. Their model consisted of three quarks, up, down, and strange, which explained much of the known particles at that time. Over the next decade more types of quarks were proposed and all of the quarks were finally experimentally found by 1995 with the discovery of the top quark at FNAL[6]. Figure 5 left panel [7] shows the current state of our picture of the quark model and the right panel shows the constituents of a He^4 atom[8]. There are 6 lepton, 6 quarks, and 4 force carriers. The particles in the first column in the left panel, generation I, are the only particles that are found in the natural world today. Columns 2 and 3, generations II and III, are produced in high energy collisions at man-made accelerators or in high energy cosmic ray interactions and do not exist normally in nature. All of the elementary particles have a corresponding antiparticle. The force carriers are the particles that are responsible for the forces between the elementary particles. The gluon is responsible for the strong force that holds quarks together in the nucleus, the photon is responsible for the force due to electromagnetic charge, and the W,Z are responsible for the weak force that governs radioactive decay and neutrino interactions. The gravitational force is not part of the Standard Model of Physics. In the right panel is a He^4 atom showing how the nucleus of an atom is made up of quarks and leptons. In the nucleus of the atom are the protons and neutrons and each are composed of three quarks bond together by the gluon force carrier.

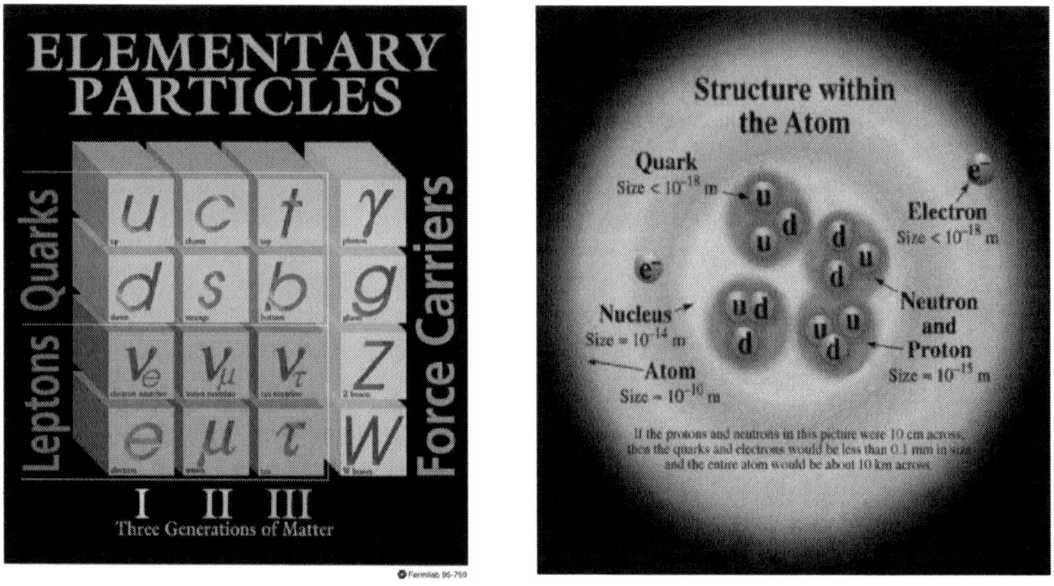

Figure 5. The left panel lists the elementary particles and the force carriers that form the Standard Model of Physics and the right panel showns how the quarks and electrons make up a nucleus, in this case He^4 atom.

The quarks are not free to roam around the nucleus but are constrained to be part of the 3 quark triplet that identifies the hadron. The only way to break up the hadron is to put the atom under extreme temperature and pressure or to collide the nuclei at very high energies in an accelerator.

Relativistic Heavy Ion Collisions

In attempting to create the Quark Gluon Plasma in the laboratory we collide two beams of heavy ions traveling at close to the speed of light. At RHIC we use Au ions accelerated to 100 GeV. A schematic representation of these collisions [9] is shown in figure 6. In the left panel the two Au ions are traveling towards each other at 99.995% of the speed of light. Due to this relativistic velocity the Au ions Lorentz contract in the direction of motion and take on a pancake shape. The consequence of this is that the quarks and gluons are more dense. In the second panel the collision takes place and some high energy collisions produce high energy particles that can be used to probe the nuclear medium. In the third panel the quarks and gluon that were part of the hadrons in the Au nucleus are released to form the high temperature and high density fire ball. In the fourth panel the resulting plasma begins to cool down and the quarks and gluons recombine to produce many particles and antiparticles. The particles that result from the recombination can be used to probe the plasma.

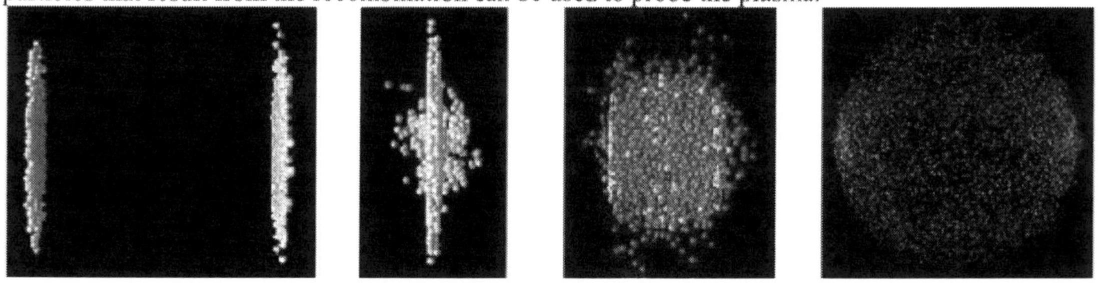

Figure 6. A heavy ion collision where in the first panel the Lorentz contracted Au ion beams a moving toward each other at 99.995% of the speed of light; in the second panel the two beams collide; in the third panel the quarks and gluons are freed; in the fourth panel the plasma is beginning to cool.

Not all heavy ion collisions are head-on as depicted in figure 6. Collisions that are head-on or near head-on are called central collisions and produced the greatest density of free quarks. When the two Au ions collide off axis with a smaller overlap, we call that a peripheral collisions. In peripheral collisions only some of the nucleons participate in the reaction and the other nucleons that don't participate are called spectator nucleons. By measuring the spectator nucleons we can determine the amount of energy available to the collision out of the total 39.4 TeV that is available. In all of the collisions, central or peripheral, not all of the participating nucleons actually collide so another term is used and that is colliding nucleons. The number of colliding nucleons is determined by a Glauber calculation.

Signatures of the Quark Gluon Plasma

The signatures of the QGP that are commonly considered are jet suppression, J/PSI suppression, collective flow, strangeness enhancement, and high pt suppression of hadrons. In jet suppression the two jets formed in the collision process travel back to back through the dense QGP. The near side jet (the jet closest to the surface of the plasma) is detected and the far side jet (the jet in the opposite direction that transverse most of the plasma) is suppressed because of the absorption and energy loss in the dense QGP. A representation of this is shown in figure 7.

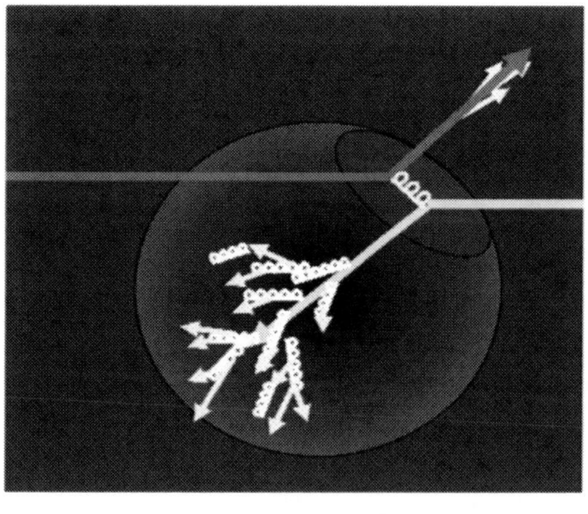

Figure 7. A descriptive example of jet suppression. The near side jet in red is detected while the far side jet in green is lost to the dense QGP medium.

In J/psi suppression we expect the color screening to suppress the bound states of charmonium (the charm-anticharm meson which is the J/psi). This is similar to Debye screening in electrostatics. Color screening is a property of quarks and the strong force much in the same way that charge is an electrostatic property of charged particles. A pictorial representation [10] of this is shown in figure 8. Normally the quark-antiquark pair bound together by the strong force but when they are in a plasma of many other quarks and antiquarks of different color they are unable to remain bound together and therefore the number of J/psi's are suppressed.

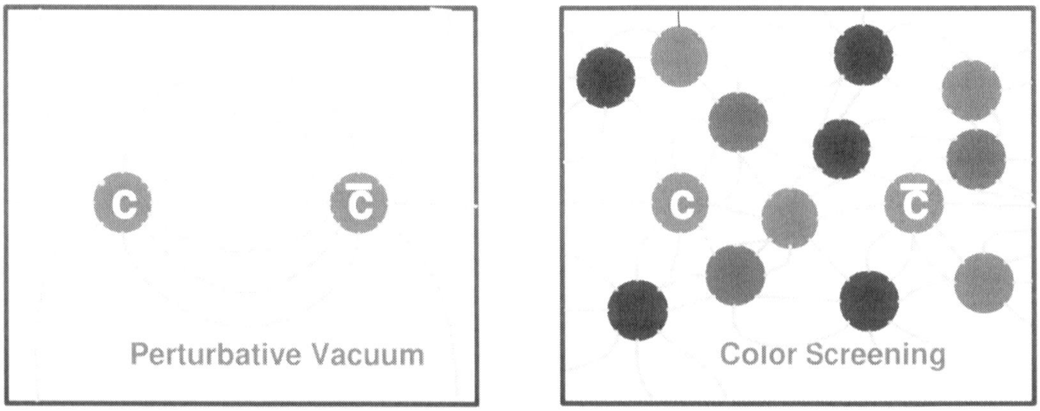

Figure 8. A description of J/psi color screening. In the left panel the J/psi quark – antiquark pair are held together by the strong force but in the right panel the quark and antiquark pair are shielded by the other quarks and the binding is broken.

Collective flow provides information on the properties of the medium, in this case the QGP, such as viscosity, equation of state, thermalization, etc. By measuring the elliptic flow we can determine whether the QGP is a strongly or weakly interacting system.

High energy quarks propagating through matter are predicted to lose energy via induced gluon radiation, with the total energy loss strongly dependent on the color charge density of the medium. This process is expected to be very sensitive to the hot dense matter created in relativistic heavy ion collision.

A Systematic Approach to the Study of the QGP

In order to study the QGP it is very important to understand the basic nucleon-nucleon interactions and nucleon-nucleus interactions so that we don't infer an effect to nucleus-nucleus collisions that is in fact due to the more simple nucleon-nucleon and nucleon-nucleus interactions. To accomplish this collider runs over the last 9 years have included Au-Au collisions at 200 GeV, d-Au collisions at 200 GeV, and p-p collisions at 200 GeV. The p-p collisions are used to understand the differences that are due to nuclear effects in the QGP in Au-Au collisions and between a superposition of independent nucleon-nucleon collisions. The analysis uses a term called the nuclear modification factor and is described by the following formula,

$$R_{AA}(p_T) = \frac{d^2N^{AA} / dp_T d\eta}{T_{AA} d^2\sigma^{NN} / dp_T d\eta}$$

where the numerator is the yield in Au-Au collisions and the denominator is the yield in p-p collisions multiplied by T_{AA}, the number of binary collisions in Au-Au collisions as calculated by the Glauber Model. In the Glauber Model the nucleons in each Au nucleus is treated as nucleon that travels in a straight line and only undergoes binary scattering. If the Au-Au collisions behave according to the Glauber Model and are just a superposition of many nucleon-nucleon collisions, then the nuclear modification factor will be unity. Departure from unity indicates nuclear matter effects either due to the QGP(sometimes referred to as hot nuclear matter) or due to nucleon-nucleus interactions(sometimes referred to as cold nuclear matter effects) and not due to the QGP. To separate out the cold nuclear matter effects another form of the nuclear modification factor is used where instead of using the Au-Au yield in the numerator, the d-Au yield is used. Here, unity also means that there are no cold nuclear matter effects and a departure from unity indicates cold nuclear matter effects. Comparing the nuclear modification factor for the Au-Au and for d-Au we can understand how much is due to the QGP.

Some important results

Over the last 9 years the RHIC community has learned much about what is formed in the collisions of Au-Au nuclei. Some of what we have learned at RHIC are that,

- There is jet suppression in Au-Au collisions
- There is a very large amount of flow
- There is evidence of shock wave dynamics
- There is excessive numbers of electrons
- There are more high momentum protons than expected.

All of these point to the creation of a new form of matter that behaves like a near perfect liquid. Unlike the original expectations of the QGP being a weakly interacting plasma of quarks and gluons, this near perfect liquid is strongly interacting.

Jet suppression has been one of the most important indicators that a new form of dense matter has been formed in the Au-Au collisions. Figure 9 shows the results for all types of runs, Au-Au, d-Au, and p-p at 200 GeV.[11] In this figure the near side jet is seen at 0 radians and the backward away side jet is at 3.14 radians. The far side jet is clearly visible for the p-p collisions and the d-Au collisions but is absent for the Au-Au collisions. This indicates that the far side jet is absorbed by the nuclear medium. The fact that the far side jet is seen for the d-Au collisions indicates cold nuclear matter effects are not the cause of the jet suppression seen in Au-Au collisions.

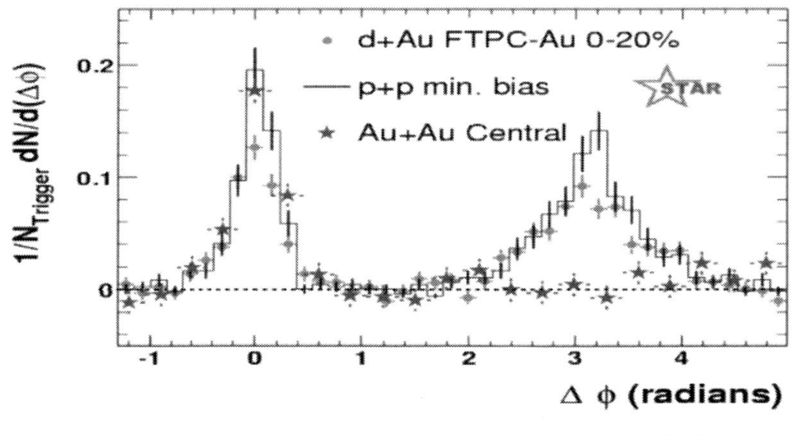

Figure 9. Jet suppression observed in the STAR experiment. The far side jet is suppressed in Au-Au collisions.

Figure 10 shows the energy dependence of jet suppression.[12] In experiments at the CERN ISR (Intersecting Storage Ring) at 31.0 GeV and at the CERN SPS (Super Proton Synchrotron) at 17.3 GeV jet suppression is not seen. This clearly indicates that the results from RHIC are unique and indicative of a new regime in high density nuclear physics.

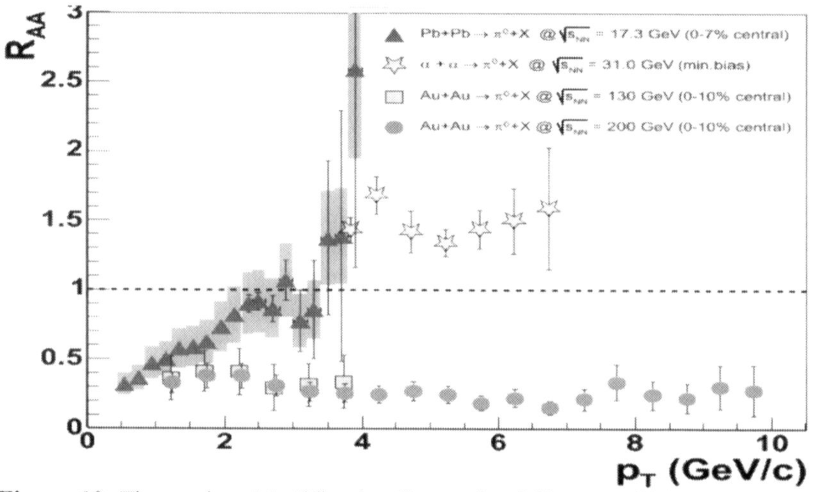

Figure 10. The nuclear Modification Factor for different collision energies.

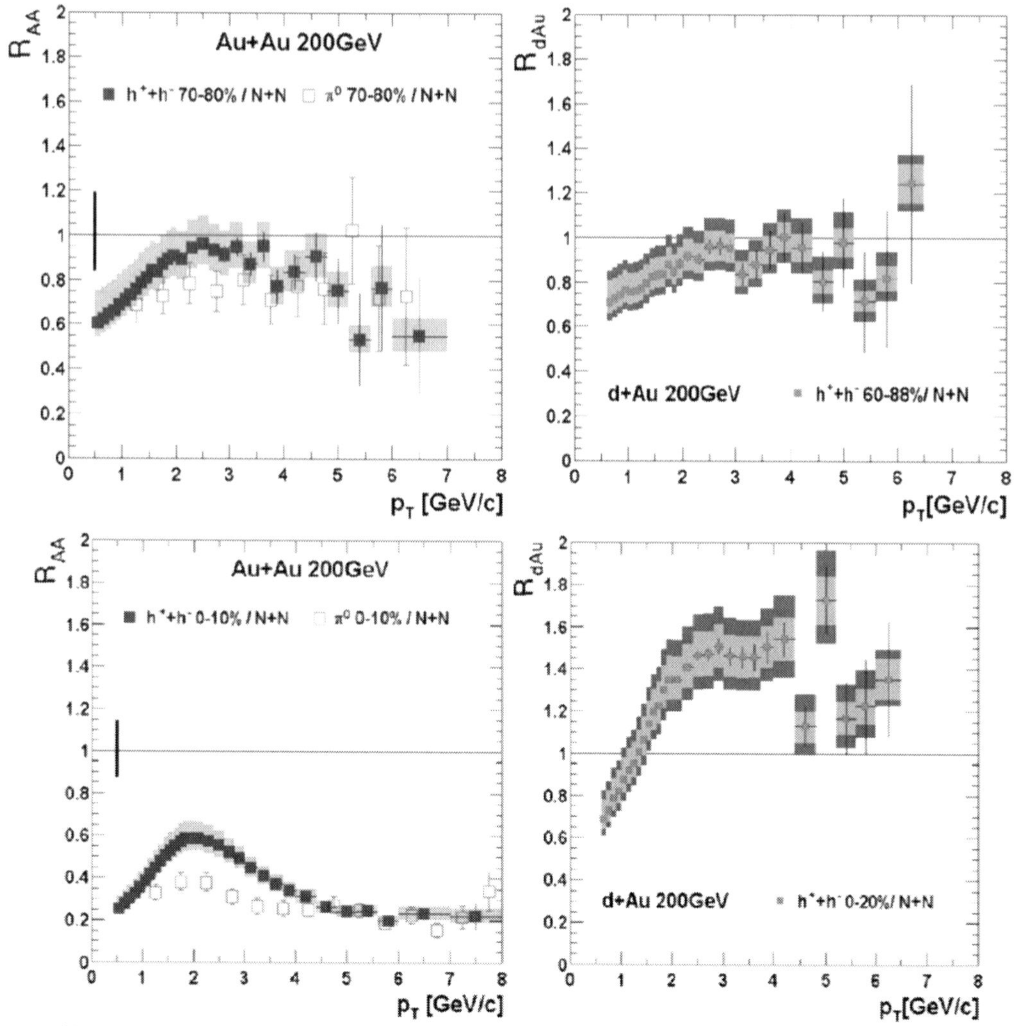

Figure 11. The Nuclear Modification Factor for different centrality bins. The top panels are for peripheral collisions and the bottom panels are for the most central collisions.

The results for the nuclear modification factor further reinforces the notion that we have observed the QGP. Figure 11 shows the nuclear modification factor for two centrality bins, peripheral and central.[13] In the top panels the peripheral bins are shown for both Au-Au(70-80%) and d-Au(60-88%) collisions. Here, because the collision is so peripheral, we don't expect to be able to produce the QGP in the Au-Au collisions and the two panels are very similar. In the bottom two panels are plotted the data for the central collisions. Here, we do expect to produce the QGP in Au-Au collisions and indeed we observe a dramatic suppression for R_{AA}. The enhancement for the d-Au data at central collisions is due to a known cold nuclear matter effect.

Another measure of the R_{AA} for direct photons and hadrons show the same suppression for the hadrons from jets but no suppression from photons. We expect hadrons from the jet will be suppressed but photons will not because photons will not interact in the QGP. This is shown in figure 12.[14][15][16]

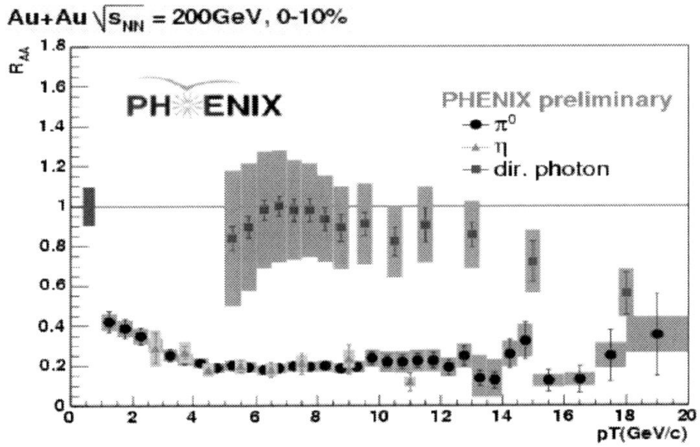

Figure 12. The Nuclear Modification Factor for direct photons and mesons.

Figure 13 is a calculation [17] of the flow parameter v_2 compared to data[18]. The large v_2 indicates that the quarks flow and the calculation fits the data only if the viscosity is very small. This is the criteria for a near perfect liquid.

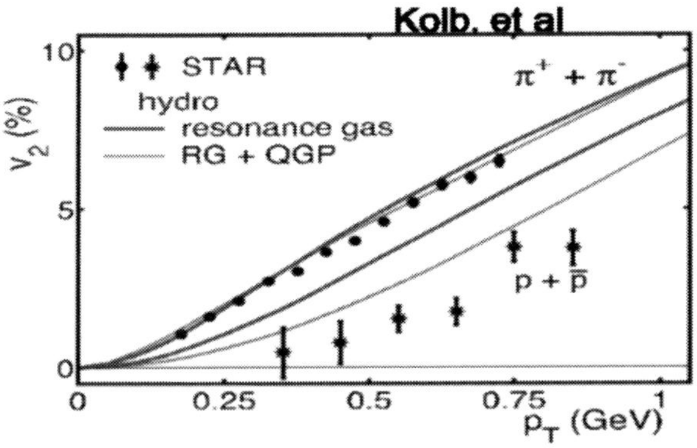

Figure 13. The flow parameter v_2 for baryons and mesons. The red and blue are calculations by Kolb, et. Al. [17]

In addition to these results we have also had some unexpected surprises. Figure 14 shows the proton to pion ratio and the antiproton to pion ratio for a variety of trigger conditions.[19] This ratio of baryons/mesons is completely unexpected. The ratio changes most rapidly when the collision process becomes more central. We do not know the reason for this result.

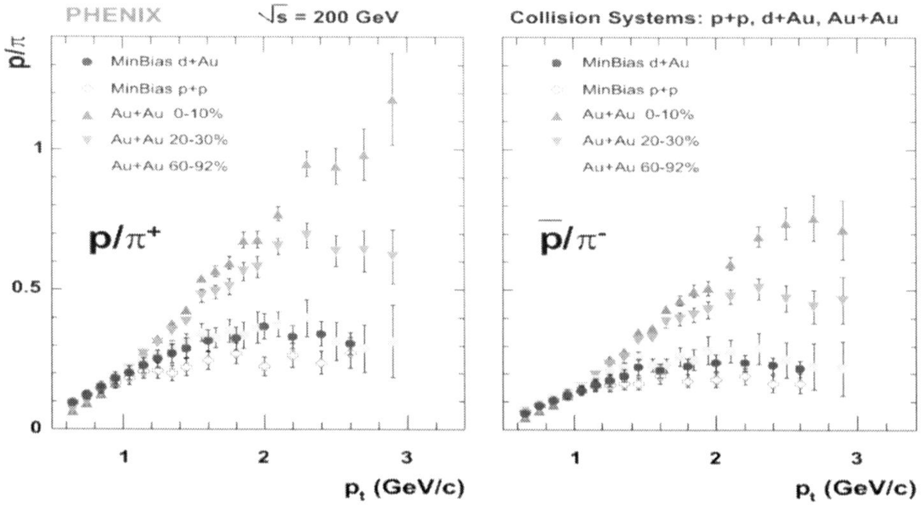

Figure 14. The baryon to pion ratio for different collision systems.

The future at RHIC and the LHC

The main effort at RHIC in the future is to further elucidate the properties of the QGP. RHIC is now in the regime where instead of discovery, it is in the determination of the details. To do this we look at different ways to probe the nuclear medium. One way to do this is to use heavy quark probes. Heavy quarks are formed in the early part of the formation of the plasma so are ideal to use as a probe of the QGP. In addition, because the heavy quarks are more massive, the theoretical calculations are more reliable. To date the identification of the heavy quarks at RHIC have been hampered by large backgrounds in the analysis and an inability to determine if the leptons have a displaced vertex from the semileptonic decay of the heavy quarks. To remedy this, the two large detectors at RHIC have embarked on an upgrade program to install new detector systems in the vertex region to accurately track the particles from the heavy ion collisions and determine which have a displaced vertex. Figure 15 illustrates the definition of a displaced vertex. The D^+ meson (contains a charm quark) decays to a muon that is detected in the muon arm. The decay length of the D^+ meson is of the order of a few hundred microns while the decay of the π meson has a decay length of the order of a few cm. By measuring the decay length we can identify the heavy quark decay (charm and bottom) from the prompt muons and the π meson decays. Figure 16 shows an example of this type of new detector.

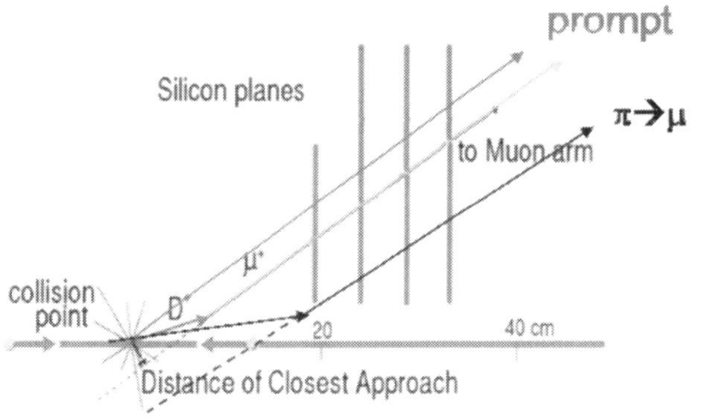

Figure 15. An illustration of the definition of a displaced vertex. The D meson travels a short distance and the decays semi-leptonically to a muon and neutrino.

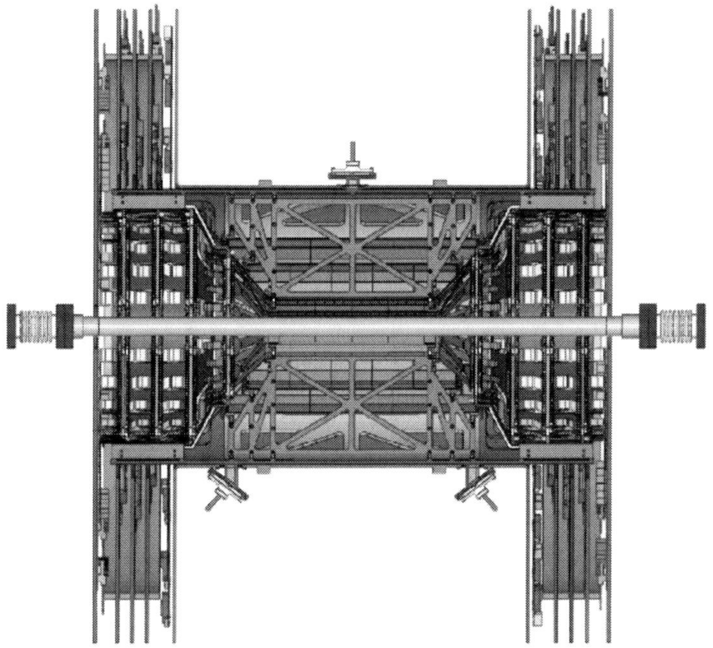

Figure 16. The new Silicon Vertex Tracker (SVT) upgrade for the PHENIX detector.

The SVT is designed to measure all of the particles emanating from the heavy ions collisions. It does this by having four layers of silicon pixel and strip detectors in the barrel region and forward regions. It has ~5.3 million channels of electronic information that must be filtered and processed for each event. The detector support structure is made of carbon fiber composite material for light weight and stable operation. The technical challenges are the need for ~ 50 micron vertex resolution, stability to ~25 micron, and a heat load ~4 kWatts.

The heavy ion program at the Large Hadron Collider (LHC) will be getting under way in the coming year. This program pushes the energy regime from 200 GeV center-of-mass/nucleon to 5.5 TeV center-of-mass/nucleon and opens the door to investigations of the QGP at higher densities and temperature. It will extend the discovery potential beyond RHIC and perhaps find a weakly interacting QGP. This ambitious program utilizes the large complex at CERN and will operate in conjunction with the proton-proton collider program. A view of the CERN complex is shown in figure 17 in the left panel and the ALICE dedicated heavy ion detector in the right panel.[20]

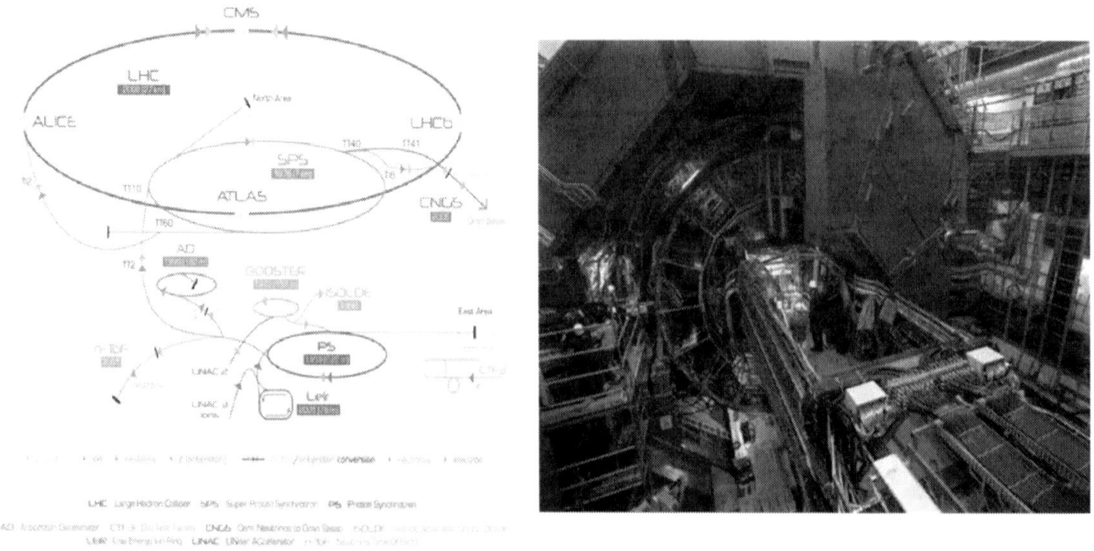

Figure 17. The CERN complex is shown in the left panel and the ALICE dedicated heavy ion detector is in the right panel.

The LHC-Heavy Ion program promises to be as exciting as the RHIC program and when the beams are operational later this year will yield many interesting results and probably a few surprises.

References

[1] Lee T D 1979 cU-TP-170, Published in *Trans. N.Y .Acad. Sci.* Ser.2 v.**40** 1980:0111
[2] Collins J C and Perry M J 1975 *Phys. Rev. Lett.* **34** 1353
[3] Shuryak E V 1980 *Phys. Rept.* **61** 71
[4] Gell-Mann M 1964 A Schematic Model of Baryons and Mesons *Physics Letters* **8** (3) 214–215
[5] Zweig G 1964 An SU(3) Model for Strong Interaction Symmetry and its Breaking *CERN Report* No.8181/Th 8419.
[6] Carithers B and Grannis P 1995 *Discovery of the Top Quark*
[7] Fermilab Visual Media Services
[8] Contemporary Physics Education Project.
[9] http://www.bnl.gov/rhic/physics.asp
[10] Zajc W A Columbia University
[11] Adams J 2005 *Nucl. Phys.* A **757** 102
[12] Adler S S et al 2003 *Phys. Rev. Lett.* **91** 072303
[13] Adler S S et al 2004 *Phys. Rev.* C **69** 034910
[14] Adler S S et al 2003 *Phys. Rev. Lett.* **91** 072301
[15] Adler S S et al 2007 *Phys. Rev.* C **76** 034904
[16] Adler S S et al 2005 *Phys. Rev.Lett.* **94** 232301
[17] Kolb P F et al 2000 *Phys. Rev.* C **62** 054909
[18] Adams J et al (STAR Collaboration) 2004 *Phys. Rev. Lett.* **92** 052302
[19] Adler S S et al 2003 *Phys. Rev. Lett.* **91** 172301
[20] CERN Photo archives

Coherent states in Quantum Information: an example of experimental manipulations

Jean-Pierre Gazeau

Astroparticules et Cosmologie (CNRS, UMR 7164)
Université Paris-Diderot (Paris 7), 75205 Paris Cedex 13, France

E-mail: gazeau@apc.univ-paris7.fr

Abstract. A part of difficulties in implementing communication in Quantum Information stems from the fragility of Shroedinger cat-like superpositions. We describe here a recent experiment in Quantum Optics proving the feasibility of a feedback-mediated quantum measurement for discriminating between optical coherent states under photodetection. The measurements validate theoretical prediction by Helstrom ("Helstrom bound"), Dolinar and Geremia. This contribution gives an account of the implementation achieved by Cook, Martin and Geremia (2007) and explains the theoretical approaches to the subject.

1. Introduction

Quantum Information Processing is about exploiting quantum mechanical features in all facets of information processing (data communication, computing). The states act as information carriers, while the communication channels are the quantum operations [Audenaert 2007]. The sender encodes information by preparing the channel into a well-defined quantum state ρ belonging to an alphabet $\mathcal{A} = \{\rho_0, \rho_1, \dots, \rho_M\}$. The receiver, following any relevant signal propagation, performs a measurement on the channel to ascertain which state was transmitted by the sender.

Quantum information theory is mainly based on superposition-basis and entanglement measurements. This requires high-fidelity implementation to be effective in the laboratory. Unfortunately, quantum measurements are "invasive" in the sense that little or no refinement is achieved by further observation of an already measured system. If the states in the sender alphabet are not orthogonal, no measurement can distinguish between overlapping quantum states without some ambiguity [von Neumann 1955, Holevo 2001, Fuchs 1996, Peres 1995, Helstrom 1976]. Then errors seem unavoidable: there exists a nonzero probability that the receiver will misinterpret the transmitted codeword[1].

However, this impossibility of discriminating non-orthogonal quantum states might represent an advantage for quantum key distribution [Bennett-Brassard 1984]. Indeed, nonorthogonality prevents an eavesdropper from acquiring information without disturbing the state. Also, in some cases it has been shown by Fuchs that the classical information capacity of a noisy channel is actually maximized by a nonorthogonal alphabet [Fuchs 1996].

Mathematically, the question of distinguishing between nonorthogonal states [Peres 1995, Fuchs 1996] is addressed by optimizing a state-determining measurement over all positive operator valued measures (POVM) [Helstrom 1976]. But arbitrary POVM's are not easy to manipulate!

[1] Even with transmission of orthogonal codewords, decoherence, energy dissipation and other imperfections deteriorate orthogonality.

The optical field produced by a laser provides a convenient quantum system for carrying information. Since optical coherent states [Glauber 1963-1] $|\alpha\rangle = e^{-|\alpha|^2/2}\sum_n \alpha^n/\sqrt{n!}|n\rangle$ are not orthogonal, one would be attempted to minimize the overlap $\langle\alpha'|\alpha\rangle = e^{i\Im(\bar{\alpha}'\alpha)}e^{-|\alpha-\alpha'|^2/2}$ by using large amplitude regimes. However, one faces to power limitation and the appearance of non-linear effects. So one is more inclined to develop optimization methods for communication processes based on small-amplitude optical coherent states and photodetection. Now, when one tries to distinguish between two non-orthogonal states through some receiver device, there exists a quantum error probability. The latter is bounded below by some minimum, named the quantum limit or Helstrom bound [Helstrom 1976] in this context.

In a recent work Cook, Martin, and Geremia (CMG)[Cook-Martin-Geremia 2007] have demonstrated that real-time quantum feedback can be used in place of a quantum superposition of the type "Schrödinger cat state", to implement an optimal quantum measurement for discriminating between optical coherent states. The aim of the present contribution is precisely to give most of the elements needed to understand the CMG experiment, both on experimental and theoretical levels. The content is mainly extracted from the seminal paper by Geremia [Geremia 2004], the CMG Nature [Cook-Martin-Geremia 2007], and the chapter 4 of the textbook [Gazeau 2009]. We give in the next section some necessary definitions concerning the use of optical coherent states in Quantum Information. The Helstrom bound for binary communication with perfect and imperfect detection is then explained. We describe in Section 3 the principles, the experimental device and the results of the CMG experiment, which proved the ability of these authors to reach the quantum limit. In Section 4 we develop the theoretical background of the CGM experiment, namely the description of the Dolinar receiver that is at the basis of the CMG experiment.

2. Binary Coherent State Communication and the Helstrom bound

2.1. Binary Coherent State Communication

Let us consider an alphabet consisting of two pure coherent states, $\rho_0 = |\Psi_0\rangle\langle\Psi_0|$, $\rho_1 = |\Psi_1\rangle\langle\Psi_1|$, corresponding to the logical "0" and "1" respectively. Without loss of generality, $|\Psi_0(t)\rangle = |0\rangle$ can be chosen as the vacuum while

$$\Psi_1(t) = \psi_1(t)\exp\left[-i(\omega t + \varphi)\right] + \text{c.c.}, \tag{1}$$

where ω is the frequency of the optical carrier and φ is (ideally) a fixed phase.

The envelope function, $\psi_1(t)$, is normalized such that

$$\int_0^\tau |\psi_1(t)|^2\, dt = \bar{n}\,, \tag{2}$$

where $\bar{n}$ is the mean number of photons to arrive at the receiver during the measurement interval, $0 \le t \le \tau$. That is, $\hbar\omega|\psi_1(t)|^2$ is the instantaneous average power of the optical signal for logical "1". By combining the incoming signal with an appropriate local oscillator, the amplitude keying with the alphabet of two coherent states $\mathcal{A} = \{|0\rangle, |\alpha\rangle\}$ can always be transformed to the phase-shift keyed alphabet, $\{|-\frac{1}{2}\alpha\rangle, |\frac{1}{2}\alpha\rangle\}$, via the unitary displacement[2], $D\left(-\frac{1}{2}\alpha\right) = \exp(-\frac{1}{2}(\alpha a^\dagger - \bar{\alpha}a))$. Similarly, if $|\Psi_0\rangle \neq |0\rangle$, a simple displacement can be used to restore $|\Psi_0\rangle$ to the vacuum state.

In the case of non-orthogonal quantum states as codewords, the receiver attempts to ascertain which state was transmitted by performing a quantum measurement, Υ, on the channel. Υ is described by an appropriate POVM represented by a complete (here countable) set of positive operators [Peres 1995] resolving the identity,

$$\sum_i \Upsilon_i = I_d \quad \Upsilon_i \ge 0\,, \tag{3}$$

where i indexes the possible measurement outcomes.

[2] We recall that $D(\beta)\,|\alpha\rangle = e^{i\Im(\bar{\alpha}\beta)}|\alpha+\beta\rangle$

An example of POVM in the euclidean plane is given by the following cyclotomic polygonal resolution of the unity:

$$\frac{2}{n}\sum_{q=0}^{n-1}\Upsilon_{\frac{2\pi q}{n}} = I_d\,, \quad \Upsilon_\theta = |\theta\rangle\langle\theta| = \begin{pmatrix} \cos^2\theta & \cos\theta\sin\theta \\ \cos\theta\sin\theta & \sin^2\theta \end{pmatrix}\,.$$

For binary communication, for which the POVM resolution of the unity reads $\Upsilon_0 + \Upsilon_1 = I_d$, the measurement by the receiver amounts to a decision between two hypotheses: H_0, that the transmitted state is ρ_0, selected when the measurement outcome corresponds to Υ_0, and H_1, that the transmitted state is ρ_1, selected when the measurement outcome corresponds to Υ_1.

2.2. *The Quantum Error Probability or Helstrom bound*

Now possibilities of errors mean that there is some chance that the receiver will select the null hypothesis, H_0 (resp. H_1), when ρ_1 is actually present (resp. ρ_0). Thus, we have in terms of conditional probabilities:

$$p(\mathrm{H}_0|\rho_1) = \mathrm{tr}[\Upsilon_0\rho_1] = \mathrm{tr}[(I_d - \Upsilon_1)\rho_1]\,, \quad p(\mathrm{H}_1|\rho_0) = \mathrm{tr}[\Upsilon_1\rho_0]\,. \tag{4}$$

The total receiver error probability is then given by

$$p[\Upsilon_0, \Upsilon_1] = \xi_0 p(\mathrm{H}_1|\rho_0) + \xi_1 p(\mathrm{H}_0|\rho_1)\,, \quad \xi_0 + \xi_1 = 1\,, \tag{5}$$

where $\xi_0 = p_0(\rho_0)$ and $\xi_1 = p_0(\rho_1)$ are the classical probabilities that the sender will transmit ρ_0 and ρ_1 respectively; they reflect the prior knowledge that enters into the hypothesis testing process implemented by the receiver, and in many cases $\xi_0 = \xi_1 = 1/2$.

Minimizing the error in receiver measurement over all possible POVM's (Υ_0, Υ_1) leads to the so-called *quantum error probability* or *Helstrom bound*,

$$P_{\mathrm{H}} \equiv \min_{\Upsilon_0,\Upsilon_1} p[\Upsilon_0, \Upsilon_1]\,, \tag{6}$$

P_{H} is the smallest physically allowable error probability, given the overlap between ρ_0 and ρ_1.

The receiver error probability (5) can be written

$$p[\Upsilon_0, \Upsilon_1] = \xi_0\mathrm{tr}\left[\Upsilon_1\rho_0\right] + \xi_1\mathrm{tr}\left[(I_d - \Upsilon_1)\rho_1\right] = \xi_1 + \mathrm{tr}\left[\Upsilon_1(\xi_0\rho_0 - \xi_1\rho_1)\right]\,,$$

and is minimized by optimizing $\min_{\Upsilon_1}\mathrm{tr}[\Upsilon_1\Gamma]$, $\Gamma \stackrel{\mathrm{def}}{=} \xi_0\rho_0 - \xi_1\rho_1$, over Υ_1 subject to $0 \leq \Upsilon_1 \leq I_d$.

Let $\Gamma = \sum_n \lambda_n|\gamma_n\rangle\langle\gamma_n|$ be the spectral decomposition of the operator Γ. One can write $\mathrm{tr}[\Upsilon_1\Gamma] = \sum_n \lambda_n\langle\gamma_n|\Upsilon_1|\gamma_n\rangle$. Then the Helstrom bound can be expressed as $P_{\mathrm{H}} = \xi_1 + \sum_{\lambda_n<0}\lambda_n$, which corresponds to the case in which Υ_1 is the projector on all eigenstates $|\gamma_n\rangle$ with negative λ_n.

For pure states, where $\rho_0 = |\Psi_0\rangle\langle\Psi_0|$ and $\rho_1 = |\Psi_1\rangle\langle\Psi_1|$, Γ has two eigenvalues of which only one is negative,

$$\lambda_- = \frac{1}{2}\left(1 - \sqrt{1 - 4\xi_0\xi_1|\langle\Psi_0|\Psi_1\rangle|^2}\right) - \xi_1 < 0\,, \tag{7}$$

and the quantum error probability is therefore [Helstrom 1976]

$$P_{\mathrm{H}} = \xi_1 + \lambda_- = \frac{1}{2}\left(1 - \sqrt{1 - 4\xi_0\xi_1|\langle\Psi_1|\Psi_0\rangle|^2}\right)\,. \tag{8}$$

Consistently, this quantity vanishes for orthogonal states.

2.3. *Helstrom bound for coherent states*

From the expansion of coherent states over the number states, $|\alpha\rangle = e^{-|\alpha|^2/2}\sum_{n=0}^{\infty}\frac{\alpha^n}{\sqrt{n!}}|n\rangle$, the overlap between $|\Psi_1\rangle$ and $|\Psi_0\rangle$ is just given in terms of the expected number of photons or average value or lower symbol of the number operator in the coherent state $|\alpha\rangle$: $\bar{n} \equiv \langle\alpha|N|\alpha\rangle$:

$$\langle\Psi_1|\Psi_0\rangle = \langle\alpha|0\rangle = e^{-|\alpha|^2/2} = e^{-\bar{n}/2}. \tag{9}$$

So, the Helstrom bound is given by

$$P_{\rm H} = \frac{1}{2}\left(1 - \sqrt{1 - 4\xi_0\xi_1 e^{-\bar{n}}}\right). \tag{10}$$

2.4. *Helstrom bound with imperfect detection*

It is further possible to evaluate the Helstrom bound for imperfect detection. Non-unit efficiency of a photodetector leads to a photon-count which is related to the ideal (efficiency $\eta = 1$) pre-measured photon distribution by a Bernoulli transformation [Loudon 1973]. Accordingly, the probability $p_n(\eta)$ to detect n-photons using a non-ideal photodetector ($\eta < 1$) is given in terms of the probability $p_m(\eta = 1)$ (using an ideal one) by

$$p_n(\eta) = \sum_{m=n}^{\infty}\binom{m}{n}\eta^n(1-\eta)^{m-n}\,p_m(\eta = 1) \tag{11}$$

Coherent states have the convenient property that sub-unity quantum efficiency is equivalent to an ideal detector masked by a beam-splitter with transmission coefficient, $\eta \leq 1$. Indeed, in the case of coherent states, we have the Poisson distribution $p_m(\eta = 1) = e^{-|\alpha|^2}\,|\alpha|^{2m}/m!$, and so, by changing m into $s = m - n$ in the summation (11) gives

$$p_n(\eta) = \frac{\eta^n|\alpha|^{2n}}{n!}\,e^{-\eta|\alpha|^2}, \tag{12}$$

which amounts to replace α by $\sqrt{\eta}\alpha$ in the expression of coherent states. Accordingly, the Helstrom bound becomes

$$P_{\rm H}(\eta) = \frac{1}{2}\left(1 - \sqrt{1 - 4\xi_0\xi_1 e^{-\bar{n}\eta}}\right). \tag{13}$$

This result and Eq. (10) indicate that there is a finite quantum error probability for all choices of $|\Psi_1\rangle$, even when an optimal measurement is performed.

In this context, three receivers have been described and compared by Geremia [Geremia 2004].

Kennedy [Kennedy 1972] proposed in 1972 a receiver based on simple photon counting to distinguish between two different coherent states. However, the Kennedy receiver error probability lies above the quantum mechanical minimum, i.e. the Helstrom bound.

Then, Dolinar [Dolinar 1973] proposed a measurement scheme capable of achieving the quantum limit. Dolinar's receiver, while still based on photon counting, approximates an optimal POVM by superposing a local feedback signal to the channel. A serious experimental drawback was that real-time adjustment of the local signal following each photon was considered (at that time) as quite impracticable.

As a result, Sasaki and Hirota [Sasaki Hirota 1996, Takeoka Sasaki 2008] later proposed an alternative receiver that applies an open-loop unitary transformation to the incoming coherent state signals to render them more distinguishable by simple photon counting.

However, the 2007 Cook-Martin-Geremia experiment has validated the Dolinar receiver.

3. Cook-Martin-Geremia experiment

In [Cook-Martin-Geremia 2007] Cook, Martin and Geremia demonstrate that shot noise can be surpassed and even the quantum limit can be approached by using real-time quantum feedback in place of the cat-state measurement. They exploit the finite duration of any real measurement and quantum states $|0\rangle$ and $|\alpha\rangle$ are realized as optical wavepackets with spatiotemporal extent.

Measurements on an optical pulse inherently persist for a time set by the pulse length τ. Photon counting generates a measurement record $\Xi_{[0,\tau]} \equiv (t_1, t_2, \ldots, t_n)$ that consists of the observed photon arrival times[3] even if is modelled using standard quantum measurement theory by viewing the total number of photon arrivals in the counting interval $[0, \tau]$ as one aggregate "instantaneous" measurement of the number operator.

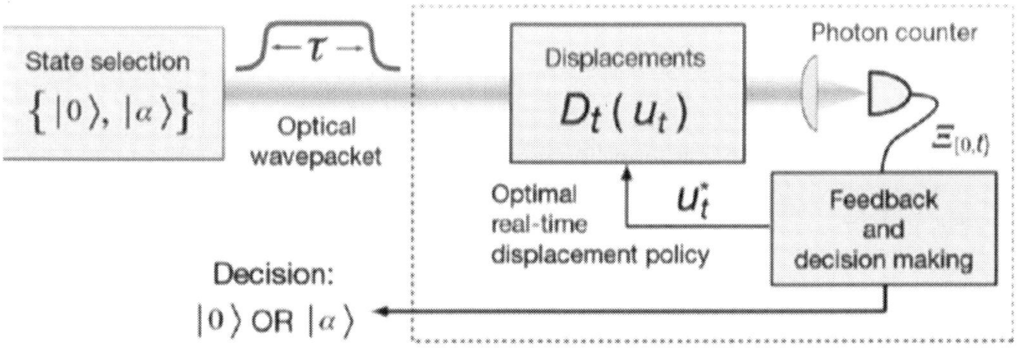

Figure 1. The CMG Measurement combines photon counting with feedback-mediated optical displacements to enact quantum-limited state discrimination between the coherent states $|0\rangle$ and $|\alpha\rangle$. Reprinted by permission from Macmillan Publishers Ltd: [Nature] (Cook, R.L., Martin, P.J., and Geremia, J.M. 446, p.774, 2007), copyright (2007)

In the closed-loop measurement as is sketched in Figure 1 and detailed in Figure 2, photon counting is combined with feedback-mediated optical displacements applied during the photon counting interval. The amplitude of the displacement u_t applied at each time t during the measurement is conditional on the accumulated measurement record $\Xi_{[0,t]}$ and based on an evolving bayesian estimate of the incoming wavepacket state.

Discrimination is performed by selecting the state $|\psi\rangle \in \{|0\rangle, |\alpha\rangle\}$ that maximizes the conditional probability $P\left(\Xi_{[0,t]} \mid \psi, u_{[0,t]}\right)$ that the measurement record $\Xi_{[0,t]}$ would be observed given the state ψ and the history of applied displacements denoted by $u_{[0,t]}$.

The feedback controller determines which state is most consistent with the accumulating record $\Xi_{[0,t]}$ and chooses the feedback amplitude at each point in time to minimize the probability of error over the remainder of the measurement interval $(t, \tau]$.

The premise behind the closed-loop measurement is **to displace the field to the vacuum in each shot and decide which state is present based on the displacement applied to cancel the field.** As the controller gains increased confidence in its guess, it is better able to perform the correct nulling displacement. The displacement magnitude $|u_t^*|$ is inversely proportional to the time-dependent decision uncertainty. The statistical optimality for the closed-loop measurement is reached by reversing state hypothesis with each detector click during the counting interval. As the measurement record accumulates, the controller eventually settles on its final (correct) decision. In Figure 3 are shown

[3] Note that the term "arrival time is not really appropriate from an experimental point of view. Time interval is more appropriate.

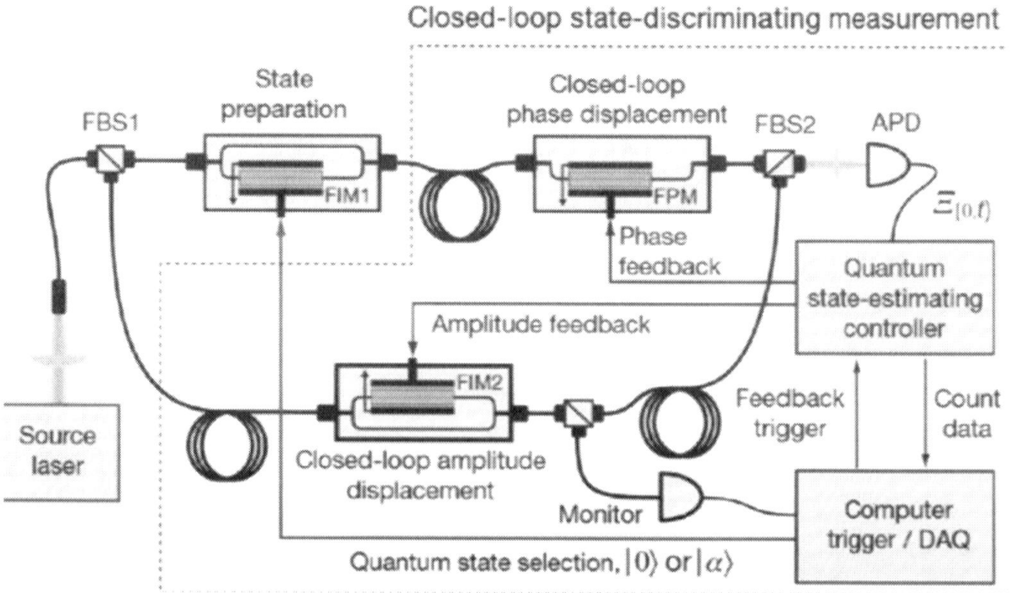

Figure 2. Diagram of the laboratory implementation of Fig. 1. Light from an external-cavity grating-stabilized diode laser operating at 852nm is coupled into a polarization maintaining (PM) fibre-optic MachZehnder interferometer. The input beam splitter (FBS1) provides two optical fields with a well defined relative phase: the upper arm of the interferometer acts as the target quantum system for state discrimination and the lower arm provides an auxiliary field used to perform closed-loop displacements at the second beam splitter (FBS2). Photon counting on the outcoupled field is implemented using a gated silicon avalanche photodiode (APD). Feedback controller is constructed from a combination of programmable waveform generators and high-speed digital signal processing electronics (feedback bandwidth 30MHz). A digital counter records the number of photon counter clicks generated in each measurement interval $[0,\tau]$, during which the feedback controller determines the feedback amplitude $u_t^*\left(n_{[0,t]}\right)$ via the accumulating count record $n_{[0,t)}$. Reprinted by permission from Macmillan Publishers Ltd: [Nature] (Cook, R.L., Martin, P.J., and Geremia, J.M. 446, p.774, 2007), copyright (2007)

the measured probability of error versus mean photon number for both direct photon counting (red squares) and the CMG closed-loop measurement interpreted using a bayesian estimator that assumes application of the optimal closed-loop control policy (blue circles) and one that accounts for experimental imperfections (green triangles). All data points were obtained from ensembles of 100,000 measurement trajectories, with error bars that reflect the sample standard deviation. The data in Fig. 3 have been adjusted to account for finite detector efficiency and optical losses. That is to say, the coherent state amplitudes in the experiment are normalized with respect to the average photon number $\bar{n}_\alpha$ measured by the APD over a t = 20μs square-envelope pulse. Owing to the nature of coherent states, it has been shown that the detection efficiency η (resulting from the combination of detector quantum efficiency η_d and optical efficiency η_e) factors out of a comparison between the shot-noise and quantum limits. For comparison, the shot-noise error and quantum limits that would correspond to ideal detection ($\eta = 1$) have been plotted. The intrinsic efficiency of the apparatus has been independently determined to be approximately $\eta \approx 0.35$.

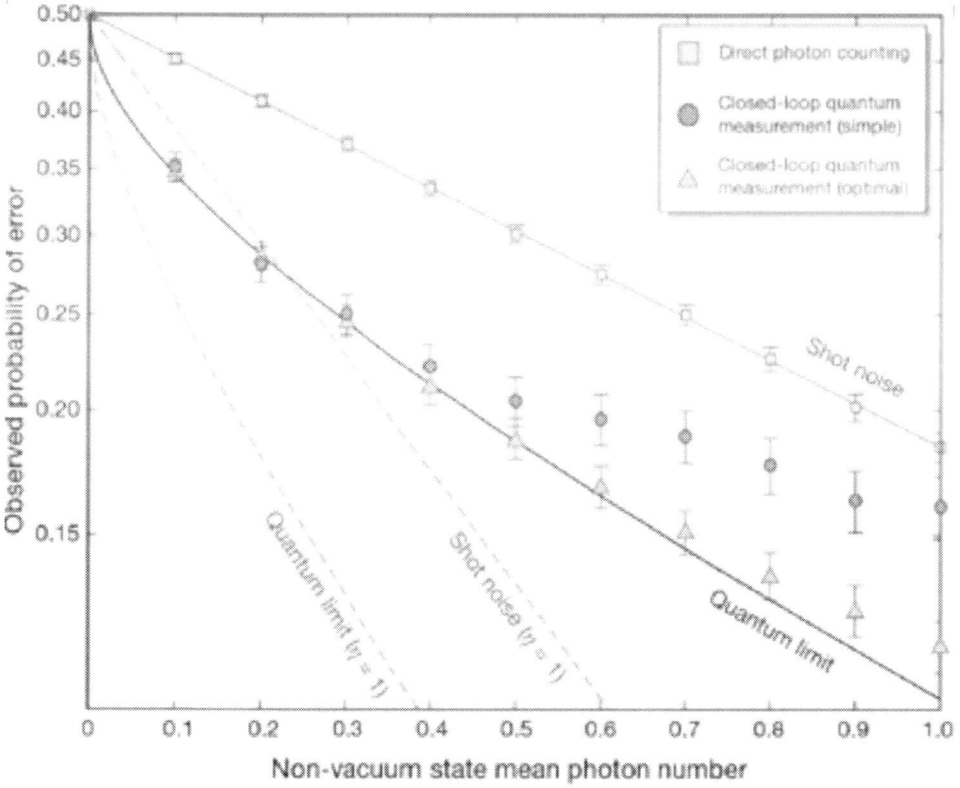

Figure 3. The measured probability of error versus mean photon number for both direct photon counting (red squares) and the CMG closed-loop measurement. Reprinted by permission from Macmillan Publishers Ltd: [Nature] (Cook, R.L., Martin, P.J., and Geremia, J.M. 446, p.774, 2007), copyright (2007)

4. The principles of the Dolinar receiver

The Dolinar receiver utilizes an adaptive strategy to implement a feedback approximation to the Helstrom POVM [Dolinar 1973]. It operates by combining the incoming signal, $\Psi(t)$, with a separate local signal,

$$U(t) = u(t) \exp\left[-i(\omega t + \phi)\right] + \text{c.c}\,. \tag{14}$$

Here $u(t)$ is the "displacement" or "feedback" amplitude.

The detector counts photons with total instantaneous mean rate,

$$\Phi(t) = |\psi(t) + u(t)|^2\,, \tag{15}$$

where we recall that $\psi(t) = 0$ (for logical "0") when the channel is in the state ρ_0, and $\psi(t) = \psi_1(t)$ (for logical "1") when the channel is in ρ_1.

4.1. Photon counting distributions

Given the alphabet $\mathcal{A} = (\rho_0, \rho_1)$, the feedback amplitude $u(t)$, a transmission coefficient η, and some subdivision $(t_0 \equiv 0, t_1, \ldots, t_n, t_{n+1} \equiv \tau)$ of the measurement time interval (or "counting interval") $[0, \tau]$, the conditional probability $w\left[t_k|\rho_i, u(t)\right]$ that a photon will arrive at time t_k and that it will be the only click during the half-closed interval, $(t_{k-1}, t_k]$ [Glauber 1963-2] is called the exponential waiting time distribution for optical coherent states. It is defined as

$$w\left[t_k|\rho_i, u(t)\right] = \eta\, \Phi(t_k)\, \exp\left(-\eta \int_{t_{k-1}}^{t_k} \Phi(t')\, dt'\right),$$

The corresponding exclusive counting densities for the measurement interval reflect the likelihood that n photon arrivals occur precisely at the times $t_1, \ldots, t_n$, given that: the channel is in the state, ρ_i, the feedback amplitude is $u(t)$, and the detector quantum efficiency is η:

$$p_\eta\left[t_1, \ldots, t_n|\rho_i, u(t)\right] = \prod_{k=1}^{n+1} w_\eta\left[t_k|\rho_i, u(t)\right].$$

They allow to evaluate, using the Bayes' rule[4], the conditional arrival time probabilities $p_\eta\left[\rho_i|t_1, \ldots, t_n, u(t)\right] = p_\eta\left[t_1, \ldots, t_n|\rho_i, u(t)\right]\, p_0(\rho_i)$.

4.2. Decision criterion of the receiver

The receiver decides between hypotheses H_0 and H_1 by selecting the one that is more consistent with the record of photon arrival times observed by the detector given the choice of $u(t)$. H_1 is selected when the ratio of conditional arrival time probabilities,

$$\Lambda = \frac{p_\eta\left[\rho_1|t_1, \ldots, t_n, u(t)\right]}{p_\eta\left[\rho_0|t_1, \ldots, t_n, u(t)\right]},$$

is greater than one; otherwise it is assumed that ρ_0 was transmitted.

By employing Bayes' rule, Λ, can be reexpressed in terms of the photon counting distributions

$$\Lambda = \frac{p_\eta\left[t_1, \ldots, t_n|\rho_1, u(t)\right] p_0(\rho_1)}{p_\eta\left[t_1, \ldots, t_n|\rho_0, u(t)\right] p_0(\rho_0)} = \frac{\xi_1\, p_\eta\left[t_1, \ldots, t_n|\rho_1, u(t)\right]}{\xi_0\, p_\eta\left[t_1, \ldots, t_n|\rho_0, u(t)\right]},$$

In terms of error probabilities, the likelihood ratio is given by

$$\Lambda = \frac{p_\eta\left[H_1|\rho_1, u(t)\right]}{p_\eta\left[H_1\rho_0, u(t)\right]} = \frac{1 - p_\eta\left[H_0|\rho_1, u(t)\right]}{p_\eta\left[H_1|\rho_0, u(t)\right]}, \quad \text{for } \Lambda > 1$$

(i.e., the receiver definitely selects H_1), and

$$\Lambda = \frac{p_\eta\left[H_0|\rho_1, u(t)\right]}{p_\eta\left[H_0|\rho_0, u(t)\right]} = \frac{p_\eta\left[H_0|\rho_1, u(t)\right]}{1 - p_\eta\left[H_1|\rho_0, u(t)\right]}, \quad \text{for } \Lambda < 1.$$

(i.e., the receiver definitely selects H_0).

The minimization over $u(t)$ of the Dolinar receiver error probability,

$$P_D[u(t)] = \xi_0 p_\eta\left[H_1|\rho_0, u(t)\right] + \xi_1 p_\eta\left[H_0|\rho_1, u(t)\right], \tag{16}$$

can be accomplished by employing the technique of dynamic programming [Bertsekas 2000], which leads to the control policy,

$$u_1^*(t) = -\psi_1(t)\left(1 + \frac{\mathcal{J}[u_1^*(t)]}{1 - 2\mathcal{J}[u_1^*(t)]}\right)$$

[4] Bayes' theorem relates the conditional and marginal probabilities of events A and B, where B has a non-vanishing probability $P(A|B) = \frac{P(B|A)\,P(A)}{P(B)}$. More generally, let $\Omega = \bigcup_i A_i$, $A_i \cap A_j = \emptyset$ for $i \neq j$, be a partition of the event space. Then we have $P(A_i|B) = \frac{P(B|A_i)\,P(A_i)}{\sum_j P(B|A_j)\,P(A_j)}$ for any A_i in the partition.

for $\Lambda > 1$, where $p_\eta[\mathrm{H}_0|\rho_1, u_1^*(t)] = 0$ and

$$\mathcal{J}[u_1^*(t)] = \xi_1 p_\eta[\mathrm{H}_1|\rho_0, u_1^*(t)] = \frac{1}{2}\left(1 - \sqrt{1 - 4\xi_0\xi_1 e^{-\eta\bar{n}(t)}}\right).$$

Here we recall that $\bar{n}(t) = \int_0^t |\psi_1(t')|^2\, dt'$ is the average number of photons expected to arrive at the detector by time t when the channel is in the state ρ_1. Conversely, the optimal control takes the form,

$$u_0^*(t) = \psi_1(t)\left(\frac{\mathcal{J}[u_0^*(t)]}{1 - 2\mathcal{J}[u_0^*(t)]}\right)$$

for $\Lambda < 1$, where $p_\eta[\mathrm{H}_1|\rho_0, u_0^*(t)] = 0$ and

$$\mathcal{J}[u_0^*(t)] = \xi_0 p_\eta[\mathrm{H}_0|\rho_1, u_0^*(t)] = \frac{1}{2}\left(1 - \sqrt{1 - 4\xi_0\xi_1 e^{-\eta\bar{n}(t)}}\right).$$

4.3. Dolinar hypothesis testing procedure

The Hamilton-Jacobi-Bellman solution leads to a conceptually simple procedure for estimating the state of the channel. The receiver begins at $t = 0$ by favoring the hypothesis that is more likely based on the prior probabilities[5], $p_0(0) = \xi_0$ and $p_1(0) = \xi_1$. Assuming that $\xi_1 \geq \xi_0$ (for $\xi_0 > \xi_1$, the opposite reasoning applies), the Dolinar receiver always selects H_1 during the initial measurement segment. The probability of deciding on H_0 is exactly zero prior to the first photon arrival such that an error only occurs when the channel is actually in ρ_0.

To see what happens when a photon does arrive at the detector, it is necessary to investigate the behavior of $\Lambda(t)$ at the boundary between two measurement segments. Substituting the optimal control policy, $u^*(t)$, which alternates between $u_1^*(t)$ and $u_0^*(t)$, into the photon counting distribution leads to

$$p(t_1, \ldots, t_n|\rho_i) = \eta^n \prod_{k=0}^{n+1} \Phi_i[u_{k|2}(t_{k-1}, t_k)] \times \exp\left(-\eta \int_{t_{k-1}}^{t_k} \Phi_i[u_{k|2}(t'_{k-1}, t'_k)]\, dt'\right),$$

where $k|2$ stands for $k \mod 2$. This expression can be used to show that the limit of $\Lambda(t)$ approaching a photon arrival time, t_k, from the left is the reciprocal of the limit approaching from the right,

$$\lim_{t \to t_k^-} \Lambda(t) = \left[\lim_{t \to t_k^+} \Lambda(t)\right]^{-1}. \tag{17}$$

That is, if $\Lambda > 1$ such that H_1 is favored during the measurement interval ending at t_k, the receiver immediately swaps its decision to favor H_0 when the photon arrives. Evidently, the optimal control policy, $u^*(t)$, engineers the feedback such that the photon counter is least likely to observe additional clicks if it is correct based on its best knowledge of the channel state at that time. Each photon arrival invalidates the current hypothesis and the receiver completely reverses its decision on every click. **This result implies that H_1 is selected when the number of photons, n, is even (or zero) and H_0 when the number of photons is odd.**

It is also asserted [Dolinar 1973] that, despite the discontinuities in the conditional probabilities, $p_\eta[\mathrm{H}_1|\rho_0, u^*(t)]$ and $p_\eta[\mathrm{H}_0|\rho_1, u^*(t)]$, at the measurement segment boundaries, the total Dolinar receiver error probability, $P_\mathrm{D}(\eta, t) = \frac{1}{2}\left(1 - \sqrt{1 - \xi_0\xi_1 e^{-\eta\bar{n}(t)}}\right)$, evolves smoothly, i.e. we have $\lim_{t \to t_k^-} \mathcal{J}[u^*(t)] = \lim_{t \to t_k^+} \mathcal{J}[u^*(t)]$ at the boundaries.

Recognizing that $\bar{n}(\tau) = \bar{n}$ leads to the final Dolinar receiver error,

$$P_\mathrm{D}(\eta) = \frac{1}{2}\left(1 - \sqrt{1 - 4\xi_0\xi_1 e^{-\eta\bar{n}}}\right)$$

which is equal to the Helstrom bound for all values of the detector efficiency, $0 < \eta \leq 1$.

[5] If $\xi_0 = \xi_1$, then neither hypothesis is *a priori* favored and the Dolinar receiver is singular with $P_\mathrm{D} = \frac{1}{2}$.

5. Conclusion

Quantum feedback can be viewed as manipulating the outcome statistics of the number operator N. In the absence of feedback, the detailed measurement record consisting of photon arrival times $\Xi_{[0,T]} = (t_1, t_2, \ldots, t_n)$ provides no more information than the total number n: Poisson processes are stationary in time, but with feedback, the significance of each click depends on when it occurs, even though the field is described by some coherent state at each point in time. The optimal feedback policy applies displacements in a manner that extracts as much information out of each photon arrival as possible. It is in this manner that shot noise can be surpassed to achieve the fundamental quantum limit over a non-trivial range of $|\alpha|$. Furthermore, this procedure appears as less demanding on the measurement resources needed to achieve optimal statistics than a direct implementation of a cat state. At no point in time has a superposition between optical coherent states been generated, yet the optimal result has been effectively achieved by exploiting the time-dependence of the measurement.

It should be added that recently Wittmann et al [Wittmann et al 2008] have experimentally realized a new quantum measurement that detects binary optical coherent states with fewer errors than the homodyne and the Kennedy receiver for all amplitudes of the coherent states. Although the scheme is not capable of achieving the Helstrom bound the implementation discriminates binary coherent states with an error probability lower than the optimal Gaussian receiver, namely the homodyne receiver. For more details on the theoretical background, see [Takeoka Sasaki 2008]

References

[Audenaert 2007] Audenaert K M R 2007 Mathematical Aspects of Quantum Information Theory. In J. P. Gazeau et al. editors. *Physics and Theoretical Computer Science* (Amsterdam: IOS Press) 3

[Bennett-Brassard 1984] Bennett C H and Brassard G 1984 in *Proceedings of the IEEE International Conference on Computers, Systems abd Signal Processing* (New York: IEEE) 175

[Bertsekas 2000] Bertsekas D P 2000 *Dynamic Programming and Optimal Control* **1** (Belmont MA: Athena Scientific)

[Cook-Martin-Geremia 2007] Cook R L, Martin P J, and Geremia J M 2007: Optical coherent state discrimination using a closed-loop quantum measurement *Nature* **446** 774

[Dolinar 1973] S. Dolinar D 1973 (Quaterly Progress Report Tech. Rep. 111 Research Laboratory of Electronics MIT)

[Dolinar 1976] Dolinar S 1976 (Ph.D. thesis Massachusets Institute of Technology)

[Fuchs 1996] Fuchs C A 1996 *Distinguishability and Accessible Information in Quantum Theory* (Ph.D. thesis University of New Mexico)

[Gazeau 2009] Gazeau J P 2009 *Coherent States in Quantum Physics* (Berlin: Wiley-VCH, Berlin)

[Geremia 2004] Geremia J M 2004: Distinguishing between optical coherent states with imperfect detection *Phys. Rev A* **70** 062303-1

[Glauber 1963-1] Glauber R J 1963: The quantum theory of optical coherence, *Phys. Rev.* **130** 2529

[Glauber 1963-2] Glauber R J 1963: Coherent and incoherent states of radiation field, *Phys. Rev.* **131** 2766

[Helstrom 1976] Carl W. Helstrom C V 1976 *Quantum Detection and Estimation Theory* (New York: Academic Press)

[Holevo 2001] Holevo A S 2001 *Statistical Structure of Quantum Theory* (Berlin: Springer-Verlag)

[Kennedy 1972] Kennedy 1972 *Quarterly Progress Report* (Tech. Rep. 110, Research Laboratory of Electronics MIT)

[Loudon 1973] Loudon R 1973 *The Quantum Theory of Light* (Oxford: Oxford Univ. Press)

[Nielsen-Chuang 2000] Nielsen M A and Chuang I L 2000 *Quantum Computation and Quantum Information* (Cambridge: Cambridge University Press)

[Peres 1995] Peres A 1995 *Quantum Theory: Concepts and Methods* (Dordrecht: Kluwer Academic Publishers)

[Preskill] Preskill J 2009 *Quantum Computation and Information*, PH219 Lecture Notes (Caltech)

[Sasaki Hirota 1996] Sasaki M and Hirota O 1996: Optimum decision scheme with a unitary control process for binary quantum-state signals *Phys. Rev A* **54** 2728

[Takeoka Sasaki 2008] Takeoka M, and Sasaki M 2008: Discrimination of the binary coherent signal: Gaussian-operation limit and simple non-Gaussian near optimal receivers *Phys. Rev A* **78** 022320-1

[von Neumann 1955] von Neumann J 1955 *Mathematical Foundations of Quantum Mechanics* English translation by R.T. Byer (Princeton: Princeton University Press)

[Wittmann et al 2008] Wittmann C, Takeoka M, Cassemiro K N, Sasaki M, Leuchs G, and Andersen U L 2008: Demonstration of near-Optimal Discrimination of Optical Coherent States. *Preprint quant-ph 0809.4953*

Graph model of the Heisenberg–Weyl algebra

P Blasiak[1], A Horzela[1], G H E Duchamp[2], K A Penson[3] and A I Solomon[3,4]

[1] H. Niewodniczański Institute of Nuclear Physics, Polish Academy of Sciences
ul. Radzikowskiego 152, PL 31342 Kraków, Poland

[2] Université Paris-Nord, LIPN, CNRS UMR 7030
99 Av. J.-B. Clement, F-93430 Villetaneuse, France

[3] Université Pierre et Marie Curie, LPTMC, CNRS UMR 7600
Tour 24 - 2ième ét., 4 pl. Jussieu, F 75252 Paris Cedex 05, France

[4] The Open University, Physics and Astronomy Department
Milton Keynes MK7 6AA, United Kingdom

E-mail: pawel.blasiak@ifj.edu.pl, ghed@lipn-univ.paris13.fr, andrzej.horzela@ifj.edu.pl, penson@lptmc.jussieu.fr, a.i.solomon@open.ac.uk

Abstract. We consider an algebraic formulation of Quantum Theory and develop a combinatorial model of the Heisenberg–Weyl algebra structure. It is shown that by lifting this structure to the richer algebra of graph operator calculus, we gain a simple interpretation involving, for example, the natural composition of graphs. This provides a deeper insight into the algebraic structure of Quantum Theory and sheds light on the intrinsic combinatorial underpinning of its abstract formalism.

1. Introduction

Quantum Theory seen in action is an interplay of mathematical ideas and physical concepts. From a present-day perspective its formalism and structure is founded on the theory of Hilbert space [1, 2, 3]. According to a few basic postulates, the physical notions of transformations and measurements on a system are described in terms of operators. In this way the algebra of operators constitutes the proper mathematical framework within which quantum theories are built. The structure of this algebra is determined by two operations, the addition and multiplication of operators; this lies at the root of all fundamental aspects of Quantum Theory [4].

However, the physical content of Quantum Theory transcends the abstract mathematical formalism. It is provided by the correspondence rules assigning operators to physical quantities. This is always an *ad hoc* procedure invoking concrete representations of the operator algebra chosen to best reflect the physical concepts related to the phenomena under investigation. The most common structure in Quantum Theory is the Heisenberg–Weyl algebra. This describes the algebraic relation between the *position* and *momentum* operators, equally the *creation* and *annihilation* operators, which provide our link to the most fundamental physical concepts. Accordingly, we take the Heisenberg–Weyl algebra as the central point of our study.

Interest in combinatorial representations of mathematical entities stems from a wealth of concrete models they provide. Their convenience comes from simplicity, which, being based on the elementary notion of enumeration, directly appeals to intuition, often rendering invaluable interpretations illustrating abstract mathematical constructions [5, 6, 7]. This makes the combinatorial perspective particularly attractive in quantum physics, given the latter's active pursuit of a proper understanding of fundamental phenomena.

In this paper we develop a combinatorial representation of the operator algebra of Quantum Theory which is based on the Heisenberg–Weyl algebra. We recast it in the language of graphs with a simple composition rule and show how, from this perspective, abstract algebraic structures gain an intuitive meaning. In some respects this draws on the Feynman idea of representing physical processes as diagrams, familiar as a bookkeeping tool in the perturbation expansions of quantum field theory [8, 9]. The combinatorial approach, however, has much more to offer if applied to the overall structure of Quantum Theory seen from the algebraic point of view. We will show that the process of lifting to a more structured algebra of graphs gives the abstract operator calculus a straightforward interpretation, reflecting natural operations on graphs. This provides an interesting insight into the algebraic counterpart of the theory and sheds light on the intrinsic combinatorial structures which lie behind its abstract formalism.

2. Quantum Theory as an Algebra of Operators

The usual setting for Quantum Theory consists of specifying a Hilbert space H, whose vectors are the states of the system, and identifying operators on those states with physically relevant quantities. Operators acting on H naturally form an algebra with addition and multiplication, which we denote by $\mathcal{O}$. The most interesting structures in $\mathcal{O}$ are, of course, those generated by operators having a physical interpretation. They usually originate from considering some observables of interest along with operations causing changes in the state of a system. Accordingly, one takes a hermitian operator, say N, representing some observable and defines a basis in H related to states with definite values of the corresponding physical quantity. The eigenvectors of N are given by $N|n\rangle = n|n\rangle$, numbering the chosen eigenbasis $|n\rangle$ in H ($n = 0, 1, 2, \ldots$). One is then interested in describing processes which change the state of the system, *e.g.* time evolution, interactions and other transformations. For that purpose it is convenient to introduce annihilation a and creation $a^\dagger$ operators which shift the basis vectors by one, *i.e.* $[a, N] = a$ and $[a^\dagger, N] = -a^\dagger$. Conventionally, these operators are required to satisfy the canonical commutation relation

$$[a, a^\dagger] = 1\,, \tag{1}$$

constituting the Heisenberg–Weyl algebra structure [10, 11] which has became the hallmark of noncommutativity in Quantum Theory [4]. The operators defined above play the role of elementary processes altering the system by changing its state with respect to the chosen physical characteristic, *i.e.* they cause a jump between the eigenstates $|n\rangle$ according to the rule $a\,|n\rangle = \sqrt{n}\,|n-1\rangle$ and $a^\dagger\,|n\rangle = \sqrt{n+1}\,|n+1\rangle$. We shall assume that any change of state can be obtained by the action of some combination of such creation and annihilation acts, making the operators a and $a^\dagger$ convenient building blocks describing the transformations of a system.

The creation and annihilation operators can be used to represent elements of the algebra $\mathcal{O}$. Indeed, each operator can be seen as an element of the free algebra generated by a and $a^\dagger$, *i.e.* written as a linear combination of words in generators. This procedure is, however, ambiguous due to the commutation relation Eq. (1) which yields different representations of the same operator [12]. To solve this problem, the order of a and $a^\dagger$ has to be fixed. Conventionally this is done by choosing the normally ordered form in which all annihilators stand to the right of creators [13, 14]. Consequently, each operator $A \in \mathcal{O}$ can be uniquely written in the normally ordered form such as[1]

$$A = \sum_{r,s} \alpha_{rs}\, a^{\dagger\, r} a^{s}. \qquad (2)$$

In this way elements of the operator algebra $\mathcal{O}$ are represented in terms of the *ladder* operators a and $a^\dagger$, and interpreted as combinations of the elementary acts of annihilation and creation. Eq. (2) will be the starting point of our combinatorial representation of the algebra $\mathcal{O}$.

3. Graphs and their Algebra

Considering combinatorial realizations of operator algebras, we shall specify two classes of graphs $\mathfrak{g}$ and $\mathfrak{g}_1$, the latter being the shadow of the former under a suitable *forgetful* procedure. We shall employ the convenient notion of graph composition to show how these structures are naturally made into algebras, providing the representation of the algebra $\mathcal{O}$.

3.1. Graphs & Composition

A graph is a collection of vertices connected by lines with internal structure determined by some construction rules. For our purposes, we consider a specific class of graphs defined in the following way.

3.1.1. Vertices & Lines. The basic building blocks of the graphs are vertices • attached to two sorts of lines, those coming into, and those going out of, the vertex, having loose ends marked with grey ▲ and white △ arrows respectively. A generic one-vertex graph $\Gamma^{(r,s)}$ is characterized by two numbers r and s counting incoming and outgoing lines respectively, see Fig. 1. We shall denote by $\mathfrak{g}_1$ the class comprising all such one-vertex graphs, and by $\emptyset$ the empty, or void, graph (no vertices, no lines). In a further construction we shall assume that all lines attached to vertices are distinguishable.

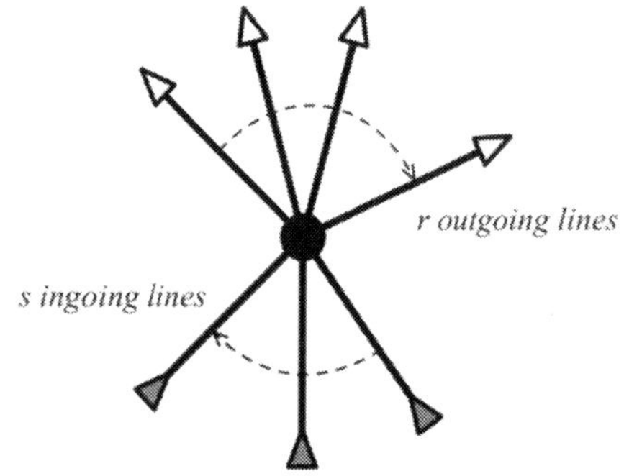

Figure 1. A generic one-vertex graph $\Gamma^{(r,s)} \in \mathfrak{g}_1$.

[1] We do not specify limits of summation and constraints on coefficients since it does not affect the algebraic considerations and can be introduced at each step if needed.

3.1.2. Construction Rules. A multi-vertex graph Γ is a set of vertices with additional structure introduced by joining some of the outgoing lines to the incoming ones. The requirement that the original direction of lines is preserved results in a *directed* structure of graphs indicated by black arrows ▲ on the inner lines. We further restrict the class of graphs we consider to those without cycles, *i.e.* we exclude graphs with closed paths. An example of a multi-vertex graph is shown in Fig. 2.

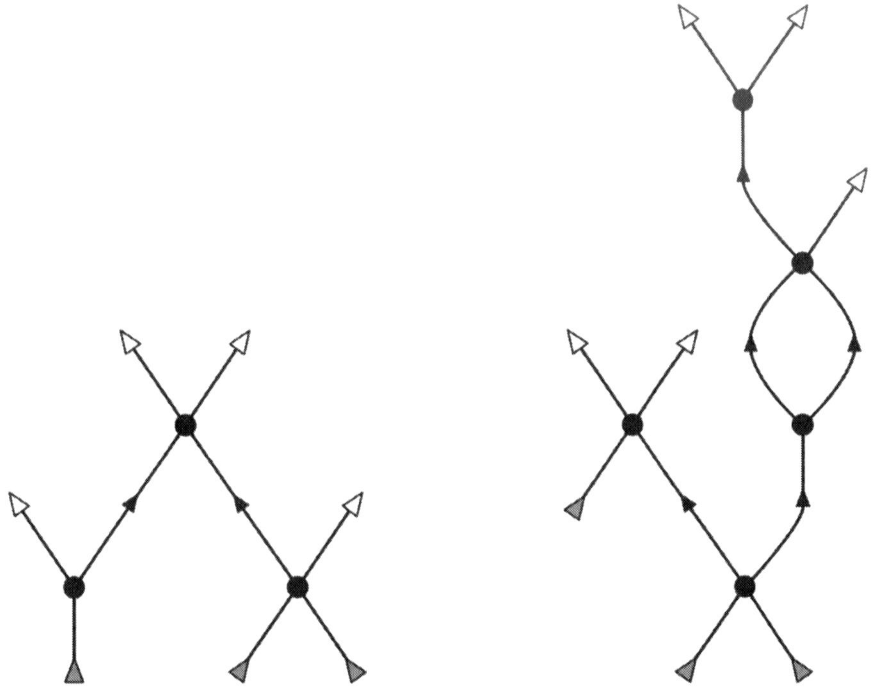

Figure 2. Example of a multi-vertex graph (8 vertices and 9 outgoing lines, 6 incoming lines, 5 inner lines) built of two kinds of vertices: $\Gamma^{(2,2)}$ (X shape) and $\Gamma^{(2,1)}$ (Y shape).

The rules specified above define the class of graphs denoted by $\mathfrak{g}$. In a less formal manner, we can describe these graphs as having an inner structure determined by directed connections between vertices and a characteristic set of outer lines marked with grey ▲ and white △ arrows at the loose ends. These graphs can be seen as a kind of process, with the vertices being intermediate steps. This observation can be developed further with the help of the convenient notion of *graph composition*.

3.1.3. Composition. Two graphs can be composed by joining some of the incoming lines (grey arrows) of the first one with some of the outgoing lines (white arrows) of the second one. This operation is *inner* in $\mathfrak{g}$ since it preserves the direction of the lines and does not introduce cycles. Observe that two graphs can be composed in many ways, *i.e.* as many as there are possible choices of pairs of lines (grey arrows from the first one and white arrows from the second one) which are joined, see Fig. 2. Note also that composing graphs in reverse order yields different results.

The notion of graph composition allows for an iterative definition, *i.e.* any element of $\mathfrak{g}$ can be constructed starting from the void graph by successive composition with one-vertex graphs.

Journal of Physics: Conference Series **213** (2010) 012014 doi:10.1088/1742-6596/213/1/012014

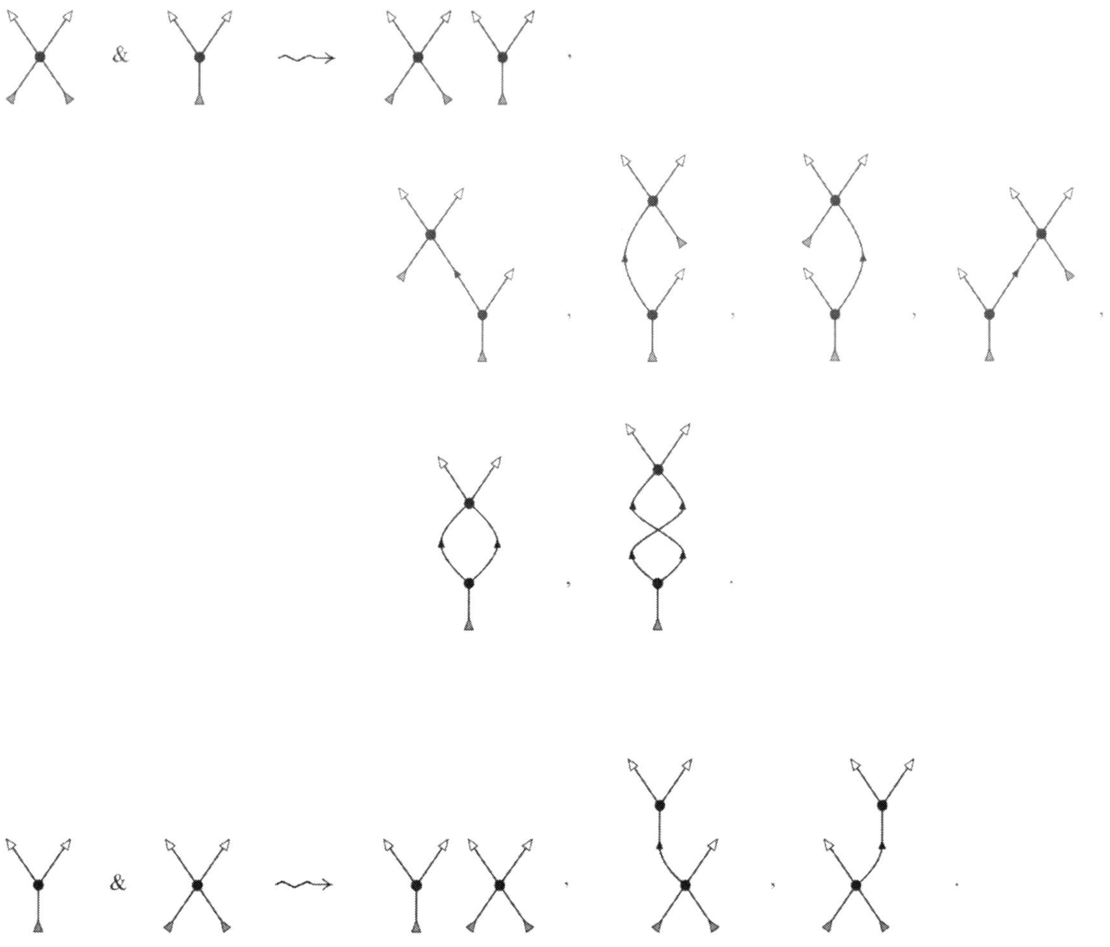

Figure 3. Two one-vertex graphs composed in different order. Note that distinguishability of the lines is taken into account.

Consequently, the class of one-vertex graphs $\mathfrak{g}_1 \subset \mathfrak{g}$ can be seen as representing basic processes or events happening one after another and constituting a composite process – a multi-vertex graph.

3.2. Equivalence of Graphs

In many cases one is not interested in the inner structure of a graph and needs only to focus on the outer lines. This is equivalent to considering the graph's one-vertex equivalents obtained by replacing all inner vertices and lines by a single vertex and keeping all the outer lines untouched, *i.e.* $\Gamma \xrightarrow{\sim} \Gamma^{(r,s)}$ where Γ is the graph with r outgoing and s incoming lines. For example, for the graph in Fig. 2 this gives $\Gamma \xrightarrow{\sim} \Gamma^{(9,6)}$. The mapping $\mathfrak{g} \xrightarrow{\sim} \mathfrak{g}_1$ reduces to forgetting about the inner structure of graphs and introduces an equivalence relation in $\mathfrak{g}$. Accordingly, two graphs are *equivalent* $\Gamma_1 \sim \Gamma_2$ if and only if both have the same number of incoming and outgoing lines respectively. The simplest choice of representatives of equivalence classes are the one-vertex graphs and so the quotient set $\mathfrak{g}/_\sim$ is isomorphic to the set of one-vertex graphs $\mathfrak{g}_1$.

There are two characteristic mappings between $\mathfrak{g}$ and $\mathfrak{g}_1$: the canonical projection map described above and the inclusion map $\mathfrak{g}_1 \subset \mathfrak{g}$, *i.e.*

$$\mathfrak{g} \underset{\supset}{\overset{\sim}{\rightleftarrows}} \mathfrak{g}_1 \tag{3}$$

In this sense $\mathfrak{g}_1$ is a shadow of the more structured class $\mathfrak{g}$. Observe that the arrows in Diag. (3) can not be reversed, *i.e.* once the inner structure of a graph is forgotten it cannot be restored.

3.3. Algebra of Graphs

Both classes of graphs $\mathfrak{g}$ and $\mathfrak{g}_1$ can be endowed with the structure of a noncommutative algebra based on the natural concept of graph composition. An algebra requires two operations, addition and multiplication, which are constructed as follows. We define the vector space $\mathcal{G}$ (over $\mathbb{C}$) generated by the basis set $\mathfrak{g}$, by

$$\mathcal{G} = \left\{ \sum\nolimits_i \alpha_i \, \Gamma_i : \; \alpha_i \in \mathbb{C}, \; \Gamma_i \in \mathfrak{g} \right\}. \tag{4}$$

Addition in $\mathcal{G}$ has the usual form

$$\sum\nolimits_i \alpha_i \, \Gamma_i + \sum\nolimits_i \beta_i \, \Gamma_i = \sum\nolimits_i (\alpha_i + \beta_i) \, \Gamma_i. \tag{5}$$

The less trivial part in the definition of an algebra $\mathcal{G}$ concerns multiplication, which by bilinearity

$$\sum\nolimits_i \alpha_i \, \Gamma_i * \sum\nolimits_j \beta_j \, \Gamma_j = \sum\nolimits_{i,j} \alpha_i \beta_j \, \Gamma_i * \Gamma_j, \tag{6}$$

only requires definition on the basis set $\mathfrak{g}$. Recalling the notion of graph composition, the definition which suggests itself is (compare with Fig. 3)

$$\Gamma_i * \Gamma_j = \sum \textit{all compositions of } \Gamma_i \textit{ with } \Gamma_j \; . \tag{7}$$

Note that all the terms in the sum are distinct with coefficient equal to one. The multiplication thus defined is noncommutative and produces $\mathcal{G}$ an associative algebra with unit (void graph).

Imposing an algebraic structure on $\mathfrak{g}_1$ roughly follows the above scheme. Accordingly, one defines the vector space

$$\mathcal{G}_1 = \left\{ \sum\nolimits_{i,j} \alpha_{i,j} \, \Gamma^{(i,j)} : \; \alpha_{i,j} \in \mathbb{C}, \; \Gamma^{(i,j)} \in \mathfrak{g}_1 \right\}, \tag{8}$$

with addition defined analogously to Eq. (5). Multiplication again reduces to defining it on the basis set $\mathfrak{g}_1$, but the obstacle here is that the composition rule is not closed within the class $\mathfrak{g}_1$, *i.e.* it produces two-vertex graphs which belong to $\mathfrak{g}$. This however can be overcome by applying the forgetful mapping $\Gamma \xrightarrow{\sim} \Gamma^{(r,s)}$ to the result. Specifically, multiplication of two graphs in $\mathfrak{g}_1$ follows Diag. (3) and consists of:

1) treating graphs as elements of $\mathfrak{g}$,
2) multiplying them according to Eq. (7),
3) forgetting the inner structure of the resulting two-vertex graphs in the sum.

Note that, contrary to Eq. (7), some of the resulting terms may be equal and their sum may involve nontrivial integer coefficients. Grouping terms with respect to the number of joined lines yields the explicit formula

$$\Gamma^{(r,s)} * \Gamma^{(k,l)} = \sum_{i=0}^{min\{k,s\}} \binom{s}{i} \binom{k}{i} i! \; \Gamma^{(r+k-i,s+l-i)}. \tag{9}$$

For example: $\Gamma^{(2,1)} * \Gamma^{(2,2)} = \Gamma^{(4,3)} + 2\,\Gamma^{(3,2)}$ and $\Gamma^{(2,2)} * \Gamma^{(2,1)} = \Gamma^{(4,3)} + 4\,\Gamma^{(3,2)} + 2\,\Gamma^{(2,1)}$, see Fig. 3. In this way, multiplication in the richer structure $\mathcal{G}$ is naturally projected onto $\mathcal{G}_1$. The resulting combinatorial algebra $\mathcal{G}_1$ is associative and noncommutative. Again, similarly to $\mathfrak{g}$ and $\mathfrak{g}_1$ in Diag. (3), both algebras are related by

$$\mathcal{G} \underset{\supset}{\overset{\sim}{\rightleftarrows}} \mathcal{G}_1 \tag{10}$$

Hence, the algebra $\mathcal{G}_1$ is an image of the more structured algebra of graphs $\mathcal{G}$.

4. Graph representation of operator algebra

The structures described above are examples of algebras having concrete representations based on the natural concept of graph composition. It appears that both are intimately related to the algebra of operators $\mathcal{O}$. As suggested by the similarity of elements in $\mathcal{O}$ and $\mathcal{G}_1$, see Eqs. (2) and (8), we make a correspondence of the basis sets

$$a^{\dagger\,r} a^s \longleftrightarrow \Gamma^{(r,s)} \tag{11}$$

establishing the isomorphism of the vector spaces. This would not be surprising if it were not for the fact that this mapping also preserves multiplication in both algebras. Indeed, multiplication of the basis elements in $\mathcal{O}$ gives

$$a^{\dagger\,r} a^s\, a^{\dagger\,k} a^l = \sum_{i=0}^{min\{k,s\}} \binom{s}{i}\binom{k}{i} i!\;\; a^{\dagger\,r+k-i} a^{s+l-i}, \tag{12}$$

which is the result of transforming $a^s a^{\dagger\,k}$ to the normally ordered form using the commutator $[a^s, a^{\dagger\,k}] = \sum_{i=1}^{min\{k,s\}} \binom{s}{i}\binom{k}{i} i!\; a^{\dagger\,k-i} a^{s-i}$, see *e.g.* [15]. It suffices to compare Eqs. (9) and (12) to show that Eq. (11) establishes an isomorphism between the algebras $\mathcal{O}$ and $\mathcal{G}_1$. In other words, both algebras are essentially the same, *i.e.* they have all elements and operations equivalent. We can thus enjoy the advantages of concrete realization of the abstract operator algebra in terms of graphs. For example, instead of multiplying operators in $\mathcal{O}$ one can do it in $\mathcal{G}_1$ simply by composing graphs. The crucial role in this procedure is played by the more structured algebra of graphs $\mathcal{G}$ where all these operations have a simple interpretation. Accordingly, Diag. (10) can be complemented to

$$\mathcal{G} \underset{\supset}{\overset{\sim}{\rightleftarrows}} \mathcal{G}_1 \overset{1:1}{\longleftrightarrow} \mathcal{O} \tag{13}$$

In this way, the algebra $\mathcal{O}$ gains a combinatorial representation via $\mathcal{G}_1$ and can be seen as reflecting the natural processes taking place in $\mathcal{G}$.

5. Discussion

We have considered the Heisenberg–Weyl algebra within the operator structure of Quantum Theory and exploited a convenient representation of operators by normally ordered expressions in ladder operators a and $a^\dagger$. This allowed us to identify the operator algebra $\mathcal{O}$ with the combinatorial algebra $\mathcal{G}_1$ constructed as the projection of the algebra of graphs $\mathcal{G}$. The main result of the paper shows that these combinatorial structures provide us with a simple interpretation of the operator calculus in terms of the natural composition of graphs.

Accordingly, the operator algebra can be seen as an image (via $\mathcal{G}_1$) of the algebra $\mathcal{G}$ which is a categorical version of the algebra $\mathcal{O}$ [16, 17]. This may be illustrated by redrawing Diag. (13) as

The diagram indicates the existence of the fundamental graph structure $\mathcal{G}$ of which the algebraic structure of Quantum Theory is just the reflection. That is, all objects in the theory, as well as its calculus, can be seen, via simple inclusion, as elements and operations present in the richer algebra of graphs described by natural composition rules. At any time it is possible to return to the coarser level of an algebra of operators, simply by forgetting about the inner structure of graphs. As a result, the algebra of graphs suggests itself as a fundamental combinatorial level of Quantum Theory. The process described here, of lifting the theory to the richer structure $\mathcal{G}$, is motivated by the natural interpretation of graphs as processes transforming quantities or objects, which is an attractive concept from the physical point of view. Moreover, the major advantage of the combinatorial representation of the algebra $\mathcal{O}$, presented above, is that the abstract operator calculus can be seen intuitively as a straightforward composition of graphs.

Acknowledgments

We wish to thank Philippe Flajolet for important discussions on the subject. Most of this research was carried out in the Mathematisches Forschungsinstitut Oberwolfach (Germany) and the Laboratoire d'Informatique de l'Université Paris-Nord in Villetaneuse (France) whose warm hospitality is greatly appreciated. The authors acknowledge support from the Polish Ministry of Science and Higher Education grant no. N202 061434 and the Agence Nationale de la Recherche under programme no. ANR-08-BLAN-0243-2.

References

[1] Isham C J 1995 *Lectures on Quantum Theory: Mathematical and Structural Foundations* (London: Imperial College Press)
[2] Peres A 2002 *Quantum Theory: Concepts and Methods* (New York: Kluwer Academic Publishers)
[3] Ballentine L E 1998 *Quantum Mechanics: A Modern Development* (Singapore: World Scientific)
[4] Dirac P A M 1982 *The Principles of Quantum Mechanics* 4th ed (New York: Oxford University Press)
[5] Flajolet P and Sedgewick R 2008 *Analytic Combinatorics* (Cambridge: Cambridge University Press)
[6] Bergeron F, Labelle G and Leroux P 1998 *Combinatorial Species and Tree-like Structures* (Cambridge: Cambridge University Press)
[7] Aigner M 2007 *A Course in Enumeration* Graduate Texts in Mathematics (Berlin: Springer-Verlag)
[8] Bjorken J D and Drell S D 1993 *Relativistic Quantum Fields* (New York: McGraw-Hill)
[9] Mattuck R D 1992 *A Guide to Feynman Diagrams in the Many-Body Problem* 2nd ed (New York: Dover Publications)
[10] Artin M 1991 *Algebra* (Prentica Hall)
[11] Bourbaki N 1998 *Algebra* vol I (Springer)
[12] Cahill K E and Glauber R J 1969 *Phys.Rev.* **177** 1857–1881
[13] Klauder J R and Skagerstam B S 1985 *Coherent States. Application in Physics and Mathematical Physics* (Singapore: World Scientific)
[14] Blasiak P, Horzela A, Penson K A, Solomon A I and Duchamp G H E 2007 *Am. J. Phys.* **75** 639–646 arXiv:0704.3116 [quant-ph]
[15] Blasiak P, Duchamp G H E, Horzela A, Penson K A and Solomon A I 2008 *J. Phys. A: Math. Theor.* **41** 415204 (8pp)
[16] Baez J and Dolan J 2001 *Mathematics Unlimited - 2001 and Beyond* (Berlin: Springer Verlag) pp 29–50 arXiv:math.QA/0004133 [math.QA]

[17] Morton J 2006 *Theory and Applications of Categories* **16** arXiv:math/0601458 [math.QA]

Journal of Physics: Conference Series **213** (2010) 012015 doi:10.1088/1742-6596/213/1/012015

Self-adjoint symmetry operators connected with the magnetic Heisenberg ring

Bernd Fiedler

Eichelbaumstr. 13, D-04249 Leipzig, Germany. URL: http://www.fiemath.de/

E-mail: bfiedler@fiemath.de

Abstract. In [1] we defined symmetry classes, commutation symmetries and symmetry operators in the Hilbert space $\mathcal{H}$ of the 1D spin-1/2 Heisenberg magnetic ring with N sites and investigated them by means of tools from the representation theory of symmetric groups $\mathcal{S}_N$ such as decompositions of ideals of the group ring $\mathbb{C}[\mathcal{S}_N]$, idempotents of $\mathbb{C}[\mathcal{S}_N]$, discrete Fourier transforms of $\mathcal{S}_N$, Littlewood-Richardson products.

In the present paper we consider symmetry operators $a \in \mathbb{C}[\mathcal{S}_N]$ on $\mathcal{H}$ which fulfil

$$\overline{a}^* = a, \tag{S}$$

where $\overline{a}$ denotes the complex conjugate of a and the element $a^* \in \mathbb{C}[\mathcal{S}_N]$ of $a = \sum_p a_p p$ is defined by $a^* = \sum_p a_p p^{-1}$. Such symmetry operators are *self-adjoint* because of the relation $\langle au|v\rangle = \langle u|\overline{a}^* v\rangle$ which holds true for all $a \in \mathbb{C}[\mathcal{S}_N]$ and $u, v \in \mathcal{H}$. They yield consequently *observables* of the quantum mechanical Heisenberg model. We prove the following results:

(i) Let $G \subseteq \mathcal{S}_N$ be an arbitrary subgroup and χ be an irreducible character of G then $\chi := \frac{\chi(\mathrm{id})}{|G|} \sum_{p \in G} \chi(p)\, p$ is an idempotent of $\mathbb{C}[\mathcal{S}_N]$ which has property (S). This leads to a big manifold of observables. In particular every commutation symmetry belonging to G yields a 1-dimensional character of G from which we can build one of the idempotents χ described here.

(ii) Let $\mathcal{R} \subseteq \mathbb{C}[\mathcal{S}_N]$ be a minimal right ideal of $\mathbb{C}[\mathcal{S}_N]$. Then the set of all generating (primitive) idempotents of $\mathcal{R}$ contains one and only one idempotent e which satisfies $\overline{e}^* = e$. This is a result by H. Weyl [2] for which we give a new proof.

(iii) Every idempotent e with property (S) can be decomposed into primitive idempotents $e = f_1 + \ldots + f_k$ which have also property (S) and satisfy $f_i \cdot f_j = 0$ if $i \neq j$. We give a computer algorithm for the calculation of such decompositions.

In big group rings $\mathbb{C}[\mathcal{S}_N]$ computer calculations are only possible by the use of a diskrete Fourier transform $D : \mathbb{C}[\mathcal{S}_N] \to \bigoplus_\lambda \mathbb{C}^{d_\lambda \times d_\lambda}$. We present two types of algorithms which allow the calculation of the matrix $B = D_\lambda(a^*)$ from a matrix $A = D_\lambda(a)$ of an $a \in \mathbb{C}[\mathcal{S}_N]$ without to determine the inverse Fourier transform of A. One of these algorithms uses as precomputed data permutation sets $\mathcal{P}_\lambda$ which define bases $\{p \cdot z_\lambda | p \in \mathcal{P}_\lambda\}$ of the minimal two-sided ideals $\mathcal{Z}_\lambda$ belonging to the partitions $\lambda \vdash N$. (z_λ is the central-primitive generating idempotent of $\mathcal{Z}_\lambda$). The sets $\mathcal{P}_\lambda$ can be used for any discrete Fourier transform D.

In our investigations we use computer calculations by means of the Mathematica packages PERMS and HRing.

1. The Heisenberg model, symmetry operators, symmetry classes, commutation symmetries

We summarize essential concepts of the one-dimensional Heisenberg model of a magnetic ring (see e.g. [3, 4, 5, 6] or [1]). First we define the Hilbert space of this model.

We denote by $\widehat{N}$ the set $\widehat{N} := \{1, \ldots, N\}$ of the integers $1, 2, \ldots, N$ and by $\widehat{K}^{\widehat{N}}$ the set of all functions $\sigma : \widehat{N} \to \widehat{K}$.

Definition 1.1 *We assign to every function $\sigma \in \widehat{K}^{\widehat{N}}$ the sequence $|\sigma\rangle := |\sigma(1), \sigma(2), \ldots, \sigma(N)\rangle$ of its values over the set $\widehat{N}$. Then the Hilbert space of a ring model with N nodes and a spin-alphabet of K letters (see [6]) is the set of all formal complex linear combinations $\mathcal{H} := \mathcal{L}_{\mathbb{C}}\left\{ |\sigma\rangle \mid \sigma \in \widehat{K}^{\widehat{N}} \right\}$ of the $|\sigma\rangle$ in which the set $\mathcal{B} := \left\{ |\sigma\rangle \mid \sigma \in \widehat{K}^{\widehat{N}} \right\}$ of all $|\sigma\rangle$ is considered a set of linearly independent elements. We equip $\mathcal{H}$ with the scalar product*

$$\langle\sigma|\sigma'\rangle := \delta_{\sigma,\sigma'} \quad and \quad \langle u|v\rangle := \sum_{\sigma\in\widehat{K}^{\widehat{N}}} u_\sigma \overline{v}_\sigma \quad for \quad u = \sum_{\sigma\in\widehat{K}^{\widehat{N}}} u_\sigma|\sigma\rangle\,,\ v = \sum_{\sigma\in\widehat{K}^{\widehat{N}}} v_\sigma|\sigma\rangle \in \mathcal{H}\,. \tag{1}$$

Obviously, $\mathcal{H}$ is an Hilbert space of dimension $\dim \mathcal{H} = K^N$, in which $\mathcal{B}$ is an orthonormal basis. Our concepts of *symmetry operators* and *symmetry classes* (defined below) can be used in all ring models which have such a Hilbert space.

In [1] we considered the example of the 1D spin-$\frac{1}{2}$ *Heisenberg ring*. It arises for $K = 2$, where $\sigma(k) = 2$ represents an *up spin* $\sigma(k) = \uparrow$ and $\sigma(k) = 1$ a *down spin* $\sigma(k) = \downarrow$ at site k. According to [3, 4], the Hamiltonians H_F (H_A) of a 1D spin-$\frac{1}{2}$ Heisenberg *ferromagnet* (*antiferromagnet*) of N sites with periodic boundary conditions $S^\alpha_{N+1} := S^\alpha_1$, $\alpha \in \{+, -, z\}$, are defined by

$$H_F := -J \sum_{k=1}^{N} \left[\frac{1}{2}\left(S^+_k S^-_{k+1} + S^-_k S^+_{k+1}\right) + S^z_k S^z_{k+1} \right], \qquad H_A := -H_F\,, \qquad J = \text{const.} > 0 \tag{2}$$

where S^+_k, S^-_k and S^z_k are *spin flip operators*.

Symmetry classes in $\mathcal{H}$ can be defined in analogy to the definition of symmetry classes of tensors (see [2, 7, 8, 9, 1]). First we define symmetry operators.

Definition 1.2 *Let $\mathcal{H}$ be the Hilbert space of a ring model of N sites and K letters, $\mathcal{S}_N$ be the symmetric group of the permutations of $1, 2, \ldots, N$ and $\mathbb{C}[\mathcal{S}_N]$ be the complex group ring of $\mathcal{S}_N$.*

(i) *Following [10, 11] we define the action of a permutation $p \in \mathcal{S}_N$ on a basis vector $|\sigma\rangle \in \mathcal{H}$ by $p|\sigma\rangle := |\sigma \circ p^{-1}\rangle$.*

(ii) *Every group ring element $a = \sum_p a_p p \in \mathbb{C}[\mathcal{S}_N]$ acts as so-called symmetry operator on the vectors $w = \sum_\sigma w_\sigma|\sigma\rangle \in \mathcal{H}$ by*

$$aw := \sum_{p\in\mathcal{S}_N} \sum_{\sigma\in\widehat{K}^{\widehat{N}}} a_p w_\sigma p|\sigma\rangle = \sum_{p\in\mathcal{S}_N} \sum_{\sigma\in\widehat{K}^{\widehat{N}}} a_p w_\sigma |\sigma \circ p^{-1}\rangle \tag{3}$$

A symmetry operator is a linear mapping $a : \mathcal{H} \to \mathcal{H}$ which possesses the following properties:

Proposition 1.3 ([1]) (i) *It holds $p(q|\sigma\rangle) = (p \circ q)|\sigma\rangle$ for all $p, q \in \mathcal{S}_N$, where "$\circ$" is defined by $(p \circ q)(i) := p(q(i))$.*

(ii) *It holds $a(bw) = (a \cdot b)w$ for all $a, b \in \mathbb{C}[\mathcal{S}_N]$ and $w \in \mathcal{H}$, where "$\cdot$" denotes the multiplication $a \cdot b := \sum_p \sum_q a_p b_q p \circ q$ of group ring elements.*

(iii) *It holds $\langle au|v\rangle = \langle u|\overline{a}^* v\rangle$ for all $a \in \mathbb{C}[\mathcal{S}_N]$ and $u, v \in \mathcal{H}$, where $\overline{a}$ denotes the complex conjugate of a and the element $a^* \in \mathbb{C}[\mathcal{S}_N]$ of $a = \sum_p a_p p$ is defined by $a^* = \sum_p a_p p^{-1}$.*

Now we define symmetry classes in $\mathcal{H}$.

Definition 1.4 *Let $\mathcal{R} \subseteq \mathbb{C}[\mathcal{S}_N]$ be a right ideal of $\mathbb{C}[\mathcal{S}_N]$. Then*

$$\mathcal{H}_{\mathcal{R}} := \{au \in \mathcal{H} \mid a \in \mathcal{R},\, u \in \mathcal{H}\} \tag{4}$$

is called the symmetry class of $\mathcal{H}$ defined by $\mathcal{R}$.

As for tensors one can prove (see e.g. [7, Chap. V, §4] or [12, p. 115]).

Proposition 1.5 *Let $e \in \mathbb{C}[\mathcal{S}_N]$ be a generating idempotent of a right ideal $\mathcal{R} \subseteq \mathbb{C}[\mathcal{S}_N]$, i.e. $\mathcal{R} = e \cdot \mathbb{C}[\mathcal{S}_N]$. Then a $u \in \mathcal{H}$ is in $\mathcal{H}_{\mathcal{R}}$ iff $eu = u$.*

Corollary 1.6 *If e, f are generating idempotents of a right ideal $\mathcal{R} \subseteq \mathbb{C}[\mathcal{S}_N]$, then $u \in \mathcal{H}_{\mathcal{R}}$ iff $eu = fu = u$.*

Definition 1.7 *We denote by $\mathcal{J}_0$ the set $\mathcal{J}_0 := \{a \in \mathbb{C}[\mathcal{S}_N] \mid au = 0 \; \forall u \in \mathcal{H}\}$.*

If $N > K$ then $\mathcal{J}_0 \neq \{0\}$, because then the idempotent $e := \frac{1}{N!}\sum_p \mathrm{sign}(p)p$ satisfies $e|\sigma\rangle = 0$ for all $|\sigma\rangle \in \mathcal{B}$.

Proposition 1.8 ([12, p. 116]) *$\mathcal{J}_0$ is a two-sided ideal of $\mathbb{C}[\mathcal{S}_N]$.*

Every two-sided ideal of $\mathbb{C}[\mathcal{S}_N]$ has one and only one generating idempotent which is central. Let f_0 be the generating idempotent of $\mathcal{J}_0$. Then $f := id - f_0$ is also a central idempotent which is orthogonal to f_0, i.e. $f \cdot f_0 = f_0 \cdot f = 0$, and which generates a two-sided ideal $\mathcal{J} := f \cdot \mathbb{C}[\mathcal{S}_N]$ that fulfills $\mathbb{C}[\mathcal{S}_N] = \mathcal{J} \oplus \mathcal{J}_0$. $\mathcal{J}$ contains all those symmetry operators a for which $\ker a \subset \mathcal{H}$. These are the symmetry operators in which we are interestet. (Compare [12, p. 116].)

A special type of symmetries are symmetries of elements $u \in \mathcal{H}$ with respect to commutations of the nodes of the ring.

Definition 1.9 *Let $C \subseteq \mathcal{S}_N$ be a subgroup of $\mathcal{S}_N$ and $\epsilon : C \to \mathcal{S}^1$ be a homomorphism of C onto a finite subgroup in the group $\mathcal{S}^1 = \{z \in \mathbb{C} \mid |z| = 1\}$ of complex units. We say that $u \in \mathcal{H}$ possesses the commutation symmetry (C, ϵ) if $cu = \epsilon(c)u$ for all $c \in C$.*

Proposition 1.10 ([12, p. 115]) *Let (C, ϵ) be a commutation symmetry.*

(i) *Then the group ring element $\epsilon := \frac{1}{|C|}\sum_{c \in C} \epsilon(c)c$ is an idempotent of $\mathbb{C}[C] \subseteq \mathbb{C}[\mathcal{S}_N]$.*

(ii) *A $u \in \mathcal{H}$ has the symmetry (C, ϵ) iff $\epsilon^* u = u$.*

Consequently, a commutation symmetry (C, ϵ) defines the symmetry class $\mathcal{H}_{\mathcal{R}}$ of the right ideal $\mathcal{R} = \epsilon^* \cdot \mathbb{C}[\mathcal{S}_N]$.

2. Self-adjoint symmetry operators

In the present paper we are interested in symmetry operators $a \in \mathbb{C}[\mathcal{S}_N]$ which are self-adjoint with respect to the scalar product $\langle \cdot | \cdot \rangle$ of $\mathcal{H}$. Such symmetry operators can lead to *observables* of the ring model. First we find

Proposition 2.1 *Two symmetry operators $a, b \in \mathbb{C}[\mathcal{S}_N]$ satisfy $\langle au|v\rangle = \langle u|bv\rangle$ for all $u, v \in \mathcal{H}$ iff $b = \overline{a}^* + c$ with $c \in \mathcal{J}_0$.*

Proof. The relations $\langle au|v\rangle = \langle u|\overline{a}^* v\rangle$ and $\langle au|v\rangle = \langle u|bv\rangle$ for all $u, v \in \mathcal{H}$ lead to $\langle u|(\overline{a}^* - b)v\rangle = 0$ for all $u, v \in \mathcal{H}$. From this we obtain $(\overline{a}^* - b)v = 0$ for all $v \in \mathcal{H}$. □

Lemma 2.2

(i) *If $a \in \mathcal{J}$, then $\overline{a}, a^*, \overline{a}^* \in \mathcal{J}$, too.*

(ii) *If $a \in \mathcal{J}_0$, then $\overline{a}, a^*, \overline{a}^* \in \mathcal{J}_0$, too.*

Proof. We show (ii). If $au = 0$ for all $u \in \mathcal{H}$, then $\overline{a}w = \overline{a}\,\overline{u} = \overline{au} = 0$ for all $w = \overline{u} \in \mathcal{H}$, i.e. $\overline{a} \in \mathcal{J}_0$. Further, the condition $au = 0$ for all $u \in \mathcal{H}$ leads to $\langle au|v\rangle = \langle u|\overline{a}^* v\rangle = 0$ for all $u, v \in \mathcal{H}$. Thus we obtain $\overline{a}^* v = 0$ for all $v \in \mathcal{H}$, i.e. $\overline{a}^* \in \mathcal{J}_0$. Finally, a combination of $\overline{a}, \overline{a}^* \in \mathcal{J}_0$ yields $a^* = \overline{\overline{a}^*} \in \mathcal{J}_0$. □

Now it follows from Proposition 2.1 and Lemma 2.2

Proposition 2.3 *A symmetry operator $a \in \mathcal{J}$ is self-adjoint iff $\overline{a}^* = a$.*

PROOF. According to Proposition 2.1 a is self-adjoint iff $a = \overline{a}^* + c$ with $c \in \mathcal{J}_0$. But since $a, \overline{a}^* \in \mathcal{J}$ we obtain also $c = a - \overline{a}^* \in \mathcal{J}$ such that $c = 0$. □

Definition 2.4 *We say that a symmetry operator $a \in \mathbb{C}[\mathcal{S}_N]$ has property* (S) *if a satisfies the condition*

$$\overline{a}^* = a\,. \tag{S}$$

Every $a \in \mathbb{C}[\mathcal{S}_N]$ with property (S) is self-adjoint. However, an arbitrary symmetry operator $a = b + \tilde{b} \in \mathbb{C}[\mathcal{S}_N]$ with $b \in \mathcal{J}$, $\tilde{b} \in \mathcal{J}_0$ has only to fulfil $\overline{b}^* = b$ to be a self-adjoint operator. We investigate in our paper operators $a \in \mathbb{C}[\mathcal{S}_N]$ which have property (S).

Now we search for classes of symmetry operators which possess property (S).

Remark 2.5 *Obviously, Young symmetrizers $y_t \in \mathbb{C}[\mathcal{S}_N]$ do not have property* (S) *in general. For instance, one can easily check that the Young symmetrizer $y_t = [1,2,3] + [2,1,3] - [3,1,2] - [3,2,1] \in \mathbb{C}[\mathcal{S}_3]$ of the standard tableau $t = \begin{smallmatrix} 1 & 2 \\ 3 & \end{smallmatrix}$ does not fulfil* (S).

The following theorem yields a big manifold of symmetry operators with property (S).

Theorem 2.6 *Let $G \subseteq \mathbb{C}[\mathcal{S}_N]$ be a subgroup and χ be an irreducible character of G. Then*[1]

$$\chi := \frac{\chi(\mathrm{id})}{|G|} \sum_{p \in G} \chi(p)\, p \tag{5}$$

is an idempotent which satisfies (S), *too.*

PROOF. The assertion follows from [12, Prop. II.1.47], $\chi(p^{-1}) = \overline{\chi(p)}$ and $\chi(\mathrm{id}) \in \mathbb{R}$. □

Corollary 2.7 *If (G, ϵ) is a commutation symmetry, then $\epsilon : G \to \mathbb{C}$ is a 1-dimensional character of G which is irreducible. Consequently, the idempotent $\epsilon := \frac{1}{|G|} \sum_{p \in G} \epsilon(p)\, p$ of every commutation symmetry has property* (S).

PROOF. The assertion follows from [12, Prop. II.1.48]. □

Already the commutation symmetries lead to a big manifold of symmetry operators with property (S). Let us consider the $\mathcal{S}_6$. A list of all commutation symmetries belonging to subgroups of $\mathcal{S}_N$ with $N \leq 6$ is given in [12, Appendix A.1]. The $\mathcal{S}_6$ possesses 55 conjugacy classes of subgroups $G \neq \{\mathrm{id}\}$ (solvable and non-solvable). The number of solvable subgroups $G \neq \{\mathrm{id}\}$ of $\mathcal{S}_6$ is equal to 1428 (see [12, III.2.2]). On all theses subgroups one can define between 1 and 8 commutation symmetries. This yields a big number of operators with property (S). However, if we use irreducible character for the construction of operators with property (S), the number of such operators becomes still bigger.

Theorem 2.8 *Let $\mathcal{R} \subset \mathbb{C}[\mathcal{S}_N]$ be a minimal right ideal with $\mathcal{R} \not\subseteq \mathcal{J}_0$. ($\mathcal{R}$ defines a symmetry class). Then the set of (primitive) generating idempotents of $\mathcal{R}$ contains one and only one idempotent f with property* (S). *f can be formed from an arbitrary generating idempotent e of $\mathcal{R}$ by*

$$f = \mu\, e \cdot \overline{e}^*\,, \quad \mu \in \mathbb{R}\,. \tag{6}$$

Remark 2.9 *A corresponding statement holds true for minimal left ideals $\mathcal{L} \subset \mathbb{C}[\mathcal{S}_N]$. For left ideals one has to replace* (6) *by $f = \mu\, \overline{e}^* \cdot e$.*

[1] $|G|$ denotes the cardinality of G.

Remark 2.10 *A statement similar to Theorem 2.8 was proven by H. Weyl in [2, p. 295]. We give here a new proof.*

Proof. Since $\mathcal{R}$ is minimal and $\mathcal{R} \not\subseteq \mathcal{J}_0$ we have $\mathcal{R} \subseteq \mathcal{J}$. Let e be a (primitive) generating idempotent of $\mathcal{R}$. Obviously, $h := e \cdot \overline{e}^*$ is a group ring element with property (S). (Note that $(a \cdot b)^* = b^* \cdot a^*$.) We show that h is essentially idempotent.

Since $e \in \mathcal{J}$ there exists an $u \in \mathcal{H}$ such that $eu \neq 0$. Using the scalar product we obtain $\langle eu|eu \rangle \neq 0$ and $\langle u|(\overline{e}^* \cdot e)u \rangle \neq 0$. Consequently, we have $(\overline{e}^* \cdot e)u \neq 0$ and $\overline{e}^* \cdot e \neq 0$. The relation $e \cdot \overline{e}^* \neq 0$ can be proved when we start a similar consideration from $\langle \overline{e}^* v | \overline{e}^* v \rangle$, where $v \in \mathcal{H}$ is an element with $\overline{e}^* v \neq 0$.

Now we consider $0 \neq \langle (\overline{e}^* \cdot e)u | (\overline{e}^* \cdot e)u \rangle = \langle eu | (e \cdot \overline{e}^* \cdot e)u \rangle$. This leads to $e \cdot \overline{e}^* \cdot e \neq 0$. If e is a primitive idempotent then $e \cdot x \cdot e$ is proportional to e for all $x \in \mathbb{C}[\mathcal{S}_N]$. Since $e \cdot \overline{e}^* \cdot e \neq 0$ we obtain $e \cdot \overline{e}^* \cdot e = \alpha\, e$ with $\alpha \neq 0$ and $h \cdot h = e \cdot \overline{e}^* \cdot e \cdot \overline{e}^* = \alpha (e \cdot \overline{e}^*) = \alpha\, h$. Thus h ist essentially idempotent.

Next we show that $\alpha \in \mathbb{R}$. From $e \cdot \overline{e}^* \cdot e = \alpha e$ it follows $\overline{e}^* \cdot e \cdot \overline{e}^* = \overline{\alpha}\, \overline{e}^*$. If we multiply this relation from the left by e we obtain $\alpha(\overline{e}^* \cdot e) = \overline{\alpha}(\overline{e}^* \cdot e)$ and $\alpha = \overline{\alpha}$. So $f := \frac{1}{\alpha} h$ is an idempotent which has property (S) because $\alpha \in \mathbb{R}$. Finally $\mathcal{R}' := f \cdot \mathbb{C}[\mathcal{S}_N] = (e \cdot \overline{e}^*) \cdot \mathbb{C}[\mathcal{S}_N]$ is a right ideal with $\mathcal{R}' \subseteq \mathcal{R}$. Since $\mathcal{R}' \ni e \cdot \overline{e}^* \neq 0$ and $\mathcal{R}$ is minimal we obtain $\mathcal{R}' = \mathcal{R}$, i.e. f generates $\mathcal{R}$.

Now we show the uniqueness of f. Assume $\mathcal{R}$ possesses two generating idempotents f_1, f_2 with property (S). We denote by $\mathcal{H}_{\mathcal{R}}$ the symmetry class defined be $\mathcal{R}$ and by $\mathcal{H}_{\mathcal{R}}^{\perp}$ the orthogonal complement of $\mathcal{H}_{\mathcal{R}}$ with respect to $\langle \cdot | \cdot \rangle$. Every idempotent f_i satisfies $f_i x = x$ for all $x \in \mathcal{H}_{\mathcal{R}}$. Since every $x \in \mathcal{H}_{\mathcal{R}}$ has a structure $x = f_i z$ with $z \in \mathcal{H}$ we obtain $0 = \langle y | f_i z \rangle = \langle \overline{f_i}^* y | z \rangle = \langle f_i y | z \rangle$ for all $y \in \mathcal{H}_{\mathcal{R}}^{\perp}$ and all $z \in \mathcal{H}$. This leads to $f_i y = 0$ for all $y \in \mathcal{H}_{\mathcal{R}}^{\perp}$. We see that both idempotents f_1 and f_2 have the same effect on $\mathcal{H}_{\mathcal{R}}$ and $\mathcal{H}_{\mathcal{R}}^{\perp}$. Consequently, we have $f_1 z = f_2 z$ for all $z \in \mathcal{H}$, from which follows $f_1 - f_2 \in \mathcal{J}_0$. But since $\mathcal{R} \subseteq \mathcal{J}$, f_1 and f_2 lie in $\mathcal{J}$ and $f_1 - f_2 = 0$. □

3. Decomposition of self-adjoint idempotents

In this section we present an algorithm for the decomposition of an idempotent e with property (S) into pairwise orthogonal, primitive idempotents f_i which have property (S), too.

Theorem 3.1 *Every idempotent $e \in \mathcal{J}$ with $\overline{e}^* = e$ has a decomposition*

$$e = f_1 + f_2 + \ldots + f_l \tag{7}$$

into primitive idempotents $f_1, \ldots f_l$ which fulfill

$$\overline{f_i}^* = f_i \quad \textit{and} \quad f_i \cdot f_j = 0\,, \ \ i \neq j\,. \tag{8}$$

Proof. The idempotent e generates a right ideal $\mathcal{R} = e \cdot \mathbb{C}[\mathcal{S}_N] \subseteq \mathcal{J}$. The decomposition (7) can be constructed by the following

Algorithm 3.2

(i) *Form a set $\mathcal{Y}$ of primitive idempotents $y \in \mathbb{C}[\mathcal{S}_N]$ such that $\mathbb{C}[\mathcal{S}_N] = \bigoplus_{y \in \mathcal{Y}} y \cdot \mathbb{C}[\mathcal{S}_N]$.*

(ii) *Determine a $y_1 \in \mathcal{Y}$ such that $e \cdot y_1 \neq 0$. The group ring element $e \cdot y_1$ generates a minimal right ideal $\mathcal{R}_1 := e \cdot y_1 \cdot \mathbb{C}[\mathcal{S}_N]$ with $\mathcal{R}_1 \subseteq \mathcal{R}$.*

(iii) *Determine a primitive generating idempotent h_1 of $\mathcal{R}_1$ from $e \cdot y_1$.*

(iv) *Calculate the unique generating idempotent $f_1 = \mu_1\, h_1 \cdot \overline{h_1}^*$ of $\mathcal{R}_1$.*

(v) *Calculate the rest $r_1 := e - f_1$.*

(vi) *Determine further idempotents $f_2, f_3, \ldots$ by iteration of the steps (ii), ..., (v) starting with the rests $r_1, r_2, \ldots$ instead of e. Stop this iteration when a rest $r_l = 0$ was reached.*

The set $\mathcal{Y}$ could be constracted from the Young symmetrizers of all standard tableaux of $\mathbb{C}[\mathcal{S}_N]$. In step (ii) the set $\mathcal{Y}$ is always the original set $\mathcal{Y}$. We do not delete from $\mathcal{Y}$ the elements $y_1, y_2, \ldots$ found in previous steps (ii). Step (iii) can be carried out by a procedure given in [12, Prop.I.2.1], [13, Prop.1] or [9, Prop.4.1]

Now we show that Algorithm 3.2 produces the decomposition (7). Obviously, the rest $r_1 = e - f_1$ has property (S). Further it follows $e \cdot f_1 = f_1$ from $\mathcal{R}_1 \subseteq \mathcal{R}$. This leads to $r_1 \cdot f_1 = 0$. Now we can carry out the calculation

$$\forall\, x, y \in \mathcal{H}:\ 0 = \langle x | (r_1 \cdot f_1) y \rangle = \langle r_1 x | f_1 y \rangle = \langle (f_1 \cdot r_1) x | y \rangle\,, \tag{9}$$

which yield $(f_1 \cdot r_1)x = 0$ (for all $x \in \mathcal{H}$) and $f_1 \cdot r_1 = 0$ since $f_1, r_1 \in \mathcal{J}$. From $f_1 \cdot r_1 = 0$ we obtain $r_1 \cdot r_1 = e \cdot r_1 - f_1 \cdot r_1 = r_1 - 0 = r_1$, i.e. r_1 is an idempotent with property (S). Thus the start of a second iteration of (ii), ..., (v) with r_1 instead of e is correct. The result of this iteration are idempotents f_2, r_2 which fulfil $r_1 = f_2 + r_2$, $r_1 \cdot f_2 = f_2$.

Consider the symmetry classes $\mathcal{H}_{\mathcal{R}_1}$, $\mathcal{H}_{\mathcal{R}_2}$, $\mathcal{H}_{\mathcal{R}'_1}$ of the right ideals $\mathcal{R}_1 = f_1 \cdot \mathbb{C}[\mathcal{S}_N]$, $\mathcal{R}_2 = f_2 \cdot \mathbb{C}[\mathcal{S}_N]$, $\mathcal{R}'_1 = r_1 \cdot \mathbb{C}[\mathcal{S}_N]$. We obtain $\mathcal{R}_1 \perp \mathcal{R}'_1$ from (9) and $\mathcal{R}_2 \subseteq \mathcal{R}'_1$ from $r_1 \cdot f_2 = f_2$. Consequently, it holds $\mathcal{R}_1 \perp \mathcal{R}_2$, too. Now a calculation similar to (9) with f_2 instead of r_1 yields $f_1 \cdot f_2 = f_2 \cdot f_1 = 0$.

Since these considerations can be carried out in all iteration steps of Algorithm 3.2 Theorem 3.1 is correct. □

4. Use of discrete Fourier transforms

Computer calculations in big group rings $\mathbb{C}[\mathcal{S}_N]$ have high costs in calculation time and memory because of the high number of permutations in $\mathcal{S}_N$. One has to use a *discrete Fourier transform* for $\mathcal{S}_N$ to reduse these costs. (See [14] for discrete Fourier transforms of groups.)

Definition 4.1 *A discrete Fourier transform for $\mathcal{S}_N$ is an isomorphism*

$$D:\ \mathbb{C}[\mathcal{S}_N] \to \bigotimes_{\lambda \vdash N} \mathbb{C}^{d_\lambda \times d_\lambda}\ ,\quad D:\ a = \sum_{p \in \mathcal{S}_N} a_p\, p \mapsto D(a) = \begin{pmatrix} A_{\lambda_1} & 0 & 0 & 0 \\ 0 & A_{\lambda_2} & 0 & 0 \\ \vdots & \vdots & \ddots & \vdots \\ 0 & 0 & 0 & A_{\lambda_l} \end{pmatrix} \tag{10}$$

according to Wedderburn's theorem which maps the group ring $\mathbb{C}[\mathcal{S}_N]$ onto an outer direct product $\bigotimes_{\lambda \vdash N} \mathbb{C}^{d_\lambda \times d_\lambda}$ of full matrix rings $\mathbb{C}^{d_\lambda \times d_\lambda}$. We denote by D_λ the natural projection $D_\lambda : \mathbb{C}[\mathcal{S}_N] \to \mathbb{C}^{d_\lambda \times d_\lambda}$, $D_\lambda(a) = A_\lambda$, belonging to the partition $\lambda \vdash N$ of N.

The dimension d_λ of the $d_\lambda \times d_\lambda$-block matrices A_λ can be computed from λ by means of the *hook length formula* (see e.g. [15, 16, 17, 18] or [12, p.38]). According to [14, p.81] the *inverse Fourier transform* for (10) is given by formula

$$\forall\, p \in \mathcal{S}_N:\ a_p \ = \ \frac{1}{N!} \sum_{\lambda \vdash N} d_\lambda \mathrm{trace}\left\{ D_\lambda(p^{-1}) \cdot A_\lambda \right\}\,. \tag{11}$$

The group ring $\mathbb{C}[\mathcal{S}_N]$ decomposes into *minimal two-sided ideals* $\mathcal{Z}_\lambda$,

$$\mathbb{C}[\mathcal{S}_N] = \bigoplus_{\lambda \vdash N} \mathcal{Z}_\lambda = \bigoplus_{\lambda \vdash N} \mathbb{C}[\mathcal{S}_N] \cdot z_\lambda\,. \tag{12}$$

Every $\mathcal{Z}_\lambda$ is generated by a unique idempotent z_λ which is *centrally primitive*. z_λ can be calculated by means of (5) from the irreducible character χ_λ belonging to $\lambda \vdash N$. Consequently every z_λ has property (S).

The structure of block matrices in (10) reflects the decomposition (12) since $D_\lambda : \mathcal{Z}_\lambda \to \mathbb{C}^{d_\lambda \times d_\lambda}$ is an isomorphism.

Elements $a \in \mathbb{C}[\mathcal{S}_N]$ which lie in minimal left, right or two-sided ideals of $\mathbb{C}[\mathcal{S}_N]$, have only one non-vanishing block matrix in (10). In particular this holds true for all primitive idempotents. The product $a \cdot b$ of such an element a by another element b is in $\mathbb{C}[\mathcal{S}_N]$ the product of two long sums $a = \sum_p a_p p$, $b = \sum_p b_p p$ of length $N!$, however in $\bigotimes_{\lambda\vdash N} \mathbb{C}^{d_\lambda \times d_\lambda}$ only the product of two $d_\lambda \times d_\lambda$-matrices $A_\lambda \cdot B_\lambda$ which are much smaller. This leads to a considerable reduction of computation costs.

The steps (i)-(iii) of Algorithm 3.2 were taken from another ideal decomposition algorithm which we described in [12, Chap.I.2] or [13, 9]. The algorithm from [12, 13, 9] runs both in $\mathbb{C}[\mathcal{S}_N]$ and in $\bigotimes_{\lambda\vdash N} \mathbb{C}^{d_\lambda \times d_\lambda}$ without a necessity of a Fourier transformation between these two spaces during the run of the algorithm. The version in $\bigotimes_{\lambda\vdash N} \mathbb{C}^{d_\lambda \times d_\lambda}$ is much more efficient than the version in $\mathbb{C}[\mathcal{S}_N]$.

Because of the relationship between Algorithm 3.2 and the algorithm from [12, 13, 9] one can transfer Algorithm 3.2 to $\bigotimes_{\lambda\vdash N} \mathbb{C}^{d_\lambda \times d_\lambda}$ by means of the remarks in [12, p.46], [13, Sec.5], [9, pp.9-10].

Only the computation of a^* for an element $a \in \mathbb{C}[\mathcal{S}_N]$ can not be transfered to $\bigotimes_{\lambda\vdash N} \mathbb{C}^{d_\lambda \times d_\lambda}$ in a simple way. The calculation of the matrix $B = D(a^*)$ from a matrix $A = D(a)$ can not be carried out in $\bigotimes_{\lambda\vdash N} \mathbb{C}^{d_\lambda \times d_\lambda}$ without Fourier transformations between $\bigotimes_{\lambda\vdash N} \mathbb{C}^{d_\lambda \times d_\lambda}$ and $\mathbb{C}[\mathcal{S}_N]$. In the next sections we present algorithms for the efficient computation of B from A.

Remark 4.2 *For $\mathcal{S}_N$ three discrete Fourier transforms are known:*

(a) *Young's natural representation of $\mathcal{S}_N$ (see* [12, p. 51]*),*

(b) *Young's semi-normal representation of $\mathcal{S}_N$ (see* [12, p. 55]*),*

(c) *Young's orthogonal representation of $\mathcal{S}_N$ (see* [7, pp. 133-135]*).*

The DFT (b) *is the basis of the fast Fourier transform for $\mathcal{S}_N$ by M. Clausen and U. Baum* [14]*. We use* (a) *as discrete Fourier transform in our* Mathematica *package* PERMS [19]*.*

The three DFT *map permutations $p \in \mathcal{S}_N$ to following types of matrices:*

	(a)	(b)	(c)
$D(p)$	integer matrix	rational matrix	real matrix

If we want to calculate $D(a)$ of an $a \in \mathbb{C}[\mathcal{S}_N]$ whose coefficients are only integers or rational numbers then (a) *and* (b) *allow calculations not only with a certain numerical precision but with infinite precision. In Section* 5 *we will see that an infinite precision has not an advantage in any case. If a calculation lead to such big integers which a computer program can not completely decompose into primes than a strong growth of integers can arise and cause memory problems.*

5. Calculation of $D(a^*)$ from $D(a)$

Now we investigate the question how one can calculate the matrix $B = D(a^*) \in \bigotimes_{\lambda\vdash N} \mathbb{C}^{d_\lambda \times d_\lambda}$ from the matrix $A = D(a) \in \bigotimes_{\lambda\vdash N} \mathbb{C}^{d_\lambda \times d_\lambda}$ without a discrete Fourier transformation between $\bigotimes_{\lambda\vdash N} \mathbb{C}^{d_\lambda \times d_\lambda}$ and $\mathbb{C}[\mathcal{S}_N]$.

Proposition 5.1 *Every $a \in \mathbb{C}[\mathcal{S}_N]$ has a unique decomposition*

$$a = \sum_{\lambda\vdash N} a_\lambda \quad \textit{where} \quad a_\lambda = a \cdot z_\lambda \in \mathcal{Z}_\lambda \,. \tag{13}$$

The a_λ fulfil

$$a^*_\lambda = a^* \cdot z_\lambda \in \mathcal{Z}_\lambda \,. \tag{14}$$

PROOF. (13) follows from (12). Since z_λ has property (S) and commutes with every element from $\mathbb{C}[\mathcal{S}_N]$ we have furthermore $a_\lambda^* = z_\lambda^* \cdot a^* = a^* \cdot z_\lambda$. □

Because of Proposition 5.1 we can restrict us to

Problem 5.2 *Determine a matrix-valued function* $f : \mathbb{C}^{d_\lambda \times d_\lambda} \to \mathbb{C}^{d_\lambda \times d_\lambda}$ *which fulfils*

$$f(A) = B \quad \Leftrightarrow \quad \exists a \in \mathcal{Z}_\lambda : \; A = D_\lambda(a) \, \wedge \, B = D_\lambda(a^*) \tag{15}$$

Obviously, a function (15) exists for every $\mathcal{Z}_\lambda$. Every such f is a linear function defined on the whole of $\mathbb{C}^{d_\lambda \times d_\lambda}$. If we can determine f we obtain a fast algorithm for the computation of B from A without a Fourier transformation between $\bigotimes_{\lambda \vdash N} \mathbb{C}^{d_\lambda \times d_\lambda}$ and $\mathbb{C}[\mathcal{S}_N]$.

The simplest algorithm which gives a solution of Problem 5.2 is

Algorithm 5.3 *Let be given a* $(d_\lambda \times d_\lambda)$*-matrix* A *of symbols representing a natural projection* $D_\lambda(a)$ *of an arbitrary element* $a \in \mathcal{Z}_\lambda$.

(i) *Calculate* $a = (D_\lambda|_{\mathcal{Z}_\lambda})^{-1}(A)$.

(ii) *Calculate* a^* *from* a.

(iii) *Calculate* $B = D_\lambda(a^*)$.

B is a $(d_\lambda \times d_\lambda)$-matrix of linear functions of the elements of A. The mapping $f : A \mapsto B$ is the solution of Problem 5.2.

Algorithm 5.3 needs a computer algebra system such as Mathematica which allows symbolic computations. It can lead to high costs in calculation time because it carries out a Fourier transformations of a relatively small $(d_\lambda \times d_\lambda)$-matrix A to a long group ring element a of length $N!$ and a second Fourier transformation of a group ring element a^* of length $N!$ to a $(d_\lambda \times d_\lambda)$-matrix B. However these costs arise only once when we run Algorithm 5.3. When we have determined the function f then we can very fast compute B from A for all $A \in \mathbb{C}^{d_\lambda \times d_\lambda}$.

Formula (11) for the inverse discrete Fourier transform allows the following modification of Algorithm 5.3 which reduces the costs in memory once more.

Algorithm 5.4 *Let A and B have the same meaning as in Algorithm* 5.3.

(o) *Determine a decomposition* $\mathcal{S}_N = \bigcup_{s=1}^t P_s$ *of* $\mathcal{S}_N$ *into pairwise disjoint subsets* P_s. *Put* $B = 0$. *Then carry out the following steps for every* $s = 1, \ldots t$:

(i) *Calculate* $a_{(s)} = (D_\lambda|_{\mathcal{Z}_\lambda})^{-1}(A)|_{P_s} = \sum_{p \in P_s} a_p \, p$.

(ii) *Calculate* $a_{(s)}^*$ *from* $a_{(s)}$.

(iii) *Calculate* $B_{(s)} = D_\lambda(a_{(s)}^*)$.

(iv) *Calculate* $B := B + B_{(s)}$.

The steps (i), (ii), (iii) in Algorithm 5.4 need less memory than the corresponding steps in Algorithm 5.3 since Algorithm 5.4 works with short group ring elements $a_{(s)}$ with support P_s.

The costs in memory for Algorithm 5.3 and 5.4 are smaller than for the other algorithms presented here. Using Algorithm 5.3 or 5.4 one runs rarely the risk of memory problems.

A second type of algorithms solving Problem 5.2 bases on the following consideration. For every $\mathcal{Z}_\lambda$ the set $\{p \cdot z_\lambda \mid p \in \mathcal{S}_N\}$ is a generating system of the vector space $\mathcal{Z}_\lambda$. We can determine a subset $\mathcal{P}_\lambda \subset \mathcal{S}_N$ of permutations such that $\mathcal{B}_\lambda := \{p \cdot z_\lambda \mid p \in \mathcal{P}_\lambda\}$ is a basis of $\mathcal{Z}_\lambda$. If $\mathcal{P}_\lambda$ is known we can write every $a \in \mathcal{Z}_\lambda$ in the following form

$$a = a \cdot z_\lambda = \sum_{p \in \mathcal{S}_N} a_p \, p \cdot z_\lambda = \sum_{p \in \mathcal{P}_\lambda} \alpha_p \, p \cdot z_\lambda = \alpha \cdot z_\lambda \quad \text{with} \quad \alpha := \sum_{p \in \mathcal{P}_\lambda} \alpha_p \, p \,. \tag{16}$$

It holds true for a, α and z_λ

$$D_\lambda(a) = D_\lambda(\alpha)\,, \quad D_\lambda(z_\lambda) = \mathrm{Id}\,, \quad a^* = z_\lambda^* \cdot \alpha^* = \alpha^* \cdot z_\lambda\,. \tag{17}$$

Definition 5.5 (i) *Let $\alpha = \sum_{p\in\mathcal{P}_\lambda} \alpha_p\, p$ be a group ring element. We denote by $v[\alpha_p]$ the column vector in which the α_p were arranged according to the lexicographical order of permutations.*

(ii) *Let $A = (A_{ij})$ be a matrix. We denote by $v[A_{ij}]$ the column vector in which the elements A_{ij} were written line by line.*

Now we formulate our second algorithm which solves Problem 5.2.

Algorithm 5.6 *Let $A = (A_{ij})$ be a symbolic $(d_\lambda \times d_\lambda)$-matrix and $\alpha = \sum_{p\in\mathcal{P}_\lambda} \alpha_p\, p$ be a group ring element with symbolic coefficients α_p.*

(i) *Determine the coefficient matrix Φ of the equation*

$$v\left[D_\lambda(\alpha)\right] = \Phi \cdot v[\alpha_p]\,. \tag{18}$$

(ii) *Calculate the inverse matrix Φ^{-1}.*

(iii) *Determine the coefficient matrix Ψ of the equation*

$$v\left[D_\lambda(\alpha^*)\right] = \Psi \cdot v[\alpha_p]\,. \tag{19}$$

(iv) *Form the function*

$$\tilde{f} : v[A_{ij}] \mapsto \Psi \cdot \Phi^{-1} \cdot v[A_{ij}]\,. \tag{20}$$

Arrange the elements of the vectors $v[A_{ij}]$ and $\Psi\cdot\Phi^{-1}\cdot v[A_{ij}]$ in $(d_\lambda\times d_\lambda)$-matrices to obtain the function $f : \mathbb{C}^{d_\lambda\times d_\lambda} \to \mathbb{C}^{d_\lambda\times d_\lambda}$.

It is clear that Algorithm 5.6 yields the function f searched in Problem 5.2. Note that $\mathcal{P}_\lambda = \dim \mathcal{Z}_\lambda = d_\lambda^2$. So both vector $v[\alpha_p]$ and vector $v[A_{ij}]$ has length d_λ^2 and Φ and Ψ are quadratic $(d_\lambda^2 \times d_\lambda^2)$-matrices. Φ has an inverse matrix because $\mathcal{B}_\lambda$ is a basis of $\mathcal{Z}_\lambda$ and $D_\lambda|_{\mathcal{Z}_\lambda} : \mathcal{Z}_\lambda \to \mathbb{C}^{d_\lambda\times d_\lambda}$ is an isomorphism. The use of $v[\alpha_p]$ in the right-hand side is correct because α and α^* have the same coefficients. Only the places of these coefficients are different in α and α^*.

By means of Algorithm 5.6 we calculated the functions f for all partitions $\lambda \vdash 5$, $\lambda \vdash 6$ and all partitions $\lambda \vdash 7$, $\lambda \neq (4\,2\,1), (3\,2\,1^2)$. The following table shows some information concerning the calculations in $\mathcal{S}_7$.

λ	d_λ	d_λ^2	d_λ^4	time	RAM	f
$(6\,1)$	6	36	1296	0.6s	14.6MB	13KB
$(5\,2)$	14	196	38416	28.6s	33.6MB	506KB
$(5\,1^2)$	15	225	50625	31.2s	34.9MB	182KB
$(4\,3)$	14	196	38416	35.9s	34.7MB	494KB
$(4\,1^3)$	20	400	160000	278.8s	131.7MB	404KB
$(3^3\,1)$	21	441	194481	615.5s	166.9MB	2149KB
$(3\,2^2)$	21	441	194481	624.3s	166.9MB	2187KB
$(3\,1^4)$	15	225	50625	31.3s	34.6MB	182KB
$(2^3\,1)$	14	196	38416	36.1s	34.8MB	471KB
$(2^2\,1^3)$	14	196	38416	30.7s	33.6MB	433KB
$(2\,1^5)$	6	36	1296	0.6s	14.7MB	13KB

The colums of d_λ^2 and d_λ^4 shows the number of elements in $v[\alpha_p]$, $v[A_{ij}]$ and Φ, Ψ, respectively. 'time' and 'RAM' give the computation time and the memory, respectively, which the algorithm needed. 'f' shows the length of the file in which the function f was stored.

For the partitions $\lambda = (4\,2\,1), (3\,2\,1^2)$ a big memory problem arose from the use of exact integers in the calculation. It turned out, that for instance the matrix Φ for $\lambda = (4\,2\,1)$ has the determinant

-96974257961965049074024912099372004573795424708044922073363953457157880990336620245014730563217365
946027469537574499367764896357932972629577610678133530230415476788046519825483493859992633559136
303784510251434033116082118775724498148053570956977145676159650656795201937451001664676330006575
437913887075107664190746396865109859747118150041195008145445396420723507449624828927005702309095
274295858496179822968799982444105484313810897703096007568025010485731184828349379597074706284167
453662581507561255480875750071503321884946609362947141428351638033559982301650092032000000000000
000000000000000
(593 digits)

During the calculation of the inverse matrix Φ^{-1} rational numbers are generated whose numerators and denominators are such large integers as the above determinat. The computer algebra system is no longer able to find all prime factors of these integers and to reduce the rational numbers. So the integers become longer and longer and the available memory is exceeded.

We used Algorithm 5.4 for the determination of f for $\lambda = (4\,2\,1), (3\,2\,1^2)$. To this end we decomposed $\mathcal{S}_7$ into 9 subsets P_s of 560 permutations. Then the calculation had the following characteristics:

λ	d_λ	d_λ^2	d_λ^4	time	RAM	f
$(4\,2\,1)$	35	1225	1500625	4.393h	440.8MB	13974KB
$(3\,2\,1^2)$	35	1225	1500625	4.503h	423.6MB	13486KB

6. Determination of the sets $\mathcal{P}_\lambda$

In this section we present an algorithm which determines the sets $\mathcal{P}_\lambda$ for the Algorithm 5.6 (see Figure 1).

We can not simply consider the set $\{p \cdot z_\lambda \mid p \in \mathcal{S}_N\}$ und determine $\dim \mathcal{Z}_\lambda = d_\lambda^2$ linearly independent vectors in this set, because this set contains $N!$ vectors from $\mathbb{C}[\mathcal{S}_N]$ of length $N!$. This data set is very large.

The algorithm in Figure 1 calculates

$$D_\lambda(p \cdot z_\lambda) = D_\lambda(p) \tag{21}$$

for permutations $p \in \mathcal{S}_N$ and checks whether these matrices are linearly independent.

The algorithm starts with the identity permutation $p = \mathrm{id} \in \mathcal{S}_N$ and $\mathcal{P}_\lambda = \{\mathrm{id}\}$. To characterize the 1-dimensional linear space spanned by $D_\lambda(\mathrm{id})$, the algorithm uses a vector $\xi = (x_1, \ldots, x_{d_\lambda^2})$ of variables x_i.

The algorithm forms the scalar product $\tau(\xi) = \xi \cdot v[D_\lambda(\mathrm{id})]$ of the vectors $\tau(\xi)$ and $[D_\lambda(\mathrm{id})]$. $\tau(\xi)$ is a linear function. The algorithm determines one of the variables x_i from the equation $\tau(\xi) = 0$ and obtains a relation $x_s = \sigma(\text{rest of the } x_i)$. By means of this relation the variable x_s is eliminated in ξ. We obtain a vector ξ which fulfil $\xi \cdot w = 0$ exactly for the elements w of the 1-dimensional span of $v[D_\lambda(\mathrm{id})]$.

Now the algorithm investigates the next permutation p. To select p, the algorithm uses a procedure **next-permutation** which determines for a given permutation $q \in \mathcal{S}_N$ the successor $p \in \mathcal{S}_N$ according to the lexicographical order of permutations. Then the algorithm checks the condition

$$\tau(\xi) := \xi \cdot v[D_\lambda(p)] = 0\,. \tag{22}$$

If (22) is satisfied, then p is canceled. If, however, (22) is not fulfilled then a new linear equation $\tau(\xi) = 0$ arises which the algorithm uses to eliminate a further variable in ξ. ξ caracterizes now the span of two linearly independent vectors $v[D_\lambda(\mathrm{id})]$, $v[D_\lambda(p)]$. The new found permutation p ist added to the set $\mathcal{P}_\lambda$. This procedure is iterated until the cardinality of $\mathcal{P}_\lambda$ has reached the value d_λ^2.

The algorithm needs not much memory because only one matrix $D_\lambda(p)$ is stored in the computer memory in every step of the algorithm.

```
input:  λ ⊢ N
output: P_λ

begin
    ξ := (x_1, x_2, ..., x_{d_λ^2});
    p := id;
    P_λ := {id};
    τ(ξ) := ξ · v[D_λ(p)];
    solve τ(ξ) = 0  ⇒ result: x_s = σ(rest of the x_i);
    ξ := ξ ← substitute x_s = σ(rest of the x_i);
    while |P_λ| < d_λ^2 do
        p := next-permutation(p);
        τ(ξ) := ξ · v[D_λ(p)];
        if    τ(ξ) ≢ 0
        then solve τ(ξ) = 0  ⇒ result: x_s = σ(rest of the x_i);
             ξ := ξ ← substitute x_s = σ(rest of the x_i);
             P_λ := P_λ ∪ {p};
        fi;
    od;
    return P_λ;
end;
```

Figure 1. Algorithm for determining of $\mathcal{P}_\lambda$

Since the property of $\mathcal{B}_\lambda = \{p \cdot z_\lambda \mid p \in \mathcal{P}_\lambda\}$ of being a basis of $\mathcal{Z}_\lambda$ is independent on the choice of a discrete Fourier transform D for our algorithm, we obtain immediately

Proposition 6.1 *A set $\mathcal{P}_\lambda$ determined by the algorithm in Figure 1 can be used for Algorithm 5.6 under every discrete Fourier transform D.*

A halving of the calculation time of our algorithm can be obtained by the following theorem.

Theorem 6.2 *Let a set $\mathcal{P}_\lambda$ be known for a partition $\lambda \vdash N$. Then $\mathcal{B}_{\lambda'} := \{p \cdot z_{\lambda'} \mid p \in \mathcal{P}_\lambda\}$ is a basis of $\mathcal{Z}_{\lambda'}$, where $\lambda' \vdash N$ denotes that partition whose Young frame is transposed to the Young frame of λ.*

Proof. The generating idempotents z_λ, $z_{\lambda'}$ of $\mathcal{Z}_\lambda$, $\mathcal{Z}_{\lambda'}$ fulfil

$$z_\lambda := \frac{\chi_\lambda(\mathrm{id})}{N!} \sum_{p \in \mathcal{S}_N} \chi_\lambda(p)\, p\,, \quad z_{\lambda'} := \frac{\chi_{\lambda'}(\mathrm{id})}{N!} \sum_{p \in \mathcal{S}_N} \chi_{\lambda'}(p)\, p\,, \tag{23}$$

where χ_λ, $\chi_{\lambda'}$ denote the irreducible characters of $\mathcal{S}_n$ belonging to λ, λ'. It holds true $\chi_\lambda(\mathrm{id}) = \chi_{\lambda'}(\mathrm{id})$, because $\chi_\lambda(\mathrm{id}) = \sqrt{\dim \mathcal{Z}_\lambda}$, $\chi_{\lambda'}(\mathrm{id}) = \sqrt{\dim \mathcal{Z}_{\lambda'}}$ and $\dim \mathcal{Z}_\lambda = \dim \mathcal{Z}_{\lambda'}$. Further, it follows from [18, Vol. 246, p. 65, Statement 4.11], that

$$\forall\, p \in \mathcal{S}_N : \ \chi_{\lambda'}(p) = \mathrm{sign}(p) \cdot \chi_\lambda(p)\,. \tag{24}$$

Because of these facts we can write

$$z_\lambda = g_\lambda + u_\lambda \qquad z_{\lambda'} = g_\lambda - u_\lambda \,, \tag{25}$$

where

$$g_\lambda := \frac{\chi_\lambda(\mathrm{id})}{N!} \sum_{p\ \mathrm{even}} \chi_\lambda(p)\, p \,, \qquad u_\lambda := \frac{\chi_\lambda(\mathrm{id})}{N!} \sum_{p\ \mathrm{odd}} \chi_\lambda(p)\, p \,. \tag{26}$$

Now we assume that $\mathcal{B}_\lambda = \{p \cdot z_\lambda \mid p \in \mathcal{P}_\lambda\}$ is a basis of $\mathcal{Z}_\lambda$, but $\mathcal{B}_{\lambda'} := \{p \cdot z_{\lambda'} \mid p \in \mathcal{P}_\lambda\}$ is a set of linearly dependent vectors from $\mathcal{Z}_{\lambda'}$. Consequently, we can form a non-trivial vanishing linear combination of the vectors from $\mathcal{B}_{\lambda'}$.

$$\sum_{p\in\mathcal{P}_\lambda} \alpha_p\, p \cdot z_{\lambda'} = 0 \,. \tag{27}$$

Now the element $\alpha := \sum_{p\in\mathcal{P}_\lambda} \alpha_p\, p$ has a decomposition $\alpha = \gamma + \omega$, where γ and ω are sums which run only through the even or odd permutations of $\mathcal{P}_\lambda$, respectively. Now we obtain

$$0 = \alpha \cdot z_{\lambda'} = (\gamma + \omega) \cdot (g_\lambda - u_\lambda) = \underbrace{(\gamma \cdot g_\lambda - \omega \cdot u_\lambda)}_{=:\Gamma} + \underbrace{(\omega \cdot g_\lambda - \gamma \cdot u_\lambda)}_{=:\Omega} \,. \tag{28}$$

It follows from (28) that

$$\Gamma = 0 \,, \quad \Omega = 0 \,, \tag{29}$$

since Γ and Ω contain only even or odd permutations, respectively.

Now we consider the non-trivial linear combination $\beta := \gamma - \omega$ of the permutations from $\mathcal{P}_\lambda$. The product of β and z_λ yields

$$\beta \cdot z_\lambda = (\gamma - \omega) \cdot (g_\lambda + u_\lambda) = (\gamma \cdot g_\lambda - \omega \cdot u_\lambda) + (\gamma \cdot u_\lambda - \omega \cdot g_\lambda) = \Gamma - \Omega = 0 \,. \tag{30}$$

So, $\beta \cdot z_\lambda$ is a non-trivial vanishing linear combination of the vectors from $\mathcal{B}_\lambda$, i.e. $\mathcal{B}_\lambda$ would not be a basis of $\mathcal{Z}_\lambda$ in contradiction to the assumptions of Theorem 6.2 □

7. An example

We finish with an example. We consider the Young symmetrizer $y_t \in \mathbb{C}[\mathcal{S}_5]$ of the Young tableau $t = \begin{smallmatrix} 5 & 4 & 2 & 1 \\ 3 \end{smallmatrix}$, which belongs to the partition $\lambda = (4\,1)$. y_t has the length 48. It is proportional to a primitive idempotent $e = \frac{1}{30}\, y_t$ and generates a 4-dimensional minimal right ideal of $\mathbb{C}[\mathcal{S}_5]$. Now our algorithms yield

$$YS := D_\lambda(y_t) = \begin{pmatrix} 0 & 0 & 0 & 0 \\ 0 & 0 & 0 & 0 \\ -30 & 0 & 30 & 0 \\ 0 & 0 & 0 & 0 \end{pmatrix}$$

$$YS^* := D_\lambda(y_t^*) = \begin{pmatrix} 6 & 6 & -24 & 6 \\ 0 & 0 & 0 & 0 \\ -6 & -6 & 24 & -6 \\ 0 & 0 & 0 & 0 \end{pmatrix}$$

$$F := \frac{1}{1440}\, YS \cdot YS^* = \begin{pmatrix} 0 & 0 & 0 & 0 \\ 0 & 0 & 0 & 0 \\ -\frac{1}{4} & -\frac{1}{4} & 1 & -\frac{1}{4} \\ 0 & 0 & 0 & 0 \end{pmatrix}$$

D_λ was calculated by means of Young's natural representation of $\mathcal{S}_5$. F represents the primitive idempotent f with property (S) which generates the same right ideal as y_t (see Theorem 2.8). f has the length 120.

One can easily check the properties $YS \cdot YS = 30YS$, $YS^* \cdot YS^* = 30YS^*$, $F \cdot F = F$, $F \cdot YS = YS$, $\frac{1}{30} YS \cdot F = F$. The relation $F^* = F$ can be verified by means of our Mathematica package PERMS [19].

We implemented all algorithms presented in this paper in PERMS. Mathematica notebooks of the calculation for this paper can be downloaded from [20].

References

[1] Fiedler B 2008 Symmetry classes connected with the magnetic Heisenberg ring (Symmetry and Structural Properties of Condensed Matter (SSPCM 2007)) ed Lulek T, Wal A, and Lulek B *Journal of Physics: Conference Series* **104** 012035 (9pp)

[2] Weyl H 1950 *The Theory of Groups and Quantum Mechanics* Number 0-486-60269-9 in Dover Books on Mathematics (New York: Dover Publications, Inc.) 2nd edition

[3] Karbach M and Müller G 1997 Introduction to the Bethe ansatz I *Computers in Physics* **11** 36 arXiv: cond-mat/9809162

[4] Karbach M, Hu K, and Müller G 1998 Introduction to the Bethe ansatz II *Computers in Physics* **12** 565 arXiv: cond-mat/9809163

[5] Karbach M, Hu K, and Müller G 2000 Introduction to the Bethe ansatz III *Online* arXiv: cond-mat/0008018

[6] Jakubczyk P, Lulek T, Jakubczyk D, and Lulek B 2008 Construction of Kostka matrix at the level of bases (Symmetry and Structural Properties of Condensed Matter (SSPCM 2007)) ed Lulek T, Wal A, and Lulek B *Journal of Physics: Conference Series* **104** 012039 (4pp)

[7] Boerner H 1970 *Representations of Groups* (Amsterdam: North-Holland Publishing Company) 2nd revised edition .

[8] Weyl H 1939 *The Classical Groups, their Invariants and Representations* (Princeton, NJ: Princeton University Press)

[9] Fiedler B 2001 Ideal decompositions and computation of tensor normal forms *Séminaire Lotharingien de Combinatoire* **B45g** (16 pp) `http://www.mat.univie.ac.at/~slc` (*Preprint* math.CO/0211156)

[10] Jakubczyk P, Lulek T, Jakubczyk D, and Lulek B 2006 The duality of Weyl and linear extension of Kostka matrices (Symmetry and Structural Properties of Condensed Matter (SSPCM 2005)) ed Lulek T, Wal A, Barbara Lulek B *Journal of Physics: Conference Series* **30** 203–208

[11] Jakubczyk D, Lulek T, Jakubczyk P, and Lulek B 2006 Geometry and rigged strings in Bethe Ansatz (Symmetry and Structural Properties of Condensed Matter (SSPCM 2005)) ed Lulek T, Wal A, and Lulek B *Journal of Physics: Conference Series* **30** 188–196

[12] Fiedler B 1999 *An Algorithm for the Decomposition of Ideals of Semi-Simple Rings and its Application to Symbolic Tensor Calculations by Computer* (*Habilitationsschrift*) (Leipzig: Fakultät für Mathematik und Informatik Universität Leipzig Germany) `http://www.fiemath.de/publicat.htm`.

[13] Fiedler B 1998 An algorithm for the decomposition of ideals of the group ring of a symmetric group *Séminaire Lotharingien de Combinatoire* **B39e** (26 pp)
`http://www.mat.univie.ac.at/~slc/wpapers/s39fiedler.html`

[14] Clausen M and Baum U 1993 *Fast Fourier Transforms* (Mannheim Leipzig Wien Zürich: BI Wissenschafts-verlag)

[15] Boerner H 1955 *Darstellungen von Gruppen* (Berlin Göttingen Heidelberg: Springer-Verlag)

[16] Fulling S A, King R C, Wybourne B G, and Cummins C J 1992 Normal forms for tensor polynomials: I. The Riemann tensor *Class. Quantum Grav.* **9** 1151

[17] Gordon D J and Kerber A 1981 *The Representation Theory of the Symmetric Group* (Reading, Mass., London, Amsterdam, Don Mills, Ont., Sidney, Tokyo: Addison-Wesley Publishing Company)

[18] Kerber A 1971, 1975 *Representations of Permutation Groups* (vol 240, 495 of *Lecture Notes in Mathematics*) (Berlin, Heidelberg, New York: Springer-Verlag)

[19] Fiedler B 2009 *PERMS* 2.5 (18.8.2009) ©1999–2009 (Leipzig: Mathematisches Institut, Universität Leipzig)

[20] Fiedler B *Examples of calculations by means of* PERMS (Mathematica notebooks) *Internet page* `http://www.fiemath.de/pnbks.htm`.

Separable N-particle Hookean models

Jacek Karwowski and Kamil Szewc

Instytut Fizyki, Uniwersytet Mikołaja Kopernika, Grudziądzka 5, PL-87-100 Toruń, Poland

E-mail: jka@fizyka.umk.pl

Abstract. A model of N particles, where interactions between disjoint pairs are described by two-particle potentials which depend only on the distance between the particles and the remaining interaction are harmonic is considered. If the coupling constants in the quadratic form describing the harmonic part of the interaction potential fulfill certain separability conditions then the model is separable. Several systems known in the literature as the Hookean two-electron atom (harmonium) and the Hookean hydrogen molecules, appear to be special cases of this model. Some properties of this model are discussed in detail.

1. Introduction

Theory of many-electron systems, in particular quantum chemistry and solid state physics, are based on approximations which impose separability to non-separable multi-dimensional Schrödinger equations. The most important examples include the one-electron model and the Born-Oppenheimer approximation. In the resulting models some features of non-separable effects related to interactions as, for example, the electron correlation or the non-adiabatic effects, are obscured by the enforced separability. Therefore, *exactly* separable models of *interacting* particles are of particular interest. Among them, particularly important are *exactly* and *quasi-exactly* solvable models.

A system of coupled harmonic oscillators can be decoupled by a linear transformation of the coordinates which diagonalizes the matrix of the coupling constants. The resulting Schrödinger equation is separable and exactly solvable. It is commonly used to model a large number of physical systems ranging from vibrations of polyatomic molecules and the electronic shells of atoms to the structure of atomic nuclei [1].

About 20 years ago Makarewicz, in a very interesting and largely overlooked work [2], formulated a separable and exactly solvable model of N particles in which the attractive interactions are represented by the harmonic potentials and the repulsive ones by the potentials proportional to the inverse of the square of the distance between the particles. Several years ago separable models of three and four particle systems, in which one or two pairs of particles interact by arbitrary binary forces while the remaining interactions are harmonic, have been constructed and applied to a description of hydrogen molecules [3–7]. The separability of these models supplied a possibility of an exact solving a molecular problem without imposing the Born-Oppenheimer approximation.

It appears [8] that the separability is retained regardless of the number of particles if the interactions within disjoint pairs of particles are described by arbitrary two-particle potentials while the remaining interactions are harmonic and the coupling constants fulfill certain conditions referred to as the separability conditions. With a proper choice of the two-particle potentials the resulting model may be exactly [2,7,8] or quasi-exactly solvable [8–10].

In this presentation some specific properties of the separable model of N interacting particles as well as its applications in atomic and molecular physics are discussed. In particular, the parametric dependence of the solutions leading to the notion of the molecular shape (otherwise defined *a priori* on the ground of the Born-Oppenheimer model) [10,11] and the structure of the separable N-particle potentials are analyzed.

2. Calogero-Sutherland model

A separable and exactly solvable model of N interacting identical one-dimensional particles was introduced by Calogero [12] and by Sutherland [13,14] nearly 40 years ago. In the original formulation the Calogero Hamiltonian contains square inverse interactions and reads:

$$\mathsf{H} = \frac{1}{2}\sum_{j=1}^{N}\left(\mathsf{p}_j^2 + \omega^2 x_j^2\right) + \sum_{j<k}^{N}\frac{A}{(x_j - x_k)^2}. \tag{1}$$

The model of Sutherland is defined by a similar Hamiltonian:

$$\mathsf{H} = \frac{1}{2}\sum_{j=1}^{N}\mathsf{p}_j^2 + \sum_{j<k}^{N}\frac{A}{\sin^2(x_j - x_k)}. \tag{2}$$

Both models are closely related and over the decades were modified and generalized in many ways. Two different modes of generalization have been recently constructed by Khare [15] and by Enciso *et al.* [16] who extended the model by introducing to the particles some internal degrees of freedom, referred to as "spin". A review of the subject has been published recently by Polychronakos [17]. Among many applications of this model one should mention quantum field and string theory, theory of solitons, quantum Hall effect, random matrix theory, many-body problems, quantum chemistry.

3. Separable three-dimensional models

Hamiltonian of coupled harmonic oscillators

$$\mathsf{H}_{\rm osc}(\mathbf{r}_1, \mathbf{r}_2, \ldots, \mathbf{r}_N) = \sum_{i=1}^{N}\frac{\mathsf{p}_i^2}{2m_i} + \frac{1}{2}\sum_{i,j=1}^{N} a_{ij}\left|\mathbf{r}_i - \mathbf{r}_j\right|^2 \equiv \sum_{i=1}^{N}\frac{\mathsf{p}_i^2}{2m_i} + \mathbb{R}^\dagger\mathbb{A}\mathbb{R}, \tag{3}$$

where $a_{ij} = a_{ji}$, $a_{ii} = 0$, $\mathbb{R}^\dagger = \{\mathbf{r}_1, \mathbf{r}_2, \ldots, \mathbf{r}_N\}$ and

$$A_{ij} = \sum_{k=1}^{N} a_{ik}\delta_{ij} - a_{ij}\left(1 - \delta_{ij}\right)$$

can be decoupled by a linear transformation of the coordinates:

$$\mathbf{q}_i = \sum_{j=1}^{N} m_j\, \mathbf{r}_j\, C_{ji}, \tag{4}$$

where $\mathbb{A}\mathbb{C} = \mathbb{M}\mathbb{C}\Lambda$, $M_{ij} = m_i\delta_{ij}$ and $\Lambda_{ij} = \lambda_i\delta_{ij}$ [18]. The decoupled Hamiltonian reads

$$\mathsf{H}_{\rm osc}(\mathbf{q}_1, \mathbf{q}_2, \ldots, \mathbf{q}_N) = \sum_{i=1}^{N}\mathsf{h}_i(\mathbf{q}_i), \tag{5}$$

where

$$\mathsf{h}_i(\mathbf{q}_i) = \tfrac{1}{2}\mathsf{P}^2(\mathbf{q}_i) + \lambda_i\, \mathbf{q}_i^2, \tag{6}$$

The resulting Schrödinger equation is separable and exactly solvable. Numerous applications include modelling the vibrations of polyatomic molecules, the electronic shells of atoms (Moshinsky atom), the internal structure of atomic nuclei, and many more.

In the case of two interacting particles the Schrödinger equation may always be decoupled. Let

$$\mathsf{H}(\mathbf{r}_1, \mathbf{r}_2) = \frac{\mathsf{p}_1^2}{2m_1} + \frac{\mathsf{p}_2^2}{2m_2} + \mathsf{V}_{12}\left(|\mathbf{r}_1 - \mathbf{r}_2|\right). \tag{7}$$

The decoupling is performed by introducing the coordinates of the center of the mass:

$$\mathbf{r}_{12}^{+} = \frac{m_1\mathbf{r}_1 + m_2\mathbf{r}_2}{m_1 + m_2} \tag{8}$$

and the relative coordinates

$$\mathbf{r}_{12}^{-} = \mathbf{r}_1 - \mathbf{r}_2. \tag{9}$$

In the new coordinates

$$\mathsf{H}(\mathbf{r}_1, \mathbf{r}_2) = \mathsf{h}_0(\mathbf{r}_{12}^{+}) + \mathsf{h}_1(\mathbf{r}_{12}^{-}) \tag{10}$$

The first one-particle Hamiltonian

$$\mathsf{h}_0(\mathbf{r}_{12}^{+}) = \frac{\mathsf{P}^2(\mathbf{r}_{12}^{+})}{2M_{12}}, \tag{11}$$

describes the free motion of the center of the mass and the second one,

$$\mathsf{h}_1(\mathbf{r}_{12}^{-}) = \frac{\mathsf{P}^2(\mathbf{r}_{12}^{-})}{2\mu_{12}} + \mathsf{V}_{12}\left(|\mathbf{r}_{12}^{-}|\right), \tag{12}$$

describes the relative motion of the particles. Here and in the following we denote

$$M_{ab} = m_a + m_b \tag{13}$$

and

$$\mu_{ab} = \frac{m_a m_b}{m_a + m_b}, \tag{14}$$

Note, that the relative motion can also be interpreted as a motion of a quasi-particle of mass μ_{12} confined in a spherically-symmetric external potential $\mathsf{V}_{12}\left(|\mathbf{r}_{12}^{-}|\right)$.

In general, separability is not possible for more than two particles. However, several years ago, separable models describing three- and four-particle systems with specific forms of potentials have been constructed [2–6]. Very recently [8] the model has been generalized for the case of N particles. The generalization is based on the following theorem:

An N-particle Hamiltonian

$$\mathsf{H} = \sum_{i=1}^{K} \mathsf{H}_{2i-1}(\mathbf{r}_{2i-1}, \mathbf{r}_{2i}) + \sum_{i=2K+1}^{N} \frac{\mathsf{p}_i^2}{2m_i} + \mathsf{W}(\mathbf{r}_1, \mathbf{r}_2, \ldots \mathbf{r}_N), \tag{15}$$

where

$$\mathsf{H}_{2i-1}(\mathbf{r}_{2i-1}, \mathbf{r}_{2i}) = \frac{\mathsf{p}_{2i-1}^2}{2m_{2i-1}} + \frac{\mathsf{p}_{2i}^2}{2m_{2i}} + \mathsf{V}_{2i-1,2i}\left(|\mathbf{r}_{2i-1} - \mathbf{r}_{2i}|\right), \tag{16}$$

and

$$\mathsf{W}(\mathbf{r}_1, \mathbf{r}_2, \dots \mathbf{r}_N) = \frac{1}{2}\sum_{i,j=1}^{N} a_{ij} \left|\mathbf{r}_i - \mathbf{r}_j\right|^2 \tag{17}$$

may be decoupled to N one-particle Hamiltonians if the masses m_i and the coefficients a_{ij} fulfill *separability conditions*.
Out of these equations:

- One equation describes the free motion of the center of the mass.
- K equations describe the relative motion of the pairs of particles $\{1,2\}$, $\{3,4\}$, ..., $\{2K-1, 2K\}$, interacting by potentials $\mathsf{V}_{2i-1,2i}\left(\left|\mathbf{r}_{2i-1} - \mathbf{r}_{2i}\right|\right) + c_{2i-1,2i}\left|\mathbf{r}_{2i-1} - \mathbf{r}_{2i}\right|^2$, $i = 1, 2, \dots, K$, where $c_{2i-1,2i}$ are constants which depend on the parameters of the Hamiltonian (15),
- $N - K - 1$ equations describe a set of spherical harmonic oscillators in properly constructed normal coordinates.

The decoupling procedure may be seen as a simple generalization of the one for two particles. Similarly as in the two-particle case, we introduce the center of the mass and the relative coordinates for the pairs of particles

$$\begin{aligned} \{a, b\} \quad &= \quad \{1,2\},\ \{3,4\},\ \dots,\ \{2K-1, 2K\} \\ &a = 2i-1,\ \ b = 2i, \quad i = 1, 2, \dots, K, \quad K \le N/2, \end{aligned} \tag{18}$$

$$\mathbf{r}_{ab}^{+} = \frac{m_a \mathbf{r}_a + m_b \mathbf{r}_b}{m_a + m_b}, \tag{19}$$

$$\mathbf{r}_{ab}^{-} = \mathbf{r}_a - \mathbf{r}_b. \tag{20}$$

As a consequence we can extract from the N-particle Hamiltonian (15) the one-particle relative-motion Hamiltonians

$$\mathsf{h}_i(\mathbf{r}_{ab}^{-}) = \frac{\mathsf{P}^2(\mathbf{r}_{ab}^{-})}{2\mu_{ab}} + \mathsf{V}_{ab}\left(\left|\mathbf{r}_{ab}^{-}\right|\right) + c_{ab}\left|\mathbf{r}_{ab}^{-}\right|^2. \tag{21}$$

If the remaining part of the interaction potential (a quadratic form) is $\mathbf{r}_{ab}^{-}$-independent, then the one-particle Hamiltonian (21) may be separated. As one can check, the separability condition reads:

$$a_{pq} = \begin{cases} m_p m_q\, \alpha_{ab}^{cd}, & \text{if } p, q \le 2K, \\ m_p\, \alpha_{aq}, & \text{if } p \le 2K,\ q > 2K, \end{cases} \tag{22}$$

where α_{ab}^{cd} and α_{aq} are constants with

$$b = 2\left[\frac{p+1}{2}\right],\ \ d = 2\left[\frac{q+1}{2}\right],\ \ a = b-1,\ \ c = d-1,\ \ \alpha_{ab}^{ab} = 0,\ \ \alpha_{ab}^{cd} = \alpha_{cd}^{ab}$$

For example, if $N = 6$, $K = 3$, then

$$a_{13} = m_1 m_3\, \alpha_{12}^{34},\quad a_{14} = m_1 m_4\, \alpha_{12}^{34},\quad a_{23} = m_2 m_3\, \alpha_{12}^{34},\quad a_{24} = m_2 m_4\, \alpha_{12}^{34},$$

$$a_{15} = m_1 m_5\, \alpha_{12}^{56},\quad a_{16} = m_1 m_6\, \alpha_{12}^{56},\quad a_{25} = m_2 m_5\, \alpha_{12}^{56},\quad a_{26} = m_2 m_6\, \alpha_{12}^{56},$$

$$a_{35} = m_3 m_5\, \alpha_{34}^{56},\quad a_{36} = m_3 m_6\, \alpha_{34}^{56},\quad a_{45} = m_4 m_5\, \alpha_{34}^{46},\quad a_{46} = m_4 m_6 \alpha_{34}^{56}.$$

Let us assume, for simplicity, that $N = 2K$ and $a_{2i-1,2i} = 0$, $i = 1, 2, \dots, K$. The last condition does not restrict the generality since the terms $a_{2i-1,2i}\left|\mathbf{r}_{2i-1} - \mathbf{r}_{2i}\right|^2$ may always be

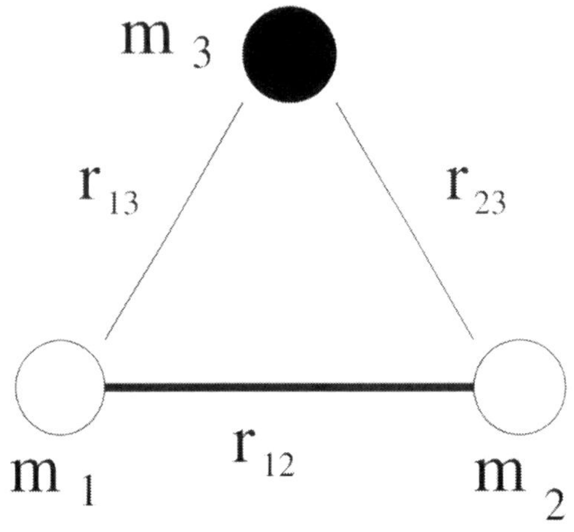

Figure 1. Hookean three-particle system.

included to $\mathsf{V}_{2i-1,2i}(|\mathbf{r}_{2i-1}-\mathbf{r}_{2i}|)$. Then, if the separability condition (21) is fulfilled, after some algebra, one gets

$$\mathsf{W}(\mathbf{r}_1,\mathbf{r}_2,\ldots\mathbf{r}_N)=\sum_{\{ab\},\{cd\}}\alpha_{ab}^{cd}M_{cd}\left[\mu_{ab}\left|\mathbf{r}_{ab}^{-}\right|^2+\frac{1}{2}M_{ab}\left|\mathbf{r}_{ab}^{+}-\mathbf{r}_{cd}^{+}\right|^2\right],\tag{23}$$

where $\{a,b\}=\{2i-1,2i\}$, $i=1,2,\ldots,K$. Consequently, if the constant c_{ab} in Eq. (21) is set equal to

$$c_{ab}=\mu_{ab}\sum_{\{cd\}}\alpha_{ab}^{cd}M_{cd}\tag{24}$$

then the residual part of W does not depend on the relative coordinates of the decoupled pairs.

4. Examples

4.1. $N=3$, $K=1$

The case of $N=3$ and $K=1$ is represented in Fig 1. It may be used to modelling a large set of systems including the Hookean two-electron atom (if $m_3>>m_1=m_2$, the negative ion of positronium (if $m_1=m_2=m_3$), and the molecular hydrogen ion (if $m_3<<m_1=m_2$) [6]. In this case

$$\begin{aligned}\mathsf{W}(\mathbf{r}_1,\mathbf{r}_2,\mathbf{r}_3)&=\alpha\left[m_1m_3\left|\mathbf{r}_{13}^{-}\right|^2+m_2m_3\left|\mathbf{r}_{23}^{-}\right|^2\right]\\&=\alpha\,\mu_{12}m_3\left|\mathbf{r}_{12}^{-}\right|^2+\alpha\,M_{12}m_3\left|\mathbf{r}_{12}^{+}-\mathbf{r}_3\right|^2.\end{aligned}\tag{25}$$

Consequently,

$$\mathsf{H}=\mathsf{h}_0(\mathbf{R}_0)+\mathsf{h}_1(\mathbf{r}_{12}^{-})+\mathsf{h}_2(\mathbf{r}_{12}^{+}-\mathbf{r}_3),\tag{26}$$

with

$$\mathbf{R}_0=\frac{m_1\mathbf{r}_1+m_2\mathbf{r}_2+m_3\mathbf{r}_3}{m_1+m_2+m_3},\tag{27}$$

$$\mathsf{h}_0(\mathbf{R}_0)=\frac{\mathsf{P}^2(\mathbf{R}_0)}{2M_{12,3}},\tag{28}$$

$$\mathsf{h}_1(\mathbf{r}_{12}^{-})=\frac{\mathsf{P}^2(\mathbf{r}_{12}^{-})}{2\mu_{12}}+\mathsf{V}\left(|\mathbf{r}_{12}^{-}|\right)+\alpha\,\mu_{12}m_3\left|\mathbf{r}_{12}^{-}\right|^2,\tag{29}$$

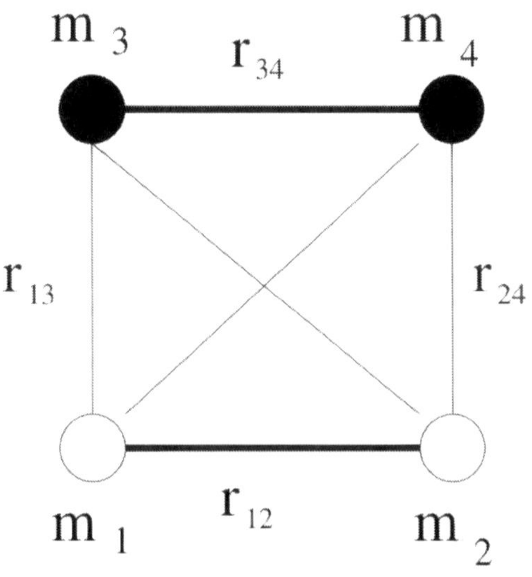

Figure 2. Hookean four-particle system.

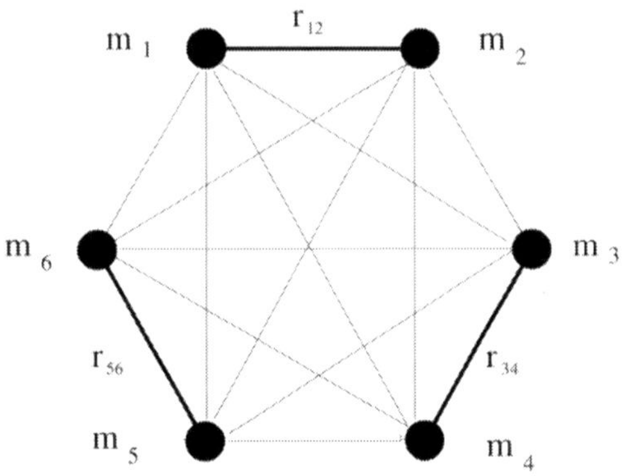

Figure 3. Hookean six-particle system. All interactions included.

$$\mathsf{h}_2(\mathbf{r}_{12}^+ - \mathbf{r}_3) = \frac{\mathsf{P}^2(\mathbf{r}_{12}^+ - \mathbf{r}_3)}{2\mu_{12,3}} + \alpha\, M_{12}\, m_3\, |\mathbf{r}_{12}^+ - \mathbf{r}_3|^2. \tag{30}$$

and $M_{12,3} = M_{12} + M_3, \quad \mu_{12,3} = \dfrac{M_{12}\, m_3}{M_{12} + m_3}$.

4.2. $N = 4$, $K = 2$

This case is represented in Fig 2 and corresponds to the Hookean H_2 molecule, to Ps_2 and to many other four-particle systems [3]. In this case

$$\begin{aligned} \mathsf{W} &= \alpha\left[m_1\, m_3\, |\mathbf{r}_{13}^-|^2 + m_1\, m_4\, |\mathbf{r}_{14}^-|^2 + m_2\, m_3\, |\mathbf{r}_{23}^-|^2 + m_2\, m_4\, |\mathbf{r}_{24}^-|^2\right] \\ &= \alpha\, \mu_{12}\, M_{34}\, \left|\mathbf{r}_{12}^-\right|^2 + \alpha\, \mu_{34}\, M_{12}\, \left|\mathbf{r}_{34}^-\right|^2 + \alpha\, M_{12}\, M_{34}\, \left|\mathbf{r}_{12}^+ - \mathbf{r}_{34}^+\right|^2, \end{aligned} \tag{31}$$

where $\alpha \equiv \alpha_{12}^{34}$.

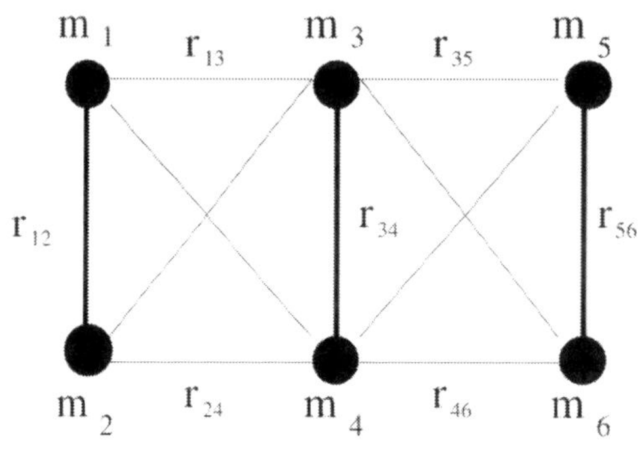

Figure 4. Hookean six-particle system. Only nearest-neighbour interactions.

4.3. $N = 6$, $K = 3$

This case is represented in Fig 3. The quadratic part W of the interaction potential has the following form:

$$\begin{aligned}
\mathsf{W} &= \alpha_{12}^{34}\left(m_1 m_3\, r_{13}^2 + m_1 m_4\, r_{14}^2 + m_2 m_3\, r_{23}^2 + m_2 m_4\, r_{24}^2\right) \\
&+ \alpha_{12}^{56}\left(m_1 m_5\, r_{15}^2 + m_1 m_6\, r_{16}^2 + m_2 m_5\, r_{25}^2 + m_2 m_6\, r_{26}^2\right) \\
&+ \alpha_{34}^{56}\left(m_3 m_5\, r_{35}^2 + m_3 m_6\, r_{36}^2 + m_4 m_5\, r_{45}^2 + m_4 m_6\, r_{46}^2\right) \qquad (32) \\
&= \mu_{12}\left(\alpha_{12}^{34} M_{34} + \alpha_{12}^{56} M_{56}\right)\left|\mathbf{r}_{12}^-\right|^2 + \alpha_{12}^{34} M_{12} M_{34}\left|\mathbf{r}_{12}^+ - \mathbf{r}_{34}^+\right|^2 \\
&+ \mu_{34}\left(\alpha_{34}^{12} M_{12} + \alpha_{34}^{56} M_{56}\right)\left|\mathbf{r}_{34}^-\right|^2 + \alpha_{12}^{56} M_{12} M_{56}\left|\mathbf{r}_{12}^+ - \mathbf{r}_{56}^+\right|^2 \\
&+ \mu_{56}\left(\alpha_{56}^{12} M_{12} + \alpha_{56}^{34} M_{34}\right)\left|\mathbf{r}_{56}^-\right|^2 + \alpha_{34}^{56} M_{34} M_{56}\left|\mathbf{r}_{34}^+ - \mathbf{r}_{56}^+\right|^2
\end{aligned}$$

If all masses, all interaction potentials and all parameters α_{ab}^{cd} are equal then, with increasing N, we get a series of "crystals" composed of pairs symmetrically distributed on a sphere.

4.4. A K-segment ladder

If the Hooke-type interactions are retained between the neighbouring pairs only, we get a ladder. The three-segment case is shown in Fig 4 If the ladder is composed of two kinds of particles with masses m and M, and $m_1 = m_2 = m_5 = m_6 = \cdots = m_{N-1} = m_N = m$ and $m_3 = m_4 = m_7 = m_8 = \cdots = m_{N-3} = m_{N-2} = M$, then

$$\begin{aligned}
\mathsf{W} &= \alpha Mm\left[|\mathbf{r}_{12}^-|^2 + 2|\mathbf{r}_{34}^-|^2 + \cdots + 2|\mathbf{r}_{N-3,N-2}^-|^2 + |\mathbf{r}_{N-1,N}^-|^2\right] \\
&+ 4\alpha Mm\left[|\mathbf{r}_{12}^+ - \mathbf{r}_{34}^+|^2 + |\mathbf{r}_{34}^+ - \mathbf{r}_{56}^+|^2 + \cdots + |\mathbf{r}_{N-3,N-2}^+ - \mathbf{r}_{N-1,N}^+|^2\right]. \qquad (33)
\end{aligned}$$

5. Three-particle systems: general solution

One most most interesting features of the separable model is a possibility of an exact analysis of the dependence of the properties of a Hookean molecule on the masses of the particles without imposing the Born-Oppenheimer approximation. Let particles $\{1\}$ and $\{2\}$ be identical fermions of mass m and let the mass of of the third particle be equal to m_3. Let us assume that the center of the mass is at the origin of our coordinate system, i.e. $\mathbf{R}_0 = 0$. Thus, Eq. (27) implies

$$m(\mathbf{r}_1 + \mathbf{r}_2) + m_3\,\mathbf{r}_3 = 0 \qquad (34)$$

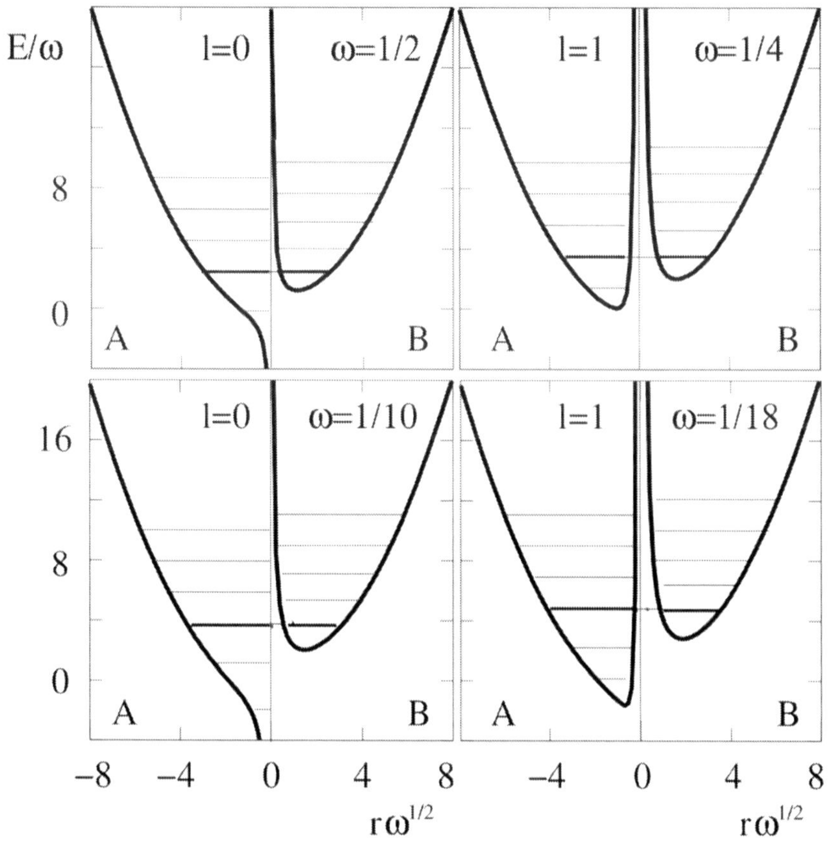

Figure 5. Shapes of the effective potential $V_{\rm eff}(r)$ and the energy levels of confined positronium (A) and of harmonium (B). The values of ω correspond to analytically solvable cases with $l = 0, 1$. The analytical energies (thick lines) of harmonium and of confined positronium are the same.

. In our case $\mu_{12} = m/2$, $M_{12} = 2m$, $M_{12,3} = 2m + m_3$,

$$\mu_{12,3} = \frac{2mm_3}{2m + m_3} \equiv \mu,$$

$$\mathbf{r}_3 - \mathbf{r}_{12}^{+} = \frac{m_3}{\mu}\mathbf{r}_3.$$

We select

$$\mathsf{V}_{12}\left(\left|\mathbf{r}_{12}^{-}\right|\right) = \frac{\zeta}{r_{12}},$$

and substitute $\omega^2 \equiv 2\alpha\, mm_3$ and $\mathbf{r} \equiv \mathbf{r}_{12}^{-}$. Since α is an arbitrary constant, we can assume that ω is a constant independent of the masses. Then Eqs. (26) – (30) yield

$$\mathsf{H} = \mathsf{h}_1 + \mathsf{h}_2, \tag{35}$$

and

$$\mathsf{h}_1 = \frac{1}{m}\mathsf{P}^2(\mathbf{r}) + \frac{\zeta}{r} + \frac{\omega^2}{4}r^2, \tag{36}$$

$$\mathsf{h}_2 = \frac{1}{2\mu}\mathsf{P}^2(\boldsymbol{\rho}) + \omega^2\rho^2, \tag{37}$$

where

$$\boldsymbol{\rho} = \frac{m_3}{\mu}\mathbf{r_3}.$$

The first Hamiltonian, h_1, describes the relative motion of particles $\{1\}$ and $\{2\}$. It is the well known Hamiltonian of either harmonium ($\zeta > 0$) or confined positronium ($\zeta < 0$) [19] and the corresponding Schrödinger equation is quasi-exactly solvable. The second Hamiltonian, h_2, describes the relative motion of particle $\{3\}$ and of the center of the mass of particles $\{1\}$ and $\{2\}$. It is the Hamiltonian of a spherical harmonic oscillator and the corresponding Schrödinger equation is exactly solvable.

The total wavefunction may be expressed as

$$\Psi(\mathbf{r}, \mathbf{r}_3) = \Phi_{nlm_l}(\mathbf{r})\, \Xi_{\nu\lambda\mu_\lambda}(\mathbf{r}_3) = (r\, r_3)^{-1}\phi_{nl}(r)\xi_{\nu\lambda}(r_3)Y_{lm_l}(\hat{\mathbf{r}})Y_{\lambda\mu_\lambda}(\hat{\mathbf{r}}_3), \tag{38}$$

where the radial functions are solutions of the following equations:

$$\left[-\frac{1}{m}\frac{d^2}{dr^2} + \mathsf{V}_{\text{eff}}(r) - E^{(1)}_{nl}\right]\phi_{nl}(r) = 0, \tag{39}$$

where

$$\mathsf{V}_{\text{eff}}(r) = \frac{l(l+1)}{mr^2} + \frac{\zeta}{r} + \frac{\omega^2 r^2}{4} \tag{40}$$

and

$$\left[-\frac{1}{2\mu}\frac{d^2}{d\rho^2} + \frac{\lambda(\lambda+1)}{2\mu\rho^2} + \omega^2\rho^2 - E^{(2)}_{\nu\lambda}\right]\xi_{\nu\lambda}(\rho) = 0. \tag{41}$$

Let us consider in more detail the ground state solutions of Eqs. (39) and (41). In the case of harmonium (39) the analytical solution exists if some specific relations between m, ζ and ω are fulfilled [10, 19]. In particular, for the ground state, the analytical solution exists if

$$\omega = \frac{\zeta^2 m^{3/2}}{2(l+1)}. \tag{42}$$

The corresponding wavefunction reads:

$$\phi(r)_{0l} \sim r^{l+1}\left[1 + \frac{m\zeta}{2(l+1)}r\right]\mathrm{e}^{-\gamma r^2} \tag{43}$$

where

$$\gamma = \frac{m^2\zeta^2}{8(l+1)} = \frac{\omega\sqrt{m}}{4}.$$

As one can show [23], if for a given set of parameters an analytical solution exists for harmonium $\zeta > 0$, it also exists for the same set of parameters but with ζ replaced by $-\zeta$, i.e. for the confined positronium. The energies for these two cases appear to be the same. More to say, the energies of harmonium and of the confined positronium are the same *only* for the states for which analytical solutions exist (see Fig. 5). This property resembles the behaviour of spectra of supersymmetric partner potentials [24,25] and reflects a specific symmetry related to the transformation $r \to -r$.

An approximate solution, valid for arbitrary values of m, ζ and ω, may be obtained by the replacement of V_{eff} by its harmonic approximation [20, 21]:

$$\mathsf{V}_{\text{eff}} = \mathsf{V}_0 + \frac{3\omega^2}{4}(r - r_0)^2, \tag{44}$$

where

$$r_0 = \left(\frac{2\zeta}{\omega^2}\right)^{1/3}. \tag{45}$$

The effective potential (44) corresponds to a shifted harmonic oscillator. Its ground state wavefunction may be expressed as

$$\phi_{00}(r) \approx \left(\frac{2D}{\pi}\right)^{1/4} \mathrm{e}^{-D(r-r_0)^2}, \tag{46}$$

where

$$D = \frac{1}{4}\sqrt{3m\omega}. \tag{47}$$

Solutions of Eq. (41) are well known. In particular,

$$\xi_{0\lambda}(r_{12}^+) \sim |r_{12}^+|^{\lambda+1} \mathrm{e}^{-c|r_{12}^+|^2}, \tag{48}$$

where

$$c = \omega\sqrt{m}\left(\frac{2m}{\mu}\right)^{3/2}. \tag{49}$$

Now, using Eqs. (34), (42) and (43), we can get the exact wavefunction of the ground state of the three-particle problem:

$$\begin{aligned} \psi(r, r_{12}^+) &= \phi_{0l}(r)\, \xi_{0\lambda}(r_{12}^+) / \left(r\, r_{12}^+\right) \\ &\sim (r_1^2 - r_2^2)^l \left(r_{12}^+\right)^{\lambda-l} \left[1 + \frac{m\zeta}{2(l+1)}\, r\right] \mathrm{e}^{-\gamma[a_+(r_1^2+r_2^2)+a_-\mathbf{r}_1\mathbf{r}_2]} \end{aligned} \tag{50}$$

where $a_+ = (2m/\mu)^{3/2} + 1$, $a_- = (2m/\mu)^{3/2} - 1$. As one can see, for $m_3 \to \infty$, i.e. for the one-center problem (Hookean helium atom with the infinite mass of the nucleus), $a_= \to 2$ and $a_- \to 0$. Consequently, the wavefunction in the limit of two-electron one-center problem becomes

$$\psi \sim (r_1^2 - r_2^2)^l \left(r_{12}^+\right)^{\lambda-l} \left[1 + \frac{m\zeta}{2(l+1)}\, r_{12}^-\right] \mathrm{e}^{-2\gamma(r_1^2+r_2^2)}. \tag{51}$$

If $m = 1$ and $\zeta = 1$ (the Coulomb repulsion of two electrons) then for $r_{12}^- \to 0$ and $l = 0$ we get the cusp condition of Kato:

$$\psi \sim 1 + \frac{1}{2} r_{12}. \tag{52}$$

6. Three-particle system: the mass dependence

An analysis of the dependence of solutions of models describing systems of several particles on the particle masses leads to a better understanding of some specific features of the Born-Oppenheimer approximation. In particular, it allows to trace the process of the emergence of the molecular structure from the atomic shape [11]. In order to study this process, let us consider the system of three particles. The density of mass operator of particle $\{3\}$ is defined as

$$\hat{\rho}(\mathbf{r}; \mathbf{r}_3) = m_3 \delta(\mathbf{r} - \mathbf{r}_3) \tag{53}$$

and for particles $\{1\}$ and $\{2\}$ as

$$\hat{\rho}(\mathbf{r}; \mathbf{r}_1, \mathbf{r}_2) = m\left[\delta(\mathbf{r} - \mathbf{r}_1) + \delta(\mathbf{r} - \mathbf{r}_2)\right], \tag{54}$$

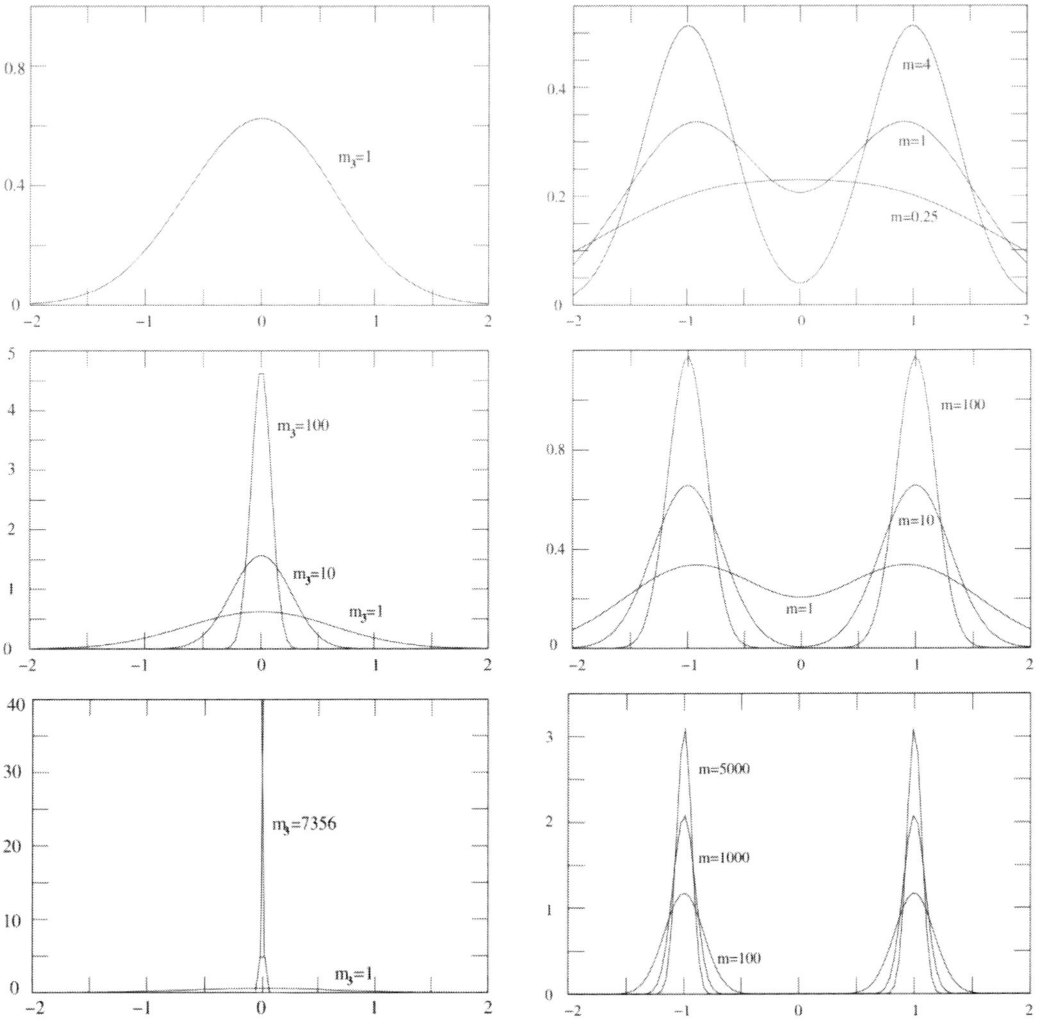

Figure 6. Left panels: Normalized to 1 density of mass of particle {3} versus z; m_3 changes from 1 (the upper panel) to 7356 (the lowest panel) while $m = 1$, $\zeta = 1$ and $\omega = 1/2$. Right panels: Normalized to 1 density of mass of particles {1} and {2}; mass m changes from 0.25 (the upper panel) to 5000 (the lowest panel) while $m_3 = 1$, $\zeta = 1$ and $\omega = 1/2$.

where

$$\mathbf{r}_1 = \frac{1}{2}\mathbf{r}_{12}^{-} - \frac{m_3}{2m_2}\mathbf{r}_3, \quad \mathbf{r}_2 = \frac{1}{2}\mathbf{r}_{12}^{-} + \frac{m_3}{2m_2}\mathbf{r}_3. \tag{55}$$

From here one can calculate the normalized to 1 density of mass distribution for particle {3}:

$$\begin{aligned} \rho(\mathbf{r})_{m_3} &= \langle\Psi(\mathbf{r}_1,\mathbf{r}_2,\mathbf{r}_3)|\delta(\mathbf{r}-\mathbf{r}_3)|\Psi(\mathbf{r}_1,\mathbf{r}_2,\mathbf{r}_3)\rangle \\ &= \langle\Phi(\mathbf{r}_{12}^{-})\,\Xi(\mathbf{r}_3)|\delta(\mathbf{r}-\mathbf{r}_3)|\Phi(\mathbf{r}_{12}^{-})\,\Xi(\mathbf{r}_3)\rangle = |\Xi(\mathbf{r})|^2 \end{aligned} \tag{56}$$

and similarly for particles {1} and {2}:

$$\rho(\mathbf{r})_m = \frac{1}{2}\langle\Phi(\mathbf{r}_{12}^{-})\,\Xi(\mathbf{r}_3)|\delta(\mathbf{r}-\mathbf{r}_1)+\delta(\mathbf{r}-\mathbf{r}_2)|\Phi(\mathbf{r}_{12}^{-})\,\Xi(\mathbf{r}_3)\rangle. \tag{57}$$

Now, using Eqs. (46) and (48), one can perform the integration and express the densities in terms of the parameters of the pertinent wavefunctions [26]. Examples of the densities, as functions of the coordinate z, are shown in Fig 6. As one can see, with increasing masses the localized nuclei emerge. At the limit of the Born-Oppenheimer model, i.e. for either $m \to \infty$ or $m_3 \to \infty$, the density distributions of the heavy particles approach the Dirac deltas. Let us note, that in the present model the internuclear distance in the Hookean H_2^+ molecule is mass-independent and equal to r_0. It is in contrast to the Coulombic molecules where the bond length scales inversely proportional to the mass of the light particle. For example, the bond length in a muonic molecule is by factor 207 shorter than in an electronic one.

7. Final remarks

The Schrödinger equation for a systems of particles in which disjoint pairs interact by arbitrary two-particle potentials while the remaining interactions are described by harmonic oscillator potentials is separable. Special cases of this model, known as the Hookean atoms and molecules, are interesting in the context of applications in atomic and molecular physics.

All equations resulting from this model may be reduced to a set of spherically-symmetric one-particle equations. For a large set of physically interesting interaction potentials the radial parts of these equations are either exactly or quasi-exactly solvable. Their analysis, apart of the academic curiosity, supplies interesting data on the nature of various schemes of separability. In particular, it allows to analyze the electron correlation problems and the nature of the Born-Oppenheimer approximation from a new perspective.

Acknowledgement

Supported by the Polish Ministry of Science and Higher Education, project No. N202 041 32/1045.

References

[1] Moshinsky M and Smirnov Y F 1996 *The harmonic oscillator in modern physics* (Amsterdam: Harwood Academic Publishers)
[2] Makarewicz J 1986 Exact solvable three-dimensional models of many-body systems *Am. J. Phys.* **54** 178-180
[3] Ludeña E V, Lopez X and Ugalde J M 2005 Non-Born-Oppenheimer treatment of the H_2 Hookean molecule *J. Chem. Phys.* **123** 024102
[4] Lopez X, Ugalde J M and Ludeña E V 2005 Extracular densities of the non-Born-Oppenheimer Hookean H_2 molecule *Chem. Phys. Letters* **412** 381-385
[5] Lopez X, Ugalde J M and Ludeña E V 2006 Exact non-Born-Oppenheimer wave function for the Hooke-Calogero model of the H_2 molecule *Eur. Phys. J.* D **37** 351-359
[6] Lopez X, Ugalde J M, Echevarría L and Ludeña E V 2006 Exact non-Born-Oppenheimer wave functions for three-particle Hookean systems with arbitrary masses *Phys. Rev.* A **74** 042504
[7] Howard I A, Amovilli C, Gidopoulos N and March N H 2007 Exactly solvable model mimicking H_2 molecule in the limit of large nuclear masses *J. Math. Chem.* **42** 603-615
[8] Karwowski J 2008 A separable model of N interacting particles *Int. J. Quantum Chem.* **108** 2253-2260
[9] Karwowski J 2008 Few-particle systems: quasi-exactly solvable models *J. Phys.: Conf. Series* **104** 012033
[10] Karwowski J and Szewc K 2008 Quasi-exactly solvable models in quantum chemistry *Collect. Czech. Chem. Commun.* **73** 1372-1390
[11] Müller-Herold U 2006 On the emergence of molecular structure from atomic shape in the $1/r^2$ harmonium model *J. Chem. Phys.* **124** 014105
[12] Calogero F 1971 Solution of the one-dimensional N-body problems with quadratic and/or inversely-quadratic pair potentials *J. Math. Phys.* **12** 419-436
[13] Sutherland B 1971 Quantum many-body problem in one dimension *J. Math. Phys.* **12** 246-250
[14] Sutherland B 1972 Exact results for a quantum many-body problem in one dimension *Phys. Rev.* A **5**, 1372-6
[15] Khare A 1996 Exact solution of an N-body problem in one dimension *J. Phys. A: Math. Gen.* **29** L45-L48
[16] Enciso A, Finkel F, Gonzáles-Lopéz A and Rodríguez M A 2005 Solvable scalar and spin models with near-neighbors interactions *Phys. Lett.* B **605** 214-22 (2005)

[17] Polychronakos A P 2006 The physics and mathematics of Calogero particles *J. Phys. A: Math. Gen.* **39** 12793-12845
[18] Thouless D J 1972 *The quantum mechanics of many-body systems*, (New York: Academic Press)
[19] Karwowski J and Cyrnek L 2004 Harmonium *Ann. Phys.* (Leipzig) **13** 181-193
[20] Taut M 1993 Two electrons in an external oscillator potential: Particular analytic solutions of a Coulomb correlation problem *Phys. Rev.* A **48** 3561-3566
[21] Mandal S, Mukherjee P K and Diercksen G H F 2003 Two electrons in a harmonic potential: an approximate analytical solution *J. Phys. B: At. Mol. Phys.* **36** 4483-4494
[22] Kato T 1957 On the eigenfunctions of many-particle systems in quantum mechanics *Commun. Pure Appl. Math.* **10** 151-177
[23] Karwowski J and Cyrnek L 2005 A class of exactly solvable Schrödinger equations *Collect. Czech. Chem. Commun.* **70** 864-880
[24] Dutt R, Khare A and Sukhatme U P 1988 Supersymmetry, shape-invariance, and exactly solvable potentials *Am. J. Phys.* **56** 163-168.
[25] Cooper F, Khare A and Sukhatme U P 1995 Supersymmetry and quantum mechanics *Physics Reports* **251** 267-385.
[26] Szewc K 2009 *Quasi-exactly solvable models of atoms and molecules*, (Thesis, Nicolaus Copernicus University, Toruń).

Hodge structures in topological quantum mechanics

Grzegorz Banaszak *, **Jan Milewski** **

* Department of Mathematical and Computer Science, Adam Mickiewicz University, ul. Umultowska 87, 61-614 Poznań, Poland

** Institute of Mathematics, Poznań University of Technology ul. Piotrowo 3A, 60-965 Poznań, Poland

E-mail: banaszak@amu.edu.pl, jsmilew@wp.pl

Abstract. We parametrize some quantum numbers in two-dimensional quantum mechanics by means of a Hodge structure. For an anyon system the statistical parameter leads to a deformation of the Hodge structure.

1. Introduction

In this paper we consider applications of Hodge structures to quantum mechanics. We consider some function spaces and show that they admit Hodge structures or their deformations. Especially the space of harmonic functions in quantum mechanics admits the Hodge structure; azimuthal and magnetic quantum numbers are weights of the Hodge structure. In the three dimensional quantum mechanics the Hodge structure is almost algebraic.

We consider in the paper theory of anyons as a special case of topological quantum mechanics (it means that wave functions are not univalent). In the case of anyons on the plane, the Hodge structure is topologically deformed.

The paper is organized as follows. In section 2 we give a notion of a Hodge structure. In section 3 we consider some examples of Hodge structures in polynomial and harmonic polynomial spaces. In section 4 there are given forms of wave function for anyons with fractional statistical parameter on a plane. In the section 5 we consider anyonic harmonic functions and we show this space admits the topologically deformed Hodge structure. In section 6 we give some concluding remarks.

2. Preliminaria

Let $\mathbf{C}^* = \mathbf{C} \setminus \{0\}$ be the multiplicative group of complex numbers not equals zero. The general form of irreducible representations of the group is given by

$$\gamma_{l,m} : \mathbf{C}^* \longrightarrow \mathbf{C}^* \ , \ \gamma(re^{i\phi}) = r^l e^{im\phi}, \ \ l \in \mathbf{R}, m \in \mathbf{Z}. \tag{1}$$

We define weights of the representation $\gamma_{l,m}$ as follows:

$$w_r = l \quad \text{the radial weight} \tag{2}$$

$$w_a = m \quad \text{the angular weight.} \tag{3}$$

A complex one-dimensional representation of the form

$$\Gamma_{p,q}(\zeta) = \zeta^p \overline{\zeta}^q \ , \quad p, q \in \mathbf{Z} \tag{4}$$

will be called an irreducible algebraic representation. Let us mention that in the case $q = 0$ the representation $\Gamma_{p,q}$ is an algebraic group representation.

The weights of the irreducible algebraic representation $\Gamma_{p,q}$ have values:

$$w_r = p + q \quad \text{the radial weight}, \quad w_a = p - q \quad \text{the angular weight}. \tag{5}$$

Observe that the difference of these weights is an even number:

$$w_r - w_a = 2q. \tag{6}$$

The irreducible representation $\gamma_{l,m}$ is an algebraic one iff $w_r = l \in \mathbf{Z}$ and $w_r - w_a = l - m \in 2\mathbf{Z}$.

If the weights w_r i w_a are integral numbers then the representation $\gamma_{l,m}$ is called almost algebraic one.

Example 1. The representation γ given by $\gamma(\zeta) = |\zeta|$ is an almost algebraic one (the radial weight equals 1 and angular equals 0). The square γ^2 of the representation γ is an algebraic representation:

$$\gamma^2(\zeta) = |\zeta|^2 = z\overline{z}. \tag{7}$$

Now let us define the notion of the Hodge structure. For an excellent servey of Hodge structures see [1].

Definition of Hodge structure. *Let V be a real vector space. A Hodge structure is a representation Γ of the group $\mathbf{C}^*$ in $V_{\mathbf{C}}$*

$$\Gamma : \mathbf{C}^* \longrightarrow \mathrm{GL}(V_{\mathbf{C}}) \tag{8}$$

which splits into one dimensional complex irreducible algebraic representations of the form

$$\Gamma = \bigoplus_{pq} m_{p,q} \Gamma_{p,q}, \tag{9}$$

where

$$m_{p,q} = m_{q,p}. \tag{10}$$

Hodge structure is strictly connected with the Hodge decomposition

$$V_{\mathbf{C}} = \bigoplus V_{p,q} \ , \quad \dim_{\mathbf{C}} V_{p,q} = m_{p,q}, \tag{11}$$

where $V_{p,q}$ are called Hodge subspaces.

If decomposition

$$\Gamma = \bigoplus n_{l,m} \gamma_{l,m}$$

has the property $n_{l,m} = n_{l,-m}$ then we say that this representation is almost algebraic Hodge structure. Complexification of Hodge (almost algebraic Hodge resp.) structure we also call Hodge (almost algebraic Hodge resp.) structure.

3. Examples of Hodge structures

Example 2. Polynomials of three variables.

The space $K_{\mathbf{R}}^{\mathbf{C}}(x,y,z)$ of complex polynomials of three real variables (x,y,z). Let us make the following substitution of variables:

$$P(u,\overline{u},z) = Q(x,y,z), \quad \text{where} \quad u := x+iy.$$

So any polynomial can be presented in the form:

$$P(u,\overline{u},z) = \sum P_{n,m} u^n \overline{u}^m z^k,$$

where coefficients are uniquely determined by the polynomial Q (or equivalently by P). The representation Γ defined in the following way

$$\Gamma(\zeta)P = P_\zeta, \quad \text{where} \quad P_\zeta(u,\overline{u},z) = P(\zeta u, \overline{\zeta u}, |\zeta| z) \tag{12}$$

is not algebraic, but is almost algebraic:

$$\Gamma = \bigoplus_{l\in\mathbf{N}_0} \bigoplus_{m=-l}^{m} n_{l,m}\gamma_{l,m}. \tag{13}$$

Monomials of the form

$$\phi_{n,m,k}(u,\overline{u},z) = u^n \overline{u}^m z^k \quad n,m,k \in \mathbf{N}_0$$

constitute a basis in $K_{\mathbf{R}}^{\mathbf{C}}(x,y,z)$. Weights of the basis monomials

$$w_r = n+m+k \;\; , \;\; w_a = n-m$$

are integers so the representation Γ in the polynomial space is an almost algebraic Hodge structure. Obvserve that the radial weight is the homogeneous degree of a homogenous polynomial.

Example 3. Spherical harmonics in R^3.

The space $H_{\mathbf{R}}^{\mathbf{C}}(x,y,z)$ of harmonic polynomials is a subspace of $K_{\mathbf{R}}^{\mathbf{C}}(x,y,z)$ and consists of polynomials which satisfy the Laplace equation

$$\Delta P = 0\,.$$

A basis of this space is a family of spherical harmonics Y_m^l, $l \in \mathbf{N_0}, m = -l, -l+1, \ldots, l$ given by the following formula [2]:

$$Y_m^l = \left[\frac{2l+1}{4\pi}(l+m)!(l-m)!\right]^{\frac{1}{2}} \sum_k \frac{(-u)^{m+k}\overline{u}^k z^{l-m-2k}}{2^{m+2k}(m+k)!k!(l-m-2k)!}, \tag{14}$$

where $u = x+iy$ is the complex variable and z is the real coordinate. The summation is over integer numbers k such that $k+m, k, l-m-2k$ are nonnegative. The space of harmonic polynomials is equivariant with respect to the representation Γ from the previous example. The restriction Γ_H has a decomposition

$$\Gamma_H = \bigoplus_{l\in\mathbf{N}_0} \bigoplus_{m=-l}^{m} \gamma_{l,m}. \tag{15}$$

Hence the space of spherical harmonics admits almost algebraic Hodge structure. The weights of the harmonic Y_m^l are $w_r = l$ and $w_a = m$. Physical interpretation of the weights of spherical harmonics in quantum mechanics: l- azimuthal quantum number, m- magnetic quantum number.

Example 4. Polynomials of even number of variables.

Let $x = (x_1, \ldots x_N)$, $y = (y_1, \ldots y_N)$ for $x_k, y_k \in \mathbf{R}$, and $z = (z_1, \ldots z_N)$, $\overline{z} = (\overline{z}_1, \ldots \overline{z}_N)$ for $z_k \in \mathbf{C}$, where $k = 1, \ldots, N$. The space $K_{\mathbf{R}}^{\mathbf{C}}(x, y)$ of complex polynomials of $2N$ real variables. Let us make the following substitution of variables: $z_k = x_k + iy_k$. So any polynomial of variables (x, y) can be expressed as a polynomial of the variables $(z, \overline{z})$. The algebraic representation

$$\Gamma_{2N} : \mathbf{C}^* \longrightarrow \mathrm{GL}_{\mathbf{C}}(K_{\mathbf{R}}^{\mathbf{C}}(x, y))$$

is given in a natural way: $\Gamma_{2N}(\zeta)P = P_\zeta$,

$$P_\zeta(z, \overline{z}) = P(\zeta z, \overline{\zeta z}), \tag{16}$$

where $\zeta z = (\zeta z_1, \ldots \zeta, z_N)$ The representation Γ has a Hodge decomposition:

$$\Gamma_{2N} = \bigoplus_{p,q \in \mathbf{N}_0} m_{p,q} \Gamma_{p,q}. \tag{17}$$

The irreducible subspaces are one dimensional vector spaces over $\mathbf{C}$ and the family of complex monomials of the form

$$\phi_{n_1,\ldots,n_N,m_1,\ldots,m_N}(z, \overline{z}) = z_1^{n_1} \ldots z_N^{n_N} \overline{z}_1^{m_1} \ldots \overline{z}_N^{m_N},$$

where $p = \sum_{i=1}^N n_i, \quad q = \sum_{j=1}^N m_j$ gives a basis consistent with this decomposition. In this case the partial weight of a homogeneous polynomial also has the meaning of the homogeneity degree.

Example 5. Harmonic polynomials of even numbers of variables.

Consider the space

$$H_{\mathbf{R}}^{\mathbf{C}}(x, y)$$

of complex harmonic polynomials of $2N$ real variables. This space is equivariant with respect to the representation Γ_{2N}. For $n = 1$ the angular weights for harmonic polynomial are $w_a = \pm w_r$, and for $N > 1$

$$w_a = -w_r, -w_r + 2, \ldots, w_r - 2, w_r$$

The space is a model of the space of harmonics for a quantum system of n identical but distinguishable softcore particles. The weights for a harmonic can be treated as quantum numbers: the angular weight w_a is the plane angular momentum m and radial weight w_r is an analog of azimuthal quantum number.

Let us introduce the following operators:

$$\hat{n}_h = \sum z_k \frac{\partial}{\partial z_k} \;, \quad \hat{n}_a = \sum \overline{z}_k \frac{\partial}{\partial \overline{z}_k} \,.$$

Hodge subspaces are eigenspaces of these operators and the eigenvalues are indices of irreducible algebraic representations.

Operators corresponding to quantum numbers l, m are

$$\hat{l} = \hat{n}_h + \hat{n}_a \;, \quad \hat{m} = \hat{n}_h - \hat{n}_a \,.$$

The number l is the homogeneity degree of a harmonic and m is the planar angular momentum.

4. Wave functions for anyons on a plane

Anyons are identical indistingushable hardcore particles. Wave functions for the system with the statistical parameter $\nu \in (0,1) \cup (1,2)$ can be considered as not univalent functions on $\mathbf{C}^N$ of the form [3]-[6]:

$$\prod_{k<l}(z_l - z_k)^{\nu} F(z,\overline{z}) \tag{18}$$

or

$$\prod_{k<l}(\overline{z}_l - \overline{z}_k)^{2-\nu} F(z,\overline{z}), \tag{19}$$

where F is a symmetric function on $\mathbf{C}^N$. In the following we assume that F is a power series of $(z,\overline{z})$ variables.

The fractional powers of the Vandermonde determinant

$$\prod_{k<l}(z_l - z_k)^{\nu} \quad , \quad \prod_{k<l}(\overline{z}_l - \overline{z}_k)^{2-\nu}$$

are the topological factors and they are no univalent factors. The theory of anyon is an example of topological quantum mechanics, because the wave function are no univalent.

Now let us decompose the space $\mathbf{C}^N$ on two mutualy orthogonal spaces: the space of mass centre motion and the relative position space. The complex *orthonormal* mass centre coordinate is

$$u_0 = \frac{1}{\sqrt{N}} \sum_{k=1}^{N} z_k. \tag{20}$$

We deal only with the relative motion space, because the topological factors depend only on relative positions. Let

$$u = (u_1, \ldots, u_M) \text{ ,where } M = N - 1 \tag{21}$$

be an orthonormal system of relative coordinates. There are such a types of relative parts of anyon wave functions

$$\Psi_I = V^{\nu}(u) F_I(u,\overline{u}) \text{ and } \Psi_{II} = V^{2-\nu}(\overline{u}) F_{II}(u,\overline{u}) \ , \tag{22}$$

where $F_I(u,\overline{u})$, $F_{II}(u,\overline{u})$ are univalent, symmetric functions in relation to transpositions of particles, it means that the composed function $F(u(z),\overline{u}(z^*))$ is symmetric or antisymmetric in relation to z-coordinates, and $V(u)$ ($V(\overline{u})$ resp.) is the Vandermonde determinant of holomorphic positions z (antiholomorphic z^* resp.).

5. Anyonic harmonics

The very important problem in the theory of anyons is to determine anyonic harmonics [7]. General forms of anyonic harmonics in the relative motion space are given by the following equations:

$$H_I = P(\overline{u}_k \frac{\partial}{\partial u_l} - \overline{u}_l \frac{\partial}{\partial u_k}) V^{n+\nu}(u) f(u) \ , \tag{23}$$

$$H_{II} = P(u_k \frac{\partial}{\partial u_l^*} - u_l \frac{\partial}{\partial u_k^*}) V^{n+2-\nu}(\overline{u}) f(\overline{u}) \ , \tag{24}$$

where P is a symmetric (antisymmetric resp.) polynomial for n even (odd resp.), f is a symmetric holomorphic polynomial.

Let us denote by l_I and l_{II} homogeneity degrees of functions H_I i H_{II}. The admissible values of planar angular momentum m_I and m_{II} are [7]:

$$m_I = l_I - 2(n_I)_{max}\,,\ l_I - 2(n_I)_{max} + 2\,,\ \ldots l_I - 2\,,\ l_I\,, \tag{25}$$

and

$$m_{II} = -(l_{II} - 2(n_{II})_{max})\,,\ -(l_I - 2(n_{II})_{max} + 2)\,,\ \ldots - (l_{II} - 2)\,,\ -l_{II}\,, \tag{26}$$

where

$$(n_I)_{max} = \left[\frac{l_I}{\binom{N}{2}} - \nu\right]\ ,\text{and}\ (n_{II})_{max} = \left[\frac{l_{II}}{\binom{N}{2}} - (2-\nu)\right]\ . \tag{27}$$

and the square bracket denotes the integer part.

For P and f homogeneous the harmonics are bihomogeneous ones.

The total homogeneity

$$l = l_f + \binom{M}{2}\mu$$

The total plane angular momentum

$$m = \pm(l - 2l_P).$$

In general quantum numbers l, m may not be integral. Hence representation corresponding to these numbers is the representation of the group $\widetilde{\mathbf{C}}^*$ which is the universal cover of $\mathbf{C}^*$:

$$\Gamma(r,\phi)H(u,\overline{u}) = H(re^{i\phi}u, \overline{re^{i\phi}u}). \tag{28}$$

The universal cover of $\mathbf{C}^*$ is the group $\tilde{\mathbf{C}}^* := \mathbf{R}_+ \times \mathbf{R}$ with the multiplicative law

$$(r_1,\phi_1)(r_2,\phi_2) := (r_1r_2, \phi_1 + \phi_2).$$

The representation Γ splits into the representations of the type $\Gamma_{\alpha,q}$ and $\Gamma_{p,\beta}$ with $\alpha = \binom{N}{2}\nu + n$, $\beta = \binom{N}{2}(2-\nu) + n$, $p, q \in \mathbf{N}$ given by

$$\Gamma_{\alpha,q}(r,\phi) = r^{\alpha+q}e^{2\pi i(\alpha-q)}\ \text{ and }\ \Gamma_{p,\beta}(r,\phi) = r^{p+\beta}e^{2\pi i(p+\beta)}.$$

We say that on the space of anyonic harmonics there exists a topologically deformed Hodge structure.

6. Conclusions

We have shown in this paper that the space of anyonic harmonics admits a topological deformation of a Hodge structure. The deformed Hodge structure is a complex representation of the universal covering of the multiplicative group $\mathbf{C}^*$. The representation in general is not a complexification of any real representation in contrary to the not deformed Hodge structure.

This structure determines only two quantum numbers l, m which is not a complete system of quantum numbers. There is an open problem how to introduce additional quantum numbers to obtain complete description of the space of quantum states for the system of anyons.

Acknowledgments

The paper was partially supported by the grant NN 201 373236.

References

[1] Peters C A and Steenbrink J H H 2008 *Mixed Hodge structures* , Ergebnisse der Mathematik **Vol 52** (Springer)
[2] Louck J D 2008 *Unitary Symmetry and Combinatorics* (World Scientific)
[3] Lerda A 1992 *Anyons Quantum Mechanics of Particles with Fractional Statistics* (Springer-Verlag)
[4] Forte S 1992 *Rev. Mod. Phys.* **64** 193
[5] Chou C 1991 *Phys. Lett.* **A 155** No. 4, 5, 245–251
[6] Basu R, Date G and Murthy M 1992 *Phys. Rev.* **B 46** No 5, 3139–3141
[7] Milewski J and Lulek T 2001 *J. Math. Phys.* **42 (3)** 1418-1426

Fourier and Schur-Weyl transforms applied to XXX Heisenberg magnet

P Jakubczyk[1], **T Lulek**[2], **D Jakubczyk**[2], **B Lulek**[2]

[1]University of Rzeszòw, Aleja Rejtana 16c, 35-959 Rzeszòw, Poland
[2]Rzeszòw University of Technology, ul. Wincentego Pola 2, 35-959 Rzeszòw, Poland

E-mail: pjakub@univ.rzeszow.pl

Abstract. Similarities and differences between Fourier and Schur-Weyl transforms have been discussed in the context of a one-dimensional Heisenberg magnetic ring with N nodes. We demonstrate that main difference between them correspond to another partitioning of the Hilbert space of the magnet. In particular, we point out that application of the quantum Fourier transform corresponds to splitting of the Hilbert space of the model into subspaces associated with the orbits of the cyclic group, whereas, the Schur-Weyl transform corresponds to splitting into subspaces associated with orbits of the symmetric group.

1. Mathematical description of the model

Let us consider a one dimensional Heisenberg magnetic ring of N nodes, each with a single node spin s and periodic boundary conditions. Such a model reveals a symmetry under collective unitary *rotations* $u \in SU(n)$ in the single-node spaces, and permutations $\sigma \in \Sigma_N$ of nodes. Mathematically, we have two sets

$$\tilde{N} = \{j = 1, 2, ..., N\}, \; \tilde{n} = \{i = 1, 2, ..., n\}, \; n = 2s + 1 \tag{1}$$

the set of N nodes of the crystal, and the set of single-node states (or the set of labels of projections $m_i = s + 1 - i$, $i \in \tilde{n}$ of the single-node spin s), respectively. A state of the whole magnet is given by assigning the spin projection i to each node j of the magnet $f : \tilde{N} \to \tilde{n}$, and it can be presented as

$$|f\rangle = |f(1)f(2)\cdots f(N)\rangle \equiv |i_1, i_2, ..., i_N\rangle, \;\; i_j \in \tilde{n}, j \in \tilde{N}. \tag{2}$$

The set

$$\tilde{n}^{\tilde{N}} = \{f : \tilde{N} \longrightarrow \tilde{n}\} \tag{3}$$

of all states of the form (2) constitutes the computational basis ($\langle f|f'\rangle = \delta_{ff'}$, $f, f' \in \tilde{n}^{\tilde{N}}$) which spans the Hilbert space of the model

$$\mathcal{H} = lc_{\mathbb{C}}\, \tilde{n}^{\tilde{N}}, \tag{4}$$

where $lc_{\mathbb{C}}\, \tilde{n}^{\tilde{N}}$ denotes linear closure of the set $\tilde{n}^{\tilde{N}}$ over the field $\mathbb{C}$ of complex numbers. The space (4) consist of all linear combinations of magnetic configurations with coefficient being complex numbers, in particular all eigenvectors of the magnet belong to this space. Moreover it is an arena of action of the Hamiltonian of the model.

In order to exploit permutational symmetry of the model, let us consider the action $A: \Sigma_N \times \tilde{n}^{\tilde{N}} \longrightarrow \tilde{n}^{\tilde{N}}$ of the symmetric group Σ_N on the basis states (2), as a purely permutational representation

$$A(\sigma)|f\rangle = \binom{f}{f \circ \sigma^{-1}}, \quad f \in \tilde{n}^{\tilde{N}}, \quad \sigma \in \Sigma_N, \tag{5}$$

where $f \circ \sigma^{-1}$ is the composition of mappings $f : \tilde{N} \mapsto \tilde{n}$ and $\sigma^{-1} : \tilde{N} \longrightarrow \tilde{N}$, so that $|f \circ \sigma^{-1}\rangle = |\ldots i_{\sigma^{-1}(j)} \ldots\rangle$, and $A(\sigma\sigma') = A(\sigma)A(\sigma')$. This action decomposes the set (3) of all magnetic configurations into orbits $\mathcal{O}_\mu$ of the symmetric group

$$\mathcal{O}_\mu = \{A(\sigma)|f\rangle \mid \sigma \in \Sigma_N\} \tag{6}$$

labelled by *weights* μ. The weight is a composition

$$\mu = (\mu_1, \mu_2, \ldots, \mu_n), \quad \sum_{i \in \tilde{n}} \mu_i = N, \tag{7}$$

where a part

$$\mu_i = |\{i_j = i \,|\, j \in \tilde{N}\}|, \quad i \in \tilde{n} \tag{8}$$

is the occupation number for the single-node state $i \in \tilde{n}$, for $f \in \mathcal{O}_\mu$. In other words, the weight μ characterises distribution of single-node states over nodes of the magnet, in such a way that the part μ_i is equal to the number of nodes with the single-node state $|i\rangle$.

The restriction

$$A\big|_{\mathcal{O}_\mu} \equiv R^{\Sigma_N:\Sigma^\mu} \tag{9}$$

of the action A to the orbit $\mathcal{O}_\mu$ gives a *transitive representation* of Σ_N with the Young subgroup

$$\Sigma^\mu = \Sigma_{\mu_1} \times \Sigma_{\mu_2} \times \ldots \times \Sigma_{\mu_n} \tag{10}$$

as the stabiliser of an initial magnetic configuration in the orbit $\mathcal{O}_\mu$.

2. Quantum Fourier transform on orbits of the cyclic group. The basis of orbits.

It is also the case that the magnet is invariant under the action of the cyclic subgroup $C_N \subset \Sigma_N$, it implies that each Σ_N-orbit (6) decomposes into C_N-orbits, in accordance with the restriction

$$R^{\Sigma_N:\Sigma^\mu} \downarrow C_N = \sum_{\kappa \in K(N)} m(\mu, \kappa)\, R^{C_N:C_\kappa}, \tag{11}$$

where $R^{C_N:C_\kappa}$ is a transitive representation of C_N, with the cyclic group $C_\kappa \lhd C_N$ being the stabiliser, so that κ is a divisor of N, and $K(N)$ is the lattice of all divisors of N. The multiplicity $m(\mu, \kappa)$ in equation (11), that is, the multiplicity of occurrence of the transitive representation $R^{C_N:C_\kappa}$ of the cyclic group in transitive representation $R^{\Sigma_N:\Sigma^\mu}$ of the symmetric group, or the number of κ-tuply rarefied C_N-orbits in the orbit $\mathcal{O}_\mu$ of the symmetric group Σ_N, is given by a combinatoric formula [1]

$$m(\mu, \kappa) = \begin{cases} \kappa \sum_{\kappa' \in K(\gcd(\frac{\mu_1}{\kappa}, \ldots, \frac{\mu_n}{\kappa}))} \tilde{\mu}(\kappa') \dfrac{(\frac{N}{\kappa\kappa'})!}{\prod_{i \in \tilde{n}} (\frac{\mu_i}{\kappa\kappa'})!} & \text{if all } \frac{\mu_i}{\kappa}, i \in \tilde{n} \text{ are integers} \\ 0 & \text{otherwise.} \end{cases} \tag{12}$$

Here, $\gcd(\frac{\mu_1}{\kappa}, \ldots, \frac{\mu_n}{\kappa})$ denotes the greatest common divisor of integers μ_i/κ, $i \in \tilde{n}$, and $\tilde{\mu} : \mathbb{Z}_{>0} \longrightarrow \{0, \pm 1\}$ is the standard Möbius function of number theory.

Thus, $A \downarrow C_N$ (the action A restricted to the subgroup C_N) decomposes each Σ_N-orbit (6) into strata labelled by $\kappa \in K(N)$, consisting of κ-tuply rarefied C_N-orbits. In this way, we achieve new set of labels

$$|\mu\kappa\mathbf{t}j\rangle, \quad \mu \models N, \quad \kappa \in K(N), \quad \mathbf{t} \in \tilde{m}(\mu,\kappa) = \{1,2,\ldots,m(\mu,\kappa)\}, \quad j \in \tilde{\overline{\kappa}} = \{1,2,\ldots,\overline{\kappa}\}, \tag{13}$$

which classify, in some different way, all basis states of the magnet. Here, $\mathbf{t} \in \tilde{m}(\mu,\kappa)$ labels orbits of C_N on $\mathcal{O}_\mu$ in the stratum κ and $j \in \tilde{\overline{\kappa}}$ labels configurations within C_N-orbits, $\overline{\kappa} = N/\kappa$ is the length of the κ-tuply rarefied C_N-orbit. We call (13) *the basis of orbits* in the set (3) of all magnetic configurations [2]. The basis of orbits specifies all *positions* of the classical counterpart of the Heisenberg magnet with the structure imposed by the cyclic group.

But, sometimes, especially if we work in the field of quantum information [3], it is more convenient to pass from the position representation to the *momentum* one, and the basis (13) allows us to do it for each C_N-orbit separately. In order to present this approach, let us observe, that each C_N-orbit spans a subspace in $\mathcal{H}$, which is invariant under C_N according to the decomposition of the transitive representation $R^{C_N:C_\kappa}$ of the cyclic group into irreps

$$R^{C_N:C_\kappa} = \sum_{k \in B/\kappa} \oplus \Gamma_k, \tag{14}$$

where $\Gamma_k = \exp(2\pi i k/N)$ is the irrep of C_N,

$$B/\kappa = \{k \in B \mid k/\kappa \in \mathbb{Z}\} \subset B \tag{15}$$

is the κ-tuply rarefied Brillouin zone, and

$$B = \{k = 0, \pm 1, \pm 2, \ldots, \left\{ \begin{array}{ll} \pm(N/2-1), N/2 & \text{for } N \text{ even} \\ \pm(N-1)/2, & \text{for } N \text{ odd} \end{array} \right\} \tag{16}$$

is the (finite) *Brillouin zone* of the (one dimensional) magnet $\tilde{N}$. The equation (14) can be written on the level of bases in the form

$$|\,\mu\,\kappa\,\mathbf{t}\,k\rangle = \sqrt{\frac{\kappa}{N}} \sum_{j=1}^{\overline{\kappa}} \exp\left(-\frac{2\pi i k j}{N}\right) |\,\mu\,\kappa\,\mathbf{t}\,j\rangle. \tag{17}$$

One can see that one dimensional subspaces in Eq. (17) which carry irreps Γ_κ, are given by the quantum Fourier transform on orbits of the cyclic group.

2.1. An example of quantum Fourier transform

For a very simple case, the Heisenberg magnet with N nodes and one spin deviation, we have

$$\mu = \{N-1, 1\}, \quad \kappa = 1, \quad \mathbf{t} = 0$$

what means that we have here one orbit of the symmetric group which consists of one stratum, and this stratum consists of just one regular orbit of the cyclic group. Thus from Eq. (17) we have

$$|k\rangle = \frac{1}{\sqrt{N}} \sum_{j=1}^{N} \exp\left(-\frac{2\pi i k j}{N}\right) |j\rangle. \tag{18}$$

The transformation (18) is exactly the quantum Fourier transform on the orthonormal basis $\{|j\rangle, \mid j \in \tilde{N}\}$ (c.f. for example [3]).

3. Schur-Weyl transform on orbits of the symmetric group. The basis of Schur-Weyl duality.

Now, we introduce another partitioning of the Hilbert space $\mathcal{H}$ by use of duality relation between the actions of the unitary and symmetric group. Namely, the space $\mathcal{H}$ is the scene of two actions of the symmetric $A : \Sigma_N \times \mathcal{H} \to \mathcal{H}$ and unitary $B : SU(n) \times \mathcal{H} \to \mathcal{H}$ groups. These actions can be decomposed into irreps, according to formulas

$$A = \sum_{\lambda \in \mathcal{D}_W(N,n)} m(A, \Delta^\lambda)\, \Delta^\lambda, \quad B = \sum_{\lambda \in \mathcal{D}_W(N,n)} m(B, D^\lambda)\, D^\lambda. \tag{19}$$

where Δ^λ and D^λ are irreducible representations of the symmetric and unitary group, respectively, $D_W(N, n)$ denotes the set of all partitions of the integer N into not more than n parts, and appropriate multiplicities, on the strength of Schur-Weyl duality [4], satisfy relations

$$m(A, \Delta^\lambda) = dim\, D^\lambda, \quad m(B, D^\lambda) = dim\, \Delta^\lambda. \tag{20}$$

Relations (20) steam from the quantum-mechanical observation that these two actions mutually commute, that is,

$$[A(\sigma), B(u)] = 0, \quad \sigma \in \Sigma_N, \quad u \in U(n), \tag{21}$$

despite the fact that both dual groups are, for $N > 2, n > 1$, highly noncommutative. In this way the Hilbert space $\mathcal{H}$ of all quantum states of the composite system decomposes into *sectors* $\mathcal{H}^\lambda$,

$$\mathcal{H} = \sum_{\lambda \in \mathcal{D}_W(N,n)} \oplus\, \mathcal{H}^\lambda \tag{22}$$

such that

$$A\big|_{\mathcal{H}^\lambda} = (dim\, D^\lambda)\, \Delta^\lambda, \quad B\big|_{\mathcal{H}^\lambda} = (dim\, \Delta^\lambda)\, D^\lambda. \tag{23}$$

Thus restriction of the action A to the sector $\mathcal{H}^\lambda$ gives $dim\, D^\lambda$ copies of irrep Δ^λ of the symmetric group, whereas restriction of the action B to the same sector $\mathcal{H}^\lambda$ gives $dim\, \Delta^\lambda$ copies of irrep D^λ of the unitary group. The duality of Schur-Weyl admits therefore the irreducible basis of the form

$$b_{irr} = \{|\lambda\, t\, y\rangle \mid \lambda \in \mathcal{D}_W(N, n),\, t \in \tilde{D}^\lambda,\, y \in \tilde{\Delta}^\lambda\},$$

where $\tilde{D}^\lambda$ and $\tilde{\Delta}^\lambda$ are some standard bases for the irrep D^λ and Δ^λ, respectively. We call this basis irreducible, because its elements transforms under the action of the symmetric or unitary group according to these irreps. It is convenient to take standard bases in the form

$$\tilde{\Delta}^\lambda = SYT(\lambda), \quad \tilde{D}^\lambda = SSWT(\lambda, \tilde{n}), \tag{24}$$

where $SYT(\lambda)$ denotes the set of all standard Young tableaux of the shape λ in the alphabet $\tilde{N}$ of nodes, and $SSWT(\lambda, \tilde{n})$ is the set of all Weyl tableaux of the shape λ in the alphabet $\tilde{n}$ of spins.

Finally, on the strength of the Schur-Weyl duality (20), and after choosing basis elements like in Eq. (24), we obtain

$$R^{\Sigma_N : \Sigma^\mu} \cong \sum_{\lambda \trianglerighteq \mu} K_{\lambda\mu}\, \Delta^\lambda, \tag{25}$$

here, the sum runs over all partitions λ greater or equal μ in dominance order, and $K_{\lambda\mu}$ denotes the Kostka number [5]. Combinatorially, $K_{\lambda\mu}$ is equal to the number of all semistandard Weyl tableaux of the shape λ and weight μ. Decomposition (25) can be written at the level of bases in a form

$$|\mu\,\lambda\,t\,y\rangle = \sum_{f\in\mathcal{O}_\mu} \langle \mu f|\lambda t y\rangle \; |\mu f\rangle, \tag{26}$$

where coefficients $\langle \mu f|\lambda t y\rangle$ form a unitary matrix, which transforms the initial basis $\mathcal{O}_\mu$ of magnetic configurations to the irreducible one of the Schur-Weyl duality. We refer to it as to the Kostka matrix at the level of bases. A method of calculation of such transformation matrix relies on representation theory technique called pattern calcullus [6] and was developed in the work [7].

One can see that states given by Eq. (26) are certain wave packets of magnetic configurations which belongs to the orbit $\mathcal{O}_\mu$ of the symmetric group. These states have strictly defined symmetry described by Weyl t and Young y tableaux, what means that they transform under the action of the symmetric and unitary groups according to these irreps.

3.1. An example of Kostka matrix at the level of bases.

For a very simple case, the Heisenberg magnet with N nodes and one spin deviation, we have

$$\mu = \{N-1, 1\},$$

so that the Schur-Weyl basis has the form

$$b_{irr} = \{|\lambda y\rangle \mid \lambda \in \{\{N\}, \{N-1,1\}\}, y \in SYT(\lambda)\}.$$

The Hilbert space $\mathcal{H}$ carries the transitive representation $R^{\Sigma_N:\Sigma_{N-1}}$, which can be decomposed into irrep of the symmetric group

$$R^{\Sigma_N:\Sigma_{N-1}} = \Delta^{\{N\}} + \Delta^{\{N-1,1\}} \tag{27}$$

where $\Delta^{\{N\}}$ is one dimensional irrep spanned on Young tableau $\boxed{1\,|\,2\,|\cdot\cdot|\,N}$, whereas $\Delta^{\{N-1,1\}}$ is the $(N-1)$-dimensional irrep spanned on Young tableaux of the form $y_{j'} = \begin{array}{|c|c|c|}\hline 1 & \cdot\cdot & \cdot \\ \hline j' \\ \cline{1-1}\end{array}$, $2 \le j' \le N$.

Eq. (27) can be written at the level of bases

$$|\lambda y\rangle = \sum_{j\in\tilde{N}} \langle j|\lambda\, y\rangle |j\rangle. \tag{28}$$

where $|j\rangle$ denotes the magnetic configurations (a states with spin deviation at the node j) and the coefficient $\langle j|\lambda\, y\rangle$ can be computed from the formula

$$\langle j|\lambda\, y\rangle = \begin{cases} \frac{1}{\sqrt{N}} & \text{for} \quad \lambda = \{N\},\ y = \boxed{1\,|\,2\,|\cdot\cdot|\,N}, \\ -\frac{1}{\sqrt{(j'-1)j'}} & \text{for} \quad \lambda = \{N-1,1\},\ 1 \le j' < j, \\ \sqrt{\frac{j'-1}{j'}} & \text{for} \qquad j = j', \\ 0 & \text{for} \qquad j' > j. \end{cases} \tag{29}$$

4. Final remarks and conclusions

Two transformations, quantum Fourier and Schur-Weyl, are very similar. The first one is strictly connected with translation symmetry of the magnet and corresponds to splitting of the Hilbert space into subspaces associated with the orbits of the cyclic group. The second, however, arises from permutational symmetry and correspond to splitting of the Hilbert space into subspaces spanned on the orbits of the symmetric group.

These two decompositions of the Hilbert space are very important, because subspaces they create are invariant under the action of the Hamiltonian so that diminishes the size of the eigenproblem by the factor N (in the case of Fourier transform) or $dim\ \lambda$ (in the case of Schur-Weyl transform).

References

[1] Lulek B 1992 *Acta Phys. Pol.* **B22** 371
[2] Milewski J Ambrozko E 2008 *J. Phys. Conf. Ser.* **104** 012040
[3] Nielsen M A Chuang I L 2000 *Quantum Computation and Quantum Information* (Cambridge: Cambridge Univ. Press.)
[4] Weyl H 1931 *Gruppentheorie und Quantummechanik* (Leipzig: Hirzel), (English translation: 1950 *The Theory of Groups and Quantum Mechanics* New York, Dover)
[5] Foulkes H O 1974 *A survey of some combinatorial aspects of symmetric functions* (Paris: Gauthier-Villars)
[6] Louck J D 2009 *Unitary symmetry and combinatorics* (Singapore: World Scientific)
[7] Lulek T Jakubczyk P Jakubczyk D 2004 *Mol. Phys.* **102** 1279

Journal of Physics: Conference Series **213** (2010) 012019 doi:10.1088/1742-6596/213/1/012019

Bethe Ansatz solutions for two qudits

M Labuz[1] **and T Lulek**[2]

[1] University of Rzeszow, Institute of Physics, Rejtana 16a, 35-959 Rzeszow, Poland

[2] Rzeszow University of Technology, The Faculty of Mathematics and Applied Physics, Powstancow Warszawy 6, 35-959 Rzeszow, Poland

E-mail: `labuz@univ.rzeszow.pl`

Abstract. We use the famous Weyl recipe to introduce Bethe Ansatz (BA) solutions for the two-magnon sector in the case of the Heisenberg magnet consisting of a small number N of nodes. In accordance with the Weyl recipe, the obvious symmetry group of the model is the translation group C_N, whereas the group of all automorphisms $\mathrm{Aut}C_N$ constitutes the hidden symmetry, and, in fact, manifests important symmetry properties of the system. With the use of the action of the $\mathrm{Aut}C_N$ on the set of eigenvalues of BA, we propose the way of drawing energy bands for the problem being discussed.

1. Introduction

Considerations concerning the one-dimensional XXX Heisenberg model of a 1/2-spin magnet [1, 2, 3] are often related to solutions obtained by using the Bethe Ansatz method [4, 5, 6, 7]. Owing to the implementation of this famous substitution one can evaluate eigenvalues and eigenfunctions of the problem being discussed for relatively small number of nodes N and spin deviations r. However, this method is not sufficient if we want to have an insight into the shape of energy bands, and so in our paper we propose the way which is helpful in drawing such bands. We focus our attention on the two-magnon sector ($r = 2$) for the magnet consisting of $N = 5$ nodes.

Let's consider the set

$$\tilde{N} = \{j = 1, 2, \ldots, N\} \tag{1}$$

of nodes of the magnet. We can think of it as a computer memory in the language of quantum computation. Furthermore, the set of spin projections on every node

$$\tilde{2} = \{+, -\} \tag{2}$$

means a single bit. Then the set

$$\tilde{2}^{\tilde{N}} = \{f : \tilde{N} \to \tilde{2}\} \tag{3}$$

constitutes *an* orthonormal basis in the Hilbert space $\mathcal{H} = \sum_{r=0}^{N} \oplus \mathcal{H}^{(r)}$, and describes the register of a classical computer with memory N. The single-node space – a qubit – can be introduced as $lc_{\mathbb{C}}\tilde{2}$, whereas $h = lc_{\mathbb{C}}\tilde{N} = \mathcal{H}^{(1)}$ describes the single-magnon space – a qudit. In the case of two qudits we have $h \otimes h \neq \mathcal{H}^{(2)}$, what means that physical states can not be considered as the tensor product of two single-magnon spaces, since decomposition

$$h \otimes h = \mathcal{H}_{\mathrm{hc}} \oplus \mathcal{H}_{\mathrm{ph}} \oplus \mathcal{H}_{\mathrm{a}} \tag{4}$$

contains hard-core, physical and antisymmetric part, respectively, where in fact $\mathcal{H}_{\mathrm{ph}} \equiv \mathcal{H}^{(2)}$.

2. The Weyl recipe

One of tools which is helpful in investigation of *a* symmetry of the Heisenberg magnet is so called the Weyl recipe [8, 9]. In general, it is a useful way in considering some algebraic structure connected with a physical system that manifests some symmetry described by a group G. According to this method, G is a group of the obvious symmetry of a physical system, while it is interesting to take into account a group $\mathrm{Aut}\, G$ of all automorphisms of G, called a group of the hidden symmetry of a set, as a group $\mathrm{Aut}\, G$ contains important properties of the symmetry of a set being investigated. In the case of the Heisenberg ring the obvious symmetry is $G = C_N$, and it constitutes a translational symmetry, so that, according to the Weyl recipe, it reflects in the hidden symmetry $\mathrm{Aut}\, C_N$.

The group $\mathrm{Aut}\, C_N$ is formed by elements τ, for which the image coincides with the group C_N, e.g.

$$\mathrm{Aut}\, C_N = \left\{ \tau : C_N \to C_N \;\middle|\; \begin{array}{c} \tau(j_1)\tau(j_2) = \tau(j_1 j_2),\ j_1, j_2 \in C_N \\ \mathrm{Im}\,\tau = C_N \end{array} \right\}. \tag{5}$$

Hence, the group of automorphisms is formed by elements τ_r satisfying relation

$$\tau_r = \begin{pmatrix} j \\ rj \bmod N \end{pmatrix}, \quad j \in \tilde{N}, \qquad \text{if} \qquad \gcd(r, N) = 1. \tag{6}$$

$\mathrm{Aut}\, C_N$ acts on the Brillouin zone B and transfers one of quasimomentum values $k \in B$ into another one $k' \in B$:

$$P : \mathrm{Aut}\, C_N \times B \to B. \tag{7}$$

Orbits of the action P

$$P = \begin{pmatrix} k \\ rk \bmod N \end{pmatrix}, \quad k \in B, \quad \text{if} \quad \tau_r \in \mathrm{Aut}\, C_N \tag{8}$$

on the Brillouin zone B

$$B = \bigcup_{\kappa \in K(N)} B^{(\kappa)} \tag{9}$$

are generalized stars of the form

$$B^{(\kappa)} = \{k \in B \,|\, \gcd(k, N) = \kappa\}. \tag{10}$$

With a concept of generalized star one can relate the structure of *a* characteristic polynomial of the $\hat{C}_N$ operator. The action of such an operator on any regular orbit of the group C_N has the form:

$$\hat{C}_N|\mathbf{t}, j\rangle = |\mathbf{t}, (j+1) \bmod N\rangle. \tag{11}$$

Then characteristic polynomial of the $\hat{C}_N$ operator can be introduced as

$$w(x) = x^N - 1 = \prod_{\kappa \in K(N)} w_N^{\kappa}(x), \tag{12}$$

where polynomials $w_N^{\kappa}(x)$ are indecomposable over the field $\mathbb{Q}$ of rational numbers, and their decomposition into linear factors are related with generalized stars $B^{(\kappa)}$ via the formula

$$w_N^{\kappa}(x) = \prod_{k \in B^{(\kappa)}} \left(x - \omega^k\right), \tag{13}$$

so that, roots of the polynomial $w_N^{\kappa}(x)$ constitute the set

$$\mathrm{RootOf}(w_N^{\kappa}(x)) = \left\{\omega^k \mid k \in B^{(\kappa)}\right\}, \tag{14}$$

with $\omega = \exp(2\pi i/N)$ being the first primary N-th root of 1. In the case of the group $\mathrm{Aut}\, C_N$ of automorphisms, it turns out that it is helpful in determination of energy bands in the energy spectrum.
It is worth to mention, that, in general, the Hamiltonian commutates with $\hat{C}_N$ operator

$$\left[\hat{H}, \hat{C}_N\right] = 0, \tag{15}$$

and so the total quasimomentum k is a motion constant, while in general it does not commute with automorphisms $\tau_r \in \mathrm{Aut}\, C_N$:

$$\left[\hat{H}, \hat{\tau}_r\right] \neq 0. \tag{16}$$

Automorphisms τ transfer quantum states one into another with different energies and quasimomenta. In next section we present our considerations for the case of Heisenberg magnet consisted of $N = 5$ nodes and $r = 2$ spin deviations.

3. An example: $N = 5$, $r = 2$

For the case of $N = 5$ nodes system, characteristic polynomial has the form:

$$w(x) = x^5 - 1 = (x-1)(x^4 + x^3 + x^2 + x + 1), \tag{17}$$

where $w' \equiv w_5^1(x) = x^4 + x^3 + x^2 + x + 1$ is the polynomial, indecomposable over the field $\mathbb{Q}$. Then we can introduce Δ

$$\Delta = \{\omega \in \mathbb{C} | w(\omega) = 0\} = \{\omega, \omega^2, \omega^3, \omega^4, \omega^5 \equiv 1\} \tag{18}$$

being the set of roots of polynomial w. Table 1 contains the action of elements of the group $\mathrm{Aut}\, C_5$ on elements of the group C_5, the action of elements of the group $\mathrm{Aut}\, C_5$ on the set of roots Δ of polynomial w and the table of multiplication group for the magnet consisted of $N = 5$, respectively.

Table 1. From the left: table of the action of elements of the group $\mathrm{Aut}\, C_5$ on elements of the group C_5, table of the action of elements of the group $\mathrm{Aut}\, C_5$ on the set of roots of polynomial w and multiplication table for the case $N = 5$ according to the rule $\tau_r \tau_{r'} = \tau_{rr' \bmod N}$.

	j				
τ	1	2	3	4	5
τ_1	1	2	3	4	5
τ_2	2	4	1	3	5
τ_3	3	1	4	2	5
τ_4	4	3	2	1	5

τ	$\omega^5 \equiv 1$	ω	ω^2	ω^3	ω^4
τ_1	1	ω	ω^2	ω^3	ω^4
τ_2	1	ω^2	ω^4	ω	ω^3
τ_3	1	ω^3	ω	ω^4	ω^2
τ_4	1	ω^4	ω^3	ω^2	ω

	τ_1	τ_2	τ_3	τ_4
τ_1	τ_1	τ_2	τ_3	τ_4
τ_2	τ_2	τ_4	τ_1	τ_3
τ_3	τ_3	τ_1	τ_4	τ_2
τ_4	τ_4	τ_3	τ_2	τ_1

Under the action of the translation group C_N in the case of $N = 5$, $r = 2$ the Hilbert space $\mathcal{H}^{(2)}$ decomposes into orbits $(\mathbf{t}, j)$ in following way:

$$\mathcal{H}^{(2)} = lc_{\mathbb{C}} \left\{ \begin{array}{cc} \mathbf{t}_1 = (1,4) & \mathbf{t}_2 = (2,3) \\ |j_1 j_2\rangle & |j_1 j_2\rangle \\ 12 & 13 \\ 23 & 24 \\ 34 & 35 \\ 45 & 14 \\ 15 & 25 \end{array} \right\} . \tag{19}$$

In general, for the two-magnon sector $r = 2$ Fourier operator expanded on the space of two-magnon states can be constructed according to the formula:

$$|B\mathbf{t}, k\rangle = \frac{1}{\sqrt{N}} \sum_j \omega^{-kj} |\mathbf{t}, j\rangle = \frac{1}{\sqrt{N}} \sum_j \omega^{(N-1)kj} |\mathbf{t}, j\rangle \ , \tag{20}$$

so that, for $N = 5$ it has the form:

$\hat{F} =$ (21)

		$k \mid \mathbf{t}$: 1		2		-2		-1		0	
$\mathbf{t}$	j	$(1,4)$	$(2,3)$	$(1,4)$	$(2,3)$	$(1,4)$	$(2,3)$	$(1,4)$	$(2,3)$	$(1,4)$	$(2,3)$
$(1,4)$	1	$5^{-\frac{1}{2}}\,\omega^4$	0	$5^{-\frac{1}{2}}\,\omega^3$	0	$5^{-\frac{1}{2}}\,\omega^2$	0	$5^{-\frac{1}{2}}\,\omega$	0	$5^{-\frac{1}{2}}$	0
	2	$5^{-\frac{1}{2}}\,\omega^3$	0	$5^{-\frac{1}{2}}\,\omega$	0	$5^{-\frac{1}{2}}\,\omega^4$	0	$5^{-\frac{1}{2}}\,\omega^2$	0	$5^{-\frac{1}{2}}$	0
	3	$5^{-\frac{1}{2}}\,\omega^2$	0	$5^{-\frac{1}{2}}\,\omega^4$	0	$5^{-\frac{1}{2}}\,\omega$	0	$5^{-\frac{1}{2}}\,\omega^3$	0	$5^{-\frac{1}{2}}$	0
	4	$5^{-\frac{1}{2}}\,\omega$	0	$5^{-\frac{1}{2}}\,\omega^2$	0	$5^{-\frac{1}{2}}\,\omega^3$	0	$5^{-\frac{1}{2}}\,\omega^4$	0	$5^{-\frac{1}{2}}$	0
	5	$5^{-\frac{1}{2}}$	0	$5^{-\frac{1}{2}}$	0	$5^{-\frac{1}{2}}$	0	$5^{-\frac{1}{2}}$	0	$5^{-\frac{1}{2}}$	0
$(2,3)$	1	0	$5^{-\frac{1}{2}}\,\omega^4$	0	$5^{-\frac{1}{2}}\,\omega^3$	0	$5^{-\frac{1}{2}}\,\omega^2$	0	$5^{-\frac{1}{2}}\,\omega$	0	$5^{-\frac{1}{2}}$
	2	0	$5^{-\frac{1}{2}}\,\omega^3$	0	$5^{-\frac{1}{2}}\,\omega$	0	$5^{-\frac{1}{2}}\,\omega^4$	0	$5^{-\frac{1}{2}}\,\omega^2$	0	$5^{-\frac{1}{2}}$
	3	0	$5^{-\frac{1}{2}}\,\omega^2$	0	$5^{-\frac{1}{2}}\,\omega^4$	0	$5^{-\frac{1}{2}}\,\omega$	0	$5^{-\frac{1}{2}}\,\omega^3$	0	$5^{-\frac{1}{2}}$
	4	0	$5^{-\frac{1}{2}}\,\omega$	0	$5^{-\frac{1}{2}}\,\omega^2$	0	$5^{-\frac{1}{2}}\,\omega^3$	0	$5^{-\frac{1}{2}}\,\omega^4$	0	$5^{-\frac{1}{2}}$
	5	0	$5^{-\frac{1}{2}}$	0	$5^{-\frac{1}{2}}$	0	$5^{-\frac{1}{2}}$	0	$5^{-\frac{1}{2}}$	0	$5^{-\frac{1}{2}}$

Hamiltonian matrix in the basis of orbits for the case being discussed has then the form:

$\hat{H} =$, (22)

		$\mathbf{t}_1 = (1,4),\ j$					$\mathbf{t}_2 = (2,3),\ j$				
$\mathbf{t}$	j	1	2	3	4	5	1	2	3	4	5
$(1,4)$	1	-2	0	0	0	0	1	0	0	0	1
	2	0	-2	0	0	0	1	1	0	0	0
	3	0	0	-2	0	0	0	1	1	0	0
	4	0	0	0	-2	0	0	0	1	1	0
	5	0	0	0	0	-2	0	0	0	1	1
$(2,3)$	1	1	1	0	0	0	-4	0	1	1	0
	2	0	1	1	0	0	0	-4	0	1	1
	3	0	0	1	1	0	1	0	-4	0	1
	4	0	0	0	1	1	1	1	0	-4	0
	5	1	0	0	0	1	0	1	1	0	-4

while in the basis of wavelets [10] Hamiltonian matrix appears in quasidiagonal form with respect to the total quasimomentum k as *a* result of the action $\hat{F}^{\dagger}\hat{H}\hat{F}$:

$\hat{F}^{\dagger}\hat{H}\hat{F} =$ (23)

		$k\vert\mathbf{t}$ 1		2		−2		−1		0	
k	$\mathbf{t}$	(1, 4)	(2, 3)	(1, 4)	(2, 3)	(1, 4)	(2, 3)	(1, 4)	(2, 3)	(1, 4)	(2, 3)
1	(1, 4)	$\hat{H}_1$		0	0	0	0	0	0	0	0
	(2, 3)			0	0	0	0	0	0	0	0
2	(1, 4)	0	0	$\hat{H}_2$		0	0	0	0	0	0
	(2, 3)	0	0			0	0	0	0	0	0
−2	(1, 4)	0	0	0	0	$\hat{H}_{-2}$		0	0	0	0
	(2, 3)	0	0	0	0			0	0	0	0
−1	(1, 4)	0	0	0	0	0	0	$\hat{H}_{-1}$		0	0
	(2, 3)	0	0	0	0	0	0			0	0
0	(1, 4)	0	0	0	0	0	0	0	0	$\hat{H}_0$	
	(2, 3)	0	0	0	0	0	0	0	0		

with blocks $\hat{H}_k$ of the form:

$$\hat{H}_1 = \begin{pmatrix} -2 & 1+\omega \\ -\omega-\omega^2-\omega^3 & -4+\omega^2+\omega^3 \end{pmatrix}, \quad \hat{H}_{-1} = \begin{pmatrix} -2 & -\omega-\omega^2-\omega^3 \\ 1+\omega & -4+\omega^2+\omega^3 \end{pmatrix},$$
$$\hat{H}_2 = \begin{pmatrix} -2 & 1+\omega^2 \\ 1+\omega^3 & -5-\omega^2-\omega^3 \end{pmatrix}, \quad \hat{H}_{-2} = \begin{pmatrix} -2 & 1+\omega^3 \\ 1+\omega^2 & -5-\omega^2-\omega^3 \end{pmatrix}, \quad \hat{H}_0 = \begin{pmatrix} -2 & 2 \\ 2 & -2 \end{pmatrix}. \tag{24}$$

Eigenvalues $E_{r'k}$ in the sector $r = 2$ $(r' \leq r)$ obtained from each block $\hat{H}_k$ (24) are introduced in Table 2.

Table 2. Eigenvalues $E_{r'k}$ for the Heisenberg ring consisting of $N = 5$ nodes.

r'	$E_{r',k}$
0	$E_{0,0} = 0$
1	$E_{1,\pm1} = -3 - \omega^2 - \omega^3$ $E_{1,\pm2} = -2 + \omega^2 + \omega^3$
2	$E_{2,0} = -4$ $E_{2,\pm1} = -3 + 2\omega^2 + 2\omega^3$ $E_{2,\pm2} = -5 - 2\omega^2 - 2\omega^3$

Finally, determination of energy bands is related to the action of elements τ of the group Aut C_5 on energy values $E_{r',k}$.

An example:
The action of the element τ_3 on the energy value $E_{r',k} = E_{2,-1} \equiv E_{2,1} = -3 + 2\omega^2 + 2\omega^3$:

$$\begin{aligned} [\tau_3(\omega)]E_{2,-1} = \; & -3 + 2(\omega^2)^3 + 2(\omega^3)^3 = -3 + 2\omega^6 + 2\omega^9 = \\ & -3 + 2\omega + 2\omega^4 = \\ & -3 + 2\omega + 2(-1-\omega-\omega^2-\omega^3) = \\ & -5 - 2\omega^2 - 2\omega^3 = E_{2,2} \equiv E_{2,-2}, \end{aligned} \tag{25}$$

where $\omega^4 + \omega^3 + \omega^2 + \omega + 1 = 0$. Results of the action of *an* each element of the group Aut C_5 on energy values $E_{r',k}$ are introduced in Table 3 and in Figure 1.

Table 3. The action of elements of the group Aut C_5 on eigenvalues spectrum of $r' = 1$ and $r' = 2$ sector for $N = 5$. Table contains only positive k values but it is also true for negative ones because of degeneracy $E_{1,1} \equiv E_{1,-1}$, $E_{1,2} \equiv E_{1,-2}$, $E_{2,1} \equiv E_{2,-1}$, $E_{2,2} \equiv E_{2,-2}$.

	$E_{r',k}$			
	$r' = 1$		$r' = 2$	
τ	$E_{1,1}$	$E_{1,2}$	$E_{2,1}$	$E_{2,2}$
τ_1	$E_{1,1}$	$E_{1,2}$	$E_{2,1}$	$E_{2,2}$
τ_2	$E_{1,2}$	$E_{1,1}$	$E_{2,2}$	$E_{2,1}$
τ_3	$E_{1,2}$	$E_{1,1}$	$E_{2,2}$	$E_{2,1}$
τ_4	$E_{1,1}$	$E_{1,2}$	$E_{2,1}$	$E_{2,2}$

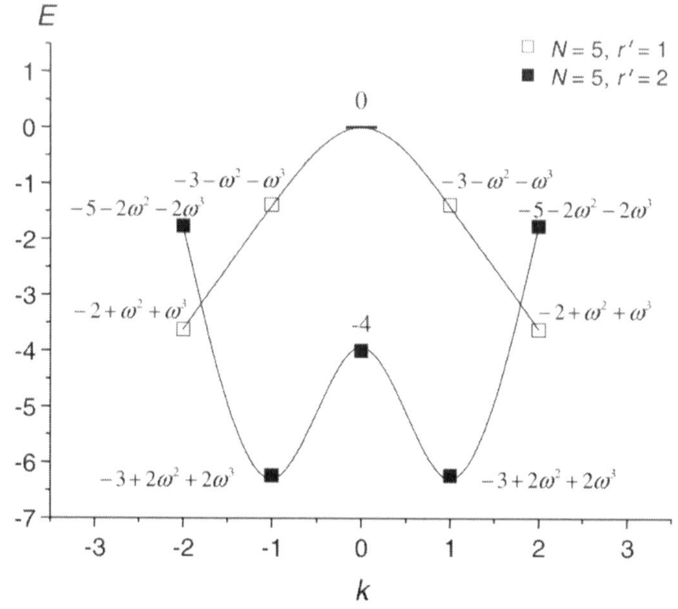

Figure 1. Energy spectrum for the case $N = 5$ in terms of ω, with energy bands for sectors $r' = 1$ and $r' = 2$.

4. Discussion

The Weyl recipe method applied in this paper for investigation of Heisenberg magnet enables us to link eigenvalues expressed in terms of roots, which results in drawing energy bands. For the case $N = 5$, $r = 2$ bands can be separated immediately because of small number of states. However, together with the growth of N one can observe an increase of the number of energy levels for a given value of $k \in B$ so that the way of connecting neighbour points of the Brillouin zone into bands becomes more and more complicated task. Then the action of the group Aut C_N makes up an important advice, and we believe that with additional considerations it can be realized for a greater number of nodes N.

References

[1] Voynarovich F 1982 *J. Phys. C: Solid State Phys.* **15** 6397
[2] Ceulemans A, Cojocaru SP and Chibotaru 2001 *Eur. Phys. J.* **B 21** 511
[3] Caspers WJ, Labuz M, Wal A, Kuzma M and Lulek T 2003 *J. Phys. A: Math. Gen.* **36** 5369

[4] Bethe H 1931 *Z. Physik* **71** 205 (in German; English translation in: Mattis DC 1993 *The Many-Body Problem* (World Sci.: Singapore) 689)
[5] Faddeev LD and Takhtajan LA 1981 *LOMI* **109** 134 (in Russian; English translation: 1984 *J. Sov. Math.* **24** 241)
[6] Caspers WJ 1989 *Spin systems* (World Sci.: Singapore)
[7] Caspers WJ, Wal A, Labuz M, Kuzma M and Lulek T 2004 *J. Math. Phys.* **45** 391-399
[8] Weyl H 1952 *Symmetry* (New Jersey: Princeton U.P)
[9] Lulek B 1993 *The recipe of Weyl, irregular orbits, and breaking of translational symmetry of the one-dimesional Heisenberg magnet* (Wydawnictwo Naukowe UAM: Poznań)
[10] Lulek B, Lulek T, Wal A and Jakubczyk P 2003 *Physica B: Phys. Cond. Matt.* **337** 375

Journal of Physics: Conference Series **213** (2010) 012020 doi:10.1088/1742-6596/213/1/012020

Algebraic Bethe Ansatz for $N = 4$

D. Jakubczyk[1], **T. Lulek**[1], **P. Jakubczyk**[2], **B. Lulek**[1]

[1]Chair of Physics, Rzeszòw University of Technology, Powstancòw Warszawy 6, 35-959 Rzeszòw, Poland

[2]Institute of Physics, University of Rzeszòw, Aleja Rejtana 16c, 35-959 Rzeszòw, Poland

E-mail: djak@prz.edu.pl

Abstract. In the present paper we explore some useful devices for discussing the exact solutions of XXX isotropic Heisenberg Hamiltonian for the single node spin equal to $\frac{1}{2}$ as the tensor product states. Our aim is to presents the monodromy matrix and the Lax operator in the contex of the Bethe Ansatz, signed from now on BA. We construct the former, for N equal to 4 nodes of the magnet, using the so called auxiliary space which is taken as a copy of $\mathbb{C}^2$. The form of this matrix in the basis of orbits of the translation group C_4 reveals its block structure of all possible nonzero elements. Each block has its meaning in the language of creation and anihilation of the magnon. This fact implies, that one can think about appropriate operators and create the theory very similar to that of the quantum oscillator.

1. Introduction

The famous BA was introduced by Bethe [1] in 1931 as an exact solution for the isotropic Heisenberg Hamiltonian with the single node spin $\frac{1}{2}$. The method used by Bethe in [1] is today often called a coordinate BA. In this paper we aim to describe the algebraic BA [2], [3], [4] which is a generalization of the former. Due to this method one can construct, for each exactly solvable model [5], its monodromy matrix, which satisfies certain commutation relations, see for example [6]. In this paper we consider the XXX Heisenberg magnet with the single node spin $\frac{1}{2}$. The monodromy matrix is a generating object for both the Hamiltonian and its eigenvectors for our system. The trace of the monodromy matrix gives the main observables such as the total spin, quasimomentum or the Hamiltonian. The elements of the monodromy matrix play the role of creation and anihilation operators, which provide the eigenvectors of the Hamiltonian when acting on the vacuum. Each such operator is a function of so called spectral parameter. This fact yields that a single exact eigenfunction is determined by the set of spectral parameters given as solutions of a system of algebraic Bethe equations.

2. Tensor product Hilbert space

Let $\tilde{N} = \{j = 1, 2, \ldots, N\}$ be the set of nodes of a magnetic ring, and $\tilde{2} = \{+, -\}$ - the set of single-node spin projections. Then the Hilbert space for such a system has a natural tensor-product form

$$\mathcal{H} = \prod_{j=1}^{N} \otimes h_j = h_1 \otimes h_2 \otimes \ldots \otimes h_j \otimes \ldots \otimes h_N, \tag{1}$$

where

$$h_j = lc_{\mathbb{C}}\tilde{2} \cong \mathbb{C}^2 \tag{2}$$

is the linear closure of the set $\tilde{2}$ over $\mathbb{C}$. The set of all mapping $f : \tilde{N} \longrightarrow \tilde{2}$ constitues $\tilde{2}^{\tilde{N}}$ magnetic configurations, which can be written in the form

$$|f> \equiv |i_1> \otimes |i_2> \otimes \ldots \otimes |i_N> \equiv |i_1, i_2, \ldots, i_N>, i_j \in \tilde{2}, j \in \tilde{N}. \tag{3}$$

Since $|i_N> \in h_N$ and $|f> \in \mathcal{H}$, the set (3) provides an orthonormal basis in $\mathcal{H}$, initial for quantum computations, i. e.

$$\mathcal{H} = lc_{\mathbb{C}}\ \tilde{2}^{\tilde{N}}. \tag{4}$$

The space $\mathcal{H}$ of all quantum states of the Heisenberg magnet, with $\dim\mathcal{H} = 2^N$, decomposes as

$$\mathcal{H} = \sum_{r=0}^{N} \oplus\ \mathcal{H}^r, \tag{5}$$

into subspaces $\mathcal{H}^r$, $\dim\mathcal{H}^r = \binom{N}{r}$, with the fixed number r of Bethe pseudoparticles (spin deviations). Stated otherwise, $\mathcal{H}^r$ is the space of all states of the magnet with the projection

$$M = \frac{N}{2} - r \tag{6}$$

of the total spin $S = \frac{N}{2} - r'$, where $0 \leq r' \leq r$, or of all states with the *weight* [7]

$$\mu = \{N - r, r\}. \tag{7}$$

From group-theorethical point of view we can express $\mathcal{H}^r$ as

$$\mathcal{H}^r = lc_{\mathbb{C}}\mathcal{O}_\mu, \tag{8}$$

where $\mathcal{O}_\mu$ is an orbit of the symmetric group Σ_N, given by the weight $\mu = \{N - r, r\}$. Let

$$\mathbf{t} = (t_1, t_2, \ldots, t_r),\ \sum_{\alpha \in \tilde{r}} t_\alpha = N \tag{9}$$

be the *relative* distribution of Bethe pseudoparticles within an orbit of the translational symmetry group C_N of the ring (C_N-orbit), being a subgroup of the symmetric group Σ_N ($C_N \subset \Sigma_N$). The vector $\mathbf{t}$ provides a label for classification of C_N-orbits in $\mathcal{O}_\mu$, so that

$$\mathcal{O}_\mu = \bigcup_{\mathbf{t} \in T} \mathcal{O}_{\mathbf{t}}, \tag{10}$$

where $T = \mathcal{O}_\mu / C_N$ is the set of orbits, which we refer to, after [8], as a *skeleton* of the configuration space $Q^{(r)} = \mathcal{O}_\mu$, for more details see [8]. The Hamiltonian of the system of interacting spin deviations in the space (1) has the following form

$$H = J \sum_{j \in \tilde{N}} (\mathbf{s}_j \cdot \mathbf{s}_{j+1} - 1/4), \tag{11}$$

which can be writed as

$$H = \frac{J}{4} \sum_{i=1}^{N} [\frac{1}{2}(\sigma_j^+ \sigma_{j+1}^- + \sigma_j^- \sigma_{j+1}^+) + \sigma_j^z \sigma_{j+1}^z - I_N], \tag{12}$$

where

$$\sigma_j = I_1 \otimes I_2 \otimes \ldots \otimes I_{j-1} \otimes \sigma \otimes I_{j+1} \otimes \ldots \otimes I_N, \tag{13}$$

σ is one of the Pauli matrices acting in h and $\sigma^+ = \frac{1}{2}\sigma_x + i\sigma_y$ ($\sigma^- = \frac{1}{2}\sigma_x - i\sigma_y$).

r	$\mathbf{t}$	j	J	J_M for + in V	J_M for - in V
0	0	1	1	1	2
1	4	1	9	17	18
		2	5	9	10
		3	3	5	6
		4	2	3	4
2	13	1	13	25	26
		2	7	13	14
		3	4	7	8
		4	10	19	20
	22	1	11	21	22
		2	6	11	12
3	112	1	15	29	30
		2	14	27	28
		3	12	23	24
		4	8	15	16
4	1111	1	16	31	32

Table 1. The basis of orbits $|r\mathbf{t}j>$ of the group C_4 in $\mathcal{H}$.

3. Some bases in $\mathcal{H} \otimes V$

Let us consider the tensor product of h_j or $\mathcal{H}$ and some auxiliary space V, namely $h_j \otimes V$ and $\mathcal{H} \otimes V$. The space V is identical with h_j, but is not related to any node of the magnetic ring $\tilde{N}$. Thus we have

$$h_j \otimes V \cong \mathbb{C}^2 \otimes V \cong (\mathbb{C}^2)^{\otimes 2}, \tag{14}$$

and

$$\mathcal{H} \otimes V \cong (\mathbb{C}^2)^{\otimes N} \otimes V \cong (\mathbb{C}^2)^{\otimes N+1}. \tag{15}$$

From now on, we will proceed with the case of the magnetic ring with 4 nodes.

As the initial basis for quantum calculations in $\mathcal{H} \otimes V$, we chose *the lexical basis* (numbered by J_M) of the simple tensor product $h^{\otimes(4+1)}$ of $4+1$ spaces h given by (2). The lexical basis for the tensor product of four h spaces, that is for the physical space $\mathcal{H}$ of our magnet, is labelled by the number J what we can see in the Table 1. Each physical state in sixteenth dimensional space $\mathcal{H}$ is doubly degenerated, since the auxiliary space V is two dimensional.

The second basis used in our calculatons is given by the structure of orbits of the translational symmetry group of the magnet C_4 (C_4-orbit) acting on the magnetic configurations given by the formula (3), and is called *the basis of orbits*. Each vector of this basis is labelled by three quantum numbers (Table 1): r, $\vec{t}$ and j, where j counts the magnetic configurations within the C_4-orbit. The only diffrence between the lexical basis and the basis of orbits is ordering of its contents.

4. The Lax and the monodromy operators

The Lax operator given, by the formula

$$L_{j,a}(\lambda) = \lambda I \otimes I_a + i \sum_{k=1}^{3} s_j^k \otimes \sigma_a^k, \tag{16}$$

acts in $\mathcal{H} \otimes V$, thus $L_{j,a}(\lambda) \in End(\mathcal{H} \otimes V)$, where a, λ and $I(I_a)$ denotes respectively the auxiliary space $V = h_a$, spectral parameter, and the identity operator in $\mathcal{H}(h_a)$. Thanks to the Dirac equivalency $P_{j,a} = \frac{1}{2}I + 2\mathbf{s_j s_a}$ between the product of two single-node spin operators $(\mathbf{s_j}, \mathbf{s_a})$ and two-node permutation oparator $P_{j,a}$, one can write $L_{j,a}(\lambda)$ in the form

$$L_{j,a}(\lambda) = (\lambda - \frac{i}{2})I \otimes I_a + iP_{j,a}, \tag{17}$$

where $P_{j,a}$ is the permutation operator between the spaces h_j and h_a.
For two different spectral parameters (λ, μ) the following relation is true

$$R_{a_1,a_2}(\lambda - \mu)L_{j,a_1}(\lambda)L_{j,a_2}(\mu) = L_{j,a_2}(\lambda)L_{j,a_1}(\lambda)R_{a_1,a_2}(\lambda - \mu), \tag{18}$$

where

$$R_{a_1,a_2}(\lambda) = \lambda I_{a_1} \otimes I_{a_2} + iP_{a_1,a_2}, \tag{19}$$

and $R_{a_1,a_2}(\lambda) \in End(V_1 \otimes V_2)$ with V_1 and V_2 being some auxiliary spaces isomorfic with V [2], [3], [4]. The equation (18) turns out to be the Yang-Baxter relation [2], [3], [4], which is equivalent with original BA equations given in [1], that is why this approach is called algebraic Bethe Ansatz (ABA). From (17) one can see that the action of $L_{j,a}(\lambda)$ in $\mathcal{H} \otimes V$ implies exchange of contents between j-th and a-th nodes. The product of two Lax operators $L_{j',a}(\lambda)$, $L_{j,a}(\lambda)$ acting in $\mathcal{H} \otimes V$ constitutes some translaton operator on the nodes (j, j', a) of the magnetic ring $\tilde{N}$. The product of three Lax operators $L_{j'',a}(\lambda)$, $L_{j',a}(\lambda)$, $L_{j,a}(\lambda)$ acting in $\mathcal{H} \otimes V$ constitutes some translaton operator on the nodes (j, j', j'', a) of the magnetic ring $\tilde{N}$. Thus the N-th product of the consecutive Lax operators, namely

$$L_{N,a}(\lambda)L_{N-1,a}(\lambda)\ldots L_{2,a}(\lambda)L_{1,a}(\lambda), \tag{20}$$

defines some translation operator on the nodes $1, 2, \ldots, N-1, N, a$ of the magnetic ring $\tilde{N}$ plus the content of the auxiliary space V. The operator given by (20) is called monodromy operator $T_a(\lambda)$ acting in $\mathcal{H} \otimes V$. The general structure of this matrix is presented in the Figure 1, but the Figure 2 presents the structure of this matrix in the basis of orbits.
$B(\lambda)$ and $C(\lambda)$ submatrices stand for creation and anihilation operator, respectively, for magnon

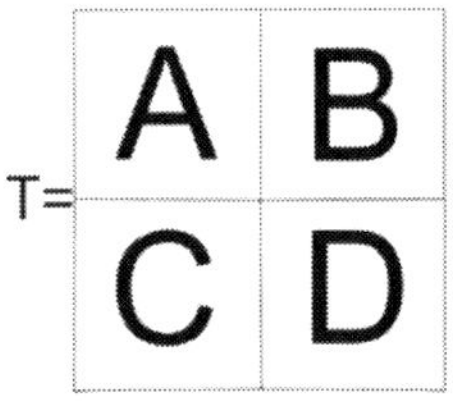

Figure 1. The general structure of the monodromy matrix.

labelled by the spectral parameter λ. The $Tr(T_a(\lambda)) = A(\lambda) + D(\lambda)$ constitutes the generating object for N constants of motion with the operator of XXX Heisenberg Hamiltonian among themselfes. For r spectral parameters $\lambda_{j'}$ $(j' = 1, 2, ..., r)$, which satisfy Bethe equation in the form

$$(\frac{\lambda_{j'} + \frac{i}{2}}{\lambda_{j'} - \frac{i}{2}})^N = \prod_{k \neq j'}^{r} \frac{\lambda_{j'} - \lambda_k + i}{\lambda_{j'} - \lambda_k - i}, \tag{21}$$

and for the vacum state Ω (consisting only of pluses) one can build the exact BA solution $\phi(\lambda_1, ..., \lambda_r)$ for the system of r magnons characterised by the set $\{\lambda\} = (\lambda_1, ..., \lambda_r)$ of spectral parameters as follows

$$\phi(\lambda_1, ..., \lambda_r) = B(\lambda_1)...B(\lambda_r)\Omega \tag{22}$$

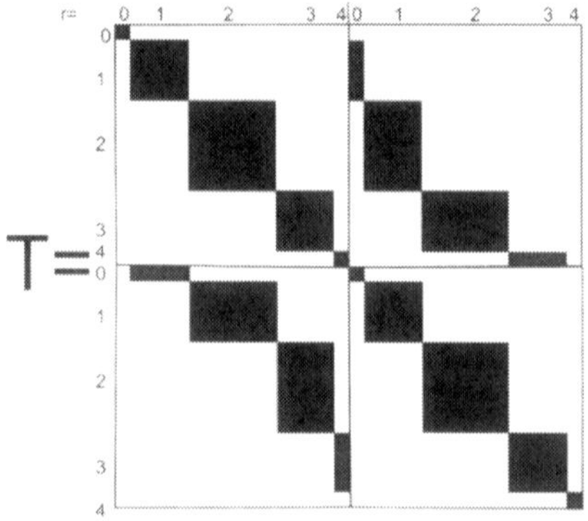

Figure 2. The structure of the monodromy matrix in the basis of orbits of the translation group C_N. The non-zero elements can occur in black boxes only.

[2], [3], [4]. It is very crucial to observe, that the set of operators $\{B(\lambda)\} = \{B(\lambda_{j'}), j' \in \tilde{r}\}$ alone, is not able to generate the complete basis in $\mathcal{H}^r$, since the vector (22) is the *highest weight vector* i. e. with the total spin $S = M = \frac{N}{2} - r$ (6). In order to obtain the remaining basis state in $\mathcal{H}^r$ one shoud use the well known operator $S^- = s_1^- + s_2^- + ... + s_N^-$ where $s_j^- = \frac{1}{2}\sigma_j^-$ $(j \in \tilde{N})$, which generates the magnon with *pseudomomentum* $p_{j'} = 0$, i. e. so called Goldstone boson. From the relation between $p_{j'}$ and $\lambda_{j'}$ i. e. $\lambda_{j'} = \frac{1}{2}cot\frac{p}{2}$ it is easy to notice that the operator $B(\lambda_{j'})$ is singular for $p_{j'}$.

5. Final remarks and conclusions

We showed some of details for construction the monodromy matrix for the magnetic ring with four nodes, using the auxiliary space. We emphasis on the fact that it is very crucial to distinguish between the physical and the nonphysical states in case when one has the auxiliary space besides the real one. We localized all possible nonzero elements of this matrix on the Picture 2. Taking the trace of the monodromy matrix over the auxiliary space one can obtain the sum of two operators acting in $\mathcal{H}$. Two remaining ones, which are placed on the antidiagonal positions of this matrix are responsible for creation and anihilation of one magnon. We presented the role of the $B(\lambda)$ and S^- operators in creation of the basis of the spaces $\mathcal{H}^r$ for $r = 0, 1, 2, 3, 4$, where the first operator is responsible for creation of the highest weight vectors $(M = S)$, and the second one creates the so called Goldstone bosons.

References

[1] Bethe H 1931 *Z. Physik* **71** 205 (in German; English translation in: Mattis D C 1993 The Many-Body Problem)(World Sci. Singapore 689-716)
[2] Faddeev L D How Algebraic Bethe Ansatz works for integrable model *hep-th/9605187v1*
[3] Faddeev L D Algebraic Aspects of Bethe-Ansatz *hep-th/9404013v1*
[4] Faddeev L D Takhtadzhyan L A 1981 Spectrum and scattering of excitations in the one-dimensional isotropic Heisenberg model *Zap. Nauch. Semin. LOMI* **109**, 134-178 (in Russian; English transl. 1984 *J. Sov. Math.* **24**, 241-267)
[5] Baxter R J 1982 *Exactly Solved Models in Statistical Mechanics* (Academic Press)
[6] Schilling A 2002 *Symmetry and Structural Properties of Condensed Matter* (World Sci., Singapore 2002) 201-223
[7] Lulek B Jakubczyk D Jakubczyk P 2004 Kinematics of the Heisenberg chain and irreducible bases of the Weyl duality *Mol. Phys.* **102** 1269-1278
[8] Lulek B Lulek T Wal A Jakubczyk P 2003 The basis of wavelets for a finite Heisenberg magnet *Physica B - Condensed Matter* **337/1** 375-387

Journal of Physics: Conference Series **213** (2010) 012021 doi:10.1088/1742-6596/213/1/012021

Thermoelectric effects in transport through a quantum dot attached to ferromagnetic electrodes

R Świrkowicz[1], M Wierzbicki[1], and J Barnaś[2,3]

[1] Faculty of Physics, Warsaw University of Technology, Koszykowa 75, 00-662 Warszawa, Poland

[2] Department of Physics, Adam Mickiewicz University, Umułtowska 85, 61-614 Poznań, Poland

[3] Institute of Molecular Physics, Polish Academy of Sciences, Smoluchowskiego 17, 60-179 Poznań, Poland

E-mail: barnas@amu.edu.pl

Abstract. We present a theoretical analysis of thermoelectric effects in a system consisting of a single-level quantum dot attached to two ferromagnetic leads. Calculations have been performed in terms of the Green function formalism based on the equation of motion method and Hartree-Fock approximation. The calculated thermopower shows an oscillatory-like behavior with increasing gate voltage and also depends on relative alignment of the magnetic moments of both electrodes and on the leads' spin polarization. External magnetic field applied to the system leads to an additional structure due to spin splitting of the dot level. Thermal conductance is also calculated and correlated with the electrical one.

1. Introduction

There is currently an increasing interest in thermoelectric and magneto-thermoelectric transport phenomena in nanoscopic systems [1,2]. Quantum confinement in such nanoscale systems as well as discrete charging with single electrons and the associated Coulomb blockade strongly affect not only charge transport properties, but also thermoelectric ones, leading to novel phenomena like for instance oscillations of the thermopower and thermal conductance with increasing gate voltage [3].

Spin-dependent thermoelectric phenomena in magnetic nanostructures also have been studied [2,4–6]. In fact, such phenomena are of particular interest as they offer a unique possibility of creating spin currents, which may be of some importance for future applications in spintronic devices [2]. Following the discovery of giant magnetoresistance in metallic magnetic multilayers, magneto-thermoelectric phenomena in such structures have been investigated, too. For instance, giant magneto-thermopower was found in some multilayer nanopillars [4], while Peltier and Seebeck effects in such structures were analyzed theoretically [5]. Some new results on thermoelectric phenomena in single-electron devices based on quantum dots have been reported as well [6]. Following the current interest, we consider in this paper thermoelectric phenomena in a system consisting of a quantum dot attached to two ferromagnetic leads. Our main interest is in variation of the thermoelectric effects with magnetic configuration of the system and with spin polarization of the leads.

2. Model

Thermopower is defined as the voltage drop ΔV created by a temperature gradient ΔT, and is quantitatively described by the Seebeck coefficient, $S = \Delta V/\Delta T$, calculated on the condition of vanishing charge current, $I = 0$ [7]. In the linear response regime, the thermopower can be expressed as [7]

$$S = -L_1/(eTL_0), \tag{1}$$

where e is the electron charge, while L_n ($n = 0, 1, 2...$) are defined as $L_n = (-1/h)\,\mathrm{Tr}\int dE\,(\partial f/\partial E)\,\mathbf{T}(E)\,(E - E_\mathrm{F})^n$. Here $f(E)$ denotes the Fermi-Dirac distribution function in the leads, h is the Planck constant, E_F is the Fermi energy of the electrodes in equilibrium, while $\mathbf{T}(E)$ describes the transmission probabilities through the system and is a 2×2 matrix in the spin space. The thermal conductance κ can be calculated in a similar way, and is given by

$$\kappa = (1/T)(L_2 - L_1^2/L_0), \tag{2}$$

with L_n defined above. Thus, the key point is to calculate the matrix $\mathbf{T}(E)$ and indirectly the required parameters L_n.

Thus, we calculate now the matrix $\mathbf{T}(E)$ for a system consisting of a single-level quantum dot attached to ferromagnetic electrodes which are characterized by the spin polarization factor p. The system is described by the Hamiltonian

$$H = H_\mathrm{e} + H_\mathrm{D} + H_\mathrm{T}, \tag{3}$$

where H_e corresponds to the left (L) and right (R) leads, H_D is the dot Hamiltonian, and H_T describes tunneling processes between the dot and electrodes. Hamiltonian H_D is taken in the following form:

$$H_\mathrm{D} = \sum_\sigma \varepsilon_\sigma n_\sigma + U n_\uparrow n_\downarrow, \tag{4}$$

with n_σ being the electron number operator, and U describing the strength of electron correlations in the dot (Hubbard parameter). The energy level becomes spin split in an external magnetic field, $\varepsilon_\sigma = \varepsilon_0 - \hat{\sigma}\Delta$ for spin-up ($\hat{\sigma} = +$) and spin-down ($\hat{\sigma} = -$) electrons, which gives rise to the splitting equal to 2Δ. The tunneling processes, in turn, are described by spin-dependent coupling parameters $\Gamma_\pm^{\mathrm{L(R)}} = \Gamma\,(1 \pm p)$ for spin-majority (upper sign) and spin-minority (lower sign) electrons. The parameter Γ characterizes the coupling strength and is treated as a constant.

In the absence of spin flip processes (eg. due to spin-orbit interaction), and when magnetic moments of the leads are collinear (parallel or antiparallel), the transmission matrix $\mathbf{T}(E)$ is diagonal and one can write $\mathrm{Tr}\,\mathbf{T}(E) = \sum_\sigma T_\sigma(E) = \sum_\sigma A_\sigma(E)\Gamma_\sigma^\mathrm{L}\Gamma_\sigma^\mathrm{R}/\left(\Gamma_\sigma^\mathrm{L} + \Gamma_\sigma^\mathrm{R}\right)$, with $A_\sigma(E)$ being the spectral function calculated in terms of the Green function formalism based on the equation of motion method and Hartree-Fock approximation. Here, $T_\sigma(E)$ is the transmission coefficient in the spin channel σ.

3. Numerical results

Thermopower S has been calculated numerically as a function of the dot's energy level ε_0 (which can be tuned by an external gate voltage) for ferromagnetic electrodes with spin polarization p, and for both parallel (P) and antiparallel (AP) configurations of the leads' magnetic moments. The influence of temperature and external magnetic field on the thermoelectric effects also has been taken into account. The results are presented in Fig. 1 for $\Gamma = 0.1$meV and $U = 2$meV.

All the calculations have been performed for the whole level position (gate voltage) range, i.e. when the dot is empty, singly occupied (Coulomb blockade regime), and doubly occupied. A typical oscillatory-like behavior of S with increasing ε_0 has been found. Maximal absolute values

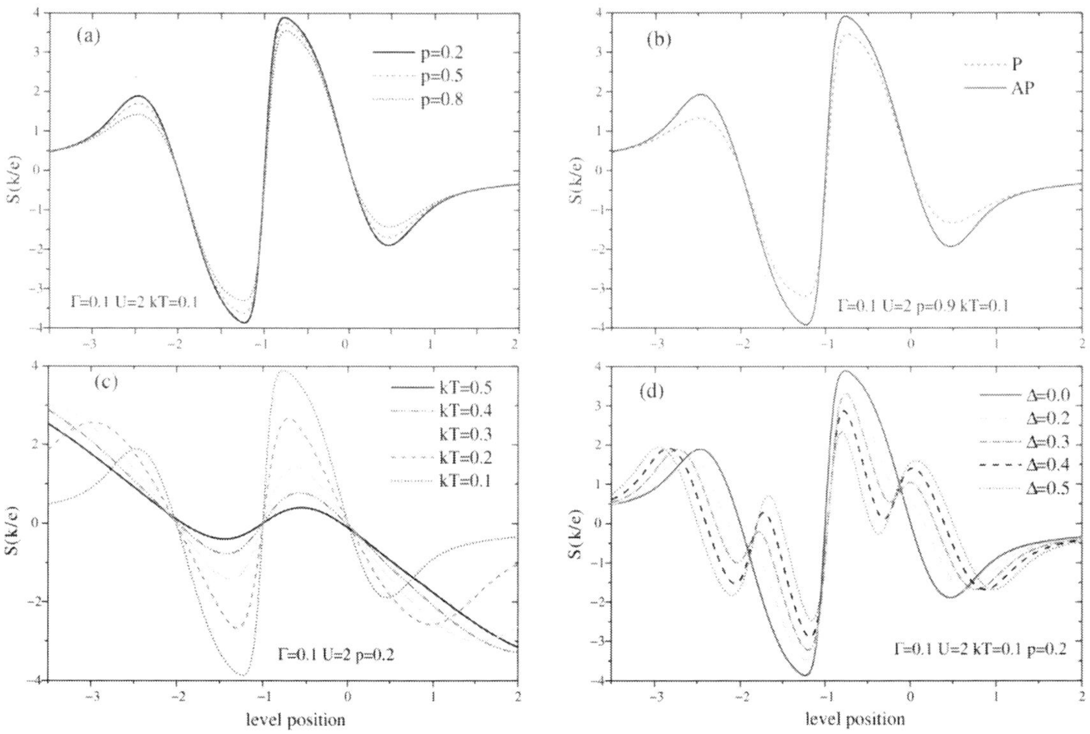

Figure 1. Thermopower S as a function of the dot's level position, calculated for indicated values of the leads' polarization p (a), for two different configurations of the leads' magnetic moments (b), and for different temperatures (c). The effects of level splitting Δ due to an external magnetic field are shown in (d). The energy parameters, like kT, U, Γ, Δ, and dot's energy level are measured in meV.

of S occur for the dot level lying in the Coulomb gap. However, the thermopower vanishes in the symmetry point $\varepsilon_0 = -U/2$ (middle of the Coulomb gap), and also at resonances $\varepsilon_0 = E_F = 0$ and $\varepsilon_0 = -U$ [8, 9]. As shown in figure 1a, the thermopower depends on the leads' polarization and decreases with increasing p. It also decreases when the magnetic configuration is changed from the antiparallel to parallel one (figure 1b). These changes in S, though not very strong, are ,however, quite remarkable, indicating that the optimal conditions for the generation of independent of spin voltage drop ΔV occur when transmissions are equal in both spin channels, which takes place in systems with non-magnetic electrodes or for ferromagnetic leads but in the antiparallel configuration of magnetic moments. In such a situation the charge current as well as spin current are simultaneously equal to zero.

Thermopower strongly varies with temperature (figure 1c). At relatively low temperatures, $kT \approx \Gamma$, well-defined maximum and minimum can be observed in the Coulomb blockade region. The blockade effects, however, become less significant with increasing T, and the curve representing S as a function of the level position becomes flatter. External magnetic field applied to the system also leads to a suppression of the thermoelectric effect for gate voltages corresponding to the Coulomb gap (figure 1d). However, an additional structure appears then due to splitting of the dot level by an external field, $\varepsilon_0 \to \varepsilon_\sigma = \varepsilon_0 \pm \Delta$. When Δ increases, the spin splitting of the dot level strongly affects transport through the dot. This is especially visible in the electrical conductance, $G = e^2 L_0$, presented in figure 2a, as well as in the thermal

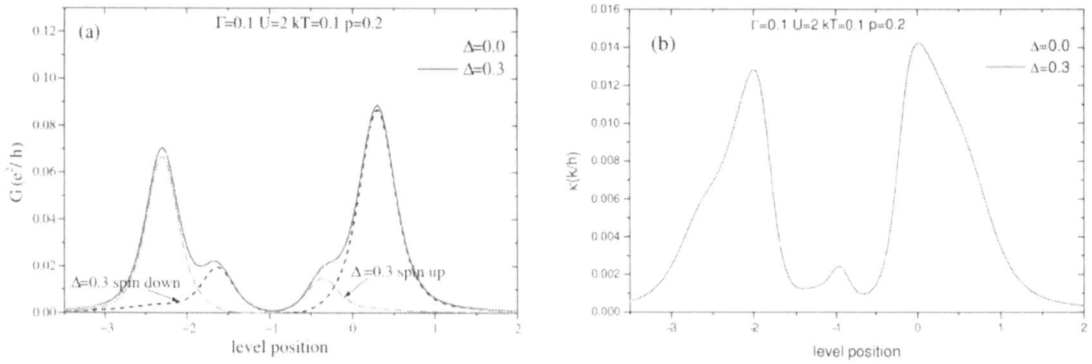

Figure 2. Linear electrical conductance G (a) and thermal conductance (b) in the parallel magnetic configuration, as a function of the dot's level position for two different values of the splitting parameter Δ. Spin resolved electrical conductance for nonzero magnetic field is also presented in (a). The energy parameters kT, U, Γ, Δ, and energy of the dot's level are measured in meV.

conductance κ shown in figure 2b. The spin resolved conductance G_σ corresponding to a given spin channel described by the transmission $T_\sigma(E)$ is also presented in figure 2a. Two well known resonance peaks, typical for the conductance of single-level quantum dots, appear in the absence of magnetic field. These peaks become split when a magnetic field is applied ($\Delta \neq 0$). Since the transmission is different in the two spin channels, the intensity of the peaks is also spin dependent, as it is clearly visible in figure 2. The influence of external field is however less pronounced at higher temperatures as the appropriate peaks are much wider and flatter, so the level splitting is not well resolved for smaller values of Δ. Similar features also appear in the thermal conductance. In the absence of magnetic field we find two peaks in the thermal conductance, whose position corresponds to the position of similar peaks in the electrical conductance. An additional peak develops between the two peaks (in the middle of the Coulomb gap). The peaks in thermal conductance are however broader, so the splitting in the external magnetic field assumed in figure 2 is not clearly resolved. The level splitting was also visible in the thermopower calculated as a function of the level position and discussed above (figure 1d).

Dimensionless figure of merit, defined as

$$ZT = \frac{GS^2}{\kappa}T, \tag{5}$$

quantitatively describes the thermoelectric efficiency of a device. Dependence of ZT on the magnetic configuration, temperature, and external field applied to the system is presented in figure 3. Note, that the thermal efficiency diminishes considerably when configuration of the moments is changed from the antiparallel to parallel one. Such a behavior is a result of two effects, namely of a relatively strong decrease of the thermopower S during the configuration change and also of an increase of thermal conductance κ (not shown here). At low temperatures ($kT \approx \Gamma$) the Coulomb blockade effects are important and heat transport is suppressed in the Coulomb gap, so the figure of merit shows two high peaks in this region of gate voltages. In turn, ZT is equal to zero in the symmetry point, $\epsilon_0 = -U/2$, and also at resonances, where the thermopower S vanishes and changes sign. Moreover, for energies far from the resonant ones, small side maxima can be observed. The picture changes dramatically with increasing temperature (figure 3b). For higher temperatures ($kT > 3\Gamma$), the intensity of both central peaks

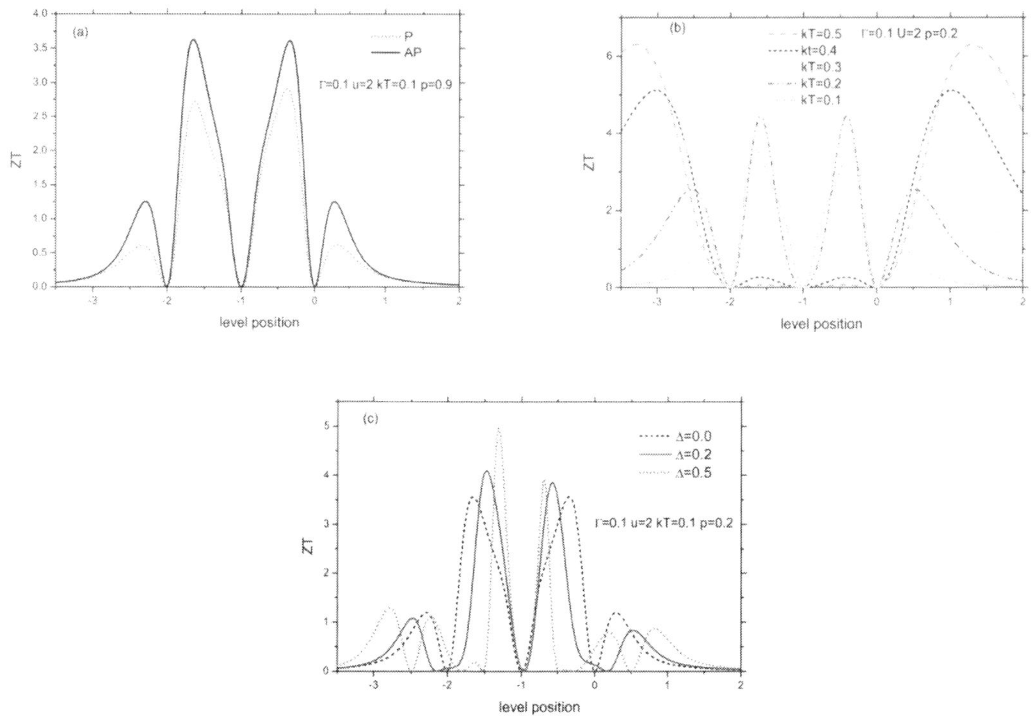

Figure 3. Figure of merit as a function of level position for a) two magnetic configurations, b) different temperatures and c) different level splitting.

rapidly drops, whereas the side maxima become enhanced and start dominating the whole dependence. Note, that for $kT > 5\Gamma$ the thermal efficiency of the device is very small for the gate voltages corresponding to the Coulomb gap. At higher temperatures the blockade effects are strongly suppressed, which leads to increase of heat transmission. The strong increase fully determines the thermoelectric efficiency of the device. Though, at high temperatures ZT shows well-developed side maxima, they are of no practical importance, as for these values of gate voltages the system is out of resonance and the electrical conductance is very low.

An external magnetic field has a strong influence on the figure of merit, as depicted in figure 3c for several values of the splitting parameter Δ. As Δ increases, the central peaks in ZT become narrower, so in strong fields the device can show the high thermoelectric efficiency only in a very narrow region of gate voltages close to the symmetry point $\epsilon_0 = -U/2$. It is due to the fact, that for large Δ the distance between split levels $\epsilon_0 + \Delta$ and $\epsilon_0 + U - \Delta$ is small and the width of the Coulomb gap is considerably reduced. One should note that the ZT curve is strongly asymmetric and moreover, its shape is changed. As energy levels are split, ZT shows additional small peaks and valleys corresponding to the situation when on of the split levels aligns with the Fermi level in the leads.

Finally, we have calculated the the voltage induced in the system due to a temperature gradient. To do this, the temperature of the right electrode is kept constant $kT_R = \Gamma$, while the temperature of the left one is changed. The induced voltage is determined under the condition that the charge current in the system vanishes. The calculated results are presented in figure 4 for two different values of gate voltages corresponding to positive and negative sign of the Seebeck

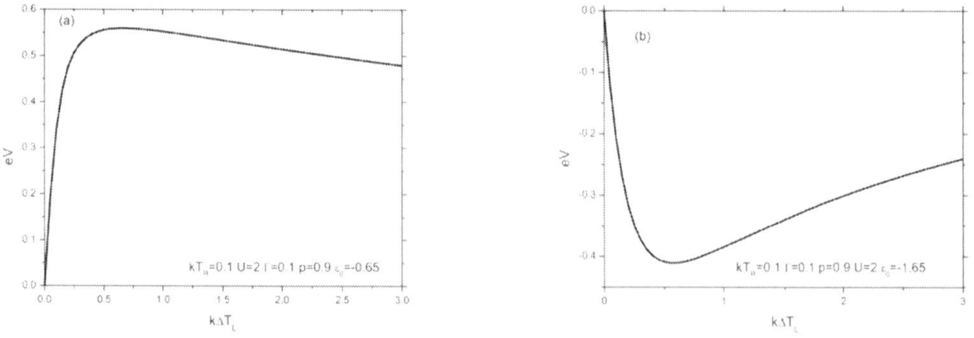

Figure 4. Voltage induced due to temperature gradient

coefficient S. The absolute values of the calculated voltage rapidly increases with increase in temperature of the left contact and then it starts to decrease slowly.

Acknowledgements This work was supported by the Polish Ministry of Science and Higher Education as a research project in years 2006 – 2009.

References

[1] Reddy P, Jang S Y, Segalman R A and Majumdar A 2007 Science **315** 1568; Hochbaum A I, Chen R, Delgado R D, Liang W, Garnett E C, Najarian M, Majumdar A and Yang P 2008 Nature (London) **451** 163
[2] Uchida K, Takahashi S, Harii K, Ieda J, Koshibae W, Ando K, Maekawa S and Saitoh E 2008 Nature **455** 778
[3] Beenakker C W J and Staring A A M 1992 Phys. Rev. B **46**; Zianni X 2007 Phys. Rev. B **75** 045344
[4] Gravier L, Serrano-Guisan S, Reuse F and Ansermet J P 2006 Phys. Rev. B **73** 024419; Phys. Rev. B **73** 052410
[5] Hatami M and Bauer G E W 2009 Phys. Rev. B **79** 174426
[6] Dubi Y and Di Ventra M 2009 Phys. Rev. B **79** 081302
[7] Mahan C D Many-Particle Physics, Plenum Press, New York 2000
[8] Koch J, von Oppen F, Oreg Y and Sela E 2004 Phys. Rev. B **70** 195107
[9] Świrkowicz R, Barnaś J and Wierzbicki M, to be published

Search for electron liquids with non-Abelian quasiparticles

Arkadiusz Wójs

TCM Group, Cavendish Laboratory, University of Cambridge,
Cambridge CB3 0HE, United Kingdom; and
Institute of Physics, Wroclaw University of Technology,
Wyb. Wyspiańskiego 27, 50-370 Wroclaw, Poland

E-mail: arkadiusz.wojs@pwr.wroc.pl

Abstract. We use exact numerical diagonalization in the search of fractional quantum Hall states with non-Abelian quasiparticle statistics. For the (most promising) states in a partially filled second Landau level, the search is narrowed to the range of filling factors $7/3 < \nu_e < 8/3$. In this range, the analysis of energy spectra and correlation functions, calculated including finite width and Landau level mixing, supports the prominent non-Abelian candidates at $\nu_e = 5/2$ (paired Moore–Read "pfafian" state) and 12/5 (clustered Read–Rezayi "parafermion" state). Outside of this range, the noninteracting composite fermion model with four attached flux quanta is validated, yielding the family of quantum liquids with fractional, but Abelian statistics. The borderline $\nu_e = 7/3$ state is shown to be adiabatically connected to the Laughlin liquid, but its short-range correlations are significantly different.

1. Introduction

The incompressible quantum liquids (IQLs) [1] continue to be the subject of extensive studies ever since the famous discovery of the fractional quantum Hall (FQH) effect [2]. The IQLs are formed by two-dimensional electrons placed in a high magnetic field B which causes them to fill a particular fraction ν of one of the lowest Landau levels (LL_n, $n = 0, 1, \dots$). The most recent storm of interest in the IQLs is motivated by the concept of "topological quantum computation" [3, 4] employing non-Abelian statistics of some of the wave functions proposed for a partially filled LL_1. The main idea is that the quantum information may be encoded in (topologically) different quantum states corresponding to the same spatial configuration of the non-Abelian quasiparticles (QPs) of a given underlying IQL. Transition between such different states would only be possible by a global transformation of the QP braiding, thereby making quantum information inherently protected from decoherence caused by any local processes (e.g., coupling to phonons or atomic spins). The best known candidate for a non-Abelian wave function is the "pfaffian" state proposed by Moore and Read [5, 6] and believed to describe the FQH state in a half-filled LL_1. Other wave functions with different complexities of braiding statistics have also been proposed [7, 8, 9, 10]. The convincing demonstration of non-Abelian statistics in a real physical system has clearly risen to the challenge of greatest importance.

The reason to search for the signatures of non-Abelian statistics in LL_1 is that, on the one hand, the partially filled LL_0 are successfully described by the composite fermion (CF) theory [11] predicting fractional but Abelian QPs and, on the other hand, that higher LLs favor ordered electron phases. Crucial recent experiments in LL_1 include confirmation [12] of the anticipated QP charge of $e/4$ for the half-filled state at $\nu_e = 5/2$ and careful measurements [13, 14] of the minute excitation gaps. In theory, some of the most recent advances are related to the role of layer width [15] and LL mixing [16, 17] in real systems, and the nature of the QPs [18, 19].

It is quite remarkable that, despite intensive studies, connection of the FQH states in LL_1 ($\nu_e = 5/2$, $7/3$, $12/5$, or $11/5$) to the few proposed wave functions remains yet to be conclusively established. In fact, in some cases it seems only tentatively assumed for the lack of other candidates. This is an urgent problem, as the anticipation of non-Abelian statistics in nature is largely driven by the connection of some of these wave functions to the particular conformal field theories. A wealth of IQLs found in various systems (electrons or CFs at different fillings of different LLs, in layers of varied width w) also invites a more general question of possible IQLs with arbitrary LL filling ν and interaction V.

This paper is an extension of our brief communication [20]. We report here on the use of large scale exact diagonalization in the search of IQLs with non-Abelian statistics. We demonstrate that non-Abelian IQLs in LL_1 can only emerge in the narrow, particle-hole symmetric range of filling factors $7/3 < \nu_e < 8/3$. In this range, the known non-Abelian candidates at $\nu_e = 5/2$ and $12/5$ are closely examined (including such previously neglected realistic effects as finite layer width and LL mixing) and found to have favorable correlation energies. Outside of this range, the LL_1 hosts the family of Abelian ground states of noninteracting CFs each carrying four magnetic flux quanta, repeating the states known from the lowest LL. The borderline $\nu = 1/3$ ground state in LL_1 is adiabatically connected to the Laughlin state of LL_0, but it has a smaller gap and distinct short-range correlations.

2. Model

We consider the systems of N spin-polarized fermions (electrons) on a Haldane sphere with unit radius and the magnetic monopole of strength $2Q(hc/e)$ inside [21]. In this geometry, LL_n is a shell of angular momentum $\ell = Q + n$, and different N-body wave functions at the same ν are (unlike on a torus) conveniently distinguished by a 'shift' γ between the LL degeneracy and $\nu^{-1}N$ (i.e., $2l = \nu^{-1}N - \gamma$). In contrast to the previous exact diagonalization studies we begin by searching the universality classes (ν, γ) of the gapped ground states with arbitrary interactions V rather than confining ourselves to the particular physical systems defined by n, w, etc.

It is a trivial fact that the many-body dynamics in a degenerate LL is completely determined by an interaction pseudopotential, defined [21] as the dependence of the pair energy V on the relative angular momentum $m = 1, 3, \ldots$. Less obviously, V_m induces particular correlations only through its deviation from a reference "harmonic pseudopotential" given by a straight line over the consecutive m's [22]. Hence, the low-energy spectra of V_m can be accurately reproduced by a suitable effective pseudopotential U_m with only a few appropriate coefficients.

In our calculations we have used $U = [U_1, U_3, U_5]$. Higher-order terms have been ignored, as they are essentially irrelevant for the dynamics in a liquid phase with short-range correlations. On the other hand, going beyond the (earlier used) U_1 and U_3 was needed for an improved description of the two lowest (electron or CF) LLs known to host IQL states. At the same time, it still allowed for useful graphical representation of the ground state properties in an effectively two-dimensional space of the (normalized, $\sum_m U_m = 1$) parameters U_m.

3. Maps of the gap for arbitrary interaction

Searching for the series of gapped ground states with particular filling factor ν and shift γ, we looked at various finite systems $(N, 2\ell)$. A few maps of the 'neutral' energy gap Δ for $\nu = 1/2$ and $\gamma = 3$ (i.e., $2\ell = 2N - 3$, as appropriate for the Moore–Read pfaffian state) are shown in Fig. 1. The gap Δ is defined as the energy difference from the ground state to the first excited state in the same spectrum as long as the ground state happens to be nondegenerate (i.e., has

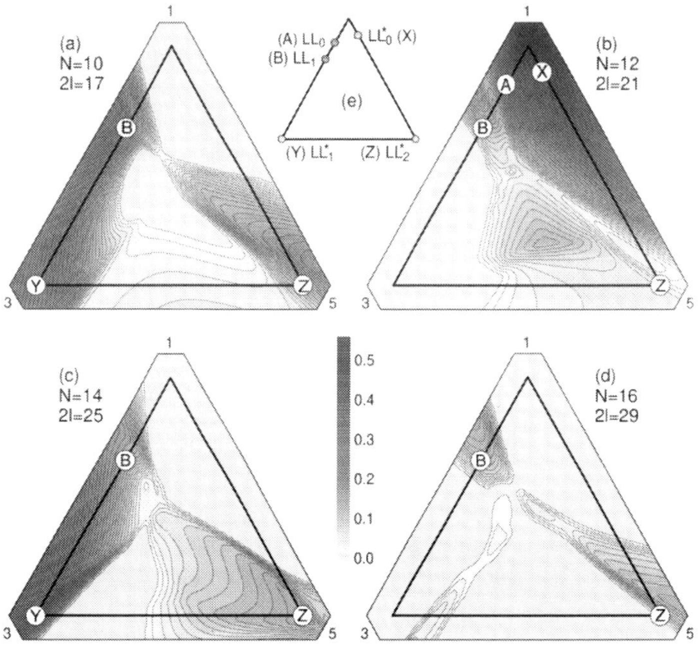

Figure 1. Ternary contour plots of the neutral excitation gap Δ for $N = 10$, 12, 14, 16 fermions in a half-filled Landau level ($\nu = 1/2$) with shell angular momentum ℓ corresponding to the shift $\gamma = 3$ of the Moore–Read "pfaffian" wave function, interacting by model pseudopotential U_m. In each plot, three corners of the big triangles correspond to $U_m = \delta_{m,\mu}$ with $\mu = 1$, 3, 5 marked in each corner. Points relevant for the actual interactions in different electron or CF Landau levels are marked on a triangle in frame (e). Different candidate incompressible states are indicated.

zero total angular momentum, $L = 0$); otherwise Δ is set to zero. The IQL candidates are the islands of positive Δ repeating regularly in the same area of the map for different values of N. Their location on the map must be compared with the actual pseudopotentials in different LLs, as indicated in frame (e). In particular, (A) and (B) mark the appropriate positions of the electron pseudopotentials in LL_0 and LL_1 (the former dominated by V_1; the latter roughly linear between $m = 1$ and 5), and (X), (Y), and (Z) mark the same for the CFs in their effective shells LL_0^*, LL_1^*, and LL_2^*.

The appearance of a significant gap around the point (B) in all maps in Fig. 1 (i.e., for each N) confirms quite definitively the earlier expectation that the Moore–Read ground state forms for a class of pseudopotentials close to that of LL_1. It also demonstrates that its accuracy depends sensitively on the fine-tuning of the leading V_m's, achieved (for example) by adjusting the layer width w [15]. Remarkably, the maps in Fig. 1 also preclude the Moore–Read state at the half-filling of other LLs (e.g., of the second CF LL, called LL_1^*, characterized by a dominant repulsion at $m = 3$ [23]). Clearly, the microscopic origin of the FQH state observed at $\nu_e = 3/8$ [24] (in the CF picture, corresponding to $\nu = 1/2$ in LL_1^*) must be different.

Analogous maps of Δ for the Laughlin filling fractions $\nu = 1/3$ and $1/5$ are shown in Fig. 2. In (a), the universality class of the Laughlin wave function ($\nu = 1/3$ and $\gamma = 3$) is shown to cover a large part of the map, including (A) and (X), and possibly also (B) and (Z). Point (B), the most interesting for the present analysis, falls just inside the island of positive gap Δ, suggesting connection of the $\nu_e = 7/3$ FQH state to the Laughlin $\nu = 1/3$ liquid in LL_1. Point (X) is relevant for the Laughlin $\nu = 1/3$ state of the CF vacancies in LL_0^*, i.e., to the robust

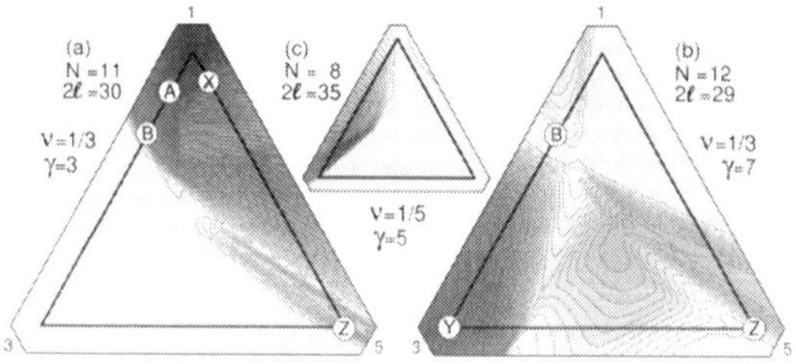

Figure 2. Similar to Fig. 1, but for Laughlin filling fractions $\nu = 1/3$ (two candidate states with $\gamma = 3$ and 7) and $\nu = 1/5$.

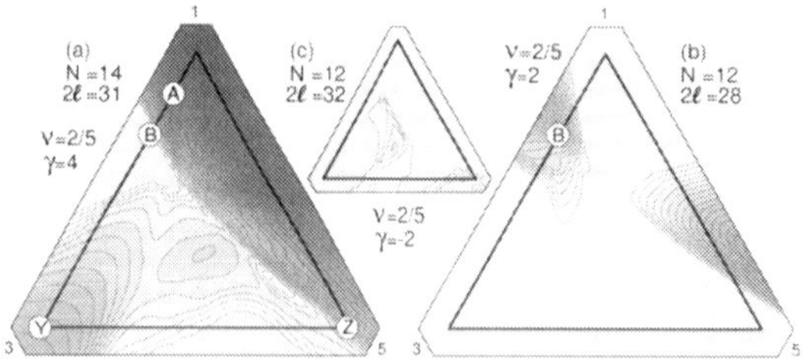

Figure 3. Similar to Fig. 1, but for the $\nu = 2/5$ states with $\gamma = 4$ (Jain), 2 (Bonderson–Slingerland), and -2 (Read–Rezayi).

Jain state at $\nu_e = 2/7$. Remarkably, the Laughlin ground state does not occur around point (Y) corresponding to the $\nu = 1/3$ filling of LL_1^*. The nature of the rather fragile FQH state observed at the corresponding fraction $\nu_e = 4/11$ [24] must therefore be different. In (b), the $\nu = 1/3$ paired (non-Laughlin) state with $\gamma = 7$, proposed earlier for both LL_1 [22] and LL_1^* [25], is tested. The gap around (Y) is quite suggestive that it may indeed describe the FQH state at $\nu_e = 4/11$ [24]. On the other hand, its relevance for LL_1 seems doubtful. In (c), any positive $U = [U_1, U_3, 0]$ yields an exact Laughlin state at $\nu = 1/5$. Our map of Δ confirms that it is true description of the FQH states in both LL_0 and LL_1 [26]. On the other hand, its relevance to the FQH effect observed in LL_1^* (i.e., at $\nu_e = 6/17$) [24] is rather unlikely.

Finally, the maps of Δ for the $\nu = 2/5$ filling have been shown in Fig. 3. In (a), the Jain series of noninteracting CF states with $\gamma = 4$ correctly represents the ground state in LL_0 (and LL_0^*). Point (B) corresponding to LL_1 lies just outside the borders of the island of $\Delta > 0$, leaving the question of relevance of the non-interacting CF picture at $\nu_e = 12/5$ open. Two other candidate ground states at $\nu = 2/5$ are the parafermion state with $\gamma = -2$ [7] and a more recently proposed state with $\gamma = 2$ [9]. Especially for the latter state, frame (b) appears suggestive of a gap emerging around (B). Clearly, the competition between these three candidate ground states in LL_1 is not convincingly resolved based on the maps of Δ alone. On the other hand, the identification of the true ground state is crucial, because the candidates with $\gamma = \pm 2$ are both non-Abelian, in contrast to the $\gamma = 4$ Jain state. More careful analysis will follow in subsequent sections.

4. Maps of the amplitudes for arbitrary interaction

Maps similar to those in Figs. 1–2 can also be used to show dependence of other parameters of the spectrum on the form of interaction. For example, in Fig. 4 we plot maps of the leading pair and triplet Haldane amplitudes for the ground states of $(\nu, \gamma) = (1/2, 3)$ and $(1/3, 3)$. The amplitudes are defined [27] as the fractions of the pairs or triplets with a given relative angular momentum m (for the pairs, $m = 1, 3, 5, \ldots$ is a measure or an average square distance; for the triplets, $m = 3$, 5, 6, ... measures an average area). These (discrete) correlation functions are particularly useful in identifying Laughlin and Moore–Read states, as they both are unique zero-energy states of simple repulsions: the former at the minimum pair angular momentum $m = 1$, the latter at the minimum triplet angular momentum $m = 3$ (hence, their corresponding amplitudes vanish exactly). The emergence and location of the islands of essentially zero amplitude in Fig. 4(c) and (d) provides additional and quite decisive support for the Laughlin $\nu = 1/3$ and Moore–Read $\nu = 1/2$ ground states (universality classes) in LL_0 and LL_1, respetively. On the other hand, other questions, such as of a Laughlin state in LL_1, remain open until the competition with other possible states (with the same ν but different γ) can be resolved.

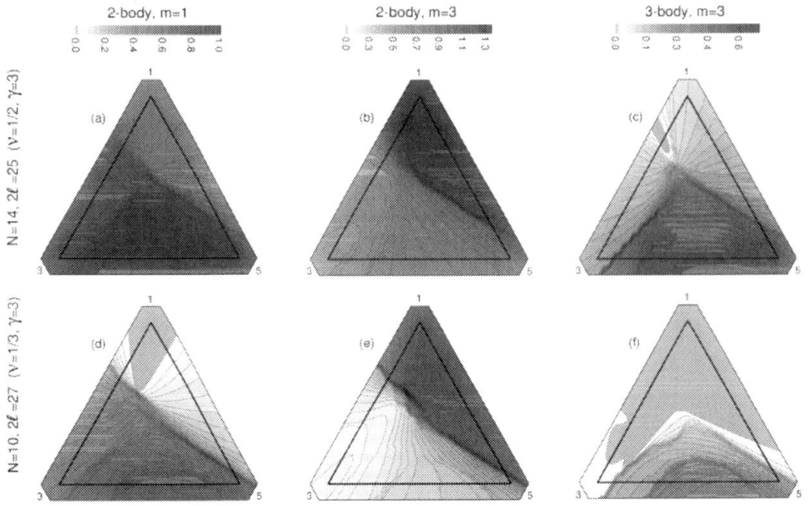

Figure 4. Ternary contour plots (similar to Fig. 1) of the pair and triplet amplitudes (labeled by the relative angular momentum m) for the universality classes of Laughlin and Moore–Read wave functions (filling factors $\nu = 1/3$ and $1/2$, respectively; shift $\gamma = 3$ in both cases).

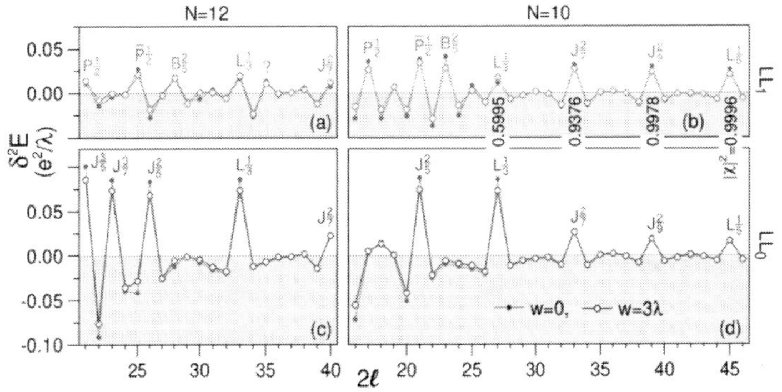

Figure 5. Symmetric second-order differences $\delta^2 E$ of the ground-state energy per particle E for $N = 10$ and 12 electrons in the lowest and second Landau level (LL_0 and LL_1), as a function of shell angular momentum ℓ, for layer widths $w = 0$ and 3λ. Candidate incompressible states are marked as '$\Phi\nu$', where $\Phi = \mathrm{P}, \overline{\mathrm{P}}, \mathrm{L}, \mathrm{J}, \mathrm{B}$ denote Moore–Read pfaffian, anti-pfaffian, Laughlin, Jain, and Bonderson–Slingerland states. Squared overlaps $|\chi|^2$ between the states repeating in both LLs at $\nu \leq 1/3$ are indicated. λ is the magnetic length.

5. Analysis of correlation energies in LL_0 and LL_1

Guided by the maps of gaps and amplitudes we now move our focus to the FQH states in a partially filled LL_1. In Fig. 5 we seek confirmation of the IQL candidates in the downward cusps of the dependence of the ground-state correlation energy per particle E on the LL degeneracy (for a fixed number of electrons N). The correlation energy E is calculated from the total Coulomb energy $\mathcal{E}$ of N electrons in a LL shell with a given ℓ by adding the energy of attraction to the uniform charge-compensating background and dividing by N. In the calculation for finite width w of a quasi-2D electron layer, the Coulomb matrix elements were computed assuming infinite-well confinement in the perpendicular direction, i.e., for the charge-density profile of $\varrho(z) = (2/w)\cos^2(\pi z/w)$. The cusps in E are most pronounced in the plots of a symmetric

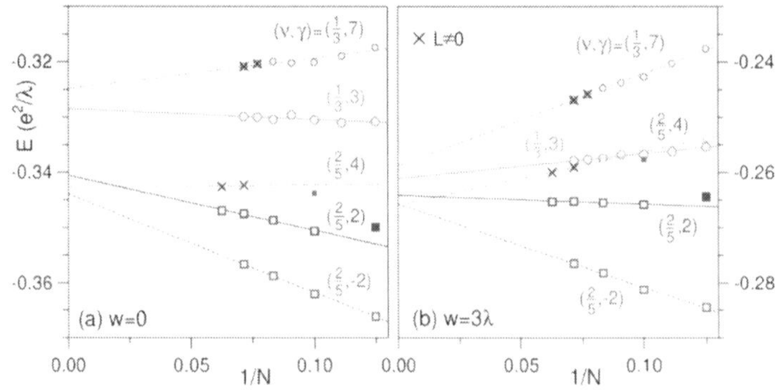

Figure 6. Size extrapolation of the ground-state correlation energies per particle E calculated for N electrons in the second Landau level (LL_1), for the layer widths $w = 0$ (a) and 3λ (b). Competing series of candidate incompressible ground states with filling factors $\nu = 1/3$ and $2/5$ are distinguished by their shifts γ.

difference $\delta^2 E_{2\ell} = E_{2\ell-1} + E_{2\ell+1} - 2E_{2\ell}$. For an IQL, its (positive) value gives the QP gap $\tilde{\Delta}$ (energy needed to create a pair of noninteracting QPs of total charge zero) times the number of QPs created per flux quantum.

Evidently, Fig. 5 complements the maps of Fig. 1-4 in the identification of the universality classes of particular IQLs. For example, it shows peaks in $\delta^2 E_{2\ell}$ which signal the Laughlin $\nu = 1/3$ and Moore–Read $\nu = 1/2$ IQLs in LL_1. But Fig. 5 also does more, by revealing the following connection between quantum statistics and the filling factor in a partially filled LL_1. At $\nu < 1/3$ the same Laughlin and Jain IQLs of filled shells of noninteracting CFs with four flux quanta occur in LL_1 and LL_0. The corresponding states in LL_1 and LL_0 have high overlaps (calculated by replacing the electron positions by the guiding centers) and similar QP gaps $\tilde{\Delta}$. This similarity, earlier pointed out in Ref. [26], is caused by a sufficiently high pseudopotential coefficient V_1 (forcing the avoidance of the $m = 1$ pair state) and a similar behavior of the pseudopotential al long range, $V_{m \geq 3}$, in the two lowest LLs. Importantly, this similarity validates the noninteracting CF model [11] with four flux quanta attached to each electron in LL_1 (in addition to LL_0 where its accuracy is well-known).

In contrast, at $\nu > 1/3$ the Jain sequence of noninteracting CF states in LL_0 is replaced in LL_1 by a different set of IQLs, including the Moore–Read pfaffian at $\nu = 1/2$, the anti-pfaffian (pfaffian's particle-hole conjugate) at the same $\nu = 1/2$ but with a different shift $\gamma = -1$, and a Bonderson–Slingerland $\nu = 2/5$ state with $\gamma = 2$. The breakdown of the noninteracting two-flux CF model in LL_1 opens, exclusively at $1/3 < \nu < 2/3$, possibility for other IQLs, including several suggested more exotic states with various non-Abelian QP statistics. The borderline $\nu = 1/3$ state, separating the Abelian from (possibly) non-Abelian states, has a moderate overlap with the Laughlin state of LL_0, despite falling into the same class of $\gamma = 3$.

Having narrowed the search for non-Abelian IQLs in LL_1 to the range of $1/3 < \nu < 2/3$, let us now try to resolve more decisively the competition between different candidate wave functions at filling factors $\nu = 1/3$ and $\nu = 2/5$ (at $\nu = 1/2$, the Moore–Read pfaffian appears to have no competition). In Fig. 6 we attempt to extrapolate to an infinite system size (to $N^{-1} \to 0$, i.e., to the planar geometry) the ground-state correlation energies per particle E. To improve convergence, the plotted energies E have been rescaled by $\sqrt{2Q\nu/N}$ so as to ensure equal units e^2/λ for each γ; here $\lambda = \sqrt{\hbar c/eB}$ is the magnetic length). In the extrapolation for $\nu = 2/5$ we only used the open squares, discarding one state aliased with the anti-pfaffian (a finite system defined by N and 2ℓ may sometimes represent states with different combinations of ν and γ)

and one apparently suffering from the small size. When choosing the IQL ground state we also checked if a given candidate series (ν, γ) consistently has a nondegenerate (i.e., one with $L = 0$) ground state. In those cases when the particular $(N, 2\ell)$ ground state had $L \neq 0$, we crossed out the corresponding data point in Fig. 6 to mark that the series containing it is unlikely to describe an IQL in the thermodynamic limit.

For $\nu = 1/3$ Fig. 6 eliminates the (paired) $\gamma = 7$ series as a viable candidate in LL_1 (it remains plausible in LL_1^* which, however, will not be further discussed here), leaving the (universality class of the) Laughlin state as the only acceptable option. For $\nu = 2/5$, the Jain series appears to extrapolate to a competitive energy, but it no longer has $L = 0$ for the larger N, and hence should probably be discarded. The remaining two candidates both consistently have $L = 0$ and both extrapolate to nearly the same E for large N (the difference in favor of the Read–Rezayi state with $\gamma = -2$ is comparable to the error of extrapolation). In realistic conditions this near degeneracy is likely to be removed by their different susceptibilities to the LL mixing. To explore this idea, we estimated the appropriate energy correction dE by including in the diagonalization additional states involving a single cyclotron excitation [16]. Specifically, in addition to the configurations with completely filled LL_0 (with both spin-$\uparrow$ and spin-$\downarrow$), N polarized electrons in LL_1, and empty higher LLs, we have also included configurations with one electron promoted from either LL_0 (with either spin) or from LL_1 to the next higher LL. It should be kept in mind that inclusion of only a single cyclotron excitation is not a rigorous treatment of the LL mixing. However, it is expected to reveal a possible difference in the susceptibility of the competing states to this process. In the calculation, we have assumed the Coulomb-to-cyclotron energy ratio of $\beta \equiv (e^2/\lambda)/(\hbar\omega_c) = 1.56$, corresponding to $B = 2.6$ T. Due to a large size of the Hilbert space, the values of dE could only be calculated for $N \leq 10$. However, we found that the corrections dE are far less size-dependent than the base energies E. It is therefore justified to apply the $N = 10$ estimates of dE to the values of E extrapolated from $N \leq 16$.

The results are presented in Tab. 1. Clearly, in addition to having the lowest E, the $\gamma = -2$ parafermion state also has the largest $|dE|$, and therefore it is quite convincingly predicted to define the universality class of the $\nu = 2/5$ ground state in LL_1. This is of considerable importance because this state is the only known candidate IQL [4] whose braiding rules are sufficiently complex to allow quantum computation (unlike, e.g., the Moore–Read state).

Let us now look at the pair correlation functions of the established IQL ground states at $\nu = 1/3$ and $2/5$ in LL_1. The results calculated for the largest available systems ($N = 14$ or 16,

Table 1. Extrapolated correlation energies per particle E (in the units of e^2/λ) in several series of candidate ground states (distinguished by shifts γ) at filling factors $\nu = 1/3$ and $2/5$ in LL_1, for layer widths $w = 0$ and 3λ, calculated without LL mixing. Also, corrections dE due to LL mixing (at $B = 2.6$ T), estimated for $N = 10$ (except for $N = 8$ for $\nu = 2/5$ and $\gamma = 4$).

(ν, γ)	$(1/3, 3)$	$(1/3, 7)$	$(2/5, -2)$	$(2/5, 2)$	$(2/5, 4)$
$w = 0$:					
E	-0.3285	-0.3249	-0.3439	-0.3406	-0.3427
dE	-0.0268	-0.0218	-0.0294	-0.0239	-0.0227
$E + dE$	-0.3553	-0.3467	-0.3733	-0.3645	-0.3654
$w = 3\lambda$:					
E	-0.2611	-0.2589	-0.2657	-0.2642	-0.2659
dE	-0.0087	-0.0077	-0.0095	-0.0085	-0.0071
$E + dE$	-0.2698	-0.2666	-0.2752	-0.2727	-0.2730

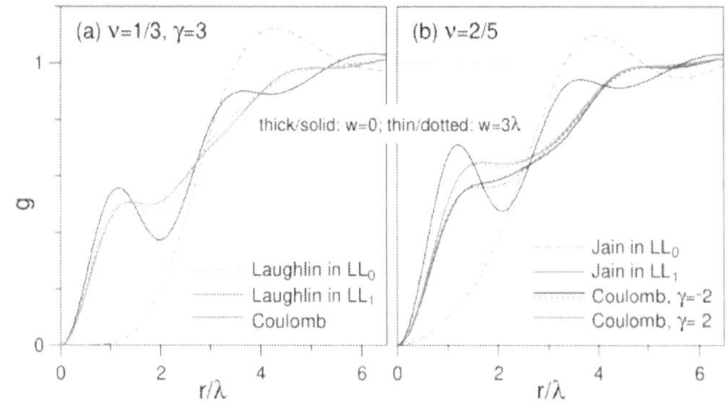

Figure 7. Pair-correlation functions of the candidate incompressible ground states at $\nu = 1/3$ (a) and $2/5$ (b) in the second Landau level (LL_1). Coulomb ground states are compared with the exact Laughlin and approximate Jain states in LL_0 and LL_1 (constructed as the ground states of a model interaction pseudopotential with pair repulsion only at $m = 1$).

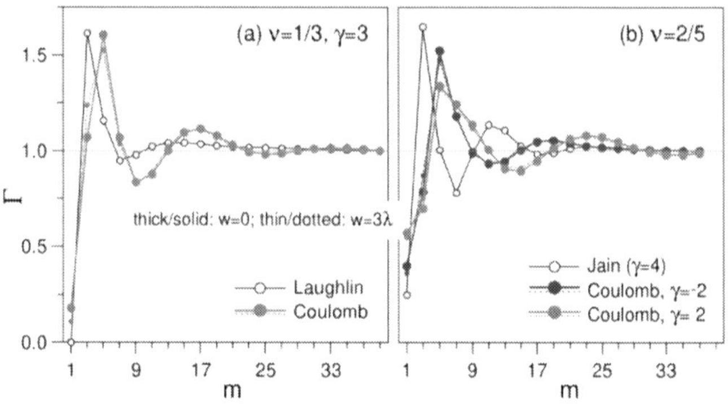

Figure 8. Renormalized pair amplitudes Γ of the same candidate states in LL_1 as in Fig. 7.

depending on a given state, as is clear from Fig. 6) are shown in Fig. 7. At $\nu = 1/3$, despite being adiabalitally connected to the Laughlin state (i.e., having the same $\gamma = 3$), the Coulomb ground state has significantly different short-range correlations. Specifically, the zigzag in $g(r)$ at $1 < r/\lambda < 2$ present in the exact Coulomb state in LL_1, caused by the particular form of the single particle wave functions and showing also in $g(r)$ for the full LL_1, gives way to a plateau in the case of the actual Coulomb ground state. (We should also clarify here that the Laughlin state in any LL is defined in a standard way as the zero energy state of the $m = 1$ pair repulsion, and its pair correlation function is drawn using the appropriate single-particle wave functions, e.g., of LL_0 or LL_1.) This agrees with only moderate squared overlaps of the $\nu = 1/3$ ground state in LL_1 with the Laughlin wave function, pointed out ourlier: 0.292 (0.501), 0.253 (0.510), 0.333 (0.549) for $N = 10$, 12, 14, for $w = 0$ (3λ). The magneto-roton band is also absent in the spectra of LL_1, and the low-energy states resembling Laughlin QEs and QHs are found at $2\ell = 3N - 3 \mp 1$, but they are not generally the lowest states in their spectra. A similar difference (removal of the characteristic zigzag) between the Jain state (approximated as the ground state of the $m = 1$ pair repulsion for $\gamma = 4$) and the competing Coulomb ground states with $\gamma = -2$ and 2 is also found at $\nu = 2/5$ in LL_1. At both filling factors dependence on w is insignificant.

Difference in short-range correlations between Laughlin and Jain states of LL_1 on one hand, and the $\nu = 1/3$ and $2/5$ Coulomb ground states of LL_1 on the other, was also earlier apparent in the leading amplitudes in Fig. 4(d,e). It is evident in the comparison of renormalized pair amplitudes Γ_m calculated for slightly larger systems (same as in Fig. 7), which have been shown in Fig. 8. Here, the coefficients Γ_m have been converted from the usual Haldane amplitudes on a sphere G_m to their planar counterparts, as described in Ref. [27]. Specifically, $\Gamma_m = 1 + (G_m - G_m^{\rm full})/G_1^{\rm full}$, where $G_m^{\rm full} = (4\ell + 1 - 2m)/\ell/(2\ell + 1)$ describes a full LL.

6. Conclusion

We have carried out extensive exact diagonalization studies (including finite layer width and LL mixing) of the fractional quantum Hall states in a partially filled second Landau level (LL_1), searching for non-Abelian incompressible quantum liquids (IQLs). We have found the range of filling factors $1/3 < \nu < 2/3$ in LL_1 in which the emergence of non-Abelian statistics is possible. Inside this range, we have demonstrated that the spin-polarized ground states at $\nu = 1/2$ and 2/5 are described by the non-Abelian "pfaffian" and "parafermion" wave functions. Outside of it, the Jain states of noninteracting CFs with four attached magnetic flux quanta repeat in both lowest LLs, thus precluding more exotic phases. The borderline $\nu = 1/3$ state is adiabatically connected to the Laughlin liquid but has a smaller gap and distinct short-range correlations.

Acknowledgment

The author thanks G. Möller, N. Cooper, S. Simon, A. Stern, and G. Gervais for many insightful comments, and acknowledges support from EU under the Marie Curie Intra-European Grant No. PIEF-GA-2008-221701 and from the Polish MNiSW under grant N-N202-1336-33.

References

[1] R. B. Laughlin, 1983 *Phys. Rev. Lett.* **50** 1395
[2] D. C. Tsui, H. L. Störmer, and A. C. Gossard, 1982 *Phys. Rev. Lett.* **48** 1559
[3] A. Yu. Kitaev, 2003 *Ann. Phys.* (N.Y.) **303** 1
[4] C. Nayak, S. H. Simon, A. Stern, M. Freedman, and S. Das Sarma, 2008 *Rev. Mod. Phys.* **80** 1083
[5] G. Moore and N. Read, 1992 *Nucl. Phys. B* **360** 362
[6] M. Greiter, X.-G. Wen, and F. Wilczek, 1991 *Phys. Rev. Lett.* **66** 3205
[7] N. Read and E. Rezayi, 1999 *Phys. Rev. B* **59** 8084
[8] G. Möller and S. H. Simon, 2008 *Phys. Rev. B* **77** 075319
[9] P. Bonderson and J. K. Slingerland, 2008 *Phys. Rev. B* **78** 125323
[10] P. Bonderson, A. Feiguin, G. Möller, and J. K. Slingerland, arXiv:0901.4965v2
[11] J. K. Jain, 1989 *Phys. Rev. Lett.* **63** 199
[12] M. Dolev, M. Heiblum, V. Umansky, Ady Stern, and D. Mahalu, 2008 *Nature* **452** 829
[13] W. Pan, J. S. Xia, H. L. Stormer, D. C. Tsui, C. Vicente, E. D. Adams, N. S. Sullivan, L. N. Pfeiffer, K. W. Baldwin, and K. W. West, 2008 *Phys. Rev. B* **77** 075307
[14] C. R. Dean, B. A. Piot, P. Hayden, S. Das Sarma, G. Gervais, L. N. Pfeiffer, and K. W. West, 2008 *Phys. Rev. Lett.* **100** 146803
[15] M. R. Peterson, Th. Jolicoeur, and S. Das Sarma, 2008 *Phys. Rev. Lett.* **101** 016807
[16] A. Wójs and J. J. Quinn, 2006 *Phys. Rev. B* **74** 235319
[17] S. H. Simon, 2008 *Phys. Rev. Lett.* **100** 116803
[18] C. Töke, N. Regnault, and J. K. Jain, 2007 *Phys. Rev. Lett.* **98** 036806
[19] B. A. Bernevig and F. D. M. Haldane, 2008 *Phys. Rev. Lett.* **100** 246802; arXiv:0810.2366
[20] A. Wójs, 2009 *Phys. Rev. B* **80** 041104(R)
[21] F. D. M. Haldane, 1983 *Phys. Rev. Lett.* **51** 605
[22] A. Wójs, 2001 *Phys. Rev. B* **63** 125312
[23] S.-Y. Lee, V. W. Scarola, and J. K. Jain, 2001 *Phys. Rev. Lett.* **87** 256803
[24] W. Pan, H. L. Störmer, D. C. Tsui, L. N. Pfeiffer, K. W. Baldwin, and K. W. West, 2003 *Phys. Rev. Lett.* **90** 016801
[25] A. Wójs, D. Wodziński, and J. J. Quinn, 2006 *Phys. Rev. B* **74** 035315
[26] N. d'Ambrumenil and A. M. Reynolds, 1988 *J. Phys. C* **21** 119
[27] A. Wójs and J. J. Quinn, 2005 *Phys. Rev. B* **71** 045324

Symmetry in finite phase plane

J. Zak [1]

Department of Physics - Technion, Israel Institute of Technology, Haifa 32000, Israel

E-mail: zak@physics.technion.ac.il

Abstract. The known symmetries in one-dimensional systems are inversion and translations. These symmetries persist in finite phase plane, but a novel symmetry arises in view of the discrete nature of the coordinate x_i and the momentum p_i : x_i and p_i can undergo permutations. Thus, if x_i assumes M discrete values, $i = 0, 1, 2, \ldots, M-1$, a permutation will change the order of the set $x_0, x_1, \ldots, x_{M-1}$ into a new ordered set. Such a symmetry element does not exist for a continuous x-coordinate in an infinite phase plane. Thus, in a finite phase plane, translations can be replaced by permutations. This is also true for the inversion operator. The new permutation symmetry has been used for the construction of conjugate representations and for the splitting of the M-dimensional vector space into independent subspaces. This splitting is exhaustive in the sense that if $M = \sqcap_i M_i$ with M_i being prime numbers, the M-dimensional space splits into $M_1, M_2, \ldots M_n$-dimensional independent subspaces. It is shown that following this splitting one can design new potentials with appropriate constants of motion. A related problem is the Weyl-Heisenberg group in the M-dimensional space which turns into a direct product of its subgroups in the M_i-dimensional subspaces. As an example we consider the case of $M = 8$.

1. Introduction

In recent years it has turned out that quantum mechanics in finite phase plane is a fascinating subject which is closely related to some notions of elementary number theory. Here are some results that depend crucially on the nature of the number M, the dimension of the vector space in the finite phase plane. It was Schwinger who in his fundamental paper on finite phase space quantum mechanics in 1960 [1], first showed that if $M = M_1M_2$ with M_1 and M_2 relatively prime, then the M-dimensional space can be subdivided into two independent subspaces of dimensions M_1 and M_2. Somewhat later Ivanovic, in 1981 [2] has proved that when M is a prime number, one can construct $M+1$ orthogonal bases with scalar products of all the vectors in different bases having the same absolute value of magnitude $\frac{1}{\sqrt{M}}$. These orthogonal bases were named mutually unbiased bases and the result was generalized in 1989 to cases when the dimension of the space M is a power of a prime number [3]. Quantum mechanics in finite phase plane can also depend on whether M is even or odd. Thus, the Wigner function for odd M can directly be obtained by replacing the integration by summation over the discrete x_i-values. This is not the case for even M, and there is a vast literature on this subject with different derivations for Wigner functions in the finite phase plane [4]. One of the reasons that the results can differ in finite and infinite phase plane is connected with the replacing of the variables x and y by the

[1] Work done with B. Simkhovich, A. Mann and M. Revzen

variables s and t in the following transformation

$$x + y = s, \quad x - y = t, \tag{1}$$

where all the variables are mod$M$. It can be checked [4] that when x and y assume all the different values from 1 till M, s and t from Eq. (1) are either both even or both odd. To put it in other words, there are, for example, no integers x and y that solve Eq. (1) for $s = 1$, $t = 2$. A related result is the action of the inversion operator on the wave function $\psi(x)$ in the finite phase plane [6]. One can check that for odd M one can derive the Wigner function in the same way as in the infinite phase plane [7]. This does not apply, however, for M even. The intuition one has acquired in infinite phase plane is, in general, not applicable in finite plane. As an example let us mention the inversion operator I, which in the infinite case is defined as follows

$$Ix = -x. \tag{2}$$

In the finite phase plane this assumes the form

$$Ix_k = -x_k = x_{-k} = x_{-k+M} \ (\text{mod}M). \tag{3}$$

For $k = M$, $Ix_M = x_M$. In the infinite case this corresponds to $Ix = x$, when $x = 0$. As was pointed out above, for the inversion operator there is a difference between the cases of M even and M odd. Thus, when M is even, it follows from Eq. (3) that

$$Ix_{M/2} = x_{M/2} \ (\text{mod}M). \tag{4}$$

This means that when M is even there are two points on the x-axis, $x_{M/2}$ and $x_M(x_0)$, which remain unchanged under inversion. For M odd only the point $x_M(x_0)$ is unchanged under inversion, which is like point $x = 0$ in the infinite case of Eq. (2). One of the conclusions of this analysis is that for M = 2, the inversion operator I turns into the unit operator. Or in other words in the $M = 2$ space there are only even functions.

In this paper, we follow closely our previous work [8]. In a finite phase plane the coordinate x and momentum p operators are replaced by the exponential operators [8].

$$\tau(\frac{2\pi}{Mc}) = \exp(i\frac{2\pi}{Mc}x) \text{ and } T(c) = \exp(\frac{-i}{\hbar}cp), \tag{5}$$

where Mc is the size of the phase plane in the x-direction and c is a constant used in the discretization of x and p

$$x_s = sc, s = 0, 1, \ldots, M-1; \ p_t = \hbar\frac{2\pi}{Mc}t, t = 0, 1, \ldots, M-1. \tag{6}$$

The discrete nature of x and p in Eq. (6) is a consequence of the finiteness of the phase plane given by the following requirement on the operators in Eq. (5)

$$\tau^M(\frac{2\pi}{Mc}) = T^M(c) = 1. \tag{7}$$

In order to make contact with Schwinger's paper [1], we denote the operators in Eq. (5) by

$$U(x) = \tau(\frac{2\pi}{Mc}) = e^{i\frac{2\pi}{Mc}x}, \quad V(p) = T(c) = e^{-\frac{i}{\hbar}pc}. \tag{8}$$

We have

$$UV = e^{i\frac{2\pi}{M}}VU. \tag{9}$$

U and V are called, according to Dirac [9], mutually conjugate operators. In Ref. [1] it was proven that when $M = M_1M_2$, with M_1 and M_2 mutually prime, the M-dimensional space can be split into M_1 and M_2-dimensional spaces, which are independent of one-another. In this paper we use a recently developed method [10] for carrying out such a splitting when $M = M_1M_2$, but without the requirement for M_1 and M_2 to be mutually prime, as was the case in Ref. [1]. This method uses permutations of the discrete coordinates, that have previously been used in the Cooley-Tukey algorithm for the fast Fourier transform [11]. These permutations have been called stride permutations [12]. We show in this paper that when used in the splitting of the M-dimensional phase space, these permutations lead to completely new periodic potentials that can exist only in the finite phase plane. Related to these potentials are special translation symmetries which are new constants of motion. We present an explicit example for the case of $M = 8 = 2^3$.

The operators U and V in Eq. (9) are generators of the Weyl-Heisenberg group in the M-dimensional phase plane [13]. The splitting of this space into its subspaces of prime dimensions M_i, $i = 1, 2, \ldots, n$ is accompanied by the writing of the Weyl-Heisenberg group in the M-dimensional space into a direct product of its subgroups in the M_i-dimensional subspaces

2. Symmetry in Finite Phase Plane

In a finite phase plane the coordinate x and momentum p assume discrete values, as given in Eq. (6). Like in the infinite case, it is meaningful to use also in the finite phase plane the x or p-representations. However, in the finite case, this opens up the possibility of a new kind of symmetry. Consider, for example, the case of dimensionality $M = 4$. According to Eq. (6) the coordinate x_s assumes 4 values. If a function is given, such that

$$\psi(x_1) = \psi(x_3) \tag{10}$$

then this function is invariant under the permutation of interchanging the values of the coordinates x_1 and x_3. In the x-representation the operator U in Eq. (8) turns into a multiplicative function of x_s, $U(x_s)$. It is easy to see that for $M = 4$

$$U^2(x_0) = U^2(x_2) = 1 \text{ and } U^2(x_1) = U^2(x_3) = -1. \tag{11}$$

This means that $U^2(x_s)$ is invariant under the permutation cycles

$$(x_0, x_2) \text{ and } (x_1, x_3). \tag{12}$$

At this point it is worthwhile to make the following remark. In quantum mechanics the group of permutations (the symmetric group) is used in the statistics of identical particles [14,15]. Here, instead of applying permutations to coordinates of identical particles, they are applied to different values of the discrete coordinate (or momentum). A general permutation S in the M-dimensional configuration space is,

$$S = \begin{pmatrix} x_0 & x_1 \ldots & x_{M-1} \\ x_{s_0} & x_{s_1} & x_{s_{M-1}} \end{pmatrix} \equiv \begin{pmatrix} x_k \\ x_{s_k} \end{pmatrix} \tag{13}$$

which means that x_k goes into x_{s_k}. This is the action of the permutation S in the configuration space. The action of S on a function $\psi(x_k)$ of x_k will respectively be [15]

$$S\psi(x_k) = \psi(S^{-1}x_k), \tag{14}$$

(in Ref. [15], pages 105, 106, a detailed discussion is given of the action in Eq. (14)) where S^{-1} is the inverse of S. As an example we consider the function in Fig. 1, and let P be the permutation

$$P = \begin{pmatrix} x_0 & x_1 & x_2 & x_3 \\ x_2 & x_3 & x_0 & x_1 \end{pmatrix} = (x_0, x_2)(x_1, x_3). \tag{15}$$

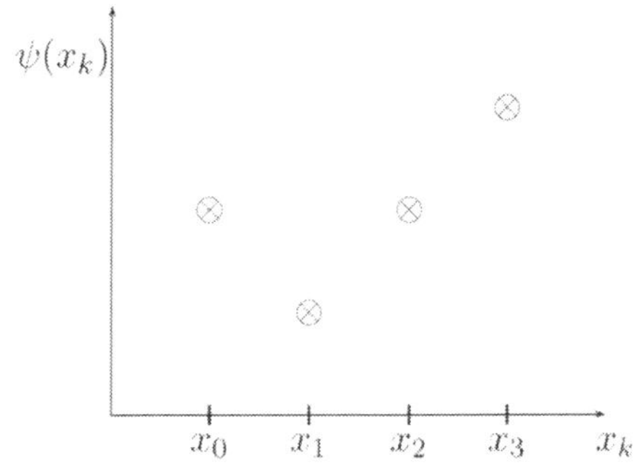

Figure 1. $\psi(x_k)$ is a function in a 4-dimensional space. $\psi(x_0) = \psi(x_2)$, but $\psi(x_1) \neq \psi(x_3)$

Under a general permutation $\psi(x_k)$ will go into a different function $\varphi(x_k)$ according to the rule of Eq. (14)

$$\varphi(x_k) = P\psi(x_k) = \psi(P^{-1}x_k). \tag{16}$$

For the example of the permutation in Eq. (15) we have

$$\varphi(x_0) = \psi(x_2), \quad \varphi(x_1) = \psi(x_3), \quad \varphi(x_2) = \psi(x_0), \quad \varphi(x_3) = \psi(x_1) \tag{17}$$

in Fig. 1. This means that $\psi(x_k)$ is invariant under the transposition (x_0, x_2), but not under (x_1, x_3).

As we know, the translation operator V in Eq. (8) acts on the wave function $\psi(x_k)$ in the following way

$$V\psi(x_k) = \psi(x_{k-1}). \tag{18}$$

By using Eqs. (14) and(18) we can define a permutation S_V in the function space which corresponds to the translation operator V

$$S_V = \begin{pmatrix} x_k \\ x_{k+1} \end{pmatrix}. \tag{19}$$

From Eqs. (14) and (19) it follows

$$S_V\psi(x_k) = \psi(x_{k-1}), \tag{20}$$

because $S_v^{-1} = \binom{x_k}{x_{k-1}}$. We have therefore replaced the translation operator V by the permutation S_V in Eq. (19) [12].

As pointed out above, the operator U in the x-representation is a multiplicative operator $U(x_s)$. The eigenfunctions $(\langle x_k|x'_l\rangle)$ of this operator in the x representation are

$$\langle x_k|x'_l\rangle = \psi_{x'_l}(x_k) = \Delta^{(M)}(x_k - x'_l). \tag{21}$$

Here $\Delta^{(M)}(x)$ is a Kornecker delta-like function which equals 1 when $x = 0$, mod$M$, and zero otherwise [8]; x_k is the coordinate and x'_l determines the eigenvalues of $U(x)$

$$\exp(ix'_l\frac{2\pi}{Mc}), \tag{22}$$

where x'_l (it assumes the same values as x_s) is given in Eq. (6). The functions in Eq. (21) form a basis in the M-dimensional space. By applying the permutation in Eq. (13) to the basis in Eq. (21), we get a new set of functions.

$$S\psi_{x'_l}(x_k) = \Delta^{(M)}(S^{-1}x_k - x'_l) = \langle S^{-1}x_k|x'_l\rangle = \langle x_k|Sx'_l\rangle, \tag{23}$$

where S is the permutation in Eq. (13). For writing the last equality the fact was used that S is a unitary operator. In the new basis functions, the operators $U(x)$ and $V(p)$ in Eq. (8) will become (we keep in mind that $V(p)$ is given by the permutation S_V in Eq. (19))

$$U_S(x_k) = SU(x_k)S^{-1} = \exp(i\frac{2\pi}{Mc}S^{-1}x_k) \tag{24}$$

$$V_S = SVS^{-1} = SS_VS^{-1}. \tag{25}$$

Again S is the general permutation in Eq. (13). It is worthwhile to make the following remark concerning the operators $U(x_k)$, $V(p)$ and S above. As was pointed out, $U(x_k)$ is a multiplicative operator in the x representation, while $V(p)$ and S are defined in Eqs. (18) and (14), respectively. In the representation of the eigenfunctions of U in Eq. (21), the operators $U(x_k)$ and $V(p)$ assume the following matrix form ($\omega = \exp(i\frac{2\pi}{M})$)

$$U_M = \begin{pmatrix} 1 & 0 & 0 & \dots & 0 \\ 0 & \omega & 0 & \dots & 0 \\ 0 & 0 & \omega^2 & \dots & 0 \\ . & . & . & . & . \\ . & & & & . \\ . & & & & . \\ 0 & . & . & . & \omega^{M-1} \end{pmatrix} \quad V_M = \begin{pmatrix} 0 & 0 & 0 & \dots & 0 & 1 \\ 1 & 0 & 0 & \dots & 0 & 0 \\ 0 & 1 & 0 & \dots & 0 & 0 \\ . & . & & & . & . \\ . & . & & & . & . \\ . & . & & & . & . \\ 0 & 0 & 0 & \dots & 1 & 0 \end{pmatrix} \tag{26}$$

U_M is diagonal, while V_M has 1 underneath the diagonal, 1 in the upper right corner and zero everywhere else. Similarly, one can represent the permutation S in Eq. (13) by the permutation matrix S_M which in the k-th column has **1** in the s_k-th row and zero otherwise [16]. As an example we write the permutation matrix P_M for the permutation P in Eq. (15)

$$P_M = \begin{pmatrix} 0 & 0 & 1 & 0 \\ 0 & 0 & 0 & 1 \\ 1 & 0 & 0 & 0 \\ 0 & 1 & 0 & 0 \end{pmatrix} \tag{27}$$

It is easy to check that V_s and U_s in Eqs. (24) and (25) satisfy the same commutation relation as the operators U or V in Eq. (9). Each of these operators U or V form a complete commuting set of operators. The same is true about U_s and V_s. When M is a prime number this also holds for any power of $U(U_s)$ or $V(V_s)$. This is, however, not true, in general, when M can be factorized into 2 or more factors, as is shown in the next section.

3. Splitting of a $M = M_1M_2$-dimensional space into M_1 and M_2-dimensional spaces

Part of this section is a review of some results of Ref. [10]. Having established the action of permutations in the finite phase plane, we show in this section how to use the new permutation symmetry in carrying out the splitting of an $M = M_1M_2$-dimensional space into M_1 and M_2-dimensional subspaces. In Ref. [1] such a splitting was carried out for the case when M_1 and M_2 are mutually prime. Here we remove this restriction on M_1 and M_2 and show how to carry out the splitting for any M_1 and M_2. Obviously, this process can be continued if either M_1 or M_2 (or both) are non-prime. Eventually, this will lead to an exhaustive splitting of the M-dimensional phase plane into a direct product of M_i-dimensional subspaces, where $M = \sqcap_i M_i$ and M_i are the prime factors of the decomposition of M. When $M = M_1 M_2$, we can use powers of the operators U and V in Eq. (8) for defining operators of two kq-representations [17]

$$U^{M_2} = \exp(i\frac{2\pi}{a}x), \quad V^{M_1} = \exp(-\frac{i}{\hbar}pa) \tag{28}$$

$$U^{M_1} = \exp(i\frac{2\pi}{b}x), \quad V^{M_2} = \exp(-\frac{i}{\hbar}pb) \tag{29}$$

where $a = M_1 c$ and $b = M_2 c$. Either of the couples of the operators in Eq. (28) or Eq. (29) forms a complete set of commuting operators. As such, each forms a quantum-mechanical representation, which can replace the x or p-representations. In a recent publication [8] it was shown that when M_1 and M_2 are relatively prime the couples of operators in Eq. (28) and (29) are mutually conjugate. This means that in the eigenstates of the operators of Eq. (28), all eigenvalues of the operators in Eq. (29) are equally probable, and vice versa, in the eigenstates of the operators of Eq. (29), all eigenvalues of the operators in Eq. (28) are equally probable. Let us use the following notation:

$$u_1 = U^{M_2}, \quad v_1 = V^{M_2}, \quad u_2 = U^{M_1}, \quad v_2 = V^{M_1}. \tag{30}$$

In references [1,8] it was shown that under the condition that M_1 and M_2 are mutually prime, the operators u_1 and v_1 are mutually conjugate in the subspace of dimension M_1. Similarly, the operators u_2 and v_2 are mutually conjugate in the subspace of dimension M_2. It should be added that the operators u_1 and v_1 commute with the operators u_2 and v_2 (no matter whether or not M_1 and M_2 are mutually prime). The use of the operators u_1, v_1 and u_2, v_2 splits the M-dimensional space into M_1 and M_2-dimensional space when M_1 and M_2 are mutually prime.

In what follows we show how to use the above introduced permutations in the finite phase plane in order to split the $M = M_1 M_2$-dimensional space into subspaces of dimension M_1 and M_2 when the latter are not necessarily mutually prime. For this we write the eigenvectors $|k, q\rangle$ [8,17] of the operators in Eq. (28) by means of the eigenvectors $|x\rangle$ of U [see Eq. (21)]

$$|k, q\rangle = \frac{1}{\sqrt{M_2}} \sum_{n=0}^{M_2-1} \exp(ikM_1 cn)|q + nM_1 c\rangle \tag{31}$$

where k and q are the quasimomentum and the quasicoordinate, respectively. They assume the following discrete values

$$k = \frac{2\pi}{Mc}f, \;\; f = 0, 1, \ldots, M_2 - 1 : \;\; q = gc, \;\; g = 0, 1, \ldots, M_1 - 1. \tag{32}$$

In order to keep in mind that k assumes M_2 values and q assumes M_1 values, let us rewrite Eq. (31) in the following way (we assume $c = 1$)

$$|k_2, q_1\rangle = \frac{1}{\sqrt{M_2}} \sum_{n_2=0}^{M_2-1} \exp(ik_2 \frac{2\pi}{M_2} n_2)|q_1 + n_2 M_1\rangle \tag{33}$$

where now k_2 is used instead of k, and also it replaces f in Eq. (32), n_2 appears instead of n and q is replaced by q_1. Similarly, we write the eigenvectors $|k_1, q_2\rangle$ of the operators in Eq. (29)

$$|k_1, q_2\rangle = \frac{1}{\sqrt{M_1}} \sum_{m_1=0}^{M_1-1} \exp(ik_1 \frac{2\pi}{M_1} m_1)|q_2 + m_1 M_2\rangle, \tag{34}$$

where

$$k_1 = 0, 1, \ldots, M_1 - 1, \quad q_2 = 0, 1, \ldots, M_2 - 1. \tag{35}$$

In Ref. [8] it was shown that for M_1 and M_2 relatively prime (e.g. $M = 6$, $M_1 = 2$, $M_2 = 3$) we have

$$|\langle k_2, q_1 | k_1, q_2 \rangle| = \frac{1}{\sqrt{M}}, \tag{36}$$

which means that the bases $|k_2, q_1\rangle$ and $|k_1, q_2\rangle$ are mutually unbiased [2,3]. However, when M_1 and M_2 are not relatively prime, the states $|k_2, q_1\rangle$ and $|k_1, q_2\rangle$ in Eqs. (33) and (34) are not mutually unbiased.

We now carry out the following transformation on the states $|k_2, q_1\rangle$ in Eq. (33) by means of the operator A, such that

$$A|q_1 + n_2 M_1\rangle = |n_2 + q_1 M_2\rangle, \tag{37}$$

where $|q_1 + n_2 M_1\rangle$ and $|n_2 + q_1 M_2\rangle$ are eigenstates $|x\rangle$ of the operator U in Eq. (8). We keep in mind that $q_1 = 0, 1, \ldots, M_1 - 1$ and $n_2 = 0, 1, \ldots, M_2 - 1$. The definition of A in Eq. (37) is based on the fact that any x can be represented uniquely by [11,12]

$$x = q_1 + n_2 M_1 \quad \text{or} \quad x' = n_2 + q_1 M_2. \tag{38}$$

The presentation of x in Eq. (38) plays an important role in the Cooley-Tukey algorithm for the fast Fourier transform [11]. As an example of the transformation A, let us consider the case of $M = 8$, $M_1 = 2$, $M_2 = 4$. The results for Eq. (38) are given in Table 1.

Table 1. x and x' in Eq. (38) for $M = 8$, $M_1 = 2$, $M_2 = 4$.

q_1	0	0	0	0	1	1	1	1
n_2	0	1	2	3	0	1	2	3
x	0	2	4	6	1	3	5	7
x'	0	1	2	3	4	5	6	7

The transformation A which transforms x into x', Eq. (38), $x' = Ax$, can be written as a permutation

$$A = \begin{pmatrix} 0 & 1 & 2 & 3 & 4 & 5 & 6 & 7 \\ 0 & 4 & 1 & 5 & 2 & 6 & 3 & 7 \end{pmatrix} = (0)(1,4,2)(3,5,6)(7). \tag{39}$$

The corresponding permutation matrix A_M is [See the permutation matrix P_M in Eq. (27) for the permutation P in Eq. (15)]:

$$A_M = \begin{pmatrix} 1 & 0 & 0 & 0 & 0 & 0 & 0 & 0 \\ 0 & 0 & 1 & 0 & 0 & 0 & 0 & 0 \\ 0 & 0 & 0 & 0 & 1 & 0 & 0 & 0 \\ 0 & 0 & 0 & 0 & 0 & 0 & 1 & 0 \\ 0 & 1 & 0 & 0 & 0 & 0 & 0 & 0 \\ 0 & 0 & 0 & 1 & 0 & 0 & 0 & 0 \\ 0 & 0 & 0 & 0 & 0 & 1 & 0 & 0 \\ 0 & 0 & 0 & 0 & 0 & 0 & 0 & 1 \end{pmatrix} \tag{40}$$

where the rows and columns are labeled from 0 to 7. It is often called stride permutation matrix [12].

Under the transformation in Eq. (37) we have

$$A|k_2, q_1\rangle = \overline{|k_2, q_1\rangle} = \frac{1}{\sqrt{M_2}} \sum_{n_2=0}^{M_2-1} \exp(ik_2 \frac{2\pi}{M_2} n_2)|n_2 + q_1 M_2\rangle. \qquad (41)$$

We keep in mind that the eigenstates $|x\rangle$ are periodic in x with the period M, because of the periodicity of U in Eq. (8) with the same period. The new states $\overline{|k_2, q_1\rangle}$, as can be checked, are mutually unbiased to the states $|k_1, q_2\rangle$ in Eq. (34)

$$\overline{\langle k_2, q_1|}k_1, q_2\rangle = \frac{1}{\sqrt{M}} \exp(ik_1 \frac{2\pi}{M_1} q_1 - ik_2 \frac{2\pi}{M_2} q_2). \qquad (42)$$

We should point out that this result holds, no matter whether or not M_1 and M_2 are relatively prime. When we take the absolute value of both sides in Eq. (42), we get the result like in Eq. (36).

With the definition in Eq. (41), it is easy to see that $\overline{|k_2, q_1\rangle}$ are eigenstates of the operators [See Eq. (28)]

$$\bar{u}_1 = \bar{U}^{M_2} \equiv AU^{M_2}A^{-1}, \;\; \bar{v}_2 = \bar{V}^{M_1} = AV^{M_1}A^{-1}, \qquad (43)$$

since $|k_2, q_1\rangle$ are eigenstates of the operators $u_1 = U^{M_2}$, $v_2 = V^{M_1}$ in Eq. (30).

One can check that the following commutation relations hold

$$\bar{u}_1 v_1 = v_1 \bar{u}_1 \exp(i\frac{2\pi}{M_1}), \;\; u_2 \bar{v}_2 = \bar{v}_2 u_2 \exp(i\frac{2\pi}{M_2}), \qquad (44)$$

where v_1 and u_2 are defined in Eq. (30) and $\bar{u}_1$ and $\bar{v}_2$ in Eq. (43). As it follows from Eq. (44), the pairs of operators $\bar{u}_1$, v_1 and u_2 $\bar{v}_2$ are mutually conjugate. In Ref. [8] it was shown that for M_1 and M_2 relatively prime, the pairs u_1 v_1 and u_2, v_2 in Eq. (30) are mutually conjugate. In Eq. (44), the M_1-power of the operators with the subscript 1 equals 1, and, similarly for the operators for the M_2-power with the subscript 2, one has

$$\bar{u}_1^{M_1} = v_1^{M_1} = u_2^{M_2} = \bar{v}_2^{M_2} = 1. \qquad (45)$$

We have therefore achieved the following goal. We have started with an M-dimensional space and the operators U and V, which are conjugate according to Eq. (9). By assuming that $M = M_1 M_2$, we have defined the operators in Eqs. (28)-(30), and consequently the operators in Eq. (43). This brought us to the operators in Eq. (44) (satisfying Eq. (45)), which split the M-dimensional space into M_1 and M_2-dimensional spaces. We keep in mind that the operators in Eq. (44) with subscript 1 commute with those with the subscript 2. Clearly, if either M_1 or M_2 (or both) are not prime, then this splitting process can be continued. Eventually, this process will come to an end, when we have split the M-dimensional space into $M_1, M_2 \ldots, M_n$ - dimensional subspaces, where $M_1, M_2 \ldots, M_n$ are the prime factors into which M splits.

A similar statement follows from this splitting with respect to the Weyl-Heisenberg group in the M-dimensional phase plane. As is known [13] this group has a single M-dimensional irreducible representation. When $M = \sqcap_i M_i$ we denote the Weyl-Heisenberg group in the M-dimensional phase plane G_M, and the corresponding subgroups in the M_i - dimensional spaces by G_{M_i}. Then our results show that G_M is a direct product [18] of its subgroups $G_{M_1}, G_{M_2}, \ldots, G_{M_n}$. The M-dimensional irreducible representation $D^{(M)}$ of G_M is therefore a direct product

$$D^{(M)} = D^{(M_1)} \times D^{(M_2)} \times \ldots D^{(M_n)} \qquad (46)$$

of the irreducible representations D^{M_i} of its subgroups. It is satisfying that this statement is consistent with the product $M = \sqcap_i M_i$.

4. Example: $M = 8$

As a first step we choose $M_1 = 2$, $M_2 = 4$. They are not mutually prime, and therefore the results of Refs. [1,8] do not apply, and we shall turn to the operators in Eqs. (30) and (43) of this paper ($a = 2c$, $b = 4c$))

$$u_1 = U^4 = e^{i\frac{2\pi}{2c}x}; \quad v_2 = V^2 = e^{-\frac{i}{\hbar}p2c} = (0,2,4,6)(1,3,5,7) \tag{47}$$

$$\bar{u}_1 = Au_1A^{-1} = e^{i\frac{\pi}{c}A^{-1}x}; \quad \bar{v}_2 = Av_2A^{-1} = (0,1,2,3)(4,5,6,7), \tag{48}$$

where we have used Eqs. (30), (39) and (43). In Figs. 2 and 3 we plot the real parts of the operators u_1 and $\bar{u}_1$.

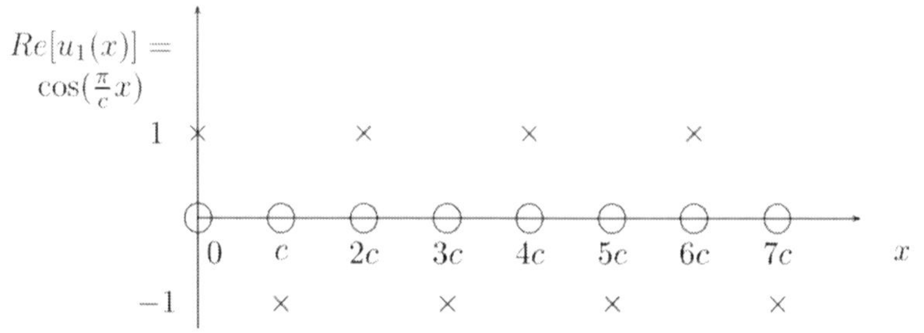

Figure 2. $Re[u_1(x)]$ as a function of x.

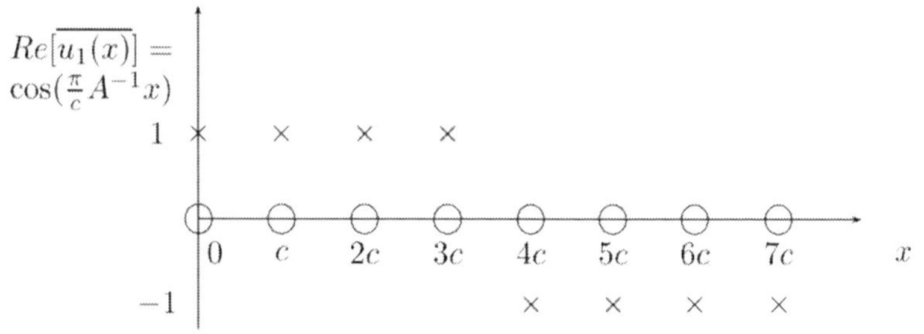

Figure 3. $Re[\overline{u_1(x)}]$ as a function of x.

There are a number of features one can see in these figures. $Re[u_1(x)]$ in Fig. 2 is a conventional periodic Bloch potential on a lattice with period $a = 2c$ for the discrete coordinate $x_s = sc \; s = 0, 1, \ldots, 7$. One can see from this figure that the operator v_2, which is a shift by the period $2c$, commutes with the potential $Re[u_1(x)]$ (it also commutes with u_1 itself in Eq. (47)). A more interesting case is the potential $Re[\overline{u_1(x)}]$ in Fig. 3. This seems to be a new kind of a potential which is invariant under the new kind of symmetry $\bar{v}_2$ in Eq. (48). $\bar{v}_2$ consists of a product of 2 independent permutations, each operating in a 4-dimensional space. From the graph in Fig. 3 one can see that $\bar{v}_2$ commutes with $Re[\overline{u_1(x)}]$ (it also commutes with $\overline{u_1(x)}$ in Eq. (48)). According to Eq. (30) we have a pair of commuting operators

$$u_2 = e^{i\frac{2\pi}{4c}x}, \quad v_1 = e^{-\frac{i}{\hbar}p4c}. \tag{49}$$

Respectively, from Eq. (44) the following commutation relations hold for $M = 8$

$$\bar{u}_1 v_1 = v_1 \bar{u}_1 \exp(i\frac{2\pi}{2}), \quad u_2 \bar{v}_2 = \bar{v}_2 u_2 \exp(i\frac{2\pi}{4}), \tag{50}$$

and the operators with index 1 commute with those of index 2.

Having in mind that v_1 in Eq. (49) is a shift by $4c$, we can read directly from Fig. 3 the first commutation relation in Eq. (50). We can see that any shift by $4c$ changes the sign of $\cos(\frac{\pi}{c}A^{-1}x)$. In order to see the same for the second relation in Eq. (50), we need the graph for $u_2(x)$. In Fig. 4 we plot the real part, $Re[u_2(x)]$, of the operator $u_2(x)$.

First, we can easily see from the graph in Fig. 4 that v_1 in Eq. (49) commutes with $Re[u_2(x)]$. The second relation in Eq. (50) contains the permutation $\bar{v}_2$ in Eq. (48), which permutes the coordinates $(0, 1, 2, 3)$ and $(4, 5, 6, 7)$ among themselves. Such permutations change the phase by $\frac{\pi}{2}$, like in Eq. (50).

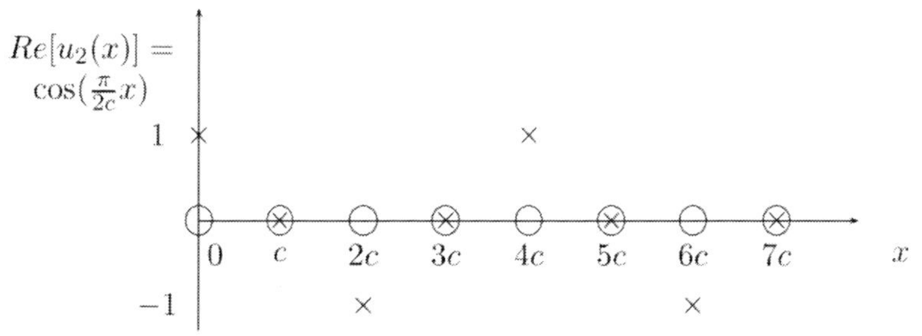

Figure 4. $Re[u_2(x)]$ as a function of x.

The commutation relations in Eq. (50) are the same as the one in Eq. (9). The only difference is that in Eq. (9) M appears (in our example $M = 8$) while in the first relation of Eq. (50) we have $M = 2$, and in the second one $M = 4$. As was pointed out in Eq. (45), the operators in Eq. (50) are of the order 2 and 4 respectively

$$\bar{u}_1^2 = v_1^2 = u_2^4 = \bar{v}_2^4 = 1. \tag{51}$$

Eqs. (50) and (51) therefore show that the two pairs of operators $\bar{u}_1$, v_1 and u_2, $\bar{v}_2$ split the 8-dimensional space for $M = 8$ into 2-dimensional and 4-dimensional spaces, respectively. The first commutation relation in Eq. (50) is in a 2-dimensional space and cannot be split any further. The second relation is in a 4-dimensional space. As was pointed out above, this relation is the same as the one in Eq. (9), for $M = 4$. We can now repeat the process and split the 4-dimensional space into two 2-dimensional spaces. This will complete the splitting of the 8-dimensional space into three 2-dimensional spaces.

We will not repeat this splitting and give directly the results. For this we write the matrices U_m and V_m in Eq. (26) for $M = 8$. For U_8 we list only its diagonal elements (the non-diagonal elements are all zero). V_8 we represent by its permutation in Eq. (19). We have ($w = \exp(i\frac{2\pi}{8})$)

$$U_8 = \begin{pmatrix} 1 & & & \\ & w & & \\ & & \ddots & \\ & & & w^7 \end{pmatrix}, \quad V_8 = (0, 1, \ldots, 7). \tag{52}$$

For giving the results of the splitting of the $M = 8$-dimensional space into three independent 2-dimensional spaces we need the operators, which are 4th powers of the ones in Eq. (52)

$$U_8^4 = \begin{pmatrix} 1 &&&&&&& \\ & -1 &&&&&& \\ && 1 &&&&& \\ &&& -1 &&&& \\ &&&& 1 &&& \\ &&&&& -1 && \\ &&&&&& 1 & \\ &&&&&&& -1 \end{pmatrix}, \quad V_8^4 = (0,4)(1,5)(2,6)(3,7). \tag{53}$$

The operators that carry out the splitting of the 8-dimensional space in 3 two-dimensional spaces are then

$$U_1 = U_8^4 = \begin{pmatrix} 1 &&&&&&& \\ & -1 &&&&&& \\ && 1 &&&&& \\ &&& -1 &&&& \\ &&&& 1 &&& \\ &&&&& -1 && \\ &&&&&& 1 & \\ &&&&&&& -1 \end{pmatrix}, \; V_1 = A^{-1}V_8^4A = (0,1)(2,3)(4,5)(6,7), \tag{54}$$

$$U_2 = AU_8^4A^{-1} = \begin{pmatrix} 1 &&&&&&& \\ & 1 &&&&&& \\ && 1 &&&&& \\ &&& 1 &&&& \\ &&&& -1 &&& \\ &&&&& -1 && \\ &&&&&& -1 & \\ &&&&&&& -1 \end{pmatrix}, \; V_2 = V_8^4 = (0,4)(1,5)(2,6)(3,7), \tag{55}$$

$$U_3 = BU_8^4B = \begin{pmatrix} 1 &&&&&&& \\ & 1 &&&&&& \\ && -1 &&&&& \\ &&& -1 &&&& \\ &&&& 1 &&& \\ &&&&& 1 && \\ &&&&&& -1 & \\ &&&&&&& -1 \end{pmatrix}, \; V_3 = CV_8^4C = (0,2)(1,3)(4,6)(5,7), \tag{56}$$

where A is given in Eqs. (39), (40), and

$$B = B^{-1} = (0)(1,2)(3)(4)(5,6)(7), \; C = C^{-1} = (0)(1)(2,4)(3,5)(6)(7). \tag{57}$$

The operators in Eqs. (54)-(56) satisfy the following commutation relations

$$U_iV_i = -V_iU_i, \quad i = 1,2,3. \tag{58}$$

This is the same commutation relation as in Eq. (9) for the operators U and V but for $M = 2$. Also, every pair of operators in one row of Eqs. (54)-(56) commutes with the 4 operators in the other 2 rows. In addition, the squares of all the operators in Eqs. (54)-(56) equal to 1, which is in agreement with Eqs. (7) and (8) for the operators U and V for $M = 2$. The conclusion is that the operators in Eqs. (54)-(56) split the 8-dimensional space in three 2-dimensional independent spaces.

This result enables us to establish a one-to-one correspondence between the operators in Eqs. (54)-(56) and the spin half Pauli matrices by the following identification

$$\sigma_{ix} = U_i, \quad \sigma_{iy} = V_i, \quad i = 1, 2, 3. \tag{59}$$

where σ_x and σ_y are the Pauli matrices [24]

$$\sigma_x = \begin{pmatrix} 0 & 1 \\ 1 & 0 \end{pmatrix}, \quad \sigma_y = \begin{pmatrix} 0 & -i \\ i & 0 \end{pmatrix}. \tag{60}$$

With the identification in Eq. (59) we accommodate 3 spin half operators in the 8-dimensional phase plane. This accommodation can be extended to n spin half particles in a 2^n-dimensional phase plane.

The Weyl-Heisenberg group in 8 dimensions G_8 has 8^3 elements (is of the order 512) [13]. The elements in each row of the Eqs. (54)-(56) are generators of the subgroups G_2 of G_8. We have 3 such subgroups which we denote by G_2^i, $i = 1, 2, 3$. These subgroups commute with one another, and therefore G_8 can be written as a direct product of 3 subgroups of order 8.

$$G_8 = G_2^{(1)} \times G_2^{(2)} \times G_2^{(3)}. \tag{61}$$

Correspondingly, the single irreducible representation of dimension 8, $D^{(8)}$ of $G^{(8)}$ can be written as a direct product in the following way

$$D^{(8)} = D^{(2)} \times D^{(2)} \times D^{(2)}, \tag{62}$$

where each $D^{(2)}$ is a 2-dimensional representation of one of the subgroups $G^{(i)}$, $i = 1, 2, 3$ in Eq. (61). The splitting of the 8-dimensional phase space, $M = 8$, into 3 independent 2-dimensional spaces, is therefore accompanied by the elegant results in Eqs. (61) and (62) for the Weyl-Heisenberg group.

References

[1] Schwinger J 1960 *Proc. Natl. Acad. Sci.* **46** 570
[2] Invanovic I D 1981 *J. Phys.* A **14** 3241
[3] Wooters W K and Fields B D 1989 *Annals of Physics* **191** 363
[4] Leonhardt U 1996 *Phys. Rev.* **A53** 2998
[5] Galetti D and De Toledo Piza A F R 1988 *Physica* **149A** 267; Cohendet O, Combe Ph, Sirugue M and Sirugue-Collin M 1988 *J. Phys.* A: Math. Gen. **21** 2875; Miquel C, Pablo Paz J and Saraceno M 2002 *Phys. Rev.* A **65** 062309
[6] We call it a related result because it is well known that in the infinite phase plane the Wigner function can be derived by using the inversion operator. See ref. [7]
[7] Royer A 1977 *Phys. Rev.* A **15** 449
[8] Mann A, Revzen M and Zak J 2005 *J. Phys.* A: Math. Gen. **38** L389; Zak J 2008 *J. Physics: Conference Series* **104** 012013
[9] Dirac P A M 1958 *The Principles of Quantum Mechanics* (New York: Oxford University Press)
[10] Simkhovich B 2009 *Master thesis* (Technion) and paper sent for publication by Simkhovich B, Mann A and Zak J
[11] Cooley J W and Tukey J W 1965 *Math. Comput.* **19** 297
[12] An M, Brodzik A K and Tolimieri R 2008 *Ideal Sequence Design in Time-Frequency Space* (Boston: Birkhauser)
[13] Zak J 1989 *Phys. Rev.* B **39** 694
[14] Landau L D and Lifshitz E M 1958 *Quantum Mechanics* (London: Pergamon Press)

[15] Wigner E P 1959 *Group Theory and its Application to the Quantum Mechanics of Atomic Spectra* (New York: Academic Press)
[16] Murnaghan F D 1963 *The Theory of Group Representations* (New York: Dover Publications)
[17] Zak J 1967 *Phys. Rev. Lett.* **19** 1385; 1970 *Phys. Today* **23** 51
[18] Hamermesh M 1964 *Group Theory and its Applications to Physical Problems* (Addison-Wesley)

Symmetry preserving and improved energy states derived from asymmetric Hartree-Fock

A K Theophilou, Iris Theophilou and S Thanos
Democritos NCSR, Institute of Materials Science,15310, Athens, Greece
E-mail: theo@ims.demokritos.gr

Abstract. Symmetry preserving Hartree-Fock (HF) states, give higher energies than those of the unrestricted HF. However, in analyzing an unrestricted HF state (UHF) in terms of states transforming according to the Irreducible Representations of the exact Hamiltonian of a many electron system, we prove that some of these states have lower energy than the initial UHF state. The case of rotational and spin symmetries are discussed.

1. Introduction

One of the main achievement of Quantum Mechanics is the calculation of the properties of atoms, molecules and solids. Most of these calculations deal with the ground state, which is not only an eigenstate of the Hamiltonian H, but it is also the minimizing state of the energy functional $E(\Psi) =< \Psi|H|\Psi >$. As it is practically impossible to obtain the exact eigenstates, even in the case of the Helium atom, which has only two electrons, many approximations appeared. One of them is the Hartree-Fock approximation, which is based on single Slater determinant minimization. A simplification of this is the Optimized Effective Potential Approximation [1, 2, 3, 4], where the space of the minimizing states is restricted to those obtained by an effective potential. Another popular method is the Kohn and Sham method [5] based on a theorem showing that there is one to one correspondence between the ground state and its density [6]. The main disadvantage of this theory is that no functionals of the exchange and correlation energy exist, which are sufficiently accurate. Many simplifications of the many electron theory have their origin in the symmetry of the Hamiltonian, resulting from the fact that it deals with identical particles. Thus, the many electron Hamiltonian is invariant under any permutation of the space and spin variables and therefore the eigenstates of the Hamiltonian belong to the irreducible representations (Ireps) of the permutation group. Fortunately only the antisymmetric Irrep appears in nature. Another symmetry which is always present, when no inhomogeneous magnetic field is applied, is spin symmetry. Thus, the energy eigenstates of atoms, molecules and solids are also eigenstates of S^2 and S_z. Other symmetries may appear, like e.g. the spherical symmetry in atoms and space group symmetries in solids. Some, molecules have point group symmetries, whereas two-atom molecules have axial symmetry. These symmetries are geometric in nature and arise because of the symmetries originating from the positions of the nuclei.

A lot of theoretical work in the early development of the many-electron theory has been developed by Löwdin [7, 8, 9, 10, 11]. Löwdin, extended HF, by projecting out of a single Slater determinant (Sladet) an eigenstates $|\Psi^S_M >$ of S^2. This state is not a single Sladet but a linear combination of such states. The resulting energy functional $< \Psi_M^S|H|\Psi_M^S >$ was next minimized with respect to the orbitals, giving energies lower than those of UHF [12, 13].

In Section 2, we shall show how from the asymmetric solutions of UHF, one can obtain solutions which belong to a definite Irrep of the symmetry group of the Hamiltonian and also have lower energies. The advantage with respect to the extended HF, is that one uses the standard UHF solutions, to improve on energy and quality. We deal with the example of spherical symmetry, whereas the problem of spin symmetry with numerical examples is treated in a sequel paper [14].

2. The Method

We consider the many electron Hamiltonian H,

$$H = T + V + Hint \tag{1}$$

where T is the kinetic energy operator, V is the external potential and H_{int} is the electron-electron interaction operator, $H_{int} = H_{int}^{\uparrow} + H_{int}^{\downarrow}$

$$H_{int}^{s} = \int d^3r \int d^3r' \frac{1}{|\mathbf{r}-\mathbf{r}'|} \psi^{s\dagger}(\mathbf{r})\psi^{s\dagger}(\mathbf{r}')\psi^{s}(\mathbf{r}')\psi^{s}(\mathbf{r}), \tag{2}$$

where $\psi^{s\dagger}(\mathbf{r})$, $\psi^{s}(\mathbf{r})$ are the fermion field creation and annihilation operators with s the spin index, which takes the values $\uparrow$ and $\downarrow$. Let us define the functional

$$E(\Phi) =< \Phi|H|\Phi >, \ |\Phi > \epsilon S_{sl}, \tag{3}$$

with S_{sl} the space of $N-particle$ Slater states $|\Phi_M >= |\varphi_1^{\uparrow}, \varphi_2^{\uparrow}, ..., \varphi_k^{\uparrow}; \varphi_1^{\downarrow}, \varphi_2^{\downarrow}, ...\varphi_l^{\downarrow} >$ corresponding to the Sladet determinants (Sladets)

$$\Phi_M(\mathbf{r}_1, ...\mathbf{r}_N; \varphi_1^{\uparrow}, \varphi_2^{\uparrow}, ..., \varphi_k^{\uparrow}; \varphi_1^{\downarrow}, \varphi_2^{\downarrow}, ...\varphi_l^{\downarrow}) = \frac{1}{\sqrt{N!}} \det\{\varphi_i^{s}(\mathbf{r}_j)\}, \tag{4}$$

where the $\varphi_i^s(\mathbf{r})$ are one-particle wave functions named as *spin orbitals* and $M = (k-l)/2$ is the eigenstate of L_z.

The above functional can be expressed as

$$E(\Phi) =< \Phi|T|\Phi > + < \Phi|V|\Phi > + E_h(\Phi) + E_x(\Phi) \tag{5}$$

where $E_h(\Phi) = E_h^{\uparrow\uparrow}(\Phi) + E_h^{\downarrow\downarrow}(\Phi) + 2E_h^{\uparrow\downarrow}$ and the suffix M has been omitted for the sake of simplicity of notation,

$$E_h^{\alpha\beta}(\Phi) = \frac{1}{2} \int d^3r \int d^3r' \frac{1}{|\mathbf{r}-\mathbf{r}'|} \rho_{\Phi}^{\alpha}(\mathbf{r})\rho_{\Phi}^{\beta}(\mathbf{r}') \tag{6}$$

is the electrostatic (Hartree) energy due to the electron density $\rho_{\Phi}(\mathbf{r}) = \rho_{\Phi}^{\uparrow}(\mathbf{r}) + \rho_{\Phi}^{\downarrow}(\mathbf{r})$ and the indices α and β take the values $\uparrow$ and $\downarrow$.

It is to be noted that the exchange energy $E_x(\Phi) = E_x^{\uparrow}(\Phi) + E_x^{\downarrow}(\Phi)$, with

$$E_x^{\alpha}(\Phi) = \frac{1}{2} \int d^3r \int d^3r' \frac{1}{|\mathbf{r}-\mathbf{r}'|} \rho_{\Phi}^{\alpha}(\mathbf{r},\mathbf{r}')\rho_{\Phi}^{\alpha}(\mathbf{r}',\mathbf{r}) \tag{7}$$

does not include a cross term. Thus the up and down terms of $E(\Phi)$, are coupled only through the Hartree term $E_h^{\uparrow\downarrow}$. In the above expressions $\rho_{\Phi}^{\alpha}(\mathbf{r},\mathbf{r}') =< \Phi|\psi^{\alpha\dagger}(\mathbf{r}) \ \psi^{\alpha}(\mathbf{r}')|\Phi >$ is the density matrix of the state $|\Phi>$ and $\rho^\alpha_\Phi(\mathbf{r}) = \rho^\alpha_\Phi(\mathbf{r},\mathbf{r})$ is the electron density of spin α. The explicit form of $\rho^\alpha_\Phi(\mathbf{r},\mathbf{r})$ is

$$\rho^\alpha_\Phi(\mathbf{r},\mathbf{r}') = \sum_{i=1}^{k_\alpha} \varphi^\alpha_i(\mathbf{r})^* \varphi^\alpha_i(\mathbf{r}'),\ k_\uparrow = k,\ k_\downarrow = l \tag{8}$$

With the above expression of $E(\Phi)$, one does not have to add the constraint that $|\Phi>$ is a Slater state. After applying the variational principle, we find that the minimizing state of $E(\Phi)$ is not an eigenstate of H, but of the Hartree-Fock Hamiltonian H^{hf},

$$H^{hf} = T + V + V_h(\Phi) + V_x(\Phi) \tag{9}$$

where T is the kinetic energy operator, V is the external potential and $V_h(\Phi)$ is the Hartree potential operator

$$V_h(\Phi) = \int d^3r \int d^3r' \frac{1}{|\mathbf{r}-\mathbf{r}'|} \rho_\Phi(\mathbf{r}')[\psi^{\uparrow\dagger}(\mathbf{r})\psi^\uparrow(\mathbf{r}) + \psi^{\downarrow\dagger}(\mathbf{r})\psi^\downarrow(\mathbf{r})] \tag{10}$$

and $V_x(\Phi)$ is the exchange operator

$$V_x(\Phi) = \int d^3r \int d^3r' \frac{1}{|\mathbf{r}-\mathbf{r}'|} \{\rho^\uparrow_\Phi(\mathbf{r},\mathbf{r}')\psi^{\uparrow\dagger}(\mathbf{r}')\psi^\uparrow(\mathbf{r}) + \rho^\downarrow_\Phi(\mathbf{r},\mathbf{r}')\psi^{\downarrow\dagger}(\mathbf{r}')\psi^\downarrow(\mathbf{r})] \tag{11}$$

Note that the HF operator H^{hf} is not linear since $V_h(\Phi)$ and $V_x(\Phi)$ depend on $|\Phi>$. Thus the HF solutions are qualitatively different with respect to symmetry from the eigenstates of the exact Hamiltonian. To see this let H be invariant under a symmetry group G, i.e., $gHg^{-1} = H$, for all $g \in G$ [15]. Then, if a state $|\Psi^\Gamma_\gamma>$ transform according to the Γ Irrep of G, then one can easily verify by direct calculation that the state $H|\Psi^\Gamma_\gamma>$ transforms according to the same Irrep since $gH|\Psi^\Gamma_\gamma> = gHg^{-1}g|\Psi^\Gamma_\gamma> = Hg|\Psi^\Gamma_\gamma>$. This is not the case with the HF operator H^{hf} since $\rho^\alpha_\Phi(\mathbf{r},\mathbf{r}')$ does not transform according to the identity Irrep of G, except in the special case that $|\Phi>$ itself belongs to the identity Irrep. To verify this statement take the expample of spherical symmetry and $l = 1$, for the case that the unfilled shell has only two particles with $m = 0$ and $m = 1$. Then $\varphi^1_m(\mathbf{r}) = \varphi^1(r)Y^1_m(\theta,\phi)$ and after introducing the explicit form of the spherical harmonic $Y^1_m(\theta,\phi)$ we have

$$\rho^\alpha_\Phi(\mathbf{r},\mathbf{r}') = \sum_{m=0}^{1} \varphi^1_m(\mathbf{r})^*\varphi^1_m(\mathbf{r}') = \varphi^1(r)\varphi^1(r')\frac{1}{2\pi}(\cos\theta\cos\theta' + \frac{1}{2}\sin\theta\sin\theta' e^{i(\phi-\phi')}) \tag{12}$$

This operator does not transform according to the identiy Irrep since, even for the case that $\mathbf{r} = \mathbf{r}'$, we get a density which depends on the angle θ and hence it has no spherical symmetry.

To see our argument about symmetry more explicitely, we consider the one-particle HF equation for a spherically symmetric external potenial $V(\mathbf{r}) = V(r)$.

$$-\frac{1}{2}\nabla^2\varphi^\alpha_i(\mathbf{r}) + V(\mathbf{r})\varphi^\alpha_i(\mathbf{r}) + V_h(r;\Phi)\varphi^\alpha_i(\mathbf{r}) + \int d^3r' \rho^\alpha_\Phi(\mathbf{r},\mathbf{r}')\varphi^\alpha_i(\mathbf{r}') = \varepsilon_i\varphi^\alpha_i(\mathbf{r}) \tag{13}$$

which determines the $\varphi^\alpha_i(\mathbf{r})$ of the minimizing Sladet. Then it is obvious that the $\varphi^\alpha_i(\mathbf{r})$ are not eigenstates of L^2 because the Hartree and exchange operators are not spherically symmetric, except in the case of filled shells, where

$$\rho^\alpha_\Phi(\mathbf{r},\mathbf{r}') = \sum_{m=-l}^{l} \varphi^l_m(\mathbf{r})^*\varphi^l_m(\mathbf{r}') = \varphi^l(r)\varphi^l(r')\frac{2l+1}{4\pi}P^l(\frac{\mathbf{r}\cdot\mathbf{r}'}{r.r'}) \tag{14}$$

One can easily verify that $\rho^{\alpha}_{\Phi}(\mathbf{r},\mathbf{r}')$ is invariant under simultaneous rotations of $\mathbf{r}$ and $\mathbf{r}'$.

From the above we conclude that the unresricted HF ground state does not transform according to a definite Irrep of the symmetry group of the exact Hamiltonian, unless one imposes the constraint that the minimization is restricted in a subspace S^{Γ}_{γ} the wavefunctions of which transform according to Irrep Γ of G. Let us define this space as S^{Γ}_{γ}. This is a subspace of the space of single $N-particle$ determinants S_{sl}.Therefore, the energy minimum of $E(\Phi^{\Gamma}_{\gamma}) =< \Phi^{\Gamma}_{\gamma}|H|\Phi^{\Gamma}_{\gamma} >$ is larger or equal to $\min E(\Phi)$ with $|\Phi >$ in S_{sl} which contains S^{Γ}_{γ}. The HF under the subspace constraint is called restricted HF (RHF) while the one without the constraint is called unrestricted HF (UHF). In the following we shall call a state asymmetric if it does not transform according to a definite Irrep of the symmetry group G of the Hamiltonian.

We shall show that one can use the asymmetric solutions of UHF to get lower energies than $E(\Phi_{UHF})$. Further, the new solutions belong to definite Irreps of the symmetry group of H, having thus the transformation properties of the exact solutions. To prove this let us denote by $|\Phi >$ the asymmetric $|\Phi_{UHF} >$ state. Then, we can express $|\Phi >$ in terms of its Irrep components $|\Psi^{\Gamma}_{\gamma} >$, which in general are not single Sladets, but linear combinations of them,

$$|\Phi >= \sum_{\Gamma,\gamma} c^{\Gamma}_{\gamma}|\Psi^{\Gamma}_{\gamma} > . \tag{15}$$

In the above expansion the $|\Psi^{\Gamma}_{\gamma} >$ are the maximum overlap states of the functional $F(\Psi'^{\Gamma}_{\gamma}) = | < \Phi|\Psi'^{\Gamma}_{\gamma} > |^2$. Thus, only one state enters from each subspace S^{Γ}_{γ}. To prove this we must show that any other state $|X^{\Gamma}_{\gamma} >$ which is orthogonal to $|\Psi^{\Gamma}_{\gamma} >$ is also orthogonal to $|\Phi >$. Since $|\Psi^{\Gamma}_{\gamma} >$ is the maximizing state of the functional $F(\Psi'^{\Gamma}_{\gamma})$ it follows that

$$\lim_{\epsilon\to 0}\frac{1}{\epsilon}\{< \Phi|\Psi^{\Gamma}_{\gamma} + \epsilon X^{\Gamma}_{\gamma} > - < \Phi|\Psi^{\Gamma}_{\gamma} >\} = 0 \tag{16}$$

and after taking the limits we find $< X^{\Gamma}_{\gamma}|\Phi >= 0$, q.e.d.

By taking into account that $H|\Psi^{\Gamma}_{\gamma} >$ is a state transforming according to the same Irrep as $|\Psi^{\Gamma}_{\gamma} >$, since $gHg^{-1} = H$, from the orthogonality relation of states belonging to different Irreps, it follows that $< \Psi^{A}_{\alpha}|H|\Psi^{\Gamma}_{\gamma} >= E^{A}\delta_{A\Gamma}\delta_{\alpha\gamma}$ and

$$E(\Phi) = \sum_{\Gamma,\gamma}|c^{\Gamma}_{\gamma}|^2 < \Psi^{\Gamma}_{\gamma}|H|\Psi^{\Gamma}_{\gamma} >= \sum_{\Gamma,\gamma}|c^{\Gamma}_{\gamma}|^2E^{\Gamma} = \sum_{\Gamma}|c^{\Gamma}|^2E^{\Gamma}, \tag{17}$$

where $|c^{\Gamma}|^2 = \sum_{\gamma}|c^{\Gamma}_{\gamma}|^2$.

We shall show that there is at least one E^{Γ} which is smaller than $E(\Phi)$, unless all $E^{\Gamma} = E(\Phi)$ for all Γ.This statement is obvious since if we consider the opposite we arrive at a contradiction. Thus, let $E^{\Gamma} \geq E(\Phi)$ and for some B, $E^{B} > E(\Phi)$. Then by taking into account that $\sum_{\Gamma}|c^{\Gamma}|^2 = 1$, we have

$$E(\Phi) = \sum_{\Gamma}|c^{\Gamma}|^2E^{\Gamma} > \sum_{\Gamma}|c^{\Gamma}|^2E(\Phi) = E(\Phi). \tag{18}$$

This inequality implies $E(\Phi) > E(\Phi)$, which is a contradiction. Hence there is at least one $|\Psi^{A}_{\alpha} >$ having energy less than that of the unrestricted HF.

In general it is not possible to judge which Irrep gives the lowest energy. Practice has shown that in most cases, the ground state is not degenerate and therefore one has to look for the one-dimensional Irreps. This is the case for molecules at their equilibrium positions because of the Jahn-Teller effect, but for atoms in some cases we have degenerate ground states.

From the above we conclude that it is possible to extract some of the correlation energy, E_c, from the usual UHF approximation. Obviously, one has to find the way to determine the $|\Psi_\gamma^\Gamma>$ from the $|\Phi>$. This can be done by the projection theorem [15]:

$$|\Psi_\gamma^\Gamma>=\frac{N_\Gamma}{N_G}\sum_g D_{\gamma a}^{\Gamma *}(g)g|\Psi>, \tag{19}$$

where $D_{\gamma a}^{\Gamma}(g)$ is the γa matrix element of the N_Γ – dim Irrep Γ of the group G the order of which is N_Γ. In most cases there are easier ways to determine the $|\Psi_\gamma^\Gamma>$. Examples for the case of spin symmetry are given in Ref [16] and details about the procedure to extract E_c from UHF are given in a sequel paper [14]. In the present paper we shall consider the case of spherical symmetry, applicable for atoms. To avoid complications, we shall omit the spin variable. Since the above comments hold also for the optimized effective potential approximation (OEP) of UHF, we shall consider this approximation for simplicity, considering that the radial parts of the orbitals are obtained by solving a Schroedinger type one-particle equation

$$-\frac{1}{2}\frac{d^2}{dr^2}u_n^l(r)+\frac{l(l+1)}{2r^2}u_n^l(r)+V_{ef}(r)u_n^l(r)=\in_n^l u_n^l(r) \tag{20}$$

restricting the effective potential $V_{ef}(r)$ to those having spherical symmetry. The orbitals are of the form

$$\varphi_{n,m}^l(\mathbf{r};v)=\frac{u_n^l(r,v)}{r}Y_m^l(\vartheta,\varphi) \tag{21}$$

where the first suffix denotes the energy levels $\in_n^l$for fixed l in ascending order.

The next step is to construct the Slater determinant with fixed L_z. As an example we take the three particle case with L_z eigenvalue equal to 0.

$$|\Phi(v)>=|\varphi_{1,0}^0,\varphi_{1,-1}^1,\varphi_{1,1}^2>. \tag{22}$$

The next step is to minimize $<\Phi(V_{ef})|H|\Phi(V_{ef})>$ with respect to the effective potential $V_{ef}(r)$. After the effective potential and the orbitals are determined, one analyzes $|\varphi_{1,0}^0,\varphi_{1,-1}^1,\varphi_{1,1}^2>$ in eigenstates of L^2. Then, after rejecting the first suffix for convenience, we have

$$|\varphi_{1,0}^0,\varphi_{1,-1}^1,\varphi_{1,1}^2>=<1,0|1,-1;2,1>|\Psi_0^1>+<2,0|1,-1;2,1>|\Psi_0^2>$$
$$+<3,0|1,-1;2,1>|\Psi_0^3> \tag{23}$$

where the $<l,m|l_1,m_1;l_2,m_2>$ are Clebsch-Gordan coefficients.

As an example, we shall calculate the corrected energy $E^1$of $|\Psi_0^1>$, the explicit form of which is

$$|\Psi_0^1>=\frac{1}{\sqrt{10}}\{\sqrt{3}|\varphi_0^0,\varphi_1^1,\varphi_{-1}^2>-2|\varphi_0^0,\varphi_0^1,\varphi_0^2>+\sqrt{3}|\varphi_0^0,\varphi_{-1}^1,\varphi_1^2>\} \tag{24}$$

In calculating the enegy $E^1=<\Psi_0^1|H|\Psi_0^1>$, one must take into account that the kinetic and external potential terms have only diagonal elements since they are one-particle operators and the Sladets of the expansion differ by two particles. Thus e.g. $<\varphi_0^0,\varphi_1^1,\varphi_{-1}^2|T|\varphi_0^0,\varphi_{-1}^1,\varphi_1^2>=0$. Further,

$$<\varphi_0^0,\varphi_1^1,\varphi_{-1}^2|T|\varphi_0^0,\varphi_1^1,\varphi_{-1}^2>=<\varphi_0^0,\varphi_0^1,\varphi_0^2|T|\varphi_0^0,\varphi_0^1,\varphi_0^2>=\sum_{l=0}^{2}<\varphi_m^l|T|\varphi_m^l> \tag{25}$$

and in general the kinetic energy does not depend on the lower indices of the orbitals since by applying the L^+ successively we can go from the lower to the higher eigenvalues of L_z. Thus

$$< \varphi^l_{m+1}|T|\varphi^l_{m+1} >= c < L^+\varphi^l_m|T|L^+\varphi^l_m >= c < \varphi^l_m|T|L^- L^+\varphi^l_m >=< \varphi^l_m|T|\varphi^l_m > . \quad (26)$$

Thus the diagonal terms will give the same energy as the minimizing state $|\Phi>$. Then,

$$E^1 =< \Psi^1_0|H|\Psi^1_0 > -\frac{1}{\sqrt{10}}\{2\sqrt{3} < \varphi^0_0, \varphi^1_1, \varphi^2_{-1}|H_{int}|\varphi^0_0, \varphi^1_0, \varphi^2_0 >$$

$$+2\sqrt{3} < \varphi^0_0, \varphi^1_0, \varphi^2_0|H_{int}|\varphi^0_0, \varphi^1_{-1}, \varphi^2_1 > +3 < \varphi^0_0, \varphi^1_1, \varphi^2_{-1}|H_{int}|\varphi^0_0, \varphi^1_{-1}, \varphi^2_1 > +c.c.\} \quad (27)$$

The off diagonal terms are of the form $< \varphi^0_0\phi_i\phi_j|H_{int}|\varphi^0_0\phi_k\phi_l >$. The explicit form of these matrix elements is

$$< \varphi^0_0, \chi_i\chi_j|H_{int}|\varphi^0_0\chi_k\chi_l >= J(\chi_i, \chi_j; \chi_k\chi_l) - J(\chi_i\chi_j; \chi_l\chi_k) \quad (28)$$

where

$$J(\chi_i, \chi_j; \chi_k\chi_l) = \int d^3r \int d^3r' \chi^*_i(\mathbf{r})\chi^*_j(\mathbf{r}')\frac{1}{|\mathbf{r}-\mathbf{r}'|}\chi_k(\mathbf{r})\chi_l(\mathbf{r}'). \quad (29)$$

The value of the off-diagonal term gives the correlation energy. This energy, can be positive or negative and there will be at least one $|\Psi^l_0>$ which will have energy less than the one obtained from the unrestricted HF, but we cannot judge, which l gives the smallest energy without doing the actual calculation.

After this example, one can consider the general case of spherical symmetry, following the procedure of the above example. In order to find the L^2 eigenstates of the expansion one can act on the Sladet state

$$|\Phi_m >= a^{l1\dagger}_{m1}...a^{li\dagger}_{mi}...a^{lN\dagger}_{mN}|0> \quad (30)$$

expressed as a single product of creation operators, and acting successively with L^+ one will get finally a single Sladet

$$|\Phi^l_l >= a^{l1\dagger}_{l1}...a^{li\dagger}_{li}...a^{lN\dagger}_{lN}|0> \quad (31)$$

Next, following the reverse procedure, one will get $|\Phi^l_{l-1} >= cL^-|\Phi^l_l>$ and by using orthogonality, can determine $|\Phi^{l-1}_{l-1}>$. A detailed account of this procedure for the case of spin symmetry is given in [16]. Finally one determines the expansion of $|\Phi_m>$ as a sum the $|\Phi^l_m>$ states by means of which the energy correction is determined.

The procedure is applicable for any symmetry. In each case some simplifications appear. Thus, in the case of the spin symmetry, which is always present in atoms, molecules and solids, one can prove certain theorems which lead to further simplifications. Details of the theory and applications to molecules are given in the paper that follows [14].

References

[1] Sharp R T and Horton G K 1953 *Phys. Rev.* **90** 317
[2] Talman J D and Shadwick W F 1976 *Phys. Rev.* A **14** 36
[3] Theophilou A K and Glushkov V N 2005 *I. J. Q. Chem.* **104** 538
[4] Theophilou A K and Glushkov V N 2006 *J. Chem. Phys.* **124** 034105
[5] Kohn W and Sham L J 1965 *Phys. Rev.* A **140** 1133
[6] Hohenberg P and Kohn W 1964 *Phys. Rev.* B **136** 864
[7] Löwdin P O 1955 *Phys. Rev.* **97** 1474
[8] Löwdin P O 1955 *Phys. Rev.* **97** 1490

[9] Löwdin P O 1955 *Phys. Rev.* **97** 1509
[10] Löwdin P O 1960 *Reviews of Modern Physics* **32** 328
[11] Löwdin P O 1962 *Reviews of Modern Physics* **34** 520
[12] Mayer I, Ladic J, Biczo G 1973 *I. J. Q. Chem.* **VII** 583
[13] Mayer I 1974 *I. J. Q. Chem.* **VIII** 893
[14] Theophilou I, Thanos S and Glushkov V (submitted to *J. Phys.: Conf. Ser.*)
[15] Tinkham M 2003 *Group Theory and Quantum Mechanics* (New York, Dover)
[16] Theophilou I, Thanos S and Theophilou A K 2007 *J. Chem. Phys.* **127** 234103

Propagation and storing of light in optically modified atomic media

Jarosław Zaremba
Institute of Physics
Nicolaus Copernicus University
ul. Grudziądzka 5/7
87 100 Toruń, Poland

E-mail: zaremba@fizyka.umk.pl

Abstract. Coherent interactions of laser light with atomic ensembles allow one to modify dispersive properties of a medium and lead to new optical phenomena. Studies of the controlled light propagation and storing in such media have recently become a dynamically developing field of research motivated both by the fundamental character of the processes and by potential applications. This article briefly reviews basic theoretical approach to the dynamics of the propagation of laser pulses in optically modified media. The method and the physical processes are discussed that allow one to slow down the group velocity of laser pulse to zero (stopping of light), to transfer the state of a light pulse to atomic coherences and to restore the pulse. The interpretation of these phenomena in the formalism of dark-state polaritons is presented. Examples of possible coherent manipulations on a stored light are also discussed.

PACS: 42.50.Gy, 03.67.-a

1. Introduction

Propagation of light pulses in atomic media is a fundamental optical phenomenon the properties of which are determined by the details of light-atoms interaction. A typical effect accompanying the propagation is the attenuation of a light beam due to the absorption of light by atoms when the radiation frequency matches the frequency of a particular atomic transition. The optical properties of a dielectric medium can be characterized by the frequency dependent electric susceptibility the imaginary part of which describes the absorptive while its real part - the dispersive properties of the medium which, via the refractive index, govern the group velocity of the pulse. Typically, in a resonance region the absorption spectrum curve presents a strong maximum while the dispersion curve is a decreasing function of frequency (steep anomalous dispersion).

In the past few decades modern optical techniques based on coherent interaction of atoms with laser radiation leading to quantum interference [1] made it possible to influence and modify this typical behaviour of a light pulse propagation by steering in a controlled way dispersive properties of an atomic sample. This is achieved by interactions of atoms with additional laser fields which "dress" a medium. As a result its dispersive properties may by strongly modified and the optical response of the medium may be significantly changed leading to new atomic and optical coherence phenomena (for a review of various phenomena of such a kind see, e.g., [2]).

In particular, it becomes possible to make an initially opaque medium transparent for resonant radiation and to control and manipulate the propagation of laser pulses modifying their group velocities and even to "stop" and then retrieve them. The phenomenon which forms the basis for such possibilities has been termed electromagnetically induced transparency (EIT) in 1990 [3] and since its observation in 1991 [4] has become a point of departure for a discovery of many interesting physical effects and potential applications ranging from nonlinear optics to quantum information [5], [6], [7].

The intention of the present lecture is to give a concise introductory presentation of basic physical ideas underlying EIT-based modifications of the group velocity of a laser pulse, of "slow" and "stopped" light and of a theoretical analysis of these processes. We refer readers for much more detailed discussion to the comprehensive reviews [8], [9], [10], [11]. It has to be stressed that much work has been published in this field and the literature of the subject is still growing.

The paper is organized as follows. In section 2 physical grounds of the EIT are presented and modification of the group velocity is discussed in a lambda system. In section 3 the idea of light storage is discussed and the picture of dark-state polaritons is introduced. More complicated situation of four level atomic system and possibilities of coherent manipulations on the stored light are presented in section 4.

2. EIT and dispersive properties of a lambda system

A generic system in which the EIT can be discussed is a three-level atomic system coupled to two co-propagating laser fields in the lambda configuration. In such a configuration two lower long living states b and c (e.g., Zeeman sublevels of a hyperfine structure in the atomic ground state) are coherently coupled to a common excited state a: a weak probe or signal field 1 of the amplitude ϵ_1 and the frequency ω_1 is tuned to the transition $b - a$ while a strong control laser field 2 of the amplitude ϵ_2 and the frequecny ω_2 drives the transition $c - a$. The coupling $b - c$ is the dipole forbidden transition. The lambda configuration is shown in Figure 1. The combined effect of the interaction with two fields is the so-called dark superposition of the lower states. The atoms in such a dark state are decoupled from the upper state, the light is not absorbed and the population is trapped in lower states [12]. Thus, if the control laser field were absent the probe beam would be resonantly absorbed, switching the control field on makes the system transparent for the probe radiation: this is the electromagnetically induced transparency effect.

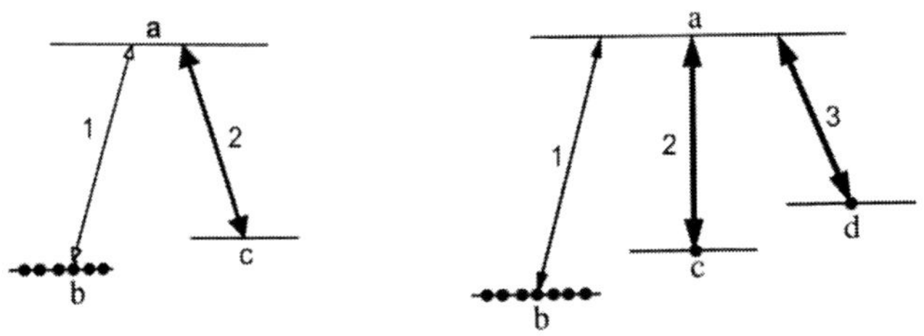

Figure 1. Lambda (left) and tripod (right) configurations. 1 denotes signal field, 2 and 3 denote control fields and black circles the initial population.

The optical response of the medium to the probe laser field is characterized by its complex susceptibility $\chi = \chi' + i\chi''$. To determine the complex susceptibility we analyze the density matrix of the three-level atom interacting with both laser fields which fulfills the von Neumann equation completed with phenomenological relaxation terms

$$i\hbar\dot{\rho} = [H, \rho] + R\rho \tag{1}$$

where $\rho = \rho(z, t)$ denotes the density matrix of the atom in the position z, R describes the relaxation processes, and H is the Hamiltonian which in the rotating wave approximation reads

$$H = \hbar\omega_a |a\rangle\langle a| + \hbar\omega_b |b\rangle\langle b| + \hbar\omega_c |c\rangle\langle c| + \left(-\hbar\Omega_1 e^{i(\omega_1 t - k_1 z)} |b\rangle\langle a| - \hbar\Omega_2 e^{i(\omega_2 t - k_2 z)} |c\rangle\langle a| + h.c.\right) \tag{2}$$

where $\Omega_1 = \frac{1}{2\hbar} d_{ba}\epsilon_1$ and $\Omega_2 = \frac{1}{2\hbar} d_{ca}\epsilon_2$ denote the Rabi frequencies of the corresponding couplings and d_{ba} and d_{ca} are the matrix elements of the electric dipole operator. We assume that Rabi frequencies are real.

Making the substitution $\rho_{ab} = \sigma_{ab}e^{-i\phi_1}$, $\rho_{ac} = \sigma_{ac}e^{-i\phi_2}$, $\rho_{cb} = \sigma_{cb}e^{-i(\phi_1-\phi_2)}$, $\rho_{ii} = \sigma_{ii}$ where $\phi_i = \omega_i t - k_i z$ one obtains the system of optical Bloch equations which describe the evolution of the slowly varying matrix elements of the density matrix σ.

Making the perturbative approximation with respect to the weak signal field 1 but treating the strong control field 2 exactly, assuming that the control field is in resonance with the corresponding atomic frequency ($\omega_2 = \omega_{ac}$) and that the atom is initially in the state b, one obtains the low intensity limit of the equations of motion in the form:

$$\dot{\sigma_{ab}} = -(\gamma_{ab} - i\Delta_1)\sigma_{ab} + i\Omega_1 + i\Omega_2\sigma_{cb} \tag{3}$$

$$\dot{\sigma_{cb}} = -(\gamma_{cb} - i\Delta_1)\sigma_{cb} + i\Omega_2\sigma_{ab} \tag{4}$$

where $\Delta_1 = \omega_1 - \omega_{ab}$ is the detuning of the signal field from atomic resonance and γ_{ik} are the constants characterizing the relaxations of the corresponding atomic coherences. The stationary solution of this system of equations allows one to find the susceptibility χ of the atomic system. To do this one calculates the polarization of the atomic system (N is the atomic density of the sample)

$$P = N\langle d\rangle = Tr(\rho d) \tag{5}$$

which, on the other hand, is related to the signal field ϵ_1 by the relation $P = \epsilon_0\chi\epsilon_1$ (ϵ_0 denotes the permittivity of the vacuum). One obtains:

$$\chi = \frac{2Nd_{ba}}{\epsilon_0\epsilon_1}\sigma_{ab} \tag{6}$$

The plots of the real and imaginary parts of the complex susceptibility as a function of a normalized detuning of the signal beam are shown in Figure 2. At resonance, the imaginary part is equal to zero which corresponds to cancellation of absorption of the signal beam, for which strong absorption would normally be expected. The medium has thus become, due to an interaction with a strong coherent control field, transparent for a resonant radiation. This is the electromagnetically induced transparency effect and the region of a weak absorption in the vicinity of resonance represents the transparency window for incoming radiation. Also, the dispersive properties of a medium illustrated by real part of susceptibility are greatly modified: a steep normal dispersion is observed inside the transparency window. Such a behaviour is completely different from that occuring in dielectric media with typical dispersive properties: the optical response of the system is significantly and qualitatively modified by coherent excitation in the lambda configuration.

Dispersive properties of a medium are characterized by the frequency dependent refractive index $n = n(\omega)$. The group velocity of a light pulse propagating in such a medium is defined as:

$$v_g = \frac{d\omega}{dk} \tag{7}$$

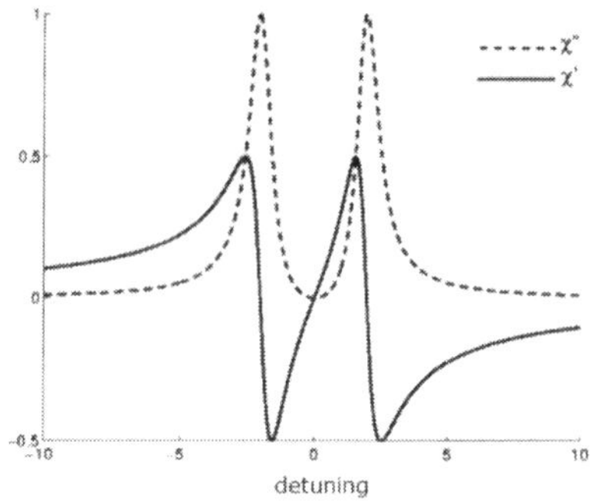

Figure 2. Absorption (imaginary part of susceptibility) and dispersion (real part of susceptibility) as a function of detuning (measured in relative units $\frac{\Delta_1}{\gamma_{ab}}$) from $a - b$ resonance.

The dispersion relation $kc = \omega n(\omega)$ leads to the following formula for the group velocity:

$$v_g = \frac{c}{n(\omega) + \omega \frac{dn(\omega)}{d\omega}} \tag{8}$$

In the region of a steep normal dispersion the derivative $\frac{dn(\omega)}{d\omega}$ can attain great values which will result in an important decrease of v_g of the pulse. Thus the pulse propagating inside the sample will be delayed with respect to the pulse propagating at the same distance in vacuum - this is the manifestation of a "light slow down" effect. When the light pulse enters the medium it exhibits a spatial compression: the front edge of the pulse is already decelerated while its back edge, still outside the medium, propagates in vacuum with the speed c much greater than the group velocity inside the medium. This picture of propagation requires that the spectral width of the pulse should be smaller than the bandwidth of the transparency window which increases proportionally to the square of the electric field intensity of the control field. However, the larger is the transparency window, the smaller the value of the derivative of the refractive index inside it.

3. Stored light and dark-state polaritons

The transparency of the medium in EIT can be modified in time via modifications of the intensity of the control field. Such a control of this phenomenon opens new possibilities of steering the propagation of light pulses. Switching off the control field makes the medium opaque and the signal pulse that has entered it cannot propagate: it is stored in the atoms of the medium. However, it should be stressed that it is not absorbed by atoms - the process is reversible and when the control field is switched on again and the transparency conditions are reestablished, the pulse reappears and is retrieved. This phenomenon is sometimes termed "stopping of light" which is an illustrative although not correct name: the pulse is rather temporarily stored in the form of an atomic coherence and can be released after some time. This effect has been demonstrated in an ultracold sample of sodium atoms [13], in a gaseous sample of warm rubidium atoms [14] and in a solid state [15].

To discuss the dynamics of propagation in EIT medium we consider a quasi one-dimensional sample of three-level atoms. The position of an atom is described by the continuous variable z and thus the atomic density matrix is z-dependent. The above Bloch equations should be completed with the Maxwell propagation equation for the signal field ϵ_1 or for the corresponding

Rabi frequency Ω_1, which reads in the slowly varying envelope approximation (SVEA, see, e.g., [16])

$$\frac{\partial \Omega_1}{\partial z} + \frac{1}{c}\frac{\partial \Omega_1}{\partial t} = -i\kappa^2 \sigma_{ba} \tag{9}$$

where $\kappa^2 = \frac{N|d_{ab}|^2\omega_1}{2\hbar\epsilon_0}$. The r.h.s term of Eq (9) is a source term which describes the polarization of the medium and must be determined from the Bloch equations. One thus obtains the coupled Bloch-Maxwell equations which are the fundamental tool to analyze the propagation phenomena in dielectric media and the solution (usually numerical) of these equations allows one to obtain a detailed information on the process.

A good insight into the dynamics of light propagation may be obtained from the formulation of its description in terms of a so-called dark-state polariton being a combination of signal field and atomic coherence. Such an approach to problems of stopped light has been introduced by Fleischhauer and Lukin [17]. Note that in their original paper the polaritons were introduced as quasi-particles in a fully quantum approach, nevertheless in the present semiclassical description the corresponding objects will also be called polaritons.

To discuss the light storage we will consider the propagation of a weak signal pulse of envelope $\epsilon_1(z,t)$ in the lambda system described in the previous chapter. For simplicity, the propagation effects for the strong control field are neglected, i.e., its envelope $\epsilon_2 = \epsilon_2(t)$ is z-indpendent and one assumes that both fields are in resonance with atomic transitions. If one additionally makes perturbative approximation with respect to the signal field (the control field is treated exactly), neglects the relaxations and makes the adiabatic approximation $\dot{\sigma}_{ab} = 0$ then from Eqs (3), (4) and (9) there follows the equation satisfied by the dark-state polariton:

$$(\frac{\partial}{\partial t} + c\, cos^2\theta \frac{\partial}{\partial z})\Psi(z,t) = 0 \tag{10}$$

The dark-state (or simply dark) polariton is defined as

$$\Psi(z,t) = \cos\theta(t)\Omega_1 - \kappa \sin\theta(t)\sigma_{bc} \tag{11}$$

where the mixing angle θ is defined as:

$$\tan\theta(t) = \frac{\kappa}{\Omega_2} \tag{12}$$

It is a shape-preserving solution of the wave equation (10)

$$\Psi(z,t) = \Psi\left(z - c\int_0^t dt' \cos^2\theta, t = 0\right) \tag{13}$$

and propagates in the sample with the group velocity

$$v_g = c\cos^2\theta = c\frac{1}{1+\tan^2\theta} \tag{14}$$

This velocity depends, via the mixing angle θ, on the intensity of the control field and can thus be controlled by varying this field in time.

The dark polariton is built of two physically different components: the signal field (term with Ω_1) and the atomic coherence (term with σ_{bc}). The relative weights of these components depend on the control field intensity and change adiabatically. Thus, although the polariton does not change its shape during the propagation, these components change considerably. When the control field is switched on ($\theta \approx 0$) the medium is transparent, the polariton is "purely optical": $\Psi(z,t) = \Omega_1(z,t)$, and the signal pulse propagates inside the sample. Switching the control field off ($\Omega_2(t) \to 0$, $\theta(t) \approx \frac{\pi}{2}$, $v_g \to 0$) makes the medium opaque, the polariton decelerates and becomes "purely atomic": $\Psi(z,t) \to -\kappa\sigma_{bc}(z,t)$ and the pulse is "stopped". At the release stage at which the control field is switched on again, the atomic component of the polariton decreases while its field component increases; the polariton becomes "optical" again which means that the signal pulse is reconstructed and retrieved. Such an evolution of the form-stable polariton explains the mechanism of the stopping of light. The signal field is mapped into the atomic coherence which is created during the switching the control field off. The pulse is thus stored inside the medium. The process of light storing is reversible and the information about the signal is read out from atomic coherence at the release stage. In this ideal scenario the coherence has not been disturbed by any relaxation, the control field at the release stage is identical with that at the storage stage and the released signal is a replica of the stored one. If the adiabatic approximation were not valid, the description of the evolution only in terms of the dark-state polariton would not be sufficient and another superposition of the field and atomic coherence called a bright-state polariton would be necessary to analyze the dynamics [18].

To describe the dynamics of the pulse storing we have used a semiclassical analysis in which the signal light has been treated as an classical electromagnetic wave. In fact, the original polariton interpretation of the process by Fleischhauer and Lukin used a second quantization formalism and a quantized light field [17]. The dark-state polariton should then be interpreted as a quasi particle corresponding to excitations of the composed atoms plus field system. Its evolution is essentially the same as that described above and the light storage and retrieval consists now in a reversible mapping of the quantum state of the signal into the atomic coherences. Such a picture forms a basis for potential interesting applications in quantum information processing: the information conveyed by photonic quantum state of the signal pulse can be stored (written down) in a "quantum memory" [18] formed by collective atomic excitations described by an atomic coherence and later retrieved (read out) by switching the control field on again and releasing the "stopped"' signal pulse.

4. Coherent processing of stored light

Light storing in the form of atomic coherences opens new possibilities of controlled modifications of the properties of the released pulse by processing the atomic medium during the storage stage. Such an additional control of the process can be obtained in more advanced systems admitting additional fields and additional active atomic states. The dynamics is then essentially enriched and various systems have been studied in the literature. Here we give an example illustarting such possibilities, namely the case of a medium of four-level atoms interacting with three laser fields in the so-called tripod configuration. Three lower states b, c, d are coherently coupled with a common excited state a: a signal field 1 drives the $b-a$ transition while two control fields 2 and 3 couple the state a with the states c and d respectively and there is no coupling among lower states. The tripod configuration is shown in Figure 1.

Light propagation and slowdown in such a system have been considered in [19], [20], [21] while its storing has recently been studied in [22] using a generalized polariton picture of the process. The medium which is opaque for the resonant signal beam may be made transparent by switching on the control fields. As in the case of a lambda system one can, by manipulations on control fields, modify dispersive properties of the medium and steer the propagation and storing of the signal pulse. However in this case two essentially new aspects appear. First, the signal is stored in two atomic coherences σ_{bc} and σ_{bd}. It becomes thus possible to manipulate the stored light by applying additional interactions which modify these two coherences and, as a consequence, to modify the properties of the released pulse. Second, the signal pulse can be stored and retrieved by different sequences of control fields. In the following we will briefly discuss examples of both cases.

The detailed analysis of pulse propagation in an atomic medium in the tripod configuration requires solutions of the Bloch-Maxwell equations. The generalized polariton description of the dynamics is also possible and allows one to obtain a clear picture of a light storage and retrieval analogous to that of a lambda case [22]. However, a single dark-state polariton is not sufficient even in the case of an adiabatic evolution. The dark-state polariton is now of the form

$$\Psi(z,t) = \cos\theta\Omega_1 - \kappa\sin\theta(\cos\phi\sigma_{bc} + \sin\phi\sigma_{bd}) \tag{15}$$

where the mixing angle θ and the angle ϕ depend on both control fields intensities

$$\sin^2\theta = \frac{\kappa^2}{\kappa^2+\Omega^2} \qquad \cos^2\theta = \frac{\Omega^2}{\kappa^2+\Omega^2} \qquad \tan\phi = \frac{\Omega_3}{\Omega_2} \qquad \Omega^2 = \Omega_2^2+\Omega_3^2 \tag{16}$$

The dark-state polariton is here a natural generalization of that for a lambda system. As in that case it is built of two physically different components: the signal field term with Ω_1 and the atomic coherence term which, however, contains now two atomic coherences σ_{bc} and σ_{bd}. For strong control fields $\theta \approx 0$ and the polariton is "purely electromagnetic" while for vanishing control fields $\theta = \pi/2$ and it becomes "purely atomic"; thus modifying the intensities of control fields in time one changes the relative weights of these two components. To describe fully the dynamics a second polariton of the form

$$Z(z,t) = \sin\phi\sigma_{bc} - \cos\phi\sigma_{bd} \tag{17}$$

is necessary, it is "purely atomic" composed solely of two atomic coherences and does not contain a field component. The two polaritons satisfy the coupled equations of motion that follow, under the assumption identical to those made in the case of a lambda system, from Bloch-Maxwell equations

$$\left(\frac{\partial}{\partial t} + c\, cos^2\theta\frac{\partial}{\partial z}\right)\Psi(z,t) = \tan^2\theta\cos\theta\Omega Z \tag{18}$$

$$\frac{\partial Z}{\partial t} = -\cos\theta\dot{\phi}\frac{1}{\Omega}\Psi \tag{19}$$

Thus, the adiabatic evolution becomes quite complicated because, in general, the polariton Z can be created inside the atomic medium even if initially, before the pulse has entered the sample, $\Psi = \Omega_1$ and $Z = 0$. It is possible that the polariton Z remains different from zero even after releasing the pulse when $\Psi = 0$ inside the sample. It means that a part of atomic coherences in which the initial pulse has been mapped can remain trapped in spite of the fact that control fields have been switched on again.

The equations of motion are decoupled and Z does not change its shape during evolution if $\phi = const$, i.e. if both control fields are identical, $\Omega_2 = \Omega_3$, or change in a proportional way. If additionally the initial value of Z is zero, which is the natural initial condition, the evolution is completely determined by the form-stable dark-state polariton $\Psi(z,t)$ only (as in the case of a lambda system) which propagates with a group velocity $v(t) = c\cos^2\theta(t)$ where $\cos\theta$ is now given by (16) and the coherences into which the pulse has been mapped are then proportional.

To illustrate the above discussion we will briefly demonstrate two examples; the detailed discussion of them has been given in [22]. In the first case the signal pulse is stored and retrieved by pairs of identical simultaneously switched-off and -on control pulses but the atomic coherences are modified in the storage stage. In the second case the signal pulse is stopped by two identical control fields but in the stage of releasing it the control fields are shifted in time.

Assume that identical control fields have been switched-off simultaneously ($\dot{\phi} = 0$) and a signal pulse has been stopped so that $Z = 0$ and two proportional coherences have been created. Then an additional interaction is switched on which modifies the coherences before the pulse is released. If the states b, c, d are Zeeman substates (e.g.,of the hyperfine structure as it is in the case of rubidium atoms typically used in experiments on light storage) such an interaction may be introduced by a pulse of a magnetic field parallel to the direction of propagation. A magnetic field pulse introduces additional phase shift δ between atomic coherences which leads to a creation of $Z \neq 0$ polariton the value of which depends on this phase shift, i.e. on the pulse duration. If the control fields used to release the signal pulse are identical to those used to store it the polariton Z does not change which means that the part of coherences constituing Z does not leave the sample. As a consequence, the coherences cannot be turned back into the pulse completely and the released signal is not a replica of the stored one. Thus a controlled modification of the relative phase of coherences changes the resulting output signal which depends on δ.

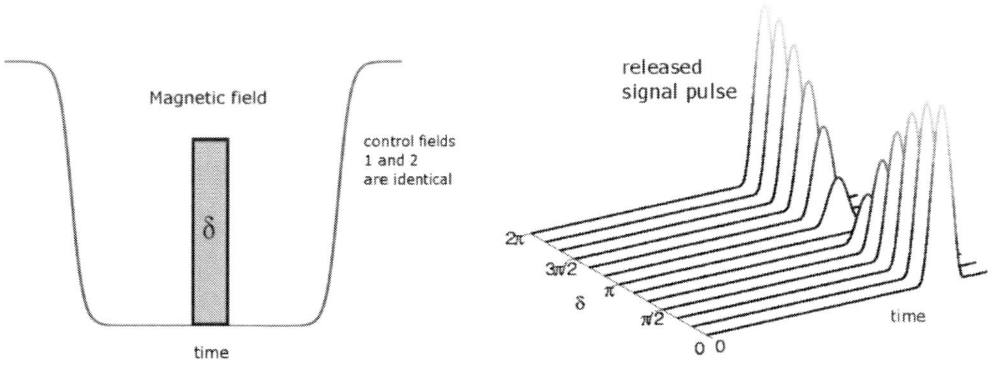

Figure 3. The schematic sequence of applied control fields and of a magnetic field pulse which introduces a phase shift δ between atomic coherences (left) and the released pulses at the end of the sample plotted for various δ (right). (In the case presented here the magnetic field modifies, for simplicity, only one of the atomic coherences; such a situation may occur, e.g., for a tripod system composed of Zeeman sublevels in rubidium - for details see [22].

In Figure 3 the shape of the released signal pulse at the end of the sample is shown for different values of δ and the interference like behavior of its height is clearly seen (the plots were obtained from numerical solutions of the Bloch-Maxwell equations the details of which are not given in the present general qualitative description, see [22]). For δ being the odd multiple of π the pulse is not released and remains trapped inside the medium in a form of atomic coherences into which it was mapped in the storing stage while for δ being a multiple of 2π it attains the maximum. Such a behavior is also reflected by the form of plots (not shown) of the spatial distribution of the corresponding Z polariton after the released pulse has left the medium: their maximum values correspond to those values of δ for which the pulse is not released. The similar idea of changing the phase of atomic coherence was discussed and verified experimentally in [23] for the simpler case of light storing in a lambda system in which the pulse was stored in a single atomic coherence. The phase of this coherence was changed and the retrieved pulse was made to interfere with a reference beam. In the tripod case the situation is different: the relative phase of two atomic coherences is changed and the interference pattern in the retrieved pulses is due to a superposition of two components inside the medium.

Our second example concerns the case in which the control fields are identical at the storing stage and shifted in time at the release stage. This situation is presented in Figure 4. The

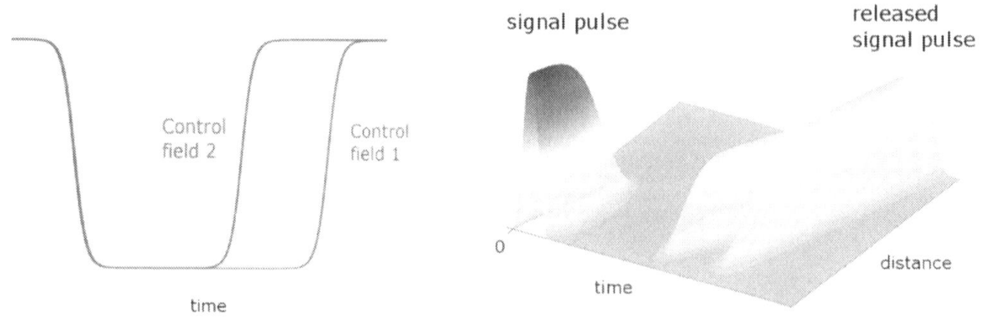

Figure 4. The schematic sequence of applied control fields in the case of the releasing fields shifted in time (left) and the corresponding pulse storing and release (right). The released pulse has a double structure; note that at the beginning its two parts have different group velocities due to the fact that the second part is triggered in the presence of both control fields.

trapped signal pulse is now released in two portions and due to $\dot{\phi} \neq 0$ the $Z \neq 0$ polariton is created which does not leave the sample. Thus a part of the stopped pulse remains stored in atomic coherences and can be further released by another pair of control pulses. It is evident that detailed properties of the pulse leaving the sample can be controlled by a changing the parameters of the storage/release procedure.

Light storing and release in a medium of atoms in the tripod configuration give rise to new interesting phenomena in the case of the input signal being a photonic quantum state. If, for example, two time-separated photons are stored and released by switching on and off two sets of control fields, the tripod medium can behave as a beam splitter in time domain in which the input and output ports concern photons separated in time. It is possible to choose conditions of photon storing and release in which a version of the Hong-Ou-Mandel interference [24] for stored light is realized, i.e., the situation in which two photons are released only either in the first or in the second release stage. Detailed analysis of this effect as well as of steering statistical properties of stored nonclassical light is given in [25] and [26]. Another interesting situation is discussed in [27] in which a possibility has been suggested of producing time-entangled photon state by transferring a part of the excitation created by a trapped photon in one atomic coherence into the other one before a photon release. Certainly investigations concerning new possibilities of controlling quantum properties of the light storage processes are important not only because of their fundamental physical interest but also in the context of quantum information processing.

5. Summary

In summary, we have briefly reviewed the basic ideas of storage and release of laser pulses in atomic media. The phenomenon is based on the possibility of modifications of dispersive properties of a medium via interactions with additional "control" light fields. Due to such interactions an initially opaque medium becomes transparent for resonant radiation, this is the electromagnetically induced transparency. By modifying the control fields in time one can dynamically modify the dispersive properties of a medium and thus manipulate the propagation of light pulses in a controlled way. In particular, the pulse can be stopped (or rather stored) inside the medium in a form of atomic coherences and later released. The simplest system in which such a scenario can be realized is a three level atomic system interacting with two laser fields in a lambda configuration and the clear physical picture of the process can be formulated in the dark-state polariton formalism. In more advanced systems which contain more laser fields and more active atomic states one has more external parameters to control the process. An example of such a system, namely the tripod configuration has been discussed and the examples of possible controlled manipulations on stored light have been presented.

Acknowledgments
I would like to express my deep and sincere gratitude and thanks to my colleagues Andrzej Raczyński, Sylwia Zielińska-Kaniasty, Małgorzata Rzepecka and Karolina Słowik with whom I have been working for many years on the subjects presented in this lecture. It would not come into being without our friendly and stimulating collaboration.

6. References

[1] Shore B W 1990 *The Theory of Coherent Atomic Excitation* (New York: John Wiley and Sons)
[2] Vitanov N V, Fleischauer M, Shore B W and Bergman K 2001 *Adv. At. Mol. Opt. Physics* vol 46 (San Diego: Academic Press) p 55
[3] Harris S E, Field J E and Imamoglu A *Phys. Rev. Lett.* **64** 1107
[4] Boller K J, Imamoglu A and Harris S E 1991 *Phys. Rev. Lett.* **66** 2593
[5] Harris S E 1997 *Phys. Today* **50** 36
[6] Marangos J P 1998 *J. Mod. Opt.* **45** 471
[7] Fleischhauer M, Imamoglu A and Marangos J P 2005 *Rev. Mod. Phys.* **77** 633
[8] Matsko A B, Kocharovskaya O,Rostovtsev Y, Welch G R, Zibrov A S and Scully M O 2001 *Adv. At. Mol. Opt. Phys* vol 46 (San Diego: Academic Press) p 191
[9] Milloni P W 2002 *J. Phys. B: At. Mol. Opt. Phys* **35** R31
[10] Lukin M D 2003 *Rev. Mod. Phys* **75** 457
[11] Andre A, Eisman M D, Walsworth R L, Zibrov A S and Lukin M D *J. Phys. B: At. Mol. Opt. Phys.* **38** S589
[12] Arimondo E 1996 *Progress in Optics* vol. 35 (Amsterdam: North-Holland) p 257
[13] Liu C, Dutton Z, Behroozi C H and Hau L V 2001 *Nature* **409** 490
[14] Phillips D F, Fleischhauer A, Mair A, Walsworth R L and Lukin M D 2001 *Phys. Rev. Lett.* **86** 783
[15] Turukhin A V, Sudarshanam V S, Shahriar M S, Musser J A, Ham B S and Hemmer P R 2002 *Phys. Rev. Lett.* **88** 023602
[16] Meystre P and Sargent III M 1990 *Elements of Quantum Optics* (Berlin: Springer-Verlag)
[17] Fleischhauer M and Lukin M D 2000 *Phys. Rev. Lett.* **84** 5094
[18] Fleischhauer M and Lukin M D 2002 *Phys. Rev. A* **65** 022314
[19] Paspalakis E and Knight P L 2002*J. Opt. B: Quantum Semiclass. Opt.* **4** S372
[20] Mazets I E 2005 *Phys. Rev. A* **71** 023806
[21] Mazets I E 2006 *JETP* **103** 365
[22] Raczyński A, Rzepecka M, Zaremba J and Zieliń ska-Kaniasty S 2006 *Opt. Commun.* **260** 73
[23] Mair A, Hager J, Phillips D F, Walsworth R L and Lukin M D 2002 *Phys. Rev. A* **65** 031802(R)
[24] Hong C K, Ou Z Y and Mandel L 1987 *Phys. Rev. Lett.* **59** 2044
[25] Raczyński A, Zaremba J and Zielińska-Kaniasty S 2007 *Phys. Rev. A* **75** 013810
[26] Raczyński A, Słowik K, Zaremba J and Zielińska-Kaniasty S 2007 *Opt. Commun.* **279** 324
[27] Wang T, Kostrun M and Yelin S F 2004 *Phys. Rev. A* **70** 053822

Asymmetry Induced Phase Transformations

J Damczyk[1], **K Ostasiewicz**[1], **Ł Radosiński** [2] **and A Radosz**[1]

[1] Institute of Physics
[2] Institute of Physical and Theoretical Chemistry
Wrocław University of Technology,
Wybrzeże Wyspiańskiego 27, 50-370 Wrocław, Poland

E-mail: jacek.damczyk@pwr.wroc.pl

Abstract. Special class of one-dimensional (1D) lattice systems with asymmetric on-site potential can exhibit a sequence of phase conversions. Whereas this result is exact in 1D, one can ask if a sequence of phase transformations may evolve into a sequence of phase transitions in higher dimensional systems, where exact results are not available. In this paper we discuss such a problem: of a presence of a sequence of phase transitions in 3D systems with asymmetric on-site potential

1. Motivation

In the thermodynamic regime, one-dimensional (1D) lattice systems with degenerated vacuum, reveal the coexistence of localized and extended excitations, kinks and phonons, respectively, that is accompanied by characteristic hump developed in a specific heat [1], [2]. Such systems in higher dimensions, 2D and/or 3D, exhibit phase transitions of second order - in fact it is the most simple manifestation of spontaneously broken (discrete) symmetry.

It has been shown recently [3], [4] that 1D or quasi one-dimensional, strongly non-linear systems, with non-degenerated vacuum exhibit interesting behavior. Such systems with local asymmetric, double-well or triple-well potential exhibit rich structure of the specific heat. In particular circumstances, of a "shape asymmetry", there appears two-peak structure in specific heat corresponding to the sequence of two phase conversions. One can ask whether such a double-peak structure of specific heat, would lead, in higher dimensions, to a sequence of the phase transitions (of the first order).

In this paper we discuss the problem of 2D and 3D system with multistable, triple-well, on-site potential where a sequence of phase transitions may occur. The paper is organized as follows. In section 2, dimensional systems revealing two phase conversions is addressed. In section 3, their three-dimensional versions are investigated, exhibiting a sequence of phase transitions.

2. Sequence of phase conversions - 1D systems

Let us consider one component system of harmonically coupled atoms with a local multistability represented by a triple-well potential, described by the Hamiltonian of the form [1], [2], [3]:

$$H = \frac{m\omega_0}{a^2}\left[\sum_l \left(\frac{1}{2}\dot{x}_l^2 + V(x_l)\right) + \frac{1}{4}\sum_{l,l'} \varphi_{ll'}(x_l - x_{l'})^2\right], \tag{1}$$

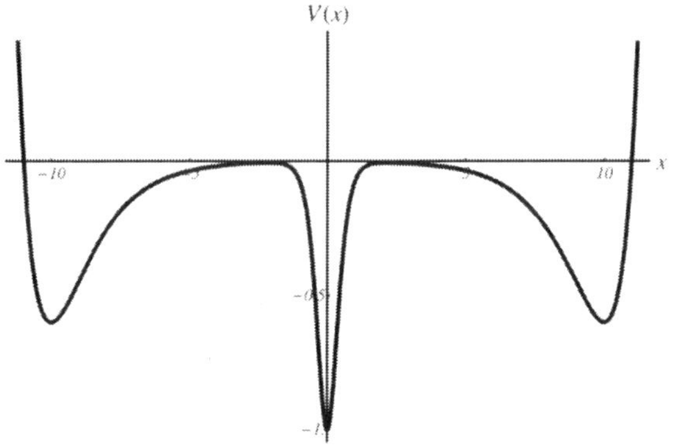

Figure 1. On-site potential $V(x)$ equ. (2) with parameters V_{DM}: $V_1 = V_2 = 0.6$, $\alpha = \beta = 0.7$, $p = 10.0$, and V_{P-T}: $U = 1.0$ $\kappa = 2.0$.

where a - denotes the characteristic inverse distance, ω_0 - characteristic frequency, and m is the characteristic mass *i.e.* mass of single atom in chain; x_l is dimensionless coordinate at l-th lattice site, $\dot{x}_l$ denotes (dimensionless) time derivative, and φ is dimensionless coupling. On-site potential $V(x)$ is composed of a central well (CW) and double-well (DW) arranged in such a way that CW is narrower and deeper than symmetric side wells (DW) (see Fig. 1). Thermodynamics of such a lattice system in one dimension, may be found by means of the transfer integral method. In continuum classical limit with nearest neighbour interaction described by coupling constant k [2], [3], [4], it exhibits interesting behavior. Namely, if we take on-site potential $V(x)$ as composed of the double-Morse potential V_{DM} as DW, and the CW as Pöshl-Teller potential V_{P-T} (see Fig. 1),

$$V(x) = V_{DM}(x) + V_{P-T}(x), \tag{2}$$

$$\begin{aligned} V_{DM}(x) = V_1\Big\{ \exp\left[-2\alpha(x+p)\right] - 2\exp\left[-\alpha(x+p)\right]\Big\} \\ +V_2\Big\{ \exp\left[-2\beta(x-p)\right] - 2\exp\left[-\beta(x-p)\right]\Big\}, \end{aligned} \tag{3}$$

$$V_{P-T}(x) = \frac{-U}{\cosh^2(\kappa x)}, \tag{4}$$

where V_1, V_2 sets depth, α, β tunes the shape, and p sets positions of minima for DM; parameter U and κ have respectively analogous meaning for CW, where position of minimum is fixed at $x = 0$. The detailed analysis of the on-site potential (2) is a bit lengthy and will be done in another paper. Here we will consider a sub-class of potentials $V(x) = V(-x)$, so that symmetric DM potentials, $V_1 = V_2$ and $\alpha = \beta$ setting symmetric DW, and observe shape transformations only by changing values of U, and κ. Such a typical $U - \kappa$ plane divided into sections of special shapes, namely: single well (SW), double well (DW), and triple well (TW), is presented in Fig. 2. For large κ, where CW is very narrow the TW shape is dominant, only for large values of U the SW class is available. In a range of moderate κ values, where specific fitting of CW shape to DM's potential barrier takes place, one can observe drastic lowering of U values, where SW class dominates. Finally, for small values of κ, where CW is crucially wide comparing to the width of DM potential ($\sim 2p$), the DW shape is available. It come from the fact, that in this interval CW is wide, leading to the shift of DM potential. For very large values of U, where the variation of CW can't be ignored, the SW domain is again available.

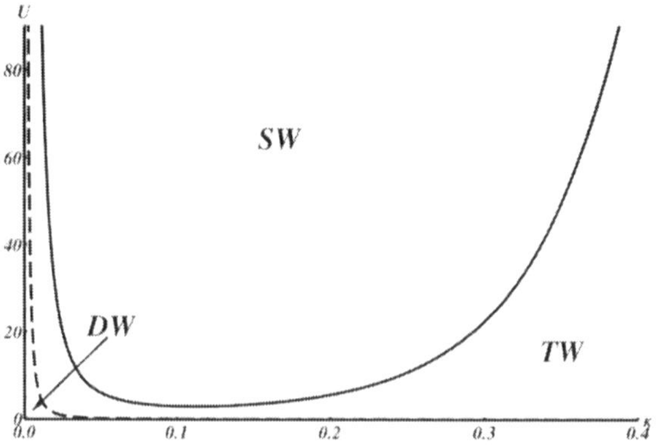

Figure 2. Plane of $U-\kappa$ parameters divided into domains of specific shapes of on-site potential (2) occurrence: single well (SW), double well (DW), and triple well (TW), with V_{DM} fixed parameters: $V_1 = V_2 = 0.6$, $\alpha = \beta = 0.7$, and $p = 10$.

The thermodynamic properties of one dimensional version of system (1) with Hamiltonian

$$H = \frac{m\omega_0}{a^2}\left[\sum_l \left(\frac{1}{2}\dot{x}_l^2 + V(x_l) + \frac{k}{2}(x_l - x_{l+1})^2\right)\right], \tag{5}$$

are described in terms of the pseudo-Schrödinger eigenvalue problem [5]:

$$\left(-\frac{1}{2m^\star}\frac{d^2}{dx^2} + V(x) + V_0(T)\right)\Psi_i(x) = E_i\Psi_i(x),$$

$$m^\star = kT^{-2}, \quad V_0(T) = -\frac{T}{2}\ln\left(\frac{2\pi T}{k}\right), \tag{6}$$

where $T = k_B\theta/\frac{m\omega_0}{a^2}$ is dimensionless temperature (k_B - Boltzmann constant, θ - temperature). Configuration part of free energy is expressed via ground state "energy" of the eigenvalue problem (6)

$$F_{con} \simeq E_0(T), \; for \; k \gg 1. \tag{7}$$

Specific heat function is presented in Fig. 3. Asymmetry of the local potential leads to a particular form of a specific heat: apart of smooth hump, sharp and intensive peak, manifesting dramatic change in the system is developed in this case. Interpretation of such a shape of a specific heat is the following. Particles occupying at low temperature range stable position in CW of local on-site potential perform radical jump to a shallower, side well of DW structure; this jump corresponds to a sharp peak of specific heat. The RTCT (Real Trajectories in Complex Time) [3], [4] approach indicates that: free energy of a more stable phase, corresponding to lower energy within CW, is increasing faster than the free energy of the less stable phase, corresponding to higher energy within DWs. Intersection of corresponding ground state "energy levels" at some temperature T_0, results in a narrow peak of specific heat. Above that temperature, DW phase is a dominant one (lower free energy). The following phase conversation is related to the hump of specific heat: particles are performing large amplitude fluctuations about central, $x = 0$, position. This picture is confirmed by means of numerical tools. First, the temperature of first phase conversion corresponds to the (avoided) crossing of the free energies of CW and DW phases ("energy level crossing", see [3], [4], and Fig. 4). The narrow peak in the specific heat arises at the temperature $T_0 \simeq 2.822$ (Fig. 3) of energy level crossing (see Fig. 4). Second, analyzing

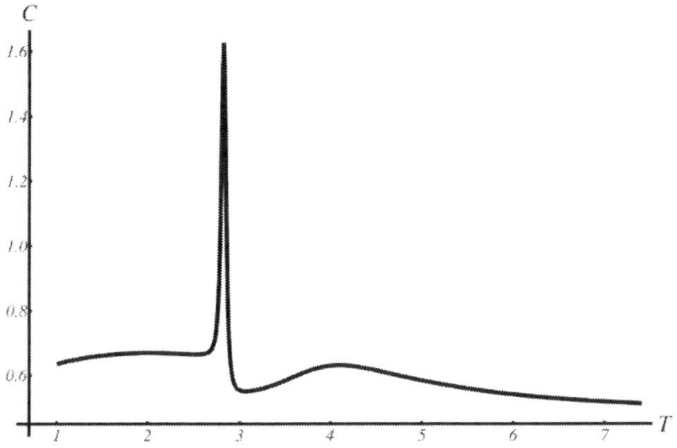

Figure 3. Specific-heat C versus temperature T for $k = 10.0$, and with on-site potential parameters as in Fig. 1.

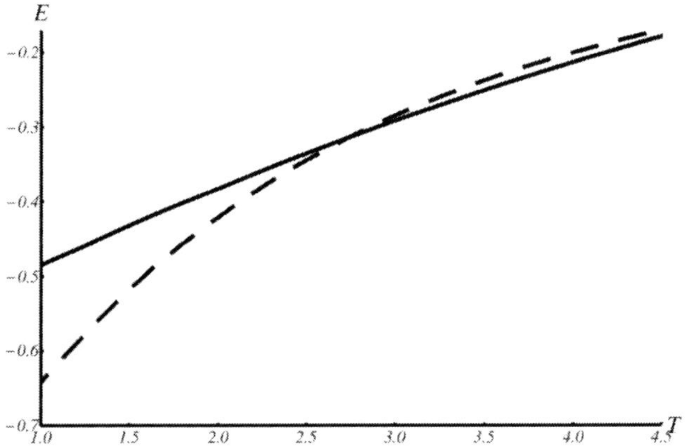

Figure 4. The ground "energy" levels crossing for on-site potential: V_{DM} – *solid line*, V_{P-T} – *dashed line* for $k =$ 10.0, and with parameters as in Fig. 1.

the specific heat structure one can notice that it is composed of three distinct contributions, see (Fig. 5). In the low temperature regime and in the high temperature regime, it reveals the shape characteristic of CW (a). In the intermediate temperatures range a hump corresponding to a double-well structure (though shifted) is formed (b). As added together, these three parts fit to the specific heat of the system (c) - sharp and narrow peak is missing as it arises when the central well and one of the side wells are present. Interpretation of the double peak structure of the specific heat becomes evident when one removes one of the side wells. Then, one only small quantitative correction are found, Fig. 6.

3. Sequence of phase conversions - 3D systems

Phase transformations manifested in specific heat of one dimensional systems would evolve into phase transitions in higher dimensions. There we shall consider 3D lattice system (1) with on-site triple-well asymmetric potential $V(x)$ of *shape asymmetry*, composed of the double-well "ϕ^4"-type potential $V_{\phi^4}(x)$, and the central, inverted *Gaussian* well $V_G(x)$ (see also [6]):

$$V(x) = V_{\phi^4}(x) + V_G(x), \tag{8a}$$

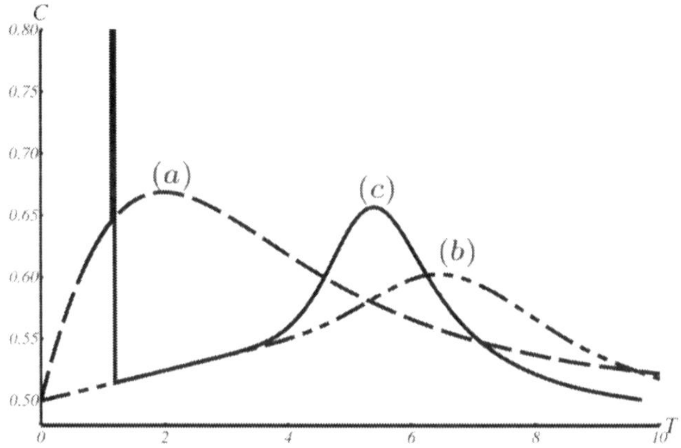

Figure 5. Specific heat C versus temperature T with distinct contributions. (**a**) – *dashed line* V_{P-T}, (**b**) – *dash-dot-dotted line* V_{DM}, (**c**) – *solid line* full $V(x)$ for $k = 10.0$, and on-site potentials parameters: V_{DM}: $V_1 = V_2 = 0.75$, $\alpha = \beta = 0.7$, $p = 10.0$, V_{P-T}: $U = 1.0$ $\kappa = 2.0$.

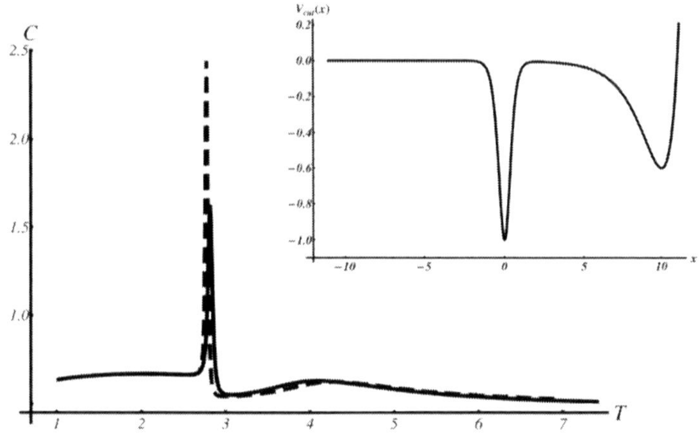

Figure 6. Specific heat C versus temperature T for: full $V(x)$ (*solid line*), and one with one removed side-well $V_{cut}(x)$ (*dashed*) for $k = 10.0$, and on-site potentials parameters: $\boldsymbol{V(x)}$ as in Fig. 1, and $\boldsymbol{V_{cut}(x)}$ as given by V_{DM}: $V_1 = 0$, $V_2 = 0.6$, $\alpha = 0$, $\beta = 0.7$, $p = 10.0$, and V_{P-T}: $U = 1.0$ $\kappa = 2.0$. Inset: Plot of the $V_{cut}(x)$ on-site potential with above parameters.

$$V_{\phi^4}(x) = -\frac{1}{2}x^2 + \frac{1}{4}x^4, \tag{8b}$$

$$V_G(x) = -\gamma \exp\Big(\frac{-x^2}{2\,\delta^2}\Big), \tag{8c}$$

$$f_0 = \sum_{l,l'} \varphi_{ll'}, \tag{8d}$$

Introduced above alternative form of $V(x)$ on-site potential, written in dimensionless units depends on two parameters $\gamma > 0$ and $\delta > 0$, which tune single central Gaussian well. The γ parameter adjust depth, with $V_G(0) = -\gamma$, and δ – sets width of central well. The invariant component of (8a), double-well $V_{\phi^4}(x)$ has fixed minima at $x_0 = \pm 1$, with depth $V_{\phi^4}(x_0) = -1/4$. Manipulating with these two parameters γ,$\delta > 0$, similarly as in the case of potential (2), on-site potential (8a) may take a form of: single well (SW), double well (DW), and triple well (TW).

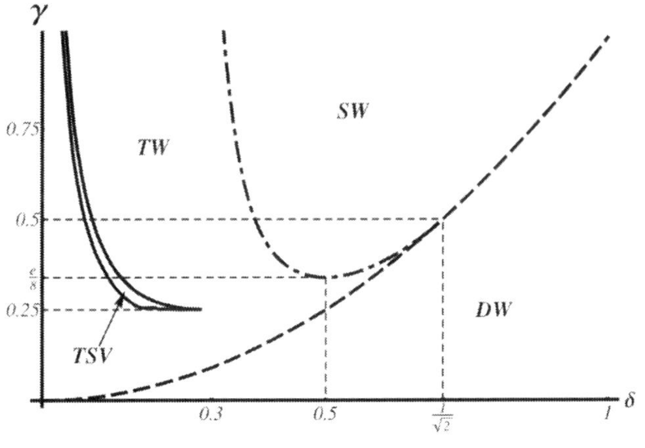

Figure 7. Plane $\gamma - \delta$ divided on regions of particular form of on-site potential equ. (8a) occurrence: SW – single well, DW – double-well, TW – triple-well, and TSV – shape asymmetry subregion of TW region. *Dashed line*: $\gamma = \delta^2$, *dash-dotted line*: $\gamma = 2\delta^4 \exp(-1 + 1/(2\delta^2))$.

The $\gamma - \delta$ plane with marked out domains of potential particular types occurrence is given in Fig. 7. We restrict our interest to the TW case; Some of its properties are following. Central well is placed, naturally in $x = 0$, while the side wells are arranged at

$$x_{min}^2 = 1 + 2\delta^4\, W_0\left(-\frac{\gamma}{2\delta^2}\, e^{\frac{-1}{2\delta^2}}\right), \tag{9}$$

and a summits of potential barrier between wells are placed in

$$x_{max}^2 = 1 + 2\delta^4\, W_{-1}\left(-\frac{\gamma}{2\delta^2}\, e^{\frac{-1}{2\delta^2}}\right), \tag{10}$$

where W_0, W_{-1} are the Lambert-W functions. The sector of TW (see Fig. 7) may be divided into two half-planes:

- $0 < \gamma < \frac{e}{8}$, where TW region is bordered only from bottom by $\gamma = \delta^2$, which comes form the condition for appearance of the central well.
- $\gamma > \frac{e}{8}$, where TW region is additionally bordered from top by $\gamma = 2\delta^4 \exp(-1 + 1/(2\delta^2))$ in interval $0 < \delta < \sqrt{\frac{1}{2}}$, which comes from the condition for existence of side wells.

In some part of TW region, on-site potential 8a belongs to the class of *shape asymmetry*, which is determined by thermodynamic properties of three dimensional version of the system (1). In fact, in this range of parameters, in 1D version, one can observe energy levels intersection, corresponding to the narrow peak appearance, followed by the moderate peak in specific heat. Considering the thermodynamic properties of three dimensional version of such a system one can find a confirmation of our hypothesis about the sequence of the phase transitions in *shape asymmetry* case. Indeed, in the mean field approximation the system shows two phase transitions and existence of an ordered phase, in the finite temperature range, see Fig. 8. Both phase transitions are of the first order. By pointing out a region of parameters γ-δ, where those two transitions are present we can identify the subregion of TW half-plane, where on-site potential (8a) takes the form of *shape asymmetry*, namely *TSV* region on Fig.7.

The detailed discussion of the asymmetry driven phase transition in model (8) will be given in a subsequent papers [7], [8].

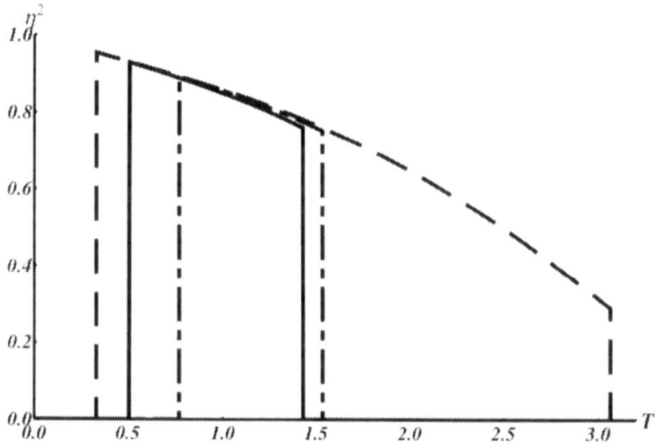

Figure 8. Square of order parameter η^2 vs temperature T for $f_0 = 20$, and on-site potential $V(x)$ equ. (8a) with parameters: $\gamma = 0.28$, $\delta = 0.185$ – *solid line*; $\gamma = 0.30$, $\delta = 0.14$ – *dashed line*; $\gamma = 0.32$, $\delta = 0.15$ – *dash-dotted line*.

4. Conclusions

We are reporting a possibility of new type of structural phase transitions. Multistable on-site potential, considered in a form of a symmetric double-well, is usually associated with the spontaneously broken symmetry phenomena. Multistable asymmetric on-site potential may lead to interesting phenomena in one-dimensional systems, e.g. resulting in a sequence of phase transformations. It is argued here that particular type of asymmetry of multistable on-site potential may result in a sequence of phase transitions in three-dimensional systems. We have proposed triple-well potential of specific shape ("large asymmetry") leading to the sequence of two consecuitive, first-order phase transitions.

References

[1] Khare A, Habib S and Saxena A 1997 *Phys. Rev. Lett.* **79** 3797
[2] Currie J, Trullinger S, Bishop A and Krumhansl J 1977 *Phys. Rev. B* **15** 5567
[3] Radosz A, Ostasiewicz K, Machnikowski P and Magnuszewski P 2001 *Phys. Rev. E* **64** 062103
[4] Radosz A, Ostasiewicz K, Magnuszewski P, Radosiński R, Damczyk J, Mituś A, Pawlik G, Kusmartsev F and Samson J 2006 *Phys. Rev. E* **73** 026127
[5] Morris J R and Gooding R J 1992 *J. Stat. Phys.* **67** 471
[6] Plakida N M 1981 *Fizika Nizkikh Temperatur* **7** 644 (in Russian)
[7] Damczyk J, Ostasiewicz K, Radosiński L and Radosz A 2009 *(submitted to Acta Phys. Pol. A)*
[8] Ostasiewicz K, Damczyk J, Radosiński L and Radosz A *(in preparation)*

Electron transport in disordered metallic nanosystems

A Paja[1]
Faculty of Physics and Applied Computer Science,
AGH-University of Science and Technology, al. Mickiewicza 30, PL 30-059 Kraków, Poland

E-mail: paja@agh.edu.pl

Abstract. We consider the electron transport through a disordered metallic system, the size of which is less than 100 nm, at least in one dimension. We first present the main ideas of the theoretical description of the electron transport in bulk amorphous metals. Then we indicate the necessary modifications we should introduce to take into account the finite size of the considered sample. Finally, we discuss some examples of metallic nanostructures and calculate their electrical conductivity.

Introduction

The notion of nanosciences and nanotechnology appear in almost every branch of modern knowledge and technology. Heinrich Rohrer, the Nobel prize winner in 1986 (jointly with Gerd Binnig), clearly formulated the definition of the nanosciences [1]. According to his view, it is the science of dealing with nano-objects, i.e. manipulating, positioning and machining nano-objects, measuring, understanding and selectively modifying their properties as well as developing new concepts in treating them. I would only add that nanotechnology is the art of the fabrication of the nanosystems.

What do we call a "nanosystem"? As everybody knows, one nanometer equals to 10^{-9} m. The diameter of the smallest atom (hydrogen) is about 10^{-10} m, i. e. 0.1 nm. People agree that nano-objects (or nanosystems) are the assemblies of atoms the size of which is less than 100 nm, at least in one direction. Do we actually need a brand new theory for description of such objects? Yes and no. No, because we already have got a beautiful theory, quantum mechanics, which proved its usefulness even for smaller objects. On the other hand, if we want to apply the quantum theory of solid state to nanosystems we have to realize that almost all the results of this theory were obtained under the assumption that the size of a sample is large enough to treat the wave vector of electron as quasi-continuous variable. It is not true if the position of the electron is confined, at least in one dimension. Thus all previous calculations should be revised. The results, differing from those obtained for bulk materials, are called "quantum size-effects". Also the considered systems gain new names: a cluster of atoms confined in each dimension is called "quantum dot", a one-dimensional nanostructure is called "nanowire" and ultra-thin layer of atoms is treated as "quantum well". In this work we will be dealing mainly with these last objects.

It is well known that the circuits of contemporary microelectronic devices are mostly made of semiconductors. The same is naturally foreseen for the oncoming nanoelectronics. The question

[1] To whom any correspondence should be addressed.

arises why we are going to deal with metallic nanosystems. Single-electron transistors (SETs) are the nanodevices very important for investigation of fundamental physical phenomena as well as for applications. A very promising branch of electronics is spintronics, where the transport of the electron spin and charge are considered on equal footing. An essential feature of spintronics devices is that orientation of electron spin should be controlled over a length scale larger than the size of the device. The required orientation of spin is achieved by means of ferromagnetic leads and they are, of course, metallic.

A good example of such device is a ferromagnetic single-electron transistor (F-SET). It is well described in Ref. [2], therefore we mention here only some its features, essential for further considerations. The drain and source electrodes are made of a ferromagnetic metal and are connected to a central island by tunnel barriers. The requirement for the electrons to be well localised on the island is an appropriate resistance of the tunnel barriers. Moreover, the interplay between the Coulomb blockade in F-SETs and spin-dependent tunnelling processes gives rise to a novel phenomenon, enhanced tunnelling magnetoresistance (TMR). The value of TMR is defined as

$$TMR = \frac{R_{\mathrm{AP}} - R_{\mathrm{P}}}{R_{\mathrm{P}}} \qquad (1)$$

where R_{AP} and R_{P} are the device resistances in the antiparallel and parallel magnetisation configuration, respectively. It has been observed that the spin relaxation time can be enhanced by orders of magnitude as the size of the F-SET is reduced to a few nanometers [3]. A method of fabrication of nano-scale Ni/Au/Ni F-SET with a single 30 nm gold central island was presented in Ref. [2].

In view of above facts it is entirely justifiable to deal with electron transport properties of metallic nanosystems. Additionally, every real sample is actually disordered so that we have to consider the influence of disorder on the electrical conductivity of these nanosystems.

2. Theory of transport in disordered media

Every real sample of condensed matter is disordered. Nevertheless, we can imagine a perfectly ordered sample as our starting point for further considerations. It would be an infinite single crystal at the temperature T=0. The absolute zero is not available but we for the moment assume that it is possible. The deviations from the ideal crystal lattice at T=0 are called "static disorder". They can take a form of isolated static defects (vacancies, foreign atoms, interstitial atoms, dislocations), or a chemical disorder visible in substitutional alloys, where different atoms occupy the sites of ideal lattice at random and, finally, a topological disorder in amorphous alloys (or metallic glasses) where atoms have got totally random positions. On the other hand, if we admit T>0 then a dynamic disorder appears. First of all, thermal vibrations of primarily perfect monoatomic lattice appear. The positions of atoms at a given moment of time are random, though they are close to their positions in perfectly ordered lattice. If the temperature reaches a definite for each solid value, T_{m}, each solid becomes a liquid (we neglect the possible effect of sublimation here). This way we move to the higher level of disorder, a topological one. If we assume that we have at least two types of atoms then we obtain a liquid alloy, the system which is disordered on the highest possible level, because it unifies both topological and chemical disorder.

The considerations on disorder are essential for our purpose: calculation of the conductivity (or resistivity) of real materials. We skip the problem of thermal vibrations in pure metals and the influence of isolated static defects on their resistivity because these subjects are treated in all textbooks on solid state physics. Calculation of the resistivity of substitutional alloys required quite new theoretical approach, the coherent potential approximation (CPA). This approach was originally proposed by Velicky et al. [4]. The main idea of this approach is the replacement of two real atomic potentials, V_{A} and V_{B}, by one complex potential Σ. It is achieved by a simple assumption that the averaged scattering vanishes for this potential. The idea of CPA turned out very fruitful. Velicky [5] applied it to the problem of the electrical resistivity and obtained the Nordheim rule. Plischke and

Mattis [6] extended the CPA method on partially ordered alloys and studied the order-disorder transition. Paja [7] calculated the electrical resistivity of partially ordered alloys as a function of their ordering parameter. Chen et al. [8] took into account the electron-phonon interaction in disordered alloys and showed that their resistivity could increase and decrease with temperature (or remain constant) in dependence on the strength of the electron-phonon coupling. This dependence is linear at low temperatures. It is suitable moment to say that the dependence of the resistivity on temperature in nanosystems may be quite different. In very thin (3 nm thick) disordered films of PdAu a logarithmic dependence is observed [9]. This dependence is of purely geometrical origin [10]. A dynamical theory of the resistivity of chemically disordered nanosystems is still waiting for its discoverer.

Much higher disorder appear in liquid and amorphous metals where no translational invariance is observed. The positions of atoms in space can be described only statistically, by means of the pair distribution function $g(r)$ which determines the probability of finding an atom at the position **r** while the another atom is at the origin of the reference frame. The pair distribution function $g(r)$ defined by the equation

$$\rho_0 \int_V g(r)\, d^3\mathbf{r} = N \tag{2}$$

is shown in Figure 1. Experimentally we can observe a structure factor $S(q)$ which is related to $g(r)$ by the equation

$$S(Q) = 1 + \rho_0 \int_V (g(r) - 1) \exp(-i\mathbf{Q}\cdot\mathbf{r})\, d^3\mathbf{r} \tag{3}$$

and is shown in Figure 2.

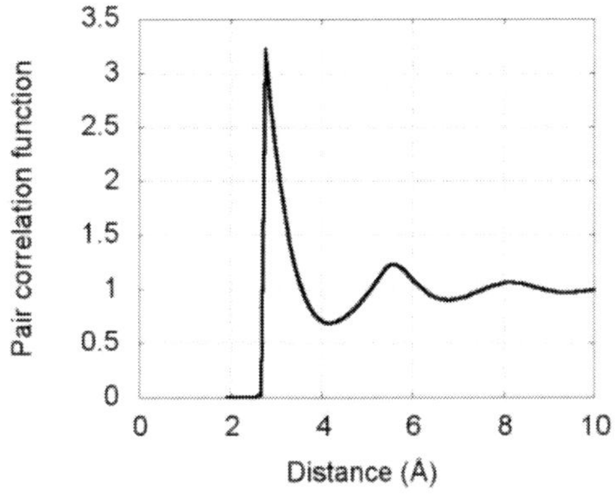

Figure 1. Pair distribution function vs. distance between atoms .

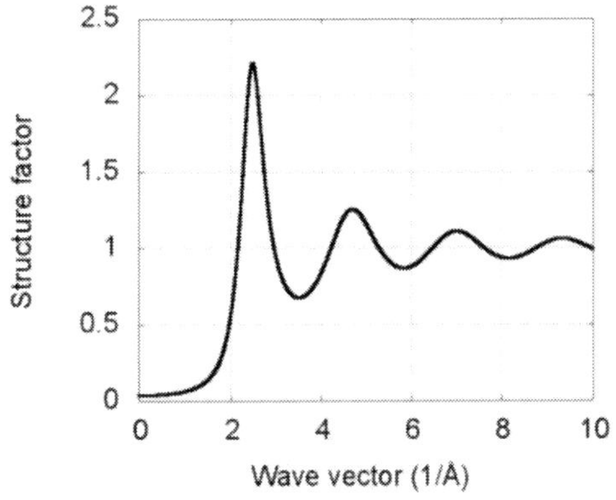

Figure 2. Structure factor vs. the value of the wave wector.

Ziman [11] viewed the process of scattering electrons from the disordered array of atomic potentials as the diffraction of the electron wave on that array (the diffraction model). His formula for the electrical resistivity ρ of liquid metal reads

$$\rho = \frac{3\pi m^2 \Omega_0}{4e^2\hbar^3 k_F^6} \int_0^{2k_F} S(q)\, |U(q)|^2 q^3 dq \tag{4}$$

where $U(q)$ is the Fourier transform of the atomic potential, Ω_0 is the atomic volume and other symbols have their usual meaning.

Faber and Ziman [12] extended the diffraction model onto liquid binary alloys. The description of the distribution of the positions of atoms requires three radial correlation functions $g_{\alpha\beta}(r)$ thus the final formula for the resistivity exploits three partial structure factors $S_{\alpha\beta}(q)$:

$$\rho = \frac{3\pi\, m^2\, \Omega_0}{4e^2\hbar^3 k_F^6} \int_0^{2k_F} \sum_\alpha \sum_\beta \left(c_\alpha c_\beta\right)^{1/2} S_{\alpha\beta}(q)\, u_\alpha(q)\, u_\beta(q)\, q^3\, dq \tag{5}$$

where c_α and c_β are the atomic concentrations of the alloy components. The same formula applies for solid amorphous alloys due to similarity of liquid and amorphous structures.

3. Electron states in confined systems

In the standard quantum theory of the solid state we usually assume that the considered sample is infinite. The finite size of the sample is taken into account by the periodic boundary conditions and the wave vector of electron, **k**, becomes a quasi-continuous variable. The dispersion relation in the free electron model

$$E(\mathbf{k}) = \frac{\hbar^2 \mathbf{k}^2}{2m^*} \tag{6}$$

(where m^* is the effective mass of the electron) allows us to define the Fermi surface

$$\frac{\hbar^2 k_x^2}{2m^*} + \frac{\hbar^2 k_y^2}{2m^*} + \frac{\hbar^2 k_z^2}{2m^*} = E_F \tag{7}$$

and to calculate the electronic density of states

$$D_3(E) = 2\frac{(2m^*)^{3/2}}{4\pi^2\hbar^3}\sqrt{E} \tag{8}$$

For 2-dimensional electron gas the Fermi sphere degenerates into a circle

$$\frac{\hbar^2 k_x^2}{2m^*} + \frac{\hbar^2 k_y^2}{2m^*} = E_F \tag{9}$$

and the density of states becomes a constant [13]

$$D_2(E) = 2\frac{m^*}{2\pi\,\hbar^2} \tag{10}$$

For 1-dimensional electron gas the Fermi surface reduces to 2 points

$$k_x = \pm\frac{\sqrt{2m^* E_F}}{\hbar} \tag{11}$$

and the density of states in this case reads

$$D_1(E) = 2\frac{\sqrt{2m^*}}{2\pi\,\hbar}\frac{1}{\sqrt{E}} \tag{12}$$

Let us consider now quasi-2-dimensional electron gas. It is a gas of electrons contained within a metallic film the thickness of which in the z-direction, L_z, is less than 100 nm (Figure 3.). If the electron is free to move in x and y directions the Schrodinger equation for this case is

$$\left(-\frac{\hbar^2}{2m^*}\Delta + V_{\text{conf}}(z)\right)\varphi(\mathbf{r}) = \varepsilon\,\varphi(\mathbf{r}) \tag{13}$$

where the confining potential $V_{conf}(z)$ is illustrated in Figure 4.

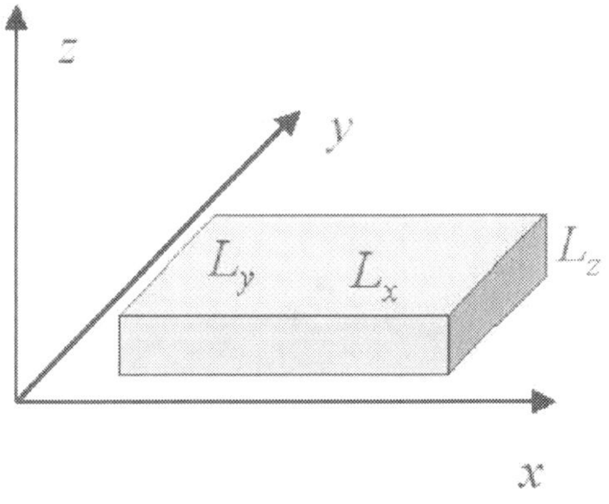

Figure 3. Positioning of a thin film in a system of coordinates.

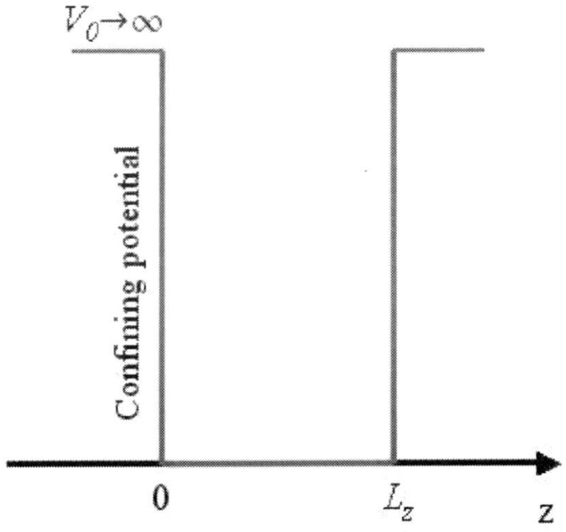

Figure 4. Confining potential along the z-axis

The Equation (13) is easy to solve and its solution is

$$\varphi_{\mathbf{k}n}(\mathbf{r}) = \frac{1}{\sqrt{A}} \exp(i\mathbf{k}_\rho \boldsymbol{\rho}) \zeta_n(z) \tag{14}$$

where

$$\boldsymbol{\rho} = (x, y); \quad \mathbf{k}_\rho = (k_x, k_y); \quad A = L_x L_y \tag{15}$$

and

$$\zeta_n(z) = \sqrt{\frac{2}{L_z}} \sin\left(\frac{nz\pi}{L_z}\right) \tag{16}$$

The eigenvalues of energy are

$$\varepsilon_{\mathbf{k}n} = \frac{\hbar^2 k_\rho^2}{2m^*} + \varepsilon_{zn} \tag{17}$$

where

$$\varepsilon_{zn} = \frac{\hbar^2}{2m^*}\left(\frac{n\pi}{L_z}\right)^2 \tag{18}$$

Thus the Fermi sphere splits into a number of parallel sheets (Fig. 5)

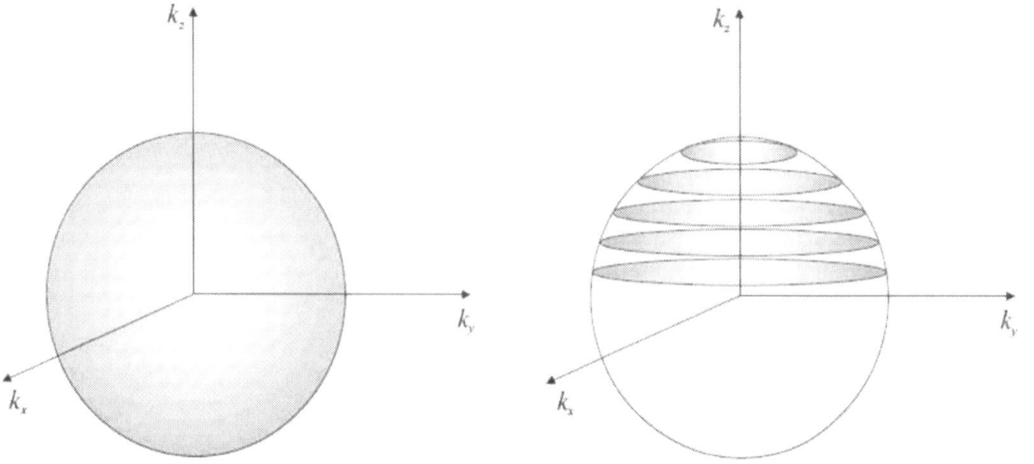

Figure 5. The Fermi sphere; left: for 3-D free electron gas; right: for quasi-2-D electron system.

The electron gas inside the Fermi sphere separates into subbands. The following inequalities hold

$$\varepsilon_{\mathbf{k}\,n} \le E_F \tag{19}$$

and

$$\frac{\hbar^2 \mathbf{k}_\rho^2}{2m^*} \le E_F - \varepsilon_{zn} \tag{20}$$

The density of states changes into a function which is constant in subsequent intervals ("a staircase").

4. Electron transport in some specific disordered metallic nanosystems

We consider a random array of scatterers represented by $S(q)$, embedded in the quantum well described above. The differential cross-section for in-plane scattering now is a function of a planar angle φ only

$$\frac{d\sigma}{d\varphi} = \frac{1}{8\pi\, k_\rho}\left(\frac{m^*}{2\pi\,\hbar^2}\right)^2 N\, S(q)\left|u_a(q)\right|^2 \tag{21}$$

where

$$q = 2k_\rho \sin\frac{\varphi}{2} \tag{22}$$

and $u_a(q)$ is a Fourier transform of a screened atomic potential $u_a(\mathbf{r})$.
The inverse transport relaxation time is

$$\frac{1}{\tau_{tr}} = 2\frac{v_F}{A}\int_0^\pi d\varphi\, \frac{d\sigma}{d\varphi}(1-\cos\varphi) \tag{23}$$

Equations (22) and (23) allow us for calculation of the resistivity ρ_n for the n-th subband

$$\rho_n = \frac{m^*}{n_c e^2 \tau_{tr}} \tag{24}$$

where n_c is the area concentration of carriers, because both k_p and n_c depend on the subband. The total Drude conductivity is a sum over all subbands

$$\sigma_D = \sum_{n=1}^{n_{max}} \sigma_n = \sum_{n=1}^{n_{max}} \frac{1}{\rho_n} \tag{25}$$

4.1. Electrical resistivity of ultrathin disordered metallic film

We at first considered the transport of exactly 2-D electron gas through a disordered array of screened Shaw pseudopotentials [14]. We assumed an alkali monoatomic amorphous film for calculations (n=1). The structure factor was obtained by means of numerical simulation. It turned out to be very similar to that of for 3-D case. The results of our calculations for five alkali metals are presented in the Table 1.

Table 1. Resistivities of alkali metals calculated for the monolayer films in comparison with the experimental values for bulk metals.

Element	Diameter of the hard sphere σ [10^{-10} m]	Screening parameter λ [10^{10} m]	Core radius R_c [10^{-10} m]	Resistivity of monolayer ρ_m [μΩcm]	Experimental bulk resistivity ρ_{exp} [μΩcm]
Li	2.70	1.36	0.66	55.4	25
Na	3.28	1.23	1.25	19.7	9.6
K	4.07	1.11	1.53	22.2	13
Rb	4.30	1.08	1.47	31.8	22
Cs	4.73	1.03	1.36	57.0	36

As we can see the resistivity of the monolayer is in each case higher than the experimental value for the bulk material (i. e. the conductivity is lower).

We next tested our model for films consisted of more than one monolayer [15]. We showed that the conductivity increases with the thickness of the film (see Figure 6).

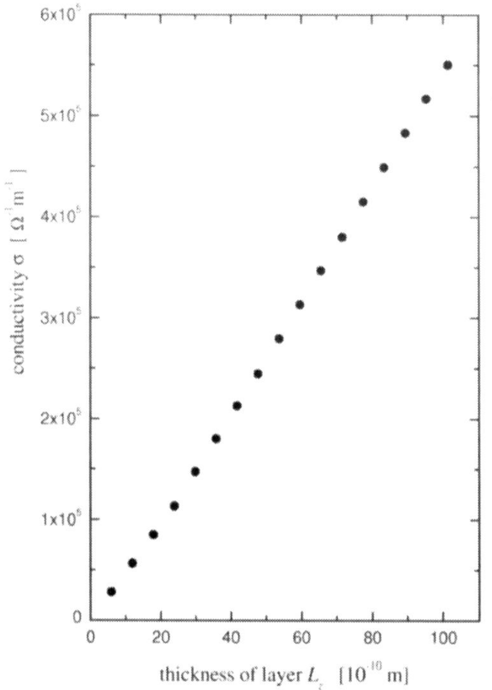

Figure 6. The conductivity σ of thin film of caesium as a function of its thickness L_z.

The conductivity should saturate as we move to bulk samples. Indeed, we have shown that in our next work.

4.2. Spin scattering of conduction electrons in quasi-two-dimensional system

We assume the current of polarised electrons going through a disordered array of N_A non-magnetic atoms doped with N_B magnetic impurities, placed at random [16]. Let spins of conduction electrons and impurities be polarised along the z-axis and in the same direction. The Hamiltonian of the conduction electron reads

$$\hat{H} = -\frac{\hbar^2}{2m^*}\Delta + U_0(\boldsymbol{\rho}) + U_{sS}(\boldsymbol{\rho}) \tag{26}$$

where the atomic potential

$$U_0(\boldsymbol{\rho}) = \sum_{l=1}^{N_A} u_{aA}(\boldsymbol{\rho} - \mathbf{R}_l) + \sum_{j=1}^{N_B} u_{aB}(\boldsymbol{\rho} - \mathbf{R}_j) \tag{27}$$

and the exchange potential

$$U_{sS}(\boldsymbol{\rho}) = -\sum_{j=1}^{N_B} J_0 \delta(\boldsymbol{\rho} - \mathbf{R}_j) s_z(\boldsymbol{\rho}) S_z(\mathbf{R}_j) \tag{28}$$

The total differential cross-section is

$$\frac{d\sigma}{d\varphi} = \frac{d\sigma_0^A}{d\varphi} + \frac{d\sigma_0^B}{d\varphi} + \frac{d\sigma_{\uparrow\uparrow}^B}{d\varphi} \tag{29}$$

where the atomic part is

$$\frac{d\sigma_0^X}{d\varphi} = \frac{1}{8\pi\, k_\rho}\left(\frac{m^*}{2\pi\,\hbar^2}\right)^2 N_X\; S_X(q) \left| u_{aX}(q) \right|^2, \qquad X = A, B \tag{30}$$

and the spin part

$$\frac{d\sigma_{\uparrow\uparrow}^B}{d\varphi} = \frac{1}{8\pi\, k_\rho}\left(\frac{m^*\, J_0}{2\pi\,\hbar^2}\right)^2 S(S+1)\, N_B\; S_B(q) \tag{31}$$

We assumed here the Coulomb screened potential

$$u_{aX}(\mathbf{\rho}) = u_X\, \frac{\exp(-\lambda_X \rho)}{\rho} \tag{32}$$

then its Fourier transform is

$$u_{aX}(q) = 2\pi\, u_X\, \frac{1}{\sqrt{\lambda_X^2 + q^2}} \tag{33}$$

We approximated the structure factor by a parabola

$$S_X(q) \approx \alpha_X q^2 \tag{34}$$

which is a good approximation in the range $[0,\, 2k_F]$.

The conductivity of the m-th sheet of the Fermi surface

$$\sigma_m(L_z,\, x) = \frac{1}{L_z}\frac{n_m e^2 \tau_m}{m^*} \tag{35}$$

depends on L_z and the concentration $x=N_B/(\,N_A+\,N_B)$ of magnetic impurities. The dependence of σ_m on L_z is through k_ρ which is different for either sheet

$$k_\rho = \sqrt{\frac{2m^* E_F}{\hbar^2} - \left(\frac{m\pi}{L_z}\right)^2} \tag{36}$$

The total conductivity of the considered film is

$$\sigma_{tot}(L_z,\, x) = \sum_{m=1}^{m_{max}} \sigma_m(L_z,\, x) \tag{37}$$

where m_{max} also depends on the system size L_z.

The results of calculations for the conductivity σ as a function of the film thickness L_z are shown in Figure 7, and as the function of concentration of magnetic impurities x in Figure 8.

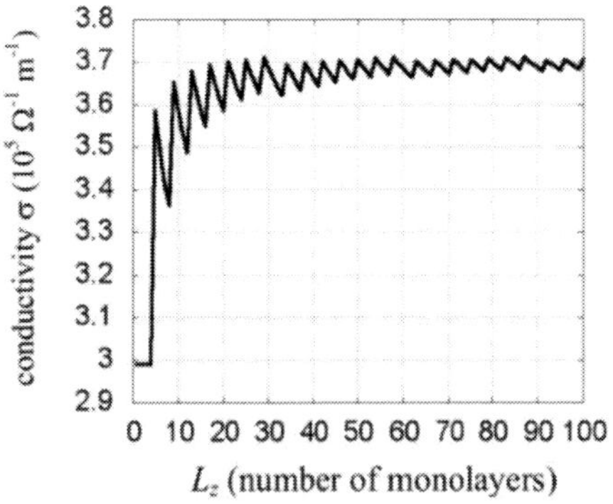

Figure 7. The conductivity σ as a function of the film thickness L_z for the concentration of magnetic impurities x=0.05

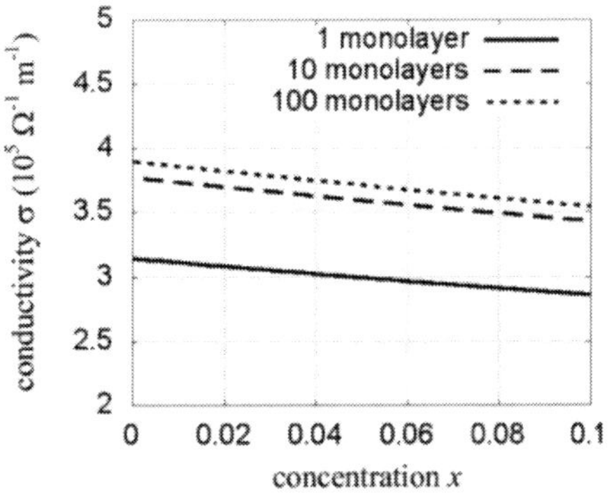

Figure 8. The conductivity σ as a function of concentration of magnetic impurities x

It is clear that the conductivity saturates for the film thickness greater than 100 monolayers.

4.3. Conductivity of a disordered magnetic monoatomic layer.

We considered a 2-D disorder array of magnetic atoms. We assume the spins can have only one of two possible orientations: "up" and "down. Thus the system can be treated as a 2-component disordered magnetic alloy [17]. Let the current go along the magnetisation axis. It can serve as an axis of quantazition. Each electron in the current takes one of the two possible orientations of its spin: parallel and antiparallel to magnetisation. There is well-known asymmetry of scattering in this case: electrons with spins parallel to that of scatterer feel only the Coulomb potential whereas those having opposite spins scatter more strongly [18].

We assume the total scattering potential

$$U(\mathbf{r}) = U_{\parallel}(\boldsymbol{\rho}) + U_{\perp}(z) \tag{38}$$

where

$$U_{\parallel}(\boldsymbol{\rho}) = \sum_{i=1}^{N} u_a(\boldsymbol{\rho} - \mathbf{R}_i)\, u_S(\mathbf{R}_i) \tag{39}$$

$U_{\perp}(z)$ is the geometrical confinement and $u_S(\mathbf{R}_i)$ is the spin part of atomic potential. The electron wave function has the form

$$\psi_s(\mathbf{r}) = \phi(\boldsymbol{\rho})\,\zeta(z)\,\chi(s), \qquad s = \pm 1 \tag{40}$$

We now can define the operator $u_S(\mathbf{R}_i)$ by its matrix elements

$$\chi(s)\, u_S(\mathbf{R}_i)\, \chi(s') = \delta_{ss'}\left(t + (1-t)\,\delta_{Ss'}\right), \qquad t \ge 1 \tag{41}$$

This definition assures us that the scattering in case of opposite spins is t times stronger. Thus the transport relaxation times for "spin-up" and "spin-down" differ substantially:

$$\frac{1}{\tau_{\rm tr}^{+}} = \frac{1}{4\pi\, k_{\rho}} \left(\frac{m}{2\pi\,\hbar^{2}}\right)^{2} n v_{\rm F} \int_{0}^{\pi} d\varphi\,(1-\cos\varphi)\, u_{\rm a}^{2}(q)\left[c_1 S_{11}(q) + 2(c_1 c_2)^{1/2} t S_{12}(q) + c_2 t^2 S_{22}(q)\right] \tag{42}$$

$$\frac{1}{\tau_{\rm tr}^{-}} = \frac{1}{4\pi\, k_{\rho}} \left(\frac{m}{2\pi\,\hbar^{2}}\right)^{2} n v_{\rm F} \int_{0}^{\pi} d\varphi\,(1-\cos\varphi)\, u_{\rm a}^{2}(q)\left[c_1 t^2 S_{11}(q) + 2(c_1 c_2)^{1/2} t S_{12}(q) + c_2 S_{22}(q)\right] \tag{43}$$

where c_1 and c_2 are the fractions of atoms polarised "up" and "down" which can be expressed by magnetisation μ

$$c_1 = \frac{1+\mu}{2}, \qquad c_2 = \frac{1-\mu}{2} \tag{44}$$

The conductivities σ^{+} and σ^{-} are given by

$$\sigma^{\pm} = \frac{n_{\pm}\, e^{2}\, \tau_{tr}^{\pm}}{m^{*} L_z} \tag{45}$$

and the total conductivity in two-band model

$$\sigma_{\rm tot} = \sigma^{+} + \sigma^{-} \tag{47}$$

The results calculated for some model set of parameters are shown in Figure 9.

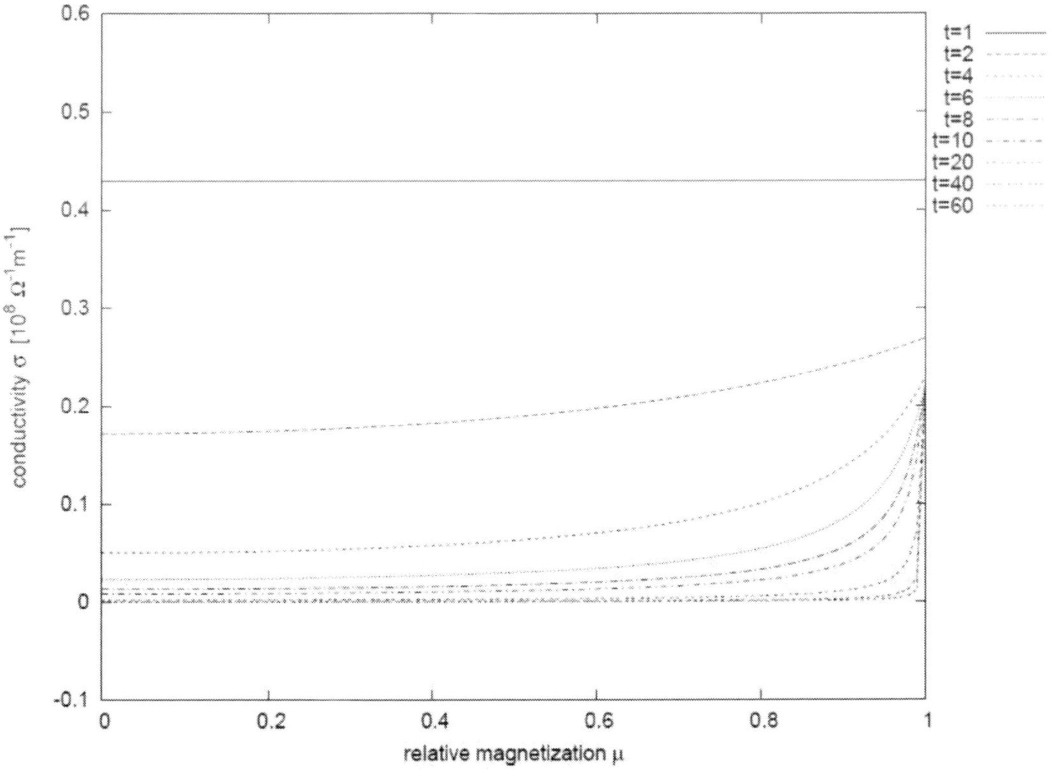

Figure 9. Conductivity σ as a function of relative magnetisation μ for various values of the asymmetry parameter t.

We can see that for asymmetry parameter t=1 the conductivity is constant as expected and for higher values of t the conductivity rises rapidly as the relative magnetisation μ approaches unity.

5. Summary

Contemporary electronics is based on circuits made of semiconducting materials. Most electronic devices exploit the transportation of the electron charge. Spintronics is a new branch of electronics which is going to manipulate the electron spin as well. Metallic nanosystems are important both for study of the behaviour of spin in the nanoscale as well as for applications. But real metallic nanosystems are disordered in most cases. In this work some specific metallic nanosystems were studied by means of methods developed for bulk disordered samples. It has been shown that the transport coefficients of such nanosystems depend on their size and differ substantially from the values for bulk materials but they approach the bulk values when their size reaches about a hundred monoatomic layers.

Acknowledgements

The author thanks dr. B. J. Spisak, Mr. P. Jarguz, MSc. Eng., and Mr. M. Ornat, MSc. Eng. for fruitful collaboration.

References

[1] Rohrer H 1996 *Microelectron. Eng.* **32** 5

[2] Yakushiji K, Ernult F, Imamura H, Yamana K, Mitani S, Takanashi K, Takahashi S, Maekawa S and Fujimori H 2005 *Nature Materials* **4** 57

[3] Petterson H, Liu R, Suyatin D and Samuelson L 2008 Assembling F-SET with AFM *Nanostructures in Electronics and Photonics* ed F Rahman (Singapore: Pan Stanford Publishing) chapter 2 pp 29-40

[4] Velicky B, Kirkpatrick S and Ehrenreich H 1968 *Phys. Rev.* **175** 747

[5] Velicky B 1969 *Phys. Rev.* **184** 614

[6] Plischke M and Mattis D 1973 *Phys. Rev.* B **7** 2430

[7] Paja A 1976 *J. Phys. C: Solid State Phys.* **9** 1445

[8] Chen A B, Weisz G and Sher A 1972 *Phys. Rev.* B **5** 2897

[9] Dolan G J and Osheroff D D 1979 . *Phys. Rev. Lett.* **43** 721

[10] Akkermans E and Montambaux G 2007 *Mesoscopic Physics of Electrons and Photons* (Cambridge: Cambridge University Press)

[11] Ziman J M 1961 *Phil. Mag.* **6** 1013

[12] Faber T E and Ziman J M 1965 *Phil. Mag.* **11** 153

[13] Heinzel T 2007 *Mesoscopic Electronics in Solid State Nanostructures* (Weinheim: WILEY-VCH Verlag GmbH&Co. KgaA)

[14] Jarguz P and Paja A 2006 *Mat. Sci. – Poland* **24** 633

[15] Spisak B J and Paja A 2006 *Mat. Sci. – Poland* **24** 605

[16] Paja A and Spisak B J 2007 *Acta Phys. Pol.* A **112** 1289

[17] Paja A and Spisak B J 2008 *Mat. Sci. – Poland* **26** 1009

[18] Camley R E and Barnas J 1989 *Phys. Rev. Lett.* **63** 664

Journal of Physics: Conference Series **213** (2010) 012028 doi:10.1088/1742-6596/213/1/012028

Local and Nonlocal Andreev Reflections in Double Barrier Junctions

RJ Wojciechowski and L Kowalewski

A. Mickiewicz University, Institute of Physics, ul. Umultowska 85, 61-614 Poznań, Poland

E-mail: `wojrysz@amu.edu.pl`

Abstract. We study spin dependent charge transport through a junction, consisting of a superconductor connected to two normal or spin-polarized electrodes. Special attention is paid to a process known as a crossed or nonlocal Andreev reflection. The nonlocal reflection occurs when an incident electron (hole) from one electrode, injected with a subgap energy onto o superconducting layer, transforms into an Andreev hole (electron) in the second lead. When two electrons (holes), forming a Cooper pair, come from the same electrode, the process of the Andreev reflection is local. The coherent spin polarized transport in ferromagnet/superconductor/ferromagnet ($F/S/F$) double barrier junctions is analyzed using the Bogolubov-de Gennes (BdG) equation with appropriate boundary conditions. We calculate and discuss probabilities of the normal, local and crossed Andreev reflections, as well as probability of the elastic co-tunneling. These processes contribute to tunneling current when the distance between the magnetic electrodes is comparable to the superconducting coherence length. The dependence of the tunneling charge transport on the strength of the exchange field in the ferromagnetic electrodes, and on the height of the tunnel barriers are presented. We also discuss the local and nonlocal spin quantum entanglements of quasiparticle pairs induced by the Andreev reflections in the junction.

1. Introduction

In this paper we have addressed the problem of charge transport through nanoscale systems where normal (N) or ferrmagnetic (F) metals (M)or half metals are attached to a superconductor. In particular, we have concentrated on the role of the local and nonlocal Andreev reflections both in single particle transport and in pairwise spin entangled transport. We have considered the charge transport through a simple model of the metal-superconductor-metal ($M/S/M$) junction. The charge transport can be decomposed into two processes. The first type,which can be named local processes, consists of the normal and Andreev reflections from the first e.g. left M/S interface. These reflections occur in the single normal- or ferromagnetic metal-superconductor junction. The nonlocal processes, occur in a region where there is no drive current. It is characteristic for multi-terminal junctions [1] and consists of the crossed Andreev (CAR) reflection and elastic cotunneling (EC) [2, 3, 4, 5, 6, 7, 8]. It has been shown [1, 8] that when the distance l between the electrodes is less than or comparable to the superconducting coherence length ξ, an injection of an electron (or hole) with a subgap energy leads to the additional nonlocal and coherent transport processes named the CAR and EC [2]. The experimental evidence for both CAR and EC has been given by Russo et al. [3] and also by Beckmann et al.([4]). If the thickness of S layer is large enough, a double tunnel junction can be regarded as two independent single M/S junctions and then the tunneling processes become incoherent [6]. In addition, the total electronic current in the $F/S/F$ junction strongly depends on the relative orientation of the magnetization of two ferromagnetic electrodes [7, 8]. The CAR is a consequence of capturing another electron (or hole) from the right electrode in order to form a Cooper pair in the superconductor. The appropriate hole (or electron) is reflected in the right electrode. The EC consists in the direct tunneling of the injected particle from one electrode to the second one, without formation of the Cooper pair [5]. The electronlike quasiparticle penetrates depth of order of the superconducting coherence length ξ before its conversion into the condensate of the Cooper pairs. The charge current $2ev_F$ in the ferromagnetic metals disappears, as an appropriate quasiparticle current J_Q in the sc layer, according with the formula $J_Q = 2ev_F \exp(-y/\xi)$, where the coherence length is given by $\xi = hv_F/(2\sqrt{\Delta^2 - E^2}$. Simultaneously, J_Q converts into an increasing supercurrent of the Cooper pairs $J_s = 2ev_F[1 - \exp(-y/\xi)]$. In this paper, we discuss how the nonlocal effects are influenced by the energy of the incident electron and the width of the superconducting layer. These effects are important for the following reasons:

(i) The spin polarized current in metallic $F/S/F$ heterostructures, due to the CAR, can be applied in spintronics by implementation of magnetoresistive devices, such as spin switches or memory elements;

(ii) The spin polarized subgap tunneling spectroscopy in double $F/S/F$ junctions provides information on the pairing symmetry in ferromagnetic superconductors;

(iii) The CAR enables realization of the solid state entanglers by creation of mobile spin entangled pairs of electrons in spatially separated nanoelectronic circuits.

2. Model and calculations

In this paper, we consider a double junction, consisting of two metallic (normal or ferromagnetic) electrodes and a superconducting layer separated from the electrodes by thin insulating interfaces. Here, to avoid unnecessary complications, we consider a model of $F/S/F$ junction with a clean superconductor. In the clean limit one can easily solve the BdG equations within the approach proposed in [9]. The wave function, describing the quasiparticle propagation across the $F/S/F$ junction, can be obtained from the extended BdG equations. In the ferromagnetic regions, these equations are reduced to the appropriate Schrödinger equations. The extended BdG equations take the following form:

$$\begin{pmatrix} H_0 - E - \sigma h_{ex}, & \Delta \\ \Delta^\dagger & -[H_0 - E + \sigma h_{ex}] \end{pmatrix} \begin{pmatrix} f_\sigma(\vec{r}) \\ g_{-\sigma}(\vec{r}) \end{pmatrix} = \check{O}\,, \tag{1}$$

where $H_0 = -\hbar^2/2m\nabla^2 - E_F + W(\vec{r})$ is the single-particle Hamiltonian, E is the quasiparticle energy measured from the Fermi energy E_F, and $\sigma = +(-1)$ denotes the up (down) spin subband. $W(\vec{r})$ stands for the interface potential which is measured by the dimensionless parameter $z = 2mW/\hbar^2 k_F$. For the ferromagnetic electrodes, we adopt the Stoner model. Thus, the exchange field in both the left ($y < -l/2$) and the right electrodes ($y > l/2$) can be defined as follows

$$h_{ex} = \begin{cases} h_0 & y < -l/2 \\ 0 & -l/2 < y < +l/2 \\ \pm h_0 & y > l/2 \end{cases} \tag{2}$$

where $+h_0$ and $-h_0$ define the exchange fields for the parallel and antiparallel configuration, respectively. The solution of eq.(1) corresponding to the injection of an electron with spin σ

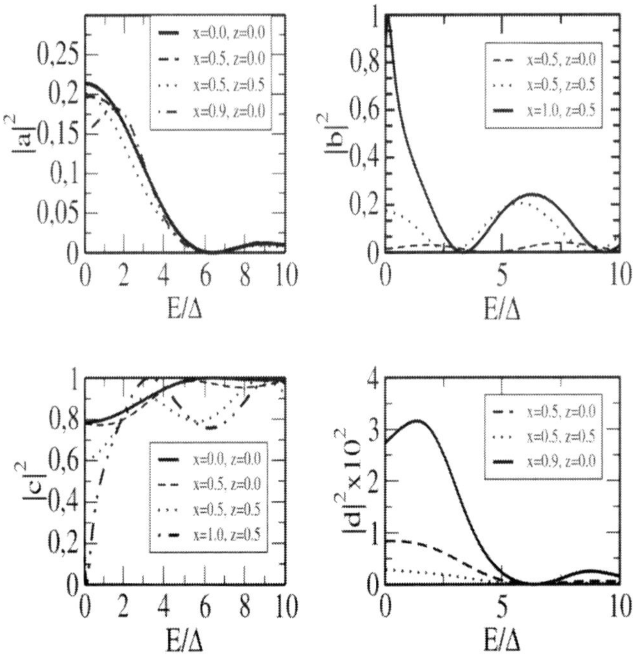

Figure 1. The probability of the four transport processes as a function of E/Δ for $lk_F = 10^3$ and for several values of x and z.

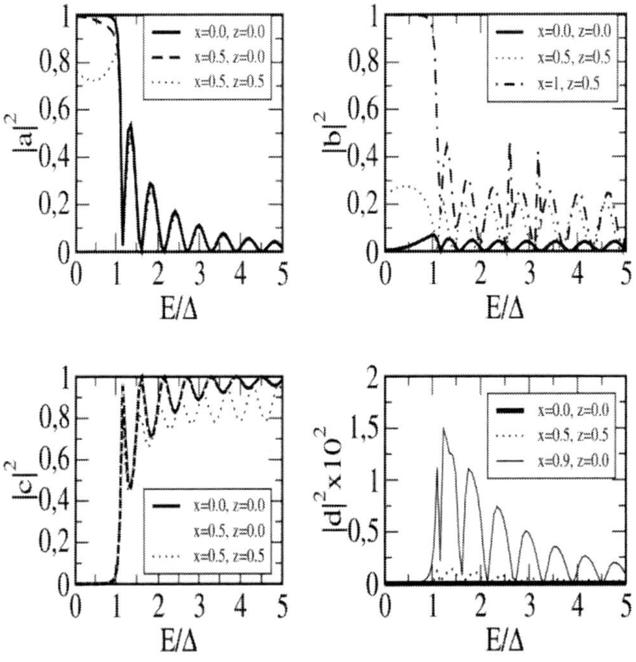

Figure 2. The probability of the four transport processes as a function of E/Δ for $lk_F = 10^4$ and for several values of $x = h_0/E_F$ and z.

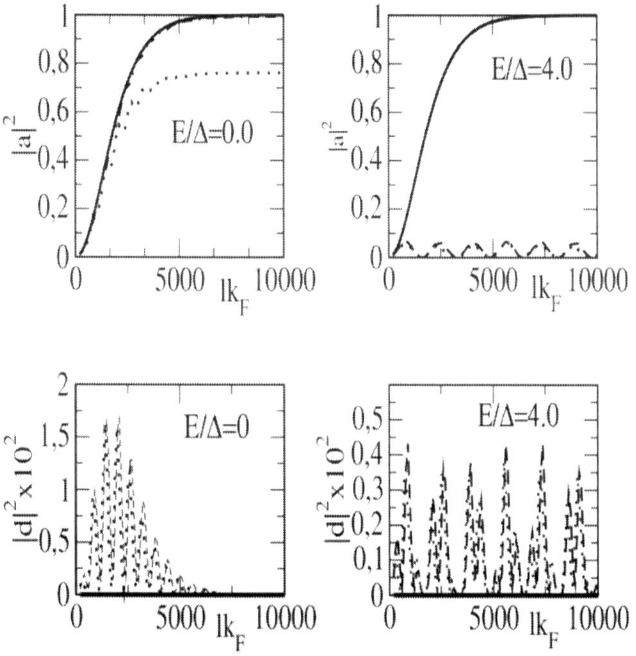

Figure 3. The probability of the ordinary ($|a|^2$) and crossed Andreev ($|d|^2$) reflections as a function of lk_F for $E/\Delta = 0.0$ (left panel) and $E/\Delta = 4.0$ (right panel)and for several values of x and z: ($x = 0.0$, $z = 0.0$)- full line, ($x = 0.5$, $z = 0.5$)- dotted line, ($x = 0.5$, $z = 0.0$)- dashed line.

(from one, e.g. the left ferromagnetic electrode) and with energy $E > 0$, and angle of incidence Θ can be written in the following form:

$$\Psi_\sigma^{(y<-l/2)}(y) = \begin{pmatrix} 1 \\ 0 \end{pmatrix} e^{ik_\sigma^+ y} + a_{-\sigma} \begin{pmatrix} 0 \\ 1 \end{pmatrix} e^{ik_{-\sigma}^- y} + b_\sigma \begin{pmatrix} 1 \\ 0 \end{pmatrix} e^{-ik_\sigma^+ y}, \tag{3}$$

$$\Psi_\sigma^{(y>l/2)}(y) = c_\sigma \begin{pmatrix} 1 \\ 0 \end{pmatrix} e^{ik_{\sigma(-\sigma)}^+ y} + d_\sigma \begin{pmatrix} 0 \\ 1 \end{pmatrix} e^{-ik_{-\sigma(\sigma)}^- y}, \tag{4}$$

where the y component of the momentum for both an electron (+) and hole(−) takes the form: $k_\sigma^\pm = \sqrt{(2m/\hbar^2)(E_F + \sigma h_0 \pm E) - \vec{k}_{||,\sigma}^2}$.The momentum parallel to the interface has the form: $|\vec{k}_{||,\sigma}| = \sqrt{(2m/\hbar^2)(E_F + \sigma h_0 + E)}\sin(\Theta)$. The probability amplitudes of the transport processes $a_\sigma(E,\Theta)$, $b_\sigma(E,\Theta)$, $c_\sigma(E,\Theta)$, $d_\sigma(E,\Theta)$ determine the ordinary Andreev and normal reflections, elastic cotunneling, and crossed Andreev reflection, respectively[5, 6, 7, 8, 10].

For $E/\Delta > 1$ the interference of incident and reflected quasiparticles in the superconducting layer leads to oscillations of all the transport coefficients and, in addition, both a_σ and d_σ vanish if the following resonance condition is fulfilled i.e. $q_\sigma^+ - q_\sigma^- = 2n\pi/l$. The momentum $q^\pm$ in the superconducting layer S is given by $q_\sigma^\pm = \sqrt{(2m/\hbar^2)(E_F \pm \sqrt{E^2 - \Delta^2}) - \vec{k}_{||,\sigma}^2}$. The coefficients $a_\sigma(E,\Theta)$, $b_\sigma(E,\Theta)$, $c_\sigma(E,\Theta)$, $d_\sigma(E,\Theta)$ are determined by the appropriate boundary conditions for the wave functions and their derivatives at both the interfaces are placed at $y = \pm l/2$. Although it is true that in experiments one does not measure directly the coefficients a, b, c, d, however, these coefficients determine the charge transport through $F/S/F$ junctions and the pairing symmetry of the superconductor. For example, the sign of the voltage measured across the second S/F tunnel barrier of the junction $F/S/F$ depends upon whether EC or CAR occurs with larger probability [3]. Similarly, in the case of half-metallic ferromagnet electrodes the process of CAR can be completely suppressed while EC will be allowed, or vice versa, depending on the mutual alignments (parallel or antiparallel) of the electrodes magnetization.

3. The Andreev entangler: spin entanglement in the F/S/F junction

As shown in the prevoius section, the coherent quasiparticle charge transport through the double $F/S/F$ junction exhibits interesting characteristic features. In particular, for $E > \Delta$, all probabilities of the transport processes (the AR, CAR, NR, and EC) oscillate with energy and the thickness of the superconducting layer as a result of the interference of incoming and outgoing particles in this layer. This coherent transport is particularly interesting because of its promising spintronic applications by implementation of magnetoresistive devices, such as spin switches, memory elements, magnetic read-out heads for computer hard drives, spin transistors etc. Additionally, as it was proved, the Andreev and normal reflected electrons (or holes) transported from the superconductor into the conducting leads can be mutually spin entangled. Pairing interaction in a superconductor is the source of spin entanglement. The pair of electrons can tunnel coherently, due to the Andreev reflections processes, from the superconductor into the same or different leads. In the case of the singlet superconductors, non-local processes, i.e. the processes which lead to tunneling of the electrons to the different electrodes, can be enhanced for the antiparallel configuration, or can be suppressed for the parallel configuration of the ferromagnetic electrodes. Entanglement of either electrons or holes depends on the sign of the voltage applied to the left lead, for instance. For negative voltage an incident spin up (down) hole in the left conductor can be Andreev reflected as spin down (up) electron either at the left (ordinary AR) or at the right (CAR) lead.

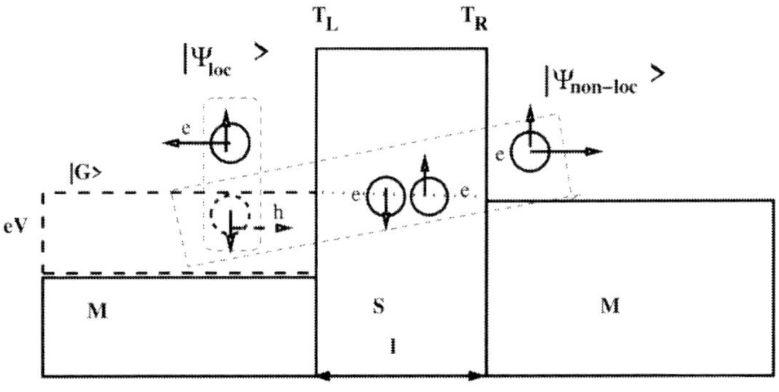

Figure 4. Sketch of the reflection processes responsible for local and non-local entanglement in the $M/S/M$ junction.

In our analysis, we consider the concept of the local and nonlocal spin-entanglement between electrons or holes, in a double barrier junction consisting of two conductors connected to a superconductor. At most, the separation between the both metallic electrodes should be comparable to the superconducting coherence length (ξ) in order to enable the crossed Andreev reflection. An application of a bias voltage $eV = \mu_S - \mu_L$, between the superconductor and the normal or ferromagnetic conductor, leads to a possible stationary tunneling transport of spin entangled pairs of electrons or holes. μ_S and μ_L stand for chemical potentials of the s-wave superconductor and the left metallic leads, respectively. Depending on the sign of the voltage applied to one of the conductors we get a stream of entangled fermions. If a negative voltage is applied to the left conductor, then an incident spin up (down) hole in the left conductor can be Andreev reflected as a spin down (up) electron either at the left conductor or at the right one. In such a way, the holes incident on the superconducting layer e.g. from the left electrode convert, due to the pairing interaction in the superconductor, their charge and spin into the opposite ones. In order to create entangled pairs of electrons, let us apply a negative voltage to the left electrode (see Fig.4). Then, the input state forming a stream of pairs of holes with opposite spins and the same energies E injected to the superconductor can be written in the form [11, 12]:

$$\mid \Psi_{in} >= \prod_{0<E<eV} a^{h\dagger}_{L\uparrow}(E) a^{h\dagger}_{L\downarrow}(E) \mid G >= \frac{i}{2} \prod_{0<E<eV} a^{h\dagger}_{L}(E) \sigma_y a^{h\dagger T}_{L}(E) \mid G > , \tag{5}$$

where $a^{h\dagger}_{L} = (a^{h\dagger}_{L\uparrow}(E) a^{h\dagger}_{L\downarrow}(E))^T$ is a matrix of the creation operators of a pair of holes with the opposite spins at the left conductor and $|G >$ denotes the Fermi sea filled up to the superconducting chemical potential in the left conductor; σ_y is the Pauli matrix. T stands for the matrix transpose. The evolution of the input state $|\Psi_{in} >$ into the output state $|\Psi_{out} >$ can be connected through unitary transformation, which can be expressd in terms of the scaterring matrix S. The scattering matrix matches the wave functions at both FS and SF interfaces in a double barrier junction [13]. In our further analysis we take into account only the Andreev reflections and completely neglect the elastic cotunneling process since such process does not lead to the entanglement of the particles. Thus, the relation between the input $\hat{a}$ and output $\hat{b}$ states can be written in a form similar to that given in [11] i.e.

$$\hat{b} = \hat{S}\hat{a} \, , \tag{6}$$

where $\hat{b}$ and $\hat{a}$ are the 8×1 matrices. $\hat{S}$ is 8×8 scattering matrix, which decribes all reflections processes in left (L) and right (R) interfaces:

$$\hat{S} = \begin{pmatrix} S^{ee}_{LL\uparrow\uparrow} & S^{ee}_{LL\uparrow\downarrow} & 0 & 0 & S^{eh}_{LL\uparrow\uparrow} & S^{eh}_{LL\uparrow\downarrow} & S^{eh}_{LR\uparrow\uparrow} & S^{eh}_{LR\uparrow\downarrow} \\ S^{ee}_{LL\downarrow\uparrow} & S^{ee}_{LL\downarrow\downarrow} & 0 & 0 & S^{eh}_{LL\downarrow\uparrow} & S^{eh}_{LL\downarrow\downarrow} & S^{eh}_{LR)\downarrow\uparrow} & S^{eh}_{LR\downarrow\downarrow} \\ 0 & 0 & S^{ee}_{RR\uparrow\uparrow} & S^{ee}_{RR\uparrow\downarrow} & S^{eh}_{RL\uparrow\uparrow} & S^{eh}_{RL\uparrow\downarrow} & S^{eh}_{RR\uparrow\uparrow} & S^{eh}_{RR\uparrow\downarrow} \\ 0 & 0 & S^{ee}_{RR\downarrow\uparrow} & S^{ee}_{RR\downarrow\downarrow} & S^{eh}_{RL\downarrow\uparrow} & S^{eh}_{RL\downarrow\downarrow} & S^{eh}_{RR\downarrow\uparrow} & S^{eh}_{RR\downarrow\downarrow} \\ S^{he}_{LL\uparrow\uparrow} & S^{he}_{LL\uparrow\downarrow} & S^{he}_{LR\uparrow\uparrow} & S^{he}_{LR\uparrow\downarrow} & S^{hh}_{LL\uparrow\uparrow} & S^{hh}_{LL\uparrow\downarrow} & 0 & 0 \\ S^{he}_{LL\downarrow\uparrow} & S^{he}_{LL\downarrow\downarrow} & S^{he}_{LR\downarrow\uparrow} & S^{he}_{LR\downarrow\downarrow} & S^{hh}_{LL\downarrow\uparrow} & S^{hh}_{LL\downarrow\downarrow} & 0 & 0 \\ S^{he}_{RL\uparrow\uparrow} & S^{he}_{RL\uparrow\downarrow} & S^{he}_{RR\uparrow\uparrow} & S^{he}_{RR\uparrow\downarrow} & 0 & 0 & S^{he}_{RR\uparrow\uparrow} & S^{he}_{RR\uparrow\downarrow} \\ S^{he}_{RL\downarrow\uparrow} & S^{he}_{RL\downarrow\downarrow} & S^{he}_{RR\downarrow\uparrow} & S^{hh}_{RR\downarrow\downarrow} & 0 & 0 & S^{hh}_{RR\downarrow\uparrow} & S^{hh}_{RR\downarrow\downarrow} \end{pmatrix} \tag{7}$$

All the elements of the scattering matrix $\hat{S}$ describe appropriate Andreev reflection processes. For example, $S^{eh}_{LL\uparrow\uparrow}$ denotes the ordinary Andreev reflection amplitude for a hole with, spin up injected into the left lead (L) and reflected as an electron with spin up in the same (L) lead.

In the tunneling limit i.e when the transparencies of the tunnel barriers ($T_{L(R)}$) fulfil the relation $T_{L(R)} << 1$, the outgoing state, to the first order of the scattering matrix elements $S^{\alpha\beta}_{....}$ can be written as a superposition of the vacuum state $|0 >$ and the local and non-local spin singlet entangled states ([11]):

$$\mid \Psi_{out} >= \sqrt{1 - T_L^2 - T_L T_R \gamma(l)} |0 > + T_L |\Psi_{loc} > + \sqrt{T_L T_R \gamma(l)} |\Psi_{non-loc} > , \tag{8}$$

where the local and non-local spin singlet states have the form

$$\mid \Psi_{loc} >= \frac{1}{\sqrt{2}} [|\uparrow (-E) >_L |\downarrow (E) >_L - |\downarrow (-E) >_L |\uparrow (E) >_L] \tag{9}$$

$$\mid \Psi_{non-loc} >= \frac{1}{\sqrt{2}}[|\uparrow (-E) >_L|\downarrow (E) >_R - |\downarrow (-E) >_L|\uparrow (E) >_R]. \tag{10}$$

$\gamma(l) = \exp(-\frac{2l}{\pi\xi})$ describes distance at which a Cooper pair can still exist. As seen from the above relations, due to the superconducting layer, the charge current of independent particles can be transformed into the (locally)(T_L) and nonlocally ($\sqrt{T_L T_R}$) Andreev reflected current of spin entangled pairs.

4. Results and conclusions

The coherent charge transport through the $F/S/F$ junction exhibits many peculiar features both for the subgap transmission and for energy greater than superconducting gap. The important role, in the transport processes through the double junction, play both the paring symmetry of a superconducting state and the magnetic orientation of the ferromagnetic electrodes [14]. For example, in the case of the spin singlet pairing and the parallel magnetic configuration of the electrodes, the ordinary and crossed Andreev reflections are suppressed and for half-metals ($x = h_0/E_F = 1$) these processes are completely blocked. In the case of antiparallel alignment, the crossed Andreev reflection is favoured while the normal Andreev reflection is suppressed (or blocked for half-metals). In the paper, we discuss the probabilities of all the components of the coherent transport processes, i.e.: the ordinary Andreev , crossed Andreev and normal reflections ($|a|^2$, $|d|^2$),$|b|^2$), as well as the direct tunneling through the double junction ($|c|^2$). The sum of these probabilities is normalized to 1. The numerical calculations were performed for a wide range of the model parameters i.e. for the strength of the exchange field x, the height of the interfacial barriers z, and the thickness of the superconducting layer l. If the energy of the incident electron is greater than the superconducting gap Δ, all these probabilities oscillate as the functions of E/Δ and l due to the interference of incoming and outgoing particles in the superconducting layer (see Figs. 1-3). If the thickness of the superconductor is smaller than the superconducting coherence length the nonlocal processes (EC and CAR) play an important role in the charge transport for the subgap energy (see Fig.1). In the opposite limit, i.e. if the thickness of the superconductor, suppression of EC and CAR is observed (see Fig.2). Observation of the changes in the tunneling current in double junctions for different widths of the superconducting layer were reported in [3]. The maxima (minima) of the elastic cotunneling correspond one to one to minima (maxima) of the ordinary Andreev, crossed Andreev and normal reflections. If the thickness $l \geq \xi$, then the non-local processes decay. When the EC dominates i.e. when the charge carriers pass from the left to the right electrode without conversion into the supercurrent, the ordinary Andreev and crossed Andreev reflections decay at the both interfaces. It is worthy mentioning that typical ranges of the coherence length are the order of $\xi = 10\mathring{A}$ and $\xi = 1.0\mu m$ for high-T$_c$ and low-temperature superconductors, respectively. The electron-beam lithography technique allows one to fabricate devices which have sizes of the order of tens nm.

In the last years, the transport phenomena in multiterminal junctions have attracted a great interest where the local and non-local entanglements, caused by the Andreev reflections, play an important role. Multiterminal junctions e.g. $N/S/N$ and $F/S/F$ are the promising examples of solid state entanglers. The controlled creation and detection of mobile entangled pairs of charge carriers, in spatially separated nanoelectronic leads, is crucial in quantum information. The spin of these mobile particles can be used as a quibit. Besides, the nonlocal spin-entangled electron pairs can be detected e.g. via the current-noise measurements in transport experiments.

Finally, it can be said that various aspects of the tunneling quasiparticle transport, especially the coherent one, through the double barrier junctions exhibits interesting features which can be used and applied in nanoelectronic physics. First, the single particle coherent transport in $F/S/F$ and $N/S/N$ junctions can be applied in spintronics. Second, the coherent pairwise tunneling transport of spin entangled pairs in spatially separated leads can be applied in quantum information. Third, it is worth mentioning that the spectroscopy of the subgap and spin polarized transport determines the pairing symmetry of Cooper pairs in the ferromagnet superconductors.

References

[1] Morten JP, Brataas A and Belzig W 2006 *Phys. Rev.* **B74** 214510
[2] Byers JM and Flatté M 1995 *Phys. Rev. Lett.* **74** 306
[3] Russo S, Kroug M, Klapwijk TM and Morpurgo AF 2005 *Phys. Rev. Lett* **95** 027002
[4] Beckmann D and v Löhneysen H 2006 *AIP Conference Proceedings 850* 875
[5] Boževič JM and Radovič Z 2002 *Phys. Rev.* **B66** 134524
[6] Dong ZC, Shen R, Zheng ZM, Xing DY and Wang ZD 2003 *Phys. Rev.* **B67** 134515
[7] Yamashita T, Imamura H, Takahashi S and Maekawa S 2003 *Phys. Rev.* **B67** 094515
[8] Yamashita T, Takahashi S and Maekawa S 2003 *Phys. Rev.* **B68** 174504
[9] Blonder GE, Tinkham M and Klapwijk TM 1982 *Phys. Rev.* **B25** 4515
[10] Kowalewski L and Wojciechowski RJ 2009 *Acta Phys. Pol.* **A115** 257
[11] Yeng Zeng Z, Zhou L, Hong J and Li B 2006 *Phys. Rev.* **74** 085312
[12] Beenakker CWJ, Emary C, Kindermann M and van Velsen LJ 2003 *Phys. Rev. Lett.* **80** 147901
[13] Beenakker CWJ 1997 *Rev. Mod. Phys.* **69** 731
[14] Benjamin C 2006 *Phys. Rev.* **B74** 1805503-1

Quadratic and cubic monocrystalline and polycrystalline materials: their stability and mechanical properties

C Jasiukiewicz, T Paszkiewicz, S Wolski

Faculty of Mathematics and Applied Physics, Rzeszów University of Technology, ul. W. Pola 1, PL-35-959 Rzeszów, Poland

E-mail: czjas@prz.edu.pl, tapasz@prz.edu.pl, wolan@prz.edu.pl

Abstract. We consider quadratic and cubic monocrystalline and polycrystalline materials. Expressions for Young's and shear moduli and Poisson's ratio are expressed in terms of eigenvalues of the stiffness tensor. Such a form is well suited for studying properties of these mechanical characteristics on sides of the stability triangles. For crystalline high-symmetry directions lines of vanishing Poisson's ratio are found. These lines demarcate regions of the stability triangle into areas of various auxeticity properties. The simplest model of polycrystalline 2D and 3D cubic materials is considered. In polycrystalline phases the region of complete auxetics is larger than for monocrystalline materials.

1. Introduction

In our recent papers we considered mechanical characteristics of 3D crystalline structures of high and middle symmetry [1] and of all 2D symmetry systems [2]. In particular we derived explicit expressions for Young's E and shear G moduli as well as for Poisson's ratio ν depending on directional cosines of angles between directions $\mathbf{n}$ of the load and the direction $\mathbf{m}$ of the lateral strain, respectively, and directions of the crystalline symmetry axes.

Our particular attention was paid to 3D and 2D cubic (i.e. cubic and quadratic) structures [3], [4]. All initially unstrained quadratic and cubic materials are characterized by three parameters s_1, s_2 and s_3 belonging to a half-infinite ($s_1 > 0$) prisms with (stability) triangles (ST) in the base lying in the (s_2,s_3)-plane.

Among mechanical characteristics, Poisson's ratio ν is particularly interesting because the phenomenon of a negative ν, i.e. solids expanding transversely to applied tensile stress, is counterintuitive. Elastic materials with negative ν are termed auxetics. Materials with negative ν for all pairs of vectors $\mathbf{n}$, $\mathbf{m}$ are *complete* auxetics. Materials are auxetics if there exist pairs of $\mathbf{n}$, $\mathbf{m}$ for which $\nu < 0$. Materials are non-auxetics if $\nu > 0$ for all pairs of vectors $\mathbf{n}$, $\mathbf{m}$. In papers [3], [4] we established regions of ST in which crystalline cubic and quadratic materials are completely auxetic, auxetic and non-auxetic. One may expect that complete auxetics are particularly promising from the standpoint of technology.

Many technically important materials are polycrystalline. In view of potential applications of crystalline auxetic materials, the question how polycrystallinity influences the auxetic properties deserves some attention.

The elastic properties of polycrystalline materials depend on the single-crystal stiffnesses (elastic constants) of the crystallites which build up the polycrystal and on the manner in which crystallites are connected. In most cases exact orientations, shapes and connections are not known. These gaps in knowledge can be overcome with the help of orientation and grain shape distribution functions. Generally, one has to resort to more or less realistic assumptions (cf. [5]). We assume that the grains have the same shape and volume and that their orientations are completely random.

2. The stability conditions for initially unstressed crystalline quadratic and cubic media

Mechanical properties of crystalline materials depend on components of the stiffness tensor **C** or on the components of the compliance tensor **S**. These tensors are mutually inverse, i.e.

$$\mathbf{CS} = \mathbf{SC} = \mathbf{I}_4, \tag{1}$$

where $\left(\mathbf{I}^{(4)}_{ij,kl}\right) = \left(\delta_{ik}\delta_{jl} + \delta_{ij}\delta_{jk}\right)/2$. The product of two fourth rank tensors A and B has components $(\mathbf{AB})_{ijkl} = \sum_{r,s=1}^{3} A_{ij,rs}B_{rs,kl}$.

In this paper we consider only 2D and 3D cubic crystalline or polycrystalline materials. For mechanically stable initially unstrained materials both **C** and **S** tensors are positive, therefore their eigenvalues c_I, s_I $(I = J, L, M)$ are positive too ([1], [2], [6])

$$c_I > 0, s_I > 0 \; (I = J, M, L). \tag{2}$$

Because of Eq. (1) $s_I = c_I^{-1}$ vanishing eigenvalues c_I $(I = J, L, M)$ signal instabilities related to phase transitions [7].

In the case of quadratic materials [2]

$$c_J = C_{11} + C_{12}, \; c_L = C_{11} - C_{12}, c_M = 2C_{66}. \tag{3}$$

In the case of cubic materials [1], [6]

$$c_J = C_{11} + 2C_{12}, \; c_L = C_{11} - C_{12}, c_M = 2C_{66}. \tag{4}$$

In Eqs. (3) and (4) we used the familiar Voigt's notation [8]: $C_{11} = C_{11,11}$, $C_{12} = C_{11,22}$ and $C_{66} = C_{12,12}$.

For cubic materials a more reasonable choice of independent parameters was proposed by Every [9] and for quadratic materials in our paper [4]

$$s_1 = C_{11} + (d-1)C_{66}, \; s_2 = \frac{C_{11} - C_{66}}{s_1}, \quad s_3 = \frac{C_{11} - C_{12} - 2C_{66}}{s_1}, \tag{5}$$

where $d = 2$ for 2D, $d = 3$ for 3D materials.

We rewrite expressions (3) and (4) using Every's parameters s_1, s_2 and s_3. In the case of quadratic materials we get

$$c_J = s_1\left(2s_2 - s_3\right), \; c_L = s_1\left(1 - s_2 + s_3\right), \; s_M = 2s_1\left(1 - s_2\right), \tag{6}$$

whereas in the case of cubic materials

$$c_J = \frac{s_1}{3}\left(10s_2 - 6s_3 - 1\right), \; c_L = 2\frac{s_1}{3}\left(1 - s_2\right), \; s_M = 2\frac{s_1}{3}\left(3s_3 - 2s_2 + 2\right). \tag{7}$$

For both cubic and quadratic materials the first of the inequalities (2) define stability regions in the form of semi-infinite prisms ($s_1 > 0$) with two distinct triangles (STs) at the base in s_2, s_3-plane.

$$s_1 > 0,\ s_2 < 1, \tag{8}$$

$$\begin{aligned} 2D &: (1 - s_2 + s_3) > 0,\ (s_2 - s_3/2) > 0, \\ 3D &: (1 + 2s_2) > |4s_2 - 3s_3 - 1|,\ (10s_2 - 6s_3) > 1. \end{aligned} \tag{9}$$

Both stability triangles are lying in the (s_2, s_3)-plane (Fig. 1). The sides of the stability triangles

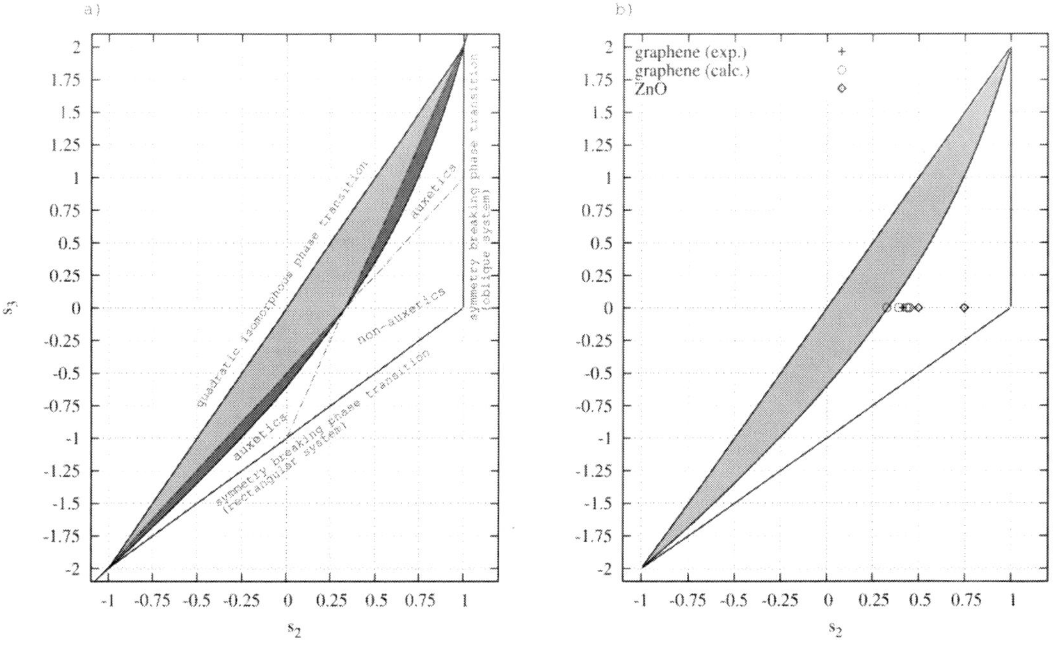

Figure 1. a) Division of the stability triangle for quadratic materials into areas of various auxetic properties. b) The shaded area is the region of complete polycrystaline auxetics. Locations of results for graphene [12], [13] and graphene-like ZnO [14] are indicated.

(STs) are defined by lines

$$\begin{aligned} 2D: &\quad s_3^{(2D,1)} = 2s_2,\ s_3^{(2D,2)} = s_2 - 1,\ s_2^{(2D,3)} = 1, \\ 3D: &\quad s_3^{(3D,1)} = 5s_2/3 - 1/6,\ s_3^{(3D,2)} = 2s_2/3 - 2/3,\ s_2^{(3d,2)} = 1. \end{aligned} \tag{10}$$

A point of the stability triangle together with the value of s_1 represents an existing or potentially existing crystalline material. For acoustic and elastic characteristics the parameter s_1 is a scaling parameter (cf. [7], [10]). In Figs.1 and 2 we show changes of the division of STs in two regions into various auxetic properties. Points of area with lighter shade represent complete monocrystaline auxetics, whereas in the area with darker shade lie complete polycrystaline auxetics.

For symmetry systems other than cubic and isotropic, as well as for the oblique symmetry system, elastic properties are defined in spaces of dimensions higher than 3. In the case of rectangular symmetry system the space of parameters can be effectively reduced to a 3D space (cf. Appendix). Monocrystalline stable *isotropic* materials are located on the interval of the lines defined by equation $s_3 = 0$ belonging to the appropriate triangle.

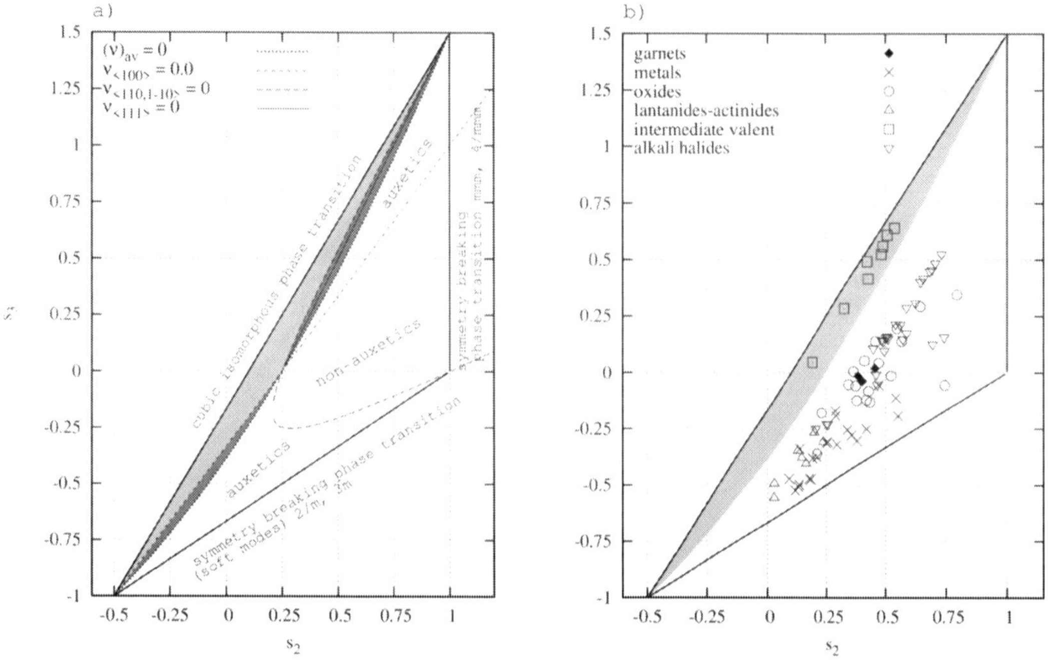

Figure 2. a) Division of the stability triangle for cubic materials into areas of various auxetic properties. b) The shaded area is the region of complete polycrystaline auxetics. Locations of several classes of elastic materials are indicated (cf. [17]).

3. Mechanical characteristics of quadratic and cubic materials

Mechanical properties of an elastic body are characterized by Young's and shear moduli and Poisson's ratio. Recent interest in crystalline materials has made the understanding these characteristics increasingly important. In papers [1] and [2] we derived expressions for them which do not depend on the choice of Cartesian coordinate system. The Young modulus $E(\mathbf{n})$ depends on the (direction) cosines that unit vector $\mathbf{n}$ of the load makes with the crystalline axes. The shear modulus G, as well as the Poisson's ratio ν), additionally depend on the direction cosines of the vector of lateral load $\mathbf{m}$.

Expressions for these characteristics can be written in terms of the eigenvalues c_J, c_L and c_M For quadratic materials we obtain

$$E_q(\mathbf{n}) = \frac{2c_J c_L c_M}{c_L c_M + c_J c_M \left(n_1^2 - n_2^2\right)^2 + 4c_J c_L n_1^2 n_2^2}, \tag{11}$$

$$G_q(\mathbf{n}, \mathbf{m}) = \frac{1}{2}\frac{c_L c_M}{c_M \left(m_1 n_1 - m_2 n_2\right)^2 + c_L \left(m_1 n_2 + m_2 n_1\right)^2}, \tag{12}$$

$$-\nu_q(\mathbf{n}, \mathbf{m}) = \frac{c_L c_M + c_J c_M \left(m_1^2 - m_2^2\right)\left(n_1^2 - n_2^2\right) + 4c_J c_L m_1 m_2 n_1 n_2}{c_L c_M + c_J c_M \left(n_1^2 - n_2^2\right)^2 + 4c_J c_L n_1^2 n_2^2}, \tag{13}$$

For cubic materials we get

$$E_c(\mathbf{n}) = \frac{3c_J c_L c_M}{\left(c_L c_M - c_J c_L + 3c_J c_M\right) + 3c_J \left(c_L - c_M\right)\sum_{i=1}^{3} n_i^4}, \tag{14}$$

$$G_c(\mathbf{n},\mathbf{m}) = \frac{c_J c_L c_M}{c_J c_M + 2c_J\,(c_L - c_m)\sum_{i=1}^{3} m_i^2 n_i^2}, \tag{15}$$

$$-\nu_c(\mathbf{n},\mathbf{m}) = \frac{c_L\,(c_M - c_J) + 3c_J\,(c_L - c_M)\sum_{i=1}^{3} m_i^2 n_i^2}{(c_L c_M - c_J c_L + 3c_J c_M) + 3c_J\,(c_L - c_M)\sum_{i=1}^{3} n_i^4}. \tag{16}$$

Note that expressions (11-16) allow one to calculate E, G and ν for a given pair of direction vectors $(\mathbf{m},\mathbf{n})$ and for chosen values of s_1, s_2 and s_3. In the case of polycrystalline materials numerical calculation provides one with analogous results. Performing calculations for a mesh of points of STs we obtained the appropriate maps [15]. These maps were calculated for $s_1 = 1$.

The direction vector of applied tension $\mathbf{n}$ and direction vector $\mathbf{m}$ in which the lateral expansion/contraction is measured are mutually perpendicular, i.e. $\mathbf{nm} = 0$. For quadratic materials

$$\begin{aligned} n_1 &= m_2 = \cos\varphi, \\ n_2 &= -m_1 = \sin\varphi, \end{aligned} \tag{17}$$

where $0 < \varphi \leq 2\pi$.

In the case of cubic materials [11]

$$\begin{aligned} n_1 &= \cos\alpha\cos\varphi\cos\theta - \sin\alpha\sin\theta, \\ n_2 &= \cos\alpha\cos\varphi\sin\theta + \sin\alpha\cos\theta, \\ n_3 &= -\cos\alpha\sin\varphi, \end{aligned} \tag{18}$$

and

$$\begin{aligned} m_1 &= \sin\varphi\cos\theta, \\ m_2 &= \sin\varphi\sin\theta, \\ m_3 &= \cos\varphi, \end{aligned} \tag{19}$$

where

$$-\pi < \alpha \leq \pi,\ 0 < \varphi \leq \pi,\ 0 < \theta \leq 2\pi. \tag{20}$$

4. Mechanical characteristics on the sides of stability triangles

For directions of high symmetry the mechanical characteristics can vanish, signaling phase transitions (cf. [7]). New phases are indicated in Fig. 1.

Consider quadratic materials. For $\mathbf{n}_1^{(q)} = \langle 11\rangle$, $\mathbf{m}_1^{(q)} = \langle 1\bar{1}\rangle$

$$E_q = \frac{2c_J c_M}{c_J + c_M},\ G_{q},= \frac{c_L}{2},\ \nu_q = \frac{c_J - c_M}{c_J + c_M}. \tag{21}$$

This means that for $E_q(\mathbf{n}_1,\mathbf{m}_1)$ vanishes on the top and on the vertical side of ST, whereas $G_q(\mathbf{n}_1,\mathbf{m}_1)$ vanishes on the bottom side of it.

For $\mathbf{n}_2 = \langle 1\,0\rangle$, $\mathbf{m}_2 = \langle 0\,1\rangle$

$$E_q = \frac{2c_J c_L}{c_J + c_L},\ G_q = \frac{c_M}{2},\ \nu_q = \frac{c_J - c_L}{c_J + c_L}. \tag{22}$$

For $\mathbf{n}_2$ and $\mathbf{m}_2$ E_q vanishes on the top and bottom side of ST, whereas G_q vanishes on the vertical side.

For $\mathbf{n}_3^{(q)} = \left\langle n_3^{(q,1)}\; n_3^{(q,2)} \right\rangle$ and $\mathbf{m}_3^{(q)} = \left\langle \bar{n}_3^{(q,2)}\; n_3^{(q,1)} \right\rangle$

$$E_q = \frac{4c_Jc_Lc_M}{2c_Lc_M + c_Jc_L + c_Jc_M}, \quad G_q = \frac{c_Lc_M}{c_L + c_M}, \quad \nu_q = \frac{c_J\left(c_L + c_M\right) - 2c_Lc_M}{c_J\left(c_L + c_M\right) + 2c_Lc_M}, \tag{23}$$

where $n_3^{(q,1)} = \cos(\pi/8)$ and $n_3^{(q,2)} = \sin(\pi/8)$. This means that $E_q(\mathbf{n}_3^{(q)})$ vanishes on all sides of ST, whereas $G_q(\mathbf{n}_3^{(q)}, \mathbf{m}_3^{(q)})$ vanishes on the bottom and vertical side of ST.

Similar results are valid for cubic materials. For $\mathbf{n}_1^{(c)} = \langle 0\,0\,1\rangle$, $\mathbf{m}_1^{(c)} = \langle 1\,1\,0\rangle$ Eqs. (14), (15) and (16) give

$$\mathbf{n}_1^{(c)}, \mathbf{m}_1^{(c)}: \quad E_c = \frac{3c_Jc_M}{2c_L + c_M}, \quad G_c = \frac{c_L}{2}, \quad \nu_c = \frac{c_J - c_M}{2c_J + c_M}, \tag{24}$$

hence $E_c(\mathbf{n}_1^{(c)}) = 0$ on the top and bottom side of ST, $G_c(\mathbf{n}_1^{(c)}, \mathbf{m}_1^{(c)}) = 0$ on the vertical side of it. For $\mathbf{n}_2^{(c)} = \langle 1\,1\,0\rangle$, $\mathbf{m}_2^{(c)} = \langle \bar{1}\,1\,0\rangle$ we obtain

$$\mathbf{n}_2^{(c)}\;\; \mathbf{m}_2^{(c)}: \quad E_c = \frac{3c_Jc_Lc_M}{c_Jc_L/2 + 3c_Jc_M/2 + c_Lc_M/2},$$
$$G_c = \frac{c_M}{2}, \quad -\nu_c = \frac{c_Jc_L/2 - 3c_Jc_M/2 + c_Lc_M}{c_Jc_L/2 + 3c_Jc_M/2 + c_Lc_M}. \tag{25}$$

For $\mathbf{n}_2^{(c)}$ and $\mathbf{m}_2^{(c)}$ Young's modulus vanishes on all sides of ST, whereas the shear modulus vanishes on the bottom side of it. Finally, for $\mathbf{n}_3^{(c)} = \langle 1\,1\,1\rangle$, $\mathbf{m}_3^{(c)} = \langle 0\,1\,\bar{1}\rangle$

$$\mathbf{n}_3^{(c)}, \mathbf{m}_3^{(c)}: \quad E_c = \frac{3c_Jc_L}{2c_J + c_L}, G_c = \frac{3c_Lc_M}{2c_L + c_M}, \quad \nu_c = \frac{c_J - c_L}{c_L - 2c_J}. \tag{26}$$

For $\mathbf{n}_3^{(c)}$ and $\mathbf{m}_3^{(c)}$ Young's modulus vanishes on the top and vertical side of ST, the shear modulus vanishes on the vertical and bottom side.

For all considered high symmetry directions Young's and shear moduli are non-negative.

5. Lines of vanishing Poisson's ratio

For a chosen direction $\mathbf{n}$ of load and lateral strain $\mathbf{m}$ the general expressions presented in Sect. 3 allow us to find the lines of vanishing ν that divide the stability triangles into regions of various auxeticity properties. These lines are particularly interesting for $\mathbf{n}$ and $\mathbf{m}$ directed along high symmetry directions.

Consider the quadratic materials. For the following pairs $(\mathbf{n}, \mathbf{m})$ the condition $\nu_q(\mathbf{m}, \mathbf{n}) = 0$ leads to equations

$$\left(\mathbf{n}_1^{(q)}, \mathbf{m}_1^{(q)}\right): \; s_3(s_2) = 2\left(s_2 - 1\right),$$
$$\left(\mathbf{n}_2^{(q)}, \mathbf{m}_2^{(q)}\right): \; s_3(s_2) = 3s_2/2 - 1/2. \tag{27}$$

For cubic materials we get

$$\left(\mathbf{n}_1^{(c)}, \mathbf{m}_1^{(c)}\right): \; s_3\left(s_2\right) = 4s_2/3 - 1,$$
$$\left(\mathbf{n}_2^{(c)}, \mathbf{m}_2^{(c)}\right): \; s_3\left(s_2\right) = \frac{7}{6}s_2 - \frac{5}{12} \pm \frac{1}{12}\sqrt{68s_2^2 + 20s_2 - 7},$$
$$\left(\mathbf{n}_3^{(c)}, \mathbf{m}_3^{(c)}\right): \; s_3\left(s_2\right) = 2s_2 - 1. \tag{28}$$

All the above lines of vanishing ν are shown in Fig. 1.

In our papers [3], [4] we introduced the division of the stability triangle into regions of complete auxeticity, auxeticity and non-auxeticity. The above regions are indicated on diagrams in Fig. 1. Inspecting them one can distinguish lines of vanishing ν studied in this section.

Mechanical characteristics of 2- and 3-d cubic polycrystalline materials
Following the assumptions made at the end of Sect. 1 to find the mechanical characteristics of polycrystalline materials one should calculate mean values of expressions defining them. Consider quadratic materials and a function F of the angle φ, then in agreement with definition (17)

$$\langle F\rangle = \frac{1}{2\pi}\int_0^{2\pi} d\varphi F(\varphi). \tag{29}$$

Calculating $\langle E_q\rangle$, $\langle G_q\rangle$ and $\langle \nu_q\rangle$ we obtained

$$\langle E_q\rangle = \frac{4c_J c_L c_M}{\sqrt{\left[2c_L c_M + c_J\left(c_L + c_M\right)\right]^2 - c_J^2\left(c_L - c_M\right)^2}}, \tag{30}$$

$$\langle G_q\rangle = \sqrt{c_L c_M}/2, \tag{31}$$

$$\langle \nu_q\rangle = 1 - \frac{4c_L c_M}{\sqrt{\left[2c_L c_M + c_J\left(c_L + c_M\right)\right]^2 - c_J^2\left(c_L - c_M\right)^2}}. \tag{32}$$

Table 1 summarizes the properties of mean values of the mechanical characteristics and eigenvalues c_I $(I = J, L, M)$ on sides of ST. These properties align with calculated maps of $\langle E_q\rangle$, $\langle G_q\rangle$ and $\langle \nu_q\rangle$ [15].

Table 1. Values of $\langle E_q\rangle$, $\langle G_q\rangle$ and $\langle \nu_q\rangle$ on the sides of ST. J – top side, M – bottom side, L – vertical side; $U = J, L, M$.

Side	c_U	$\langle E_q\rangle$	$\langle G_q\rangle$	$\langle \nu_q\rangle$
J	0	0	$\sqrt{c_M c_L}/2$	-1
L	0	0	0	1
M	0	0	0	1

In the case of quadratic polycrystalline materials the condition $\langle \nu_q\rangle = 0$ leads to equations of line

$$s_3\left(s_2\right) = \frac{3s_2^2 - 10s_2 + 3}{3s_2 - 5}, \tag{33}$$

demarcating the stability triangle to the regions of *complete* auxetics and non-auxetics. In case of cubic materials such a line is found numerically. Inspecting Fig. 2 we note that for polycrystalline 2D and 3D cubic materials the area of region of complete auxeticity is larger than for monocrystalline materials.

In the case of cubic materials and a function F of angles α, φ and θ (cf. Eqs. (18) and (19)) the mean value $\langle F\rangle$ of $F(\alpha, \varphi, \theta)$ is defined as

$$\langle F\rangle = \frac{1}{8\pi^2}\int_{-\pi}^{+\pi} d\alpha \int_0^{\pi} d\varphi \sin\varphi \int_0^{2\pi} d\theta F(\alpha, \varphi, \theta). \tag{34}$$

Since we consider the model of randomly oriented crystallites, the averaged mechanical characteristics are isotropic. This means, that if $\langle \nu\rangle > 0$, one deals with isotropic non-auxetics, whereas if $\langle \nu\rangle < 0$, a polycrystal is a complete isotropic auxetic. We conclude that in the case of considered models of polycrystals the division of the stability triangles simplifies.

6. Concluding remarks

We have shown that mechanical characteristics of elastic materials expressed in terms of eigenvalues of the stiffness tensor are well suited for the study of their general properties. The simplest model of polycrystalline 2D and 3D cubic materials indicates that some monocrystalline auxetics become complete auxetics in the polycrystalline phase.

Appendix

Elastic properties of rectangular materials are characterized by four stiffnesses C_{11}, C_{22}, C_{12}, C_{66} (cf. [2] and [16]). Initially unstrained rectangular materials are stable if

$$C_{11} > 0,\ C_{22} > 0,\ C_{66} > 0,\ C_{11}C_{22} > C_{12}^2. \tag{35}$$

Introduce four new parameters

$$s_1 = (C_{11} + C_{22})/2 + C_{66}, \tag{36}$$

$$s_2 = [(C_{11} + C_{22})/2 - C_{66}]/s_1, \tag{37}$$

$$s_3 = [(C_{11} + C_{22})/2 - C_{12} - 2C_{66}]/s_1, \tag{38}$$

$$s_4 = (C_{11} - C_{12})/2s_1. \tag{39}$$

Because of inequalities (36)-(39)

$$s_1 > 0,\quad -1 < s_2 < 1, \tag{40}$$

s_4 can be positive and negative.

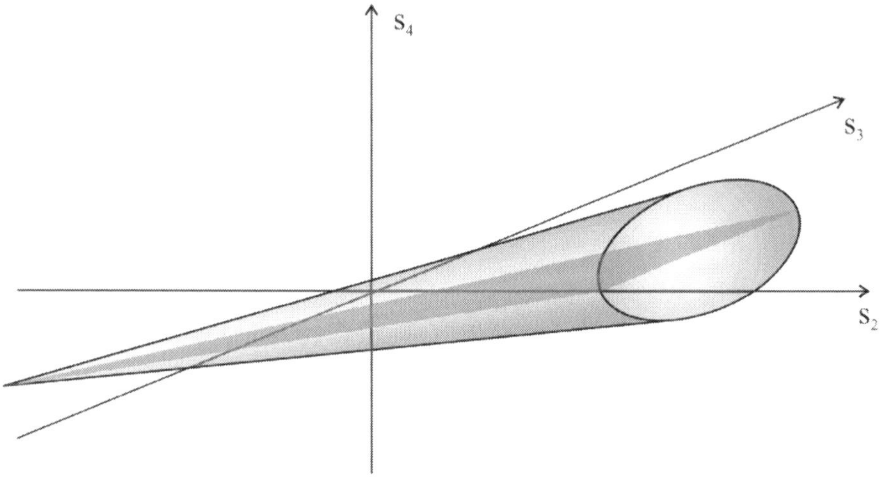

Figure 3. Rectangular materials: mechanical stability region in (s_2, s_3, s_4)-space.

Consider 3D space of parameters s_2, s_3, s_4. In planes $s_2 = \alpha$ $(-1 < \alpha < 1)$ the last of inequalities (35) takes the form

$$s_4^2 + (s_3 - d_1)^2 < d_2^2, \tag{41}$$

where $d_1(\alpha) = (3\alpha - 1)/2$, $d_2(\alpha) = (1 + \alpha)/2$. Inequality (41) defines a section of the oblique cone with the apex located at the point $(s_2 = -1, s_3 = -2, s_4 = 0)$. For $-1 < \alpha < 1$ the cross-sections of the cone perpendicular to s_2 axis are circles with radii equal to $d_2(\alpha)$ and centers $(s_2 = \alpha,\ s_3 = d_1(\alpha))$, $s_4 = 0$ (Fig. 3). For $s_4 = 0$ the cone (41) reduces to ST defined by Eq. (9).

References

[1] Paszkiewicz T, Wolski S 2007 *phys. stat. sol. (b)* **244** 978
[2] Jasiukiewicz Cz, Paszkiewicz T, Wolski S 2008 *phys. stat. sol. (b)* **244** 557
[3] Paszkiewicz T, Wolski S, 2007 *phys. stat. sol. (b)* **244**, 966
[4] Jasiukiewicz Cz, Paszkiewicz T, Wolski S 2008 *phys. stat sol. (b)* **245** 562
[5] Hirsekorn S 1990 *Textures and Microstructures* **12** 1
[6] Walpole L J 1984 *Proc. R. Soc. Lond. A* **391** 149
[7] Paszkiewicz T, Pruchnik M, Zieliński P 2001 *Eur. Phys. J. B* **24** 372
[8] Nye J F 1957 *Physical Properties of Crystals* (Oxford: Clarendon Press)
[9] Every A G 1980 *Phys. Rev.* B **122** 1746
[10] Paszkiewicz T, Pruchnik M 2001 *Eur. Phys. J. B* **24** 91
[11] Aouni N, Wheeler L 2008 *phys. stat. sol. (b)* **245** 2454
[12] L A Falkovsky, 2008 *Phys. Lett.* **A 372** 5189.
[13] K H Michel, Verberck B 2008 *Phys. Rev.* B **78** 085424.
[14] Z C Tu 2009 arXiv:091.1112v1.
[15] Jasiukiewicz Cz, Paszkiewicz T, Wolski S in preparation.
[16] Musgrave M P J 1970 *Crystal Acoustics* (San Francisco: Holden Day)
[17] Paszkiewicz T, Wolski S 2008 *J. Phys. Conf. Series* **104** 012038

Dissociation energies for small molecules from symmetry projected Hartree Fock states

Iris Theophilou[1], **S Thanos**[1] **and V N Glushkov**[2]
[1]Democritos NCSR, Institute of Materials Science,15310, Athens, Greece
[2]Departement of Physics, State University, Dniepropetrovsk, 320625, Nauchniy 13, Ukraine
E-mail: irida@ims.demokritos.gr

Abstract. The Unrestricted Hartree Fock (UHF) method gives approximate lower energy states of N-electron Hamiltonians which are not eigenstates of S^2 although the exact ones are. Starting from the Slater determinant (Sladet) of the UHF, we express the spatial part of each spin down orbital as a sum of the spatial part of a spin up orbital and a vector belonging to an orthogonal subspace. In this way, the initial UHF state is expressed as a sum of Sladets that have the spatial parts of the up and down orbitals either the same or mutually orthogonal. This expansion makes it possible to express the initial UHF state in terms of eigenstates of S^2. It is shown that at least one of these states will give lower energy than UHF. We applied our method in the dissociation range of small molecules and found that it gives more accurate results than UHF. Details of the implementation of our approach and numerical results for LiH molecule are presented.

1. Introduction

The Unrestricted Hartree-Fock approximation [1], to the energy of a N-electron system with Hamiltonian H and $S_z = M$ is defined as the minimum of the functional

$$E(\Phi'_M) = \langle \Phi'_M | H | \Phi'_M \rangle, \langle \Phi'_M | \Phi'_M \rangle = 1, \tag{1}$$

$$H = T + H_{int} + V, \tag{2}$$

$$T = -\frac{1}{2}\sum_{i=1}^{N} \nabla_i^2, \quad H_{int} = -\frac{1}{2}\sum_{i \neq j}^{N} \frac{e^2}{|r_i - r_j|}, \quad V = \sum_{i=1}^{N_u}\sum_{j=1}^{N} \frac{Z_i e}{|R_i - r_j|}. \tag{3}$$

Let $|\Phi_M\rangle$ the Sladet minimizing $E(\Phi'_M)$,

$$|\Phi_M\rangle = \left|\phi_1^{\uparrow}, \phi_2^{\uparrow}, ..., \phi_k^{\uparrow}; \phi_1^{'\downarrow}, \phi_2^{'\downarrow}, ..., \phi_l^{'\downarrow}\right\rangle, \tag{4}$$

i.e. $|\Phi_M\rangle$ is the Sladet with k spin orbitals up and l down. Thus, $|\Phi_M\rangle$ corresponds to the $N = k + l$ electron wave function with $M = (k - l)/2$. Since the spatial parts of the up and the down orbitals of $|\Phi_M\rangle$ in the UHF approximation are not always the same i.e. $\left|\left\langle \phi_i | \phi_i' \right\rangle\right| \leq 1$, $|\Phi_M\rangle$ is not in general an eigenstate of S^2[2], [3], although the exact eigenstates of the Hamiltonian of $N-$electron systems, are eigenstates of S^2.

The Spin Projected Extended Hartree Fock (EHF) is a method introduced by Löwdin [4, 5, 6, 7, 8], which consists of the variational optimization of a spin projected UHF Sladet. i.e. a determinant projected to an eigenstate of S^2. A lot of theoretical work on EHF was done [9, 10, 11, 12, 13, 14], especially in the early days of quantum many body problem theory, but the method was applied mainly to small systems [15, 16] due to the fact that "spin projection" before variation leads to rather complex minimizing equations. In our method the expansion of a Sladet in eigenstates of S^2 is done by using the minimizing UHF determinant. Orbitals could also be derived by the Optimized Effective Potential approximation [17, 18, 19, 20], where the space of the minimizing states is restricted to those obtained by an effective potential. We tested our method calculating the dissociation energy of $|\Psi_0^0\rangle$ state of LiH , but it can also be applied to large molecules irrespective of the total spin S, using the already existing UHF orbitals.

After expanding the minimizing $|\Phi_M\rangle$ in terms of eigenstates of S^2 :

$$|\Phi_M\rangle = \sum_{S=M}^{N/2} C_M^S \left|\Psi_M^S\right\rangle \tag{5}$$

we get

$$E(\Phi_M) = \langle\Phi_M| H |\Phi_M\rangle = \sum_{S=M}^{N/2} \left|C_M^S\right|^2 \left\langle\Psi_M^S\right| H \left|\Psi_M^S\right\rangle \geq \sum_{S=M}^{N/2} \left|C_M^S\right|^2 \left\langle\Psi_M^{S'}\right| H \left|\Psi_M^{S'}\right\rangle, \tag{6}$$

where

$$\left\langle\Psi_M^{S'}\right| H \left|\Psi_M^{S'}\right\rangle = \min \left\langle\Psi_M^S\right| H \left|\Psi_M^S\right\rangle. \tag{7}$$

Since

$$\sum_{S=M}^{N/2} \left|C_M^S\right|^2 = 1 \tag{8}$$

it follows that

$$E(\Phi_M) \geq \left\langle\Psi_M^{S'}\right| H \left|\Psi_M^{S'}\right\rangle. \tag{9}$$

In this way we proved that there will be at least one eigenstate of S^2 that gives lower or equal energy than that of UHF. Thus by expanding $|\Phi\rangle$ according to Eq. 5 we can find a better minimizing state for our Hamiltonian. In addition, this state has the proper transformation properties, i.e. it transforms according to the irreducible representations of the symmetry group of the Hamiltonian, which are rotations in the spin space.

According to a theorem proved in [9], a simple proof of which we give in Appendix B, if M_k and M_l are subspaces of an inner product space of dimension k and l respectively, with $k \geq l$, then an orthonormal basis $|\theta_i\rangle$ exists in M_k so that the basis vectors $|\theta_i'\rangle$ of M_l can be expressed as

$$|\theta_i'\rangle = c_i |\theta_i\rangle + c_{i+k} |\theta_{i+k}\rangle \tag{10}$$

with $\langle\theta_i|\theta_j\rangle = \delta_{ij}$ for $i, j = 1, ..., k$, $\langle\theta_{i+k}|\theta_{j+k}\rangle = \delta_{ij}$ for $i, j = 1, ..., l$ and $\langle\theta_i|\theta_{j+k}\rangle = 0$ for $i = 1, ..., k$ and $j = 1, ..., l$. In Appendix A we show how we can change our basis sets in order to obey the above relation. Thus, we can express any Sladet as

$$|\Phi_M'\rangle = \left|\theta_1^\uparrow, \theta_2^\uparrow, ..., \theta_k^\uparrow; c_1\theta_1^\downarrow + c_{1+k}\theta_{1+k}^\downarrow, c_2\theta_2^\downarrow + c_{2+k}\theta_{2+k}^\downarrow, .., c_l\theta_l^\downarrow + c_{l+k}\theta_{l+k}^\downarrow\right\rangle = \sum_{i=1}^{2^l} |\Theta_{M,i}\rangle \tag{11}$$

which is a sum of Sladets, $|\Theta_{M,i}\rangle$, where the spatial parts of the down orbitals either coincide with the up ones or are orthogonal to them. It must be noted that the choice of a different basis set changes the Sladet only by a phase factor and leaves the functional $\langle\Phi_M|H|\Phi_M\rangle$ unchanged[2]. The doubly occupied orbitals in Eq. 11 do not contribute to the total spin S^2. Thus we have to deal only with the orthogonal orbitals. It was shown[21] that when the spatial parts of the single particle wave functions are mutually orthogonal irrespective of the spin i.e., when $\left\langle\phi_i|\phi_j\right\rangle = \delta_{ij}$, then,

$$\left|\phi_1^{\uparrow},\phi_2^{\uparrow},..,\phi_k^{\uparrow};\phi_{k+1}^{\downarrow},\phi_{k+2}^{\downarrow},..,\phi_{k+l}^{\downarrow}\right\rangle = \sum_{S=(\frac{k-l}{2})}^{(\frac{k+l}{2})}\left[\frac{(2S+1)k!l!}{(S+1+\frac{(k+l)}{2})!(\frac{(k+l)}{2}-S)!}\right]\left|\Psi_M^S\right\rangle. \tag{12}$$

We can now express $|\Theta_{M,i}\rangle$ in linear combination of $\left|\Psi_M^S\right\rangle$. Then using a procedure described in a previous paper[2], which is based on the action of S^+ and S^-, we can analyze every $\left|\Psi_M^S\right\rangle$ in Sladets that will have the same spatial orbitals but different spin parts. Thus, we get:

$$|\Phi'_M\rangle = \sum_{i=1}^{2^l}|\Theta_{M,i}\rangle, \tag{13}$$

$$|\Theta_{M,i}\rangle = \sum_{S'=(\frac{k'-l'}{2})}^{(\frac{k'+l'}{2})}\left[\frac{(2S'+1)k'!l'!}{(S'+1+\frac{(k'+l')}{2})!(\frac{(k'+l')}{2}-S')!}\right]\left|\Psi_{M,i}^{S'}\right\rangle \tag{14}$$

and

$$\left|\Psi_{M,i}^{S'}\right\rangle = P_M^S|\Theta_{M,i}\rangle \tag{15}$$

where by P_M^S we denote the projection operator.

Applying the above procedure we analyzed the initial Sladet that was the minimizing state of UHF in eigenstates of S^2. Then we evaluated the energies $\left\langle\Psi_M^{S'}\right|H\left|\Psi_M^{S'}\right\rangle$ in order to find the ones that lie lower than the UHF energy $\langle\Phi_M|H|\Phi_M\rangle$.

2. Implementation to LiH.

Application of our method to the ground state $(1\sigma^2 2\sigma^2, {}^1\Sigma^+)$ of LiH was performed for internuclear distances R ranging from 2 to 8 bohrs. Molecular orbitals ϕ_i in terms of a finite set of basis functions g_p were constructed i.e. $= \sum_{p=1}^{m} c_{ip}|g_p\rangle$ $(p = 1,..,m)$. By $g_p(r)$ we denote the s-type Gaussian functions $g_p(r) = \exp\{-\zeta_p(\mathbf{r}-\mathbf{R}_p)^2\}$ where the orbital exponent ζ_p and the centers of the Gaussians $\mathbf{R}_p$ (note that these do not coincide with the position of the atomic nuclei) were determined by employing a variational procedure [20]. The dimension of the basis set used in this application is $p = 24$. Up to 4.2 bohrs internuclear distance, the spatial parts of the up orbitals almost coincide with those of the down ones and thus we have a Sladet $|\Phi_M\rangle = \left|\phi_1^{\uparrow},\phi_2^{\uparrow};\phi_1^{\downarrow},\phi_2^{\downarrow}\right\rangle$ that has $S = S_z = 0$ and therefore our method cannot give any better results since the down orbitals cannot be analyzed further in order to have the properties of Eq. 10. For internuclear distances greater than 4.2 bohrs, $\left|\left\langle\phi_1|\phi_1'\right\rangle\right| \simeq 1$ but $\left|\left\langle\phi_2|\phi_2'\right\rangle\right| < 1$. Thus we can map ϕ_2 and ϕ_2' in θ_2 and θ_2' so that $|\theta_2'\rangle = c_2|\theta_2\rangle + c_4|\theta_4\rangle$ with $\langle\theta_2|\theta_4\rangle = 0$. Note that this happens because as one increases the internuclear distance, the 3 electrons of Lithium localize closer to its nuclei and the one electron of Hydrogen acts in the same way, i.e. the molecule starts splitting in two separate atoms and therefore the 3rd electron of Lithium is not anymore very much "related" to the electron of Hydrogen and they do not make a couple.

After using the expressions of $|\theta_i\rangle$, we have $\left|\theta_1^\uparrow,\theta_2^\uparrow;\theta_1^\downarrow,\theta_2'^\downarrow\right\rangle = c_2\left|\theta_1^\uparrow,\theta_2^\uparrow;\theta_1^\downarrow,\theta_2^\downarrow\right\rangle + c_4\left|\theta_1^\uparrow,\theta_2^\uparrow;\theta_1^\downarrow,\theta_4^\downarrow\right\rangle$. The Sladet $\left|\theta_1^\uparrow,\theta_2^\uparrow;\theta_1^\downarrow,\theta_2^\downarrow\right\rangle$ has $S = S_z = 0$, and

$$\left|\theta_1^\uparrow,\theta_2^\uparrow;\theta_1^\downarrow,\theta_4^\downarrow\right\rangle = g_0^0\left|\Theta_0^0\right\rangle + g_0^1\left|\Theta_0^1\right\rangle \tag{16}$$

with $g_0^0 = g_0^1 = \frac{1}{\sqrt{2}}$, where g_M^S are Clebsh-Gordon coefficients. Thus, the above Sladet is a linear combination of two eigenfunctions of S^2 with $S = 0$ and $S = 1$. By applying S^- to $\left|\theta_1^\uparrow,\theta_2^\uparrow;\theta_1^\uparrow,\theta_4^\uparrow\right\rangle$ we get

$$\left|\Theta_0^1\right\rangle = \frac{1}{\sqrt{2}}\{\left|\theta_1^\uparrow,\theta_2^\uparrow;\theta_1^\downarrow,\theta_4^\downarrow\right\rangle + \left|\theta_1^\uparrow,\theta_2^\downarrow;\theta_1^\downarrow,\theta_4^\uparrow\right\rangle\} \tag{17}$$

From Eqs. 16,17 we have:

$$\left|\Theta_0^0\right\rangle = \frac{1}{\sqrt{2}}\{\left|\theta_1^\uparrow,\theta_2^\uparrow;\theta_1^\downarrow,\theta_4^\downarrow\right\rangle - \left|\theta_1^\uparrow,\theta_2^\downarrow;\theta_1^\downarrow,\theta_4^\uparrow\right\rangle\}. \tag{18}$$

and $\left|\theta_1^\uparrow,\theta_2^\uparrow;\theta_1^\downarrow,\theta_2'^\downarrow\right\rangle = c_2\left|\theta_1^\uparrow,\theta_2^\uparrow;\theta_1^\downarrow,\theta_2^\downarrow\right\rangle + c_4\frac{1}{\sqrt{2}}\{|\Theta_0^0\rangle + |\Theta_0^1\rangle\}$, where c_2 and c_4 can be extracted by a method presented in Appendix A. $\left|\theta_1^\uparrow,\theta_2^\uparrow;\theta_1^\downarrow,\theta_2'^\downarrow\right\rangle = C_0^0|\Psi_0^0\rangle + C_0^1|\Psi_0^1\rangle$ with

$$\left|\Psi_0^0\right\rangle = c_2\left|\theta_1^\uparrow,\theta_2^\uparrow;\theta_1^\downarrow,\theta_2^\downarrow\right\rangle + c_4\frac{1}{\sqrt{2}}\{\left|\theta_1^\uparrow,\theta_2^\uparrow;\theta_1^\downarrow,\theta_4^\downarrow\right\rangle - \left|\theta_1^\uparrow,\theta_2^\downarrow;\theta_1^\downarrow,\theta_4^\uparrow\right\rangle\} \tag{19}$$

and

$$\left|\Psi_0^1\right\rangle = c_4\frac{1}{\sqrt{2}}\{\left|\theta_1^\uparrow,\theta_2^\uparrow;\theta_1^\downarrow,\theta_4^\downarrow\right\rangle + \left|\theta_1^\uparrow,\theta_2^\downarrow;\theta_1^\downarrow,\theta_4^\uparrow\right\rangle\}. \tag{20}$$

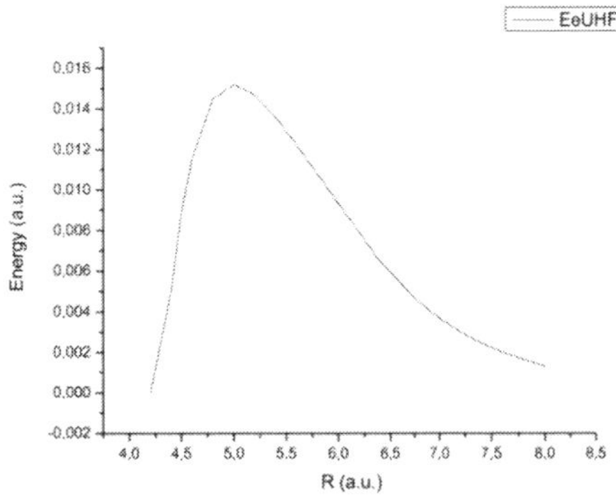

Figure 1. Electronic energy difference $\langle\Phi_0|H|\Phi_0\rangle - \langle\Psi_0^0|H|\Psi_0^0\rangle$ (in hartrees) vs. internulclear distance (in bohrs) for LiH.

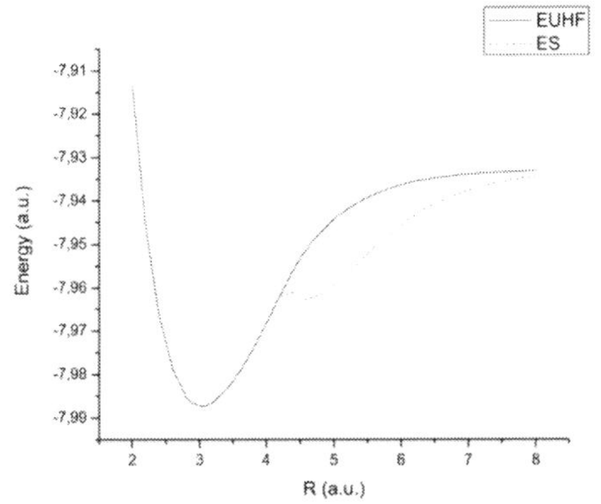

Figure 2. Total Molecular energy (in hartrees) vs. internulclear distance (in bohrs) of LiH. The solid line is the UHF energy and the dashed line the energy of $|\Psi_0^0\rangle$ state derived by our method.

We calculated $\langle\Psi_0^0|H|\Psi_0^0\rangle$,$\langle\Psi_0^1|H|\Psi_0^1\rangle$ and compared them with the UHF energy $\langle\Phi_0|H|\Phi_0\rangle$. When R is between 4.2 and 8 bohrs we found that $\langle\Psi_0^0|H|\Psi_0^0\rangle < \langle\Phi_0|H|\Phi_0\rangle < \langle\Psi_0^1|H|\Psi_0^1\rangle$. For the dissociation limit $R > 8\ bohrs$, $\langle\theta_2\ |\theta_2'\rangle \simeq 0$, the molecular energies

$\langle \Psi_0^0 | H | \Psi_0^0 \rangle$, $\langle \Psi_0^1 | H | \Psi_0^1 \rangle$ present degeneracy and converge to the energy $\langle \Phi_0 | H | \Phi_0 \rangle$ of the two separate atoms.

In Figure 1, one can see the difference between the electronic part $\langle \Phi_M | H | \Phi_M \rangle$ and $\langle \Psi_0^0 | H | \Psi_0^0 \rangle$ for $R \in [4.2, 8]$ *bohrs*. In Figure 2, the total energy of the molecule is plotted for $R \in [2, 8]$ *bohrs* extracted by $\langle \Phi_M | H | \Phi_M \rangle + Enn$ and $\langle \Psi_0^0 | H | \Psi_0^0 \rangle + Enn$ where Enn is the nuclear-nuclear interaction $Enn = \frac{Z_{Li} Z_H}{R}$. The inequality $\langle \Psi_0^1 | H | \Psi_0^1 \rangle > \langle \Phi_0 | H | \Phi_0 \rangle$ holds within this range of internuclear distances.

Appendix A. Construction of the appropriate basis having the properties of Eq. 10

Let M_k, M_l be subsaces of dimension k and l of an inner product space and $\{|\phi_i\rangle\}$, $\{|\phi_i'\rangle\}$ their orthonormal basis set respectively. We want to find a new basis set $\{|\theta_i\rangle\}$ for M_k and $\{|\theta_i'\rangle\}$ for M_l such that:

$$|\theta_i'\rangle = c_i |\theta_i\rangle + c_{i+k} |\theta_{i+k}\rangle \tag{A.1}$$

where $|\theta_i'\rangle$ belongs to M_l and $|\theta_i\rangle$ belongs to M_k and $\langle \theta_i | \theta_j \rangle = \delta_{ij}$ for $i, j = 1, ..., k$, $\langle \theta_{i+k} | \theta_{j+k} \rangle = \delta_{ij}$ for $i, j = 1, ..., l$ and $\langle \theta_i | \theta_{j+k} \rangle = 0$ for $i = 1, ..., k$ and $j = 1, ..., l$. It was shown [2], that the relation above holds when the overlap between one unit vector in M_k and one in M_l is maximum, i.e. when $\max\{|\langle \phi | \phi' \rangle|, |\phi\rangle \in M_k, |\phi'\rangle \in M_l\} = |\langle \theta | \theta' \rangle|$. As we can always choose a phase factor so that $|\langle \theta | \theta' \rangle| = \langle \theta | \theta' \rangle$, we shall omit the absolute values in the following. We have

$$|\theta\rangle = \sum_{i=1}^{k} c_i |\phi_i\rangle \tag{A.2}$$

and

$$|\theta'\rangle = \sum_{i=1}^{l} c_i' |\phi_i'\rangle \tag{A.3}$$

after using the explicit form of the coefficients c_i, c_i' we get

$$\langle \theta | \theta' \rangle = \sum_{i=1}^{k} \sum_{j=1}^{l} \langle \theta | \phi_i \rangle \left\langle \phi_i | \phi_j' \right\rangle \left\langle \phi_j' | \theta' \right\rangle \tag{A.4}$$

with $\langle \theta | \theta \rangle = \langle \theta' | \theta' \rangle = 1$. By using a vector notation

$$\langle \theta | \theta' \rangle = \langle C | A | C' \rangle \tag{A.5}$$

with $\langle C | C \rangle = \langle C' | C' \rangle = 1$, where $\langle C| = [< \theta|\phi_1 > ... < \theta|\phi_k >]$, $\langle C'| = [< \theta'|\phi_1' > ... < \theta'|\phi_l' >]$ and the matrix elements of A are $A_{ij} = \left[\left\langle \phi_i | \phi_j' \right\rangle\right]$.

After applying the maximization principle for $|C>$ and correspondingly for $|C'>$ we find

$$A |C'\rangle = \lambda |C\rangle \tag{A.6}$$

and

$$\langle C| A = \lambda' \langle C'| \tag{A.7}$$

or

$$A^\dagger |C\rangle = \lambda'^* |C'\rangle . \tag{A.8}$$

But since we have chosen $\langle \theta | \theta' \rangle$ to be real and

$$\langle \theta | \theta' \rangle = \langle C | A | C' \rangle = \lambda', \tag{A.9}$$

$$\langle\theta|\theta'\rangle = \langle\theta'|\theta\rangle = \langle C'|\, A\, |C\rangle = \lambda \tag{A.10}$$

we get that $\lambda' = \lambda'^{*} = \lambda$ and Eq. A.8 can now be written as:

$$A^{\dagger}\,|C\rangle = \lambda\,|C'\rangle\,. \tag{A.11}$$

From Eqs. A.8, A.6 we can easily have

$$AA^{\dagger}\,|C\rangle = \lambda^2\,|C\rangle \tag{A.12}$$

and

$$A^{\dagger}A\,|C'\rangle = \lambda^2\,|C'\rangle\,. \tag{A.13}$$

Setting $B = AA^{\dagger}$ we can rewrite the above Eqs. as:

$$B\,|C\rangle = \lambda^2\,|C\rangle \tag{A.14}$$

and

$$B^{\dagger}\,|C'\rangle = \lambda^2\,|C'\rangle\,. \tag{A.15}$$

Solving these eigenvalue equations we get the expansion coefficients of $|\theta\rangle$ and $|\theta'\rangle$ with respect to the "old" basis set $\{|\phi_i\rangle\}$ and $\{|\phi_i'\rangle\}$. The positive square root of the eigenvalue λ^2 corresponds to c_i of Eq. A.1. We keep just the positive root since we need $\langle\theta|\theta'\rangle = \lambda$ to be a maximum. From Eq. A.1 we also get that $c_{i+k}^2 = 1 - c_i^2 = 1 - \lambda^2$.

Appendix B. Proof of Eq. 10

Let M_k , M_l be subspaces of an inner product space of dimension k and l, with $k \succeq l$ and $\{|\phi_i\rangle\}$, $\{|\phi_i'\rangle\}$ their orthonormal basis sets respectively. We shall prove that there exists orthonormal bases for M_k and M_l such that:

$$|\theta_i'\rangle = c_i\,|\theta_i\rangle + \lambda_i'\,|u_i\rangle \tag{B.1}$$

where $|\theta_i'\rangle$ belongs to M_l and $|\theta_i\rangle$ belongs to M_k and $\langle\theta_i|\theta_j\rangle = \delta_{ij}$ for $i,j = 1,...,k$, $\langle u_i|u_j\rangle = \delta_{ij}$ for $i,j = 1,...,l$ and $\langle\theta_i|u_j\rangle = 0$ for $i = 1,...,k$ and $j = 1,...,l$. Let us construct a basis in M_k and one in M_l such that the overlap of two unit vectors in M_k and M_l respectively is maximum i.e. $\max\{|\langle\phi_i|\phi_i'\rangle|\,, |\phi_i\rangle \in M_k\,, |\phi_i'\rangle \in M_l\,\} = |\langle\theta_i|\theta_i'\rangle|$ and choose it to be real and equal to λ_i. Expanding the "new" basis sets $\{|\theta_i\rangle\}$, $\{|\theta_i'\rangle\}$ in terms of the "old" one $\{|\phi_i\rangle\}$, $\{|\phi_i'\rangle\}$ we get: $|\theta_i\rangle = \sum_{j=1}^{k} c_{ij}\,|\phi_j\rangle$ and $|\theta_i'\rangle = \sum_{j=1}^{l} c_{ij}'\,|\phi_j'\rangle$.

Thus, $|\theta_i\rangle$ and $|\theta_i'\rangle$ satisfy the equations:

$$\langle\theta_i|\theta_i'\rangle = \langle C_i|\, A\, |C_i'\rangle = \lambda_i \tag{B.2}$$

with $\langle C_i|C_i\rangle = \langle C_i'|C_i'\rangle = 1$, where $\langle C_i| = [<\theta_i|\phi_1> ... <\theta_i|\phi_k>]$, $\langle C_i'| = [<\theta_i'|\phi_1'> ... <\theta_i'|\phi_l'>]$ and $A_{mn} = [\langle\phi_m|\phi_n'\rangle]$.

But since $\langle\theta_i|\theta_i'\rangle$ is maximum with respect to $\langle C_i|$ we get that:

$$A\,|C_i'\rangle = \lambda_i\,|C_i\rangle\,. \tag{B.3}$$

Thus, we have the following equation:

$$\left\langle\theta_i|\theta_j'\right\rangle = \langle C_i|\, A\left|C_j'\right\rangle = \lambda_j\,\langle C_i|C_j\rangle = \lambda_j\,\delta_{ij}. \tag{B.4}$$

As it is obvious now every unit vector in M_k will have non zero overlap with just one unit vector in M_l and thus we can write

$$|\theta_i'\rangle = \lambda_i |\theta_i\rangle + \lambda_i' |u_i\rangle .$$

For another unit vector $|\theta_j\rangle$ in M_k which has maximum overlap with a $\left|\theta_j'\right\rangle$ in M_l we can write again:

$$\left|\theta_j'\right\rangle = \lambda_j |\theta_j\rangle + \lambda_j' |u_j\rangle .$$

$$\left\langle\theta_i'|\theta_j'\right\rangle = \lambda_i\lambda_j \langle\theta_i|\theta_j\rangle + \lambda_i\lambda_j' \langle\theta_i|u_j\rangle + \lambda_j\lambda_i' \langle u_i|\theta_j\rangle + \lambda_j'\lambda_i' \langle u_i|u_j\rangle . \tag{B.5}$$

But for $i \neq j$, $\left\langle\theta_i'|\theta_j'\right\rangle = 0$ and $\langle\theta_i|\theta_j\rangle = 0$. It also holds

$$\langle\theta_i|u_j\rangle = \frac{1}{\lambda_j'}\{\left\langle\theta_i|\theta_j'\right\rangle - \lambda_j \langle\theta_i|\theta_j\rangle\} \tag{B.6}$$

which is equal to zero since $\left\langle\theta_i|\theta_j'\right\rangle = 0$ and $\langle\theta_i|\theta_j\rangle = 0$. In the same way we can show that $\langle u_i|\theta_j\rangle = 0$. So now we deduce from Eq. B5 that $\langle u_i|u_j\rangle = 0$. Following the same procedure for every vector $|\theta_i'\rangle$ of the space we can now easily extract the Eq. B1.

Acknowledgments

The authors would like to thank Prof. A.K. Theophilou for discussions and comments on the manuscript.

References

[1] Szabo A and Ostlund N S 1989 *Modern Quantum Chemistry* (New York: Dover)
[2] Theophilou Iris, Thanos S and Theophilou A K 2007 *J. Chem. Phys.* **127** 234103
[3] Theophilou A K, Thanos S, Theophilou Iris and V N Glushkov 2008 *J. Phys.: Conf. Ser.* **104** 012020
[4] Löwdin P O 1955 *Phys. Rev.* **97** 1474
[5] Löwdin P O 1955 *Phys. Rev.* **97** 1490
[6] Löwdin P O 1955 *Phys. Rev.* **97** 1509
[7] Löwdin P O 1960 *Reviews of Modern Physics* **32** 328
[8] Löwdin P O 1962 Reviews of Modern Physics **34** 520
[9] Amos A T and Hall G G 1961 *Proc. Roy. Soc.* **A263** 483
[10] Amos A T 1962 *Mol. Phys.* **5** 91
[11] Pauncz R, J. Heer J and Löwdin P O 1962 *J. Chem. Phys.* **36** 2247
[12] Pauncz R 1962 *J. Chem. Phys.* **37** 2739
[13] Mayer I, Ladic J, Biczo G 1973 *I. J. Q. Chem.* **VII** 583
[14] Mayer I 1974 *I. J. Q. Chem.* **VIII** 893
[15] Lunell S 1968 *Phys. Rev.* **173** 85
[16] Burden F R 1972 *I. J. Q. Chem* **VI** 647
[17] Sharp R T and Horton G K 1953 *Phys. Rev.* **90** 317
[18] Talman J D and Shadwick W F 1976 *Phys. Rev.* A **14** 36
[19] Theophilou A K and Glushkov V N 2005 *I. J. Q. Chem.* **104** 538
[20] Theophilou A K and Glushkov V N 2006 *J. Chem. Phys.* **124** 034105
[21] Thanos S and Theophilou A K 2006 *J. Chem. Phys.* **124** 204109

The symmetry analysis of structural deformations related to the hydrogen implementation in borohydrides

Agnieszka Kuna, Wiesława Sikora

Faculty of Physics and Applied Computer Science, AGH – University of Science and Technology, al. Mickiewicz 30, 30-059 Cracow, Poland

E-mail: sikora@novell.ftj.agh.edu.pl

Abstract. The symmetry of magnesium borohydride Mg(BH4)2 has been analyzed. In particular, relations between space groups I41/amd(IT 141), I-4m2(IT 119) and F222(IT 22) have been derived. As a result of the analysis based on the theory of the space groups and their representations we found all transformations from the parent structure, described by the high symmetry group, to the structures with symmetries belonging to the appropriate subgroups. For a given transformation we obtained the displacements of the atoms from their initial positions, the ordering of the hydrogen atoms over the interstitial sites and the ordering of the magnetic moments. In the description the number of free parameters has been reduced to the necessary minimum.

1. Introduction

Metal hydrides are known as safe media to store hydrogen with high density. In order to use hydrogen as a fuel in the transport sector, we need not only a light, cheap, safe and reliable hydrogen storage methods, but also high hydrogen storage capacity, fast kinetics and favourable thermodynamics. A usable hydrogen storage candidate, which meets all these criteria, has yet to be discovered. In this context light metal borohydrides are attractive due to their high gravimetric and volumetric hydrogen densities compared to other complex hydrides. Magnesium borohydrides, $Mg(BH_4)_2$, is a very promising candidate for hydrogen storage materials due to its high hydrogen content -14,8 wt\% [1] and due to the low hydrogen binding enthalpy [2].

Magnesium borohydride has been known since the 1950s, but the ground-state structure of it is still under debate. The experimentally [3][4][5][6] and theoretically [7][8][9][10] proposed structures do not match. The main debated question is related to the lattice stability of the proposed structures.

Recently a project under which several computational methods have been combined to build and compute the lowest energy structure of magnesium borohydride, has been initiated by the Division Hydrogen and Energy of the Swiss Federal Laboratories for Materials Testing and Research (EMPA) [11]. It has been found that the building motif of the crystalline structure of the alkali and the alkaline-earth metal tetrahydroborates is dictated by the coordination of the metal atom. The present study is closely related to this project. In particular, we have found relations of the space groups with the structures of $Mg(BH_4)_2$ reported in [11].

2. An introduction to the symmetry analysis

The symmetry analysis based on the theory of groups and representations has been applied to the description of the magnetic ordering in crystals for the first time by E.F. Bertaut [12][13]. He obtained the symmetry-adapted ordering modes, derived from the representation analysis, by the calculation of the basis vectors of the irreducible representations. This method of analysis has been later developed by many other theoreticians, like Izyumov [14] and others.

The superstructure, as an ordering of some property of the initial crystal structure, may occur as a result of a phase transition induced by temperature, magnetic field or pressure. Each of the properties of the crystal that is localised on the atom sites may be described by a Wannier function S defined by a set of equivalent positions. It may be a scalar (the change of probability of site occupations), a polar vector (displacements of atoms from the equilibrium positions), an axial vector (the ordering of magnetic moments), a tensor (the ordering of the quadrupole moments). The presentation of this function in the commonly used frame of coordinates related to the crystallographic system takes advantage of the translation symmetry only. The other symmetry relations are lost in this description. As a consequence, the description of many crystal properties becomes unnecessarily complicated. The presentation of the model structures in terms of the basis vectors (BV) $\vec{\Psi}_{\lambda}^{k_l,\nu}$ of the irreducible representations of the initial symmetry group rather than in the frame of the crystallographic system (x,y,z), properly reflects the symmetry of the problem and provides the simplest, i.e. requiring the lowest number of independent parameters, form of the description of the structure. Thus, the Wannier function is expressed as a linear combination of the BVs:

$$\vec{S} = \sum_{l,\nu,\lambda} c_{\lambda}^{k_l,\nu} \vec{\Psi}_{\lambda}^{k_l,\nu} \tag{1}$$

where l, ν and λ - correspond, respectively, to the vectors **k**, to the irreducible representations (IRs) of the group and to the dimensions of the IRs

The symmetry group G(**k**) of the **k** vectors is a subgroup of the space group G. Consequently, the set of equivalent positions in the group G, the so-called orbit in G, may split to independent sets of equivalent positions in G(**k**). Thus, one orbit in the group G can lead to two or more orbits in the subgroup G(**k**). The symmetry analysis uncovers the relations between the old sets of the equivalent positions (in the parent group) and the new sets of the equivalent positions (in the resulting subgroup).

The form of the basis vectors and the information which of the representations are relevant for the phase transition under consideration are directly given by the theory of groups and representations. Here we derive this information using the computer programme MODY [15] based on this It is important to note that the basis vectors have the same translational properties as the Bloch functions. Therefore, the basis vectors may be defined on positions of a given orbit in the elementary cell of the crystal as well as in the elementary cell translated by a lattice vector **t**, which just corresponds to a multiplication by $e^{ik_l t}$. Let us note that the set of parameters $c_{\lambda}^{k_l,\nu}$ is restricted by several conditions. In particular, the resulting magnetic moments related to all atoms should be real. The conditions influence the set of equations satisfied by $c_{\lambda}^{k_l,\nu}$. As a consequence the number of the independent free parameters is reduced and strictly determined. Finally, the model contains clearly defined minimum number of the free parameters and supplies strictly defined relations between the quantities localised on different crystal sites and describing the considered property. Each choice of these free parameters uniquely determines one of the possible models of the new structure that may appear after the phase transition.

The representation τ_ν and the coefficients $c_{\lambda}^{k_l,\nu}$ uniquely determine the symmetry of the structure, independently on the kind of the property taken into account. The type of the phase transition and the property under consideration is encoded in the form of the basis vectors.

3. Calculations:

On the request of the EMPA group we investigated whether there are some relations between the space groups: $I4_1$/amd(141), I-4m2(119) and F222(22). We performed the analysis using the programme MODY[15]. The programme gives relations between the parent groups and their subgroups. It also calculates the basis vectors of the IRs and other characteristics of the groups which can be used for deriving the ordering modes. The results of the analysis are shown in Tables 1-7.

The subgroups of the parent group $I4_1/amd(141)$ related to the IRs τ_j, j=1,2,...,10 and ordering parameters $c_{\lambda}{}^{k}{}_{l}{}^{,\nu}$ for **k**=(0,0,0), are listed in Table 1. As we can see, the destination group I-4m2(119), relevant for the case under consideration, corresponds to the representation τ_8

Table 1. Destination group - $I4_1/amd$ (141), Rep.- representations; Param. – free parameter $c_{\lambda}^{k_l,\nu}=c^j$;

	Rep.	Param	Destination groups
Dimension 1	τ_1	c^1	$I4_1/amd$ (141)
	τ_2	c^2	$I4_122$ (98)
	τ_3	c^3	$I4_1/a$ (88)
	τ_4	c^4	$I4_1md$ (109)
	τ_5	c^5	Imma (74)
	τ_6	c^6	I-42d (122)
	τ_7	c^7	Fddd (70)
	$\boldsymbol{\tau_8}$	$\mathbf{c^8}$	**I-4m2 (119)**
Dimension 2	τ_9	[16]	P-1 (2)
			C2/m (12)
			C2/c (15)
	τ_{10}		Ima2 (46)
			Fdd2 (43)
			Cc (9)

Next we had to take as parent group – symmetry I-4m2 group. Destination groups for parent (119) are reported in Table 2. and now we can see, that group F222 (22) is a subgroup

In the next step we took I-4m2(119) as the parent group. The destination groups are collected in Table 2. As one can see, F222(22) appears as a subgroup for τ_4.

Table 2. Destination group - I-4m2 (119), Rep.- representations; Param. – free parameter $c_{\lambda}^{k_l,\nu}=c^j$;

	Rep.	Param	Destination groups
Dim 1	τ_1	c^1	I-4m2 (119)
	τ_2	c^2	I-4 (82)
	τ_3	c^3	Imm2 (44)
	$\boldsymbol{\tau_4}$	$\mathbf{c^4}$	**F222 (22)**
Dim2	τ_5	(c^5, c^6)	P1 (1)
		$(c^7, \pm ic^7)$	C2 (5)
		$(c^8, \pm c^8)$	Cm (8)

The most important results contained in Tables 1 and 2 may be summarized as:

$I4_1$/amd (141) → τ_8 →I-4m2 (119) → τ_4 → F222 (22)

The splitting relations for the subgroups I-4m2 and F222 for all types of $Mg(BH_4)_2$ atoms are shown in Table 3. The Wyckoff positions in the group $I4_1$/amd: Mg - 4a and 4b, B - 16h and H - two sets of 16h and 32i. It should be mentioned that the F-type cell is 2 times larger as the I-type cell. This leads to the doubling of the multiplicity of corresponding Wyckoff positions.

Table 3. Symmetry analysis – splitting of Wyckoff positions

atoms	Wyckoff positions		
	$I4_2$/amd (141)	**I-4m2 (119)**	**F222 (22)**
	4a	2a	4a
Mg		2c	4d
	4b	2b	4b
		2d	4c
B	16 h	$8i_1$	$16k_1$
		$8i_2$	$16k_2$
	1. 16 h	1. $8i_1$	1. $16k_1$
		1. $8i_2$	1. $16k_2$
	2. 16 h	2. $8i_1$	2. $16k_1$
H		2. $8i_2$	2. $16k_2$
	32 i	16 j_1	3. $16k_1$
			3. $16k_2$
		16 j_2	3. $16k_3$
			3. $16k_4$

All transformations of the parent structure corresponding to the group $I4_1$/amd(141) leading to the structures of lower symmetry, for **k**=(0,0,0), have been analysed. The active representations of the invariable lattice **k**=(0,0,0) for different positions and for different types of orderings (modes) are shown in the Table 4. The designations **S**, **P** and **A** of the mode types correspond, respectively, to scalar (the change of probability of site occupations), polar (displacement of atoms from the equilibrium positions in the high symmetry structure) and axial (ordering of the magnetic moments). The column corresponding to τ_8 is boldfaced because this case is most relevant for our discussion – it leads to I-4m2 destination group.

Table 4. Active representations for all types of modes - $I4_1$/amd (141)

position	type of mode	representation									
		dimension 1								dim 2	
		τ_1	τ_2	τ_3	τ_4	τ_5	τ_6	τ_7	$\boldsymbol{\tau_8}$	τ_9	τ_{10}
4a (0,0,0)	S	1	-	-	-	-	-	-	**1**	-	-
4b (0,0, ½)	P	-	-	-	1	1	-	-	-	1	1
	A	-	-	1	-	-	1	-	-	1	1
	S	1	-	-	1	1	-	-	**1**	1	1
16h (0,z,y)	P	2	1	1	2	2	1	1	**2**	3	3
	A	1	2	2	1	1	2	2	**1**	3	3
	S	1	1	1	1	1	1	1	**1**	2	2
32i (x,y,z)	P	3	3	3	3	3	3	3	**3**	6	6
	A	3	3	3	3	3	3	3	**3**	6	6

Similar results for the group I-4m2(119) are collected in Table 5. Here the result for τ_4 are most important – according to this representation the transformation to the F222 group is performed.

Table 5. Active representations for all types of modes - I-4m2 (119)

Position	type of mode	Representation				
		dimension 1				dim2
		τ_1	τ_2	τ_3	τ_4	τ_5
2a (0,0,0) 2b (0,0,1/2) 2c (0,1/2,1/4) 2d (0,1/2,3/4)	S	1	-	-	-	-
	P	-	-	1	-	1
	A	-	1	-	-	1
8i (x,0,z)	S	1	-	1	-	1
	P	2	1	2	**1**	3
	A	1	2	1	**2**	3
16j (x,y,z)	S	1	1	1	**1**	2
	P	3	3	3	**3**	6
	A	3	3	3	**3**	6

Using MODY for calculation of basis vectors, and corresponding mixing coefficients (Table 1) we can easily calculate many details concerning all types of modes. An example, corresponding to the position 16h (possible for B and H atoms) in 141 symmetry group, is shown in Table 6. The y and z quantities characterizing the 16h positions are taken from sets of data presented in [8]. ΔP_n ,$\mathbf{R}_n$, $\mathbf{M}_n$ characterize respectively change of probability of sites occupation, displacements of atoms from their equilibrium positions in the high symmetry structure and magnetic moments localized on these positions.

From our point of view the most important is the polar type displacement of the atoms.

Table 6. $I4_1/amd$ (141) **k**=(0,0,0), τ_8

Wyckoff position 16 h (0, y, z)	Type of modes	Results of calculations
h1 :(0,000 0,485 0,230)	S	For atoms: h1, h4, h5, h8, h9, h12, h13,
h2 :(0,000 0,015 0,020)	c^8=C	h16: ΔP_{hj}= +C
h3 :(0,000 0,985 0,020)		For atoms: h2, h3, h6, h7, h10, h11, h14,
h4 :(0,000 0,515 0,230)		h15: ΔP_{hj}= -C
h5 :(0,515 0,000 0,770)	P_1-version	For atoms: h1, h2, h9, h10: $\mathbf{R}_{hj}$= (0,A,0)
h6 :(0,515 0,500 0,480)	$c^{8,1}$=A	For atoms: h3, h4, h11, h12: $\mathbf{R}_{hj}$= -(0,A,0)
h7 :(0,485 0,500 0,480)		For atoms: h5, h7,h13, h15: $\mathbf{R}_{hj}$= -(A,0,0)
h8 :(0,485 0,000 0,770)		For atoms: h6, h8, h14, h16: $\mathbf{R}_{hj}$= (A,0,0)
h9 :(0,500 0,985 0,730)		
h10:(0,500 0,515 0,520)	P_2-version	For atoms: h1, h2, h3, h4, h9, h10, h11,
h11:(0,500 0,485 0,520)	$c^{8,2}$=B	h12: $\mathbf{R}_{hj}$= (0, 0,B)
h12:(0,500 0,015 0,730)		For atoms: h5, h6, h7, h8, h13, h14, h15,
h13:(0,015 0,500 0,270)		h16: $\mathbf{R}_{hj}$= - (0, 0,B)
h14:(0,015 0,000 0,980)	A	For atoms: h1, h3, h9, h11: $\mathbf{M}_{hj}$= (M,0,0)
h15:(0,985 0,000 0,980)	c^8=M	For atoms: h2, h4, h10, h12: $\mathbf{M}_{hj}$= - (M,0,0)
h16:(0,985 0,500 0,270)		For atoms: h5, h6, h13, h14: $\mathbf{M}_{hj}$= - (0,M,0)
		For atoms: h7, h8, h15, h16: $\mathbf{M}_{hj}$= (0,M,0)

Likewise, Table 7 presents results for the position 8i in 119 group, for **k** =(0,0,0) and τ_4.

Table 7. I-4m2 (119) k=(0,0,0), τ_4

Wyckoff position 8 i (x, 0, z)	Type of modes	Results of calculations
i1: (0,248 0,000 0,728)	S	No active
i2: (0,752 0,000 0,728)	P	For atoms: i1, i5: $\mathbf{R}_{ij}$= (0,C,0)
i3: (0,000 0,752 0,272)	c^4=C	For atoms: i2, i6: $\mathbf{R}_{ij}$= - (0,C,0)
i4: (0,000 0,248 0,272)		For atoms: i3, i7: $\mathbf{R}_{ij}$= (-C,0,0)
i5: (0,748 0,500 0,228)		For atoms: i4, i8: $\mathbf{R}_{ij}$= - (-C,0,0)
i6: (0,252 0,500 0,228)	A_1-version	For atoms: i1, i5: $\mathbf{M}_{ij}$= (A,0,0)
i7: (0,500 0,252 0,772)	$c^{4,1}$=A	For atoms: i2, i6: $\mathbf{M}_{ij}$= - (A,0,0)
i8: (0,500 0,748 0,772)		For atoms: i3, i7: $\mathbf{M}_{ij}$= -(0,A,0)
		For atoms: i4, i8: $\mathbf{M}_{ij}$= (0,A,0)
	A_2-version	For atoms: i1, i2, i5, i6: $\mathbf{M}_{ij}$= (0,0,B)
	$c^{4,2}$=B	For atoms: i3, i4, i7, i8: $\mathbf{M}_{ij}$= - (0,0,B)

4. Summary

Using a computer programme MODY [15] we have derived all types of modes for all Wyckoff positions and for all active representations both for $I4_1/amd$ (141) and for I-4m2 (119) space groups. Tables 1 - 7 indicate that the deformation of the structure from $I4_1/amd$ to F222 is going in two steps. In each of them the Mg positions does not change while the B and H atoms move. This leads to the displacements and deformations of the BH4 thetrahedra surrounding the Mg atoms. The comparison of the predictions of the symmetry analysis will be compared with the predictions of the structure stability calculations and experimental data in a future.

5. Acknowledgments

The work was partially supported by Polish ministry of Science and Higher Education and CMA European Network for Excellence.

References

[1] George L, Drozd V, Saxena S K, Bardaji E G and Fichtner M 2009 Structural Phase Trahsitions of $Mg(BH_4)_2$ under Pressure, *J.Phys. Chem. C* **113** 486-492
[2] Her J-H, Stephens P W, Gao Y, Soloveichik G L, Rijssenbeek J, Andrus M and Zhao J-C 2007 Structure of unsolvated magnesium borohydride $Mg(BH_4)_2$, *Acta Cryst. B* **63**: 561-568
[3] Riktor M D, Sørby M H, Chlopek K, Fichtner M, Buchter F, Züttel A and Hauback B C 2007 In situ synchrotron diffraction studies of phase transitions and thermal decomposition of Mg(BH4)2 and Ca(BH4)2, *J. Mater. Chem.* **17** 4939–4942
[4] Her J-H, Stephens P W, Gao Y, Soloveichik G L, Rijssenbeek J, Andrus M, and Zhao J-C 2007 Structure of unsolvated magnesium borohydride Mg(BH4)2, *Acta Cryst. B* **63** 561–568
[5] Cerny R, Filinchuk Y, Hagermann H, and Yvon K 2007 Magnesium borohydride: Synthesis and crystal structure, *Angew. Chem.* **119** 5867–5869
[6] George L, Drozd V, Bardaji E G and Fichtner M and Saxena S K, 2009 Structural phase transitions of Mg(BH4)2 under pressure *Journal of Physical Chemistry: C* **113** 486-492
[7] Ozolins V, Majzoub E H and Wolverton C 2008 First-principles prediction of a ground state crystal structure of magnesium borohydride. *Phys. Rev. Lett.* **100** 135501
[8] Voss J, Hummelshj J S, Lodziana Z, and Vegge T 2009 Structural stability and decomposition of Mg(BH4)2 isomorphs- an ab initio free energy study, *J. Phys.: Condens. Matter* **21**
[9] Dai B, Sholl D S and Johnson J K 2008 First-principles study of experiemental and hypothetical

Mg(BH4)2 crystal structures, *J. Phys. Chem. C* **112** 4391–4395

[10] Zhou, Xiang-Feng; Qian, Quang-Rui; Zhou, Jian; Xu, Bo; Tian, Yongjun; Wang, Hui-Tian 2009 Crystal structure and stability of magnesium borohydride from first principles. *Physical Review B* **79** Issue **21**, id. 212102

[11] Caputo R, Tekin A, Sikora W, Zuttel A 2009 First-principles determination of the ground-state structure of $Mg(BH_4)_2$ *Chemical Physics Letters* **480** 203–209

[12] Bertaut E F in: Rado G T, Suhl H (Ed.) 1963 *Treatise on Magnetism, 3*, Academic Press, New York (Chapter. 4)

[13] Bertaut E F 1971 Magnetic structure analysis and group theory *J. Phys.Colloques* **32C1** 462-470

[14] Izyumov Yu A, Syromyatnikov V N 1990 *Phase Transitions and Crystal Symmetry* (Dordrecht: Kluwer Academic Publishers) Chapter 2

[15] Sikora W, Białas F, Pytlik L 2004 MODY: a program for calculation of symmetry – adapted functions for ordered structures in crystals, J. *Appl. Cryst.* **37** 1015-1019 (Program available from: http://novell.ftj.agh.edu.pl/sikora/mody.htm)

[16] IZOTROPY (by Stokes and Hutch) http://www.physics.byu.edu/~stokesh/isotropy.html

Induced magnetic moment in graphene with a nonmagnetic impurity

M Inglot[1] and V K Dugaev[1,2]

[1]Department of Physics, Rzeszów University of Technology, Al. Powstańców Warszawy 6, 35-959 Rzeszów, Poland

[2]Department of Physics and CFIF, Instituto Superior Técnico, Technical University of Lisbon, Av. Rovisco Pais, 1049-001 Lisbon, Portugal

E-mail: ming@prz.edu.pl

Abstract. We consider a two-dimensional crystalline monolayer of carbon atoms with a single non-magnetic impurity. Using the Weyl Hamiltonian to describe electronic energy spectrum near the Dirac points, we calculate the wave function and energy of impurity states, as well as the induced magnetic moment associated with polarization of the electron system. We present a phase diagram of the localized magnetic state as a function of the chemical potential μ and perturbation related to impurity V_0.

1. Introduction

Graphene has been attracting a lot of attention recently due to its unusual electronic structure and, correspondigly, due to its unique transport properties, which differ substantially from those of traditional metals and semiconductors [1, 2, 3]. The key characteristic features of graphene are the exact two-dimensionality of its crystal structure and massless relativistic energy spectrum near the Fermi surface. In the clean limit of undoped graphene, the Fermi surface consists of a pair of nonequivalent isolated points and the electronic spectrum in the vicinity of these points should be described by Dirac or Weyl equation for „massless" particles. It is now commonly believed that these unusual properties of graphene make it an excellent candidate for various device applications in nanoelectronics and/or spintronics. [4].

Although there is great interest in magnetic properties of graphene, these are still not well understood. The main problem is that the usual doping with magnetic impurities like Mn or Fe can be ineffective because the magnetic atoms change dramatically the structure of electronic states due to bonding with carbon atoms. However, the possibility of doping with magnetic impurities, which can lead to ferromagnetism in graphene, has been discussed in several recent publications [5, 6, 7]. The main interest now is related to the possibility of magnetism induced by the presence of non-magnetic defects [8, 9, 10]. Recently, the possibility of electron correlation - induced magnetism has bean also discused in the case of graphene dots [14]. Strong electron correlation effect is an additional factor which enhances the tendency to magnetization of electron system.

In this work we consider formation of a local magnetic moment in graphene, induced by a non-magnetic impurity. Assuming exchange coupling between the localized spin and electronic states we calculate magnetic polarization of the system and analyze conditions for the appearance of a local magnetic moment.

2. Model of graphene with a single impurity

The energy spectrum of graphene is described by the following tight-binding Hamiltonian [3],

$$\mathcal{H}_{tb} = \sum_i \varepsilon_A c^\dagger_{Ai} c_{Ai} + \sum_n \varepsilon_B c^\dagger_{Bn} c_{Bn} + t \sum_{<i,n>} (c^\dagger_{Ai} c_{Bn} + c^\dagger_{Bn} c_{Ai}) \tag{1}$$

where the creation and annihilation operators correspond to electrons localized in atomic sites of two nonequivalent sublattices A and B, with the hopping between the nearest neighbors of different sublattices described by the parameter t. The indices i and n refer to sites in the sublattices A and B, respectively. The onsite energy in graphene is the same for both sublattices, $\varepsilon_A = \varepsilon_B$.

By transforming Hamiltonian (1) (with $\varepsilon_A = \varepsilon_B$) to the momentum representation, one reproduces the energy spectrum of graphene, which has two nonequivalent Dirac points in the k space, with a linear energy spectrum near these points. The Dirac points are known to be the most interesting features of the electronic structure, because the Fermi level of graphene is usually located at these points or in their close vicinity.

In our calculations we restrict ourselves to the energy spectrum near the Dirac points. The corresponding Hamiltonian is known in quantum field theory as the 2D Weyl Hamiltonian of massless relativistic particles,

$$\mathcal{H}_0 = \begin{pmatrix} 0 & v\hbar\,(k_x - ik_y) \\ v\hbar\,(k_x + ik_y) & 0 \end{pmatrix} = v\hbar\boldsymbol{\tau}\cdot\mathbf{k}, \tag{2}$$

where v is the velocity of electrons, which in graphene is $v \simeq 10^6\, m/s$, $\boldsymbol{\tau}$ are the Pauli matrices acting in the sublattice space (matrices of pseudospin), and we put the energy onset in the Dirac points, $\varepsilon_A = \varepsilon_B = 0$.

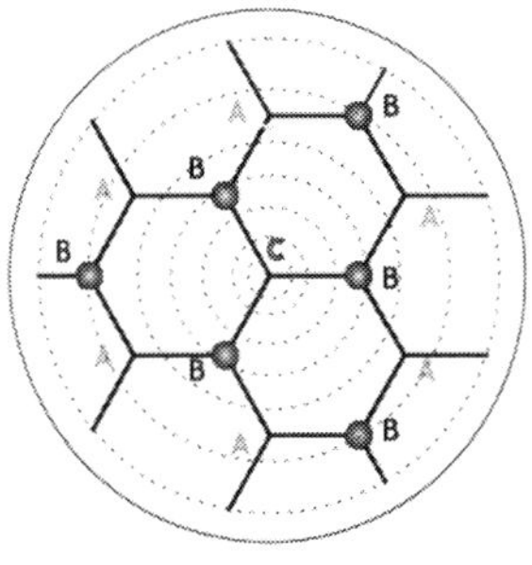

Figure 1. Schematic view of an impurity C, which substitutes a carbon atom in the sublattice A of graphene.

In the following we assume that one carbon atom in the sublattice A is replaced by an impurity atom C (located at $\mathbf{r} = \mathbf{0}$), as shown schematically in Fig. 1. This replacement can be formally included in the Hamiltonian (1) as a term with onsite energy ε_C at the site $i = 0$, so that the perturbation is $V_0 = (\varepsilon_C - \varepsilon_A)c^\dagger_{A0}c_{A0}$. After reducing to the Weyl Hamiltonian, the corresponding perturbation term in the polar coordinates (ρ, ϕ)

$$\mathcal{H}_{int} = \begin{pmatrix} V_0\,\delta(\rho) & 0 \\ 0 & 0 \end{pmatrix}. \tag{3}$$

3. Localized electronic states

Our task now is to find impurity states described by the Hamiltonian $\mathcal{H} = \mathcal{H}_0 + \mathcal{H}_{int}$, with $\mathcal{H}_0$ and $\mathcal{H}_{int}$ defined by Eqs. (2) and (3), respectively. The corresponding Schrödinger equation, $(\mathcal{H} - \varepsilon)\Psi(\rho, \varphi) = 0$, gives us the energy and eigenfunction of the localized electronic states.

Solving the Schrödinger equation for $\rho \neq 0$ we find a general solution in the form of a superposition of the Bessel functions $J_m(\xi)$ and $Y_m(\xi)$, where $\xi = \rho\varepsilon/\hbar v)$ and m is the magnetic quantum number. It turns out that only the following pseudospinor form of the solution gives us the localized states near the Dirac point:

$$\Psi_m(\varrho, \varphi) = N \begin{pmatrix} e^{im\varphi}\, Y_m(\xi) \\ i\, e^{i(m+1)\varphi}\, Y_{m+1}(\xi) \end{pmatrix}, \tag{4}$$

where N is the normalization factor.

Energy of the localized states can be found by integrating the Schrödinger equation with the wavefunction (4) over a small area near the point $\rho = 0$. Within the lattice model, this corresponds to integration over $\rho < a_0$, where a_0 is the lattice parameter. This way we account for the boundary condition at $\rho = 0$ with the δ-like perturbation localized at this point. The ground state corresponds to the choice $m = 0$ or $m = -1$. Both states give us the same energy of the localized state,

$$\varepsilon_0 = \frac{2\pi\,\hbar^2 v^2}{V_0 \ln\left(a_0|\varepsilon_0|/\hbar v\right)} \simeq \frac{2\pi\,\hbar^2 v^2}{V_0 \ln\left(2\pi a_0 \hbar v/|V_0|\right)}. \tag{5}$$

As we see from the solution (5), the energy of the localized state $\varepsilon_0 \to 0$ as the perturbation $|V_0|$ increases. For the parameters typical of graphene, the sign of logarithm in (5) is negative, and therefore the sign of the energy state ε_0 is opposite to that of perturbation potential. In other words, the attractive (negative) potential, $V_0 < 0$, gives us a state with the energy $\varepsilon_0 > 0$ and *vice versa*. This is just the opposite behavior as compared to the impurity state in traditional semiconductors – if we assume that graphene is similar to the gapless semiconductors. An important point is the above mentioned degeneracy of localized states, which is related to the pseudospinor wave function in graphene.

We have also checked the result (5) for the energy of impurity state by calculating the exact T-matrix of scattering from the impurity described by the matrix perturbation (3). Both calculation methods give exactly the same results.

4. Induced magnetic moment

The impurity state in graphene is occupied by an electron when the Fermi level is above the impurity level (5). Hence, in our consideration we assume that the location of the Fermi level is an independent parameter which is not related with the single impurity state under consideration. In reality, this assumption is well justified because position of the Fermi level in graphene is usually related with various defects or simply can be controlled by a gate voltage.

The single electron localized at the impurity has the magnetic moment μ_B. In the following consideration we assume that the Coulomb interaction does not allow for two or more electrons to occupy the same impurity state, which means that the Hubbard energy is sufficiently large.

Magnetic moment of the localized electron polarizes the electron system, inducing magnetic moment whose magnitude depends on the magnetic polarizability of the electron gas in graphene. Assuming exchange coupling of electrons in graphene with the localized electrons we calculate the induced magnetic moment. The Hamiltonian of such an exchange interaction is

$$H_{int}^m = \frac{1}{2}\, g_c \mu_z(\mathbf{r})\sigma_z, \tag{6}$$

where g_c is the coupling constant, $\mu_z(\mathbf{r})$ is the spatial distribution of magnetization associated with the wavefunction profile of the localized electron in Eq. (4),

$$\mu_z(\mathbf{r}) = g\mu_B|\Psi(\mathbf{r})|^2, \tag{7}$$

and σ_z is the Pauli matrix for electron spin. We take the quantization axis along the spin orientation of the localized electron.

The induced magnetic moment $\mathbf{M}(\mathbf{r})$ of electrons in graphene can be calculated using the quantum field theoretical method, used for calculation of magnetic polarizability [11]. Accordingly, one can present it as a loop Feynman diagram, which gives

$$M_z(\mathbf{r}) = ig_c g\mu_B \operatorname{Tr}\int d^2\mathbf{r}' \int \frac{d\varepsilon}{2\pi}\, \sigma_z\, G_0\left(\varepsilon, \mathbf{r}-\mathbf{r}'\right) \sigma_z G_0\left(\varepsilon, \mathbf{r}'-\mathbf{r}\right)\, \mu_z(\mathbf{r}'), \tag{8}$$

where $G_0(\varepsilon, \mathbf{r})$ is the Green function of electrons in graphene in the energy-coordinate representation. The calculation of the Green function using the Hamiltonian (2) gives

$$G_0(\varepsilon, \pm\mathbf{r}) = -\frac{i\varepsilon}{4\hbar^2 v^2} H_0^{(1)}\left(\frac{r\varepsilon}{\hbar v}\right) \pm \boldsymbol{\tau}\cdot\mathbf{r}\frac{\varepsilon}{4r\hbar^2 v^2} H_1^{(1)}\left(\frac{r\varepsilon}{\hbar v}\right), \tag{9}$$

where $H_{0,1}^{(1)}$ are the Hankel functions [12]. Using Eqs. (7)-(9) we find

$$M_z(\mathbf{r}) = -\frac{g_c g\mu_B \zeta}{4\pi\hbar v}\int d^2\mathbf{r}'\, \frac{\mu_z(\mathbf{r}')}{|\mathbf{r}-\mathbf{r}'|^3}, \tag{10}$$

where

$$\zeta = \int_0^\infty d\xi\, \xi^2\left[K_0^2(\xi) + K_1^2(\xi)\right], \tag{11}$$

and

$$K_m(\xi) = \frac{i^{m+1}\pi}{2} H_m^{(1)}(i\xi) \tag{12}$$

is the the modified Bessel function [12].

Using Eqs. (10)-(12) one can write the induced magnetic moment in the following form:

$$M_z(\rho) = \frac{g_c g^2\mu_B^2\pi}{8\hbar v}\int_0^\infty \frac{\rho'|\psi(\rho')|^2\mathcal{E}\left(\varsigma\right)}{(\rho+\rho')(\rho-\rho')^2}\, d\rho' \tag{13}$$

where $\mathcal{E}\left(\varsigma\right)$ is the elliptic integral of the second kind and $\varsigma = 2\sqrt{\rho\rho'}/(\rho+\rho')$. This expression was used for the numerical calculations, and the results are presented in Fig. 2.

We have also calculated the total induced magnetic moment

$$M_0(V_0, g_c) = 2\pi\int_0^\infty M_z(\rho)\, \rho d\rho \tag{14}$$

The dependence of M_0 on the strength of impurity potential $|V_0|$ is found to be rather weak, whereas the main factor which may enhance M_0 is the magnitude of the coupling parameter g_c.

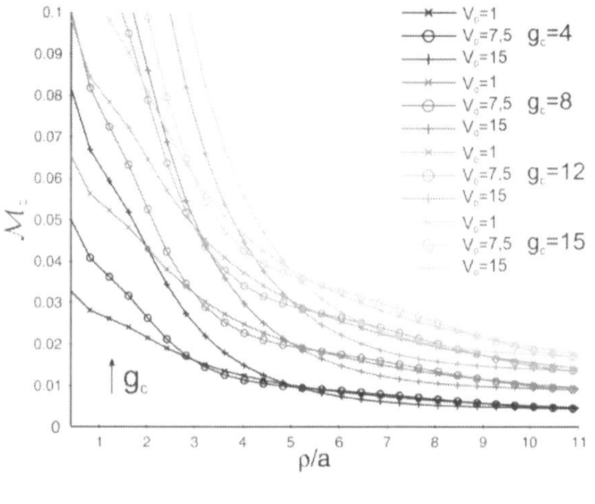

Figure 2. The density of induced magnetic moment as a function of the distance ρ for different coupling parameters.

5. Magnetic coupling of the localized spin with induced magnetic moment

The magnetic interaction of the localized spin with induced magnetic density $M_z(\rho)$ leads to the renormalization of the impurity energy level [13]. This interaction can be written as

$$E_{int} = -2\pi g_c \int_0^\infty M_z(\varrho)|\Psi(\varrho)|^2 \rho \, d\rho. \tag{15}$$

We have calculated numerically the interaction (15) and found the corresponding shift of the energy level. Our results are presented in Fig. 3.

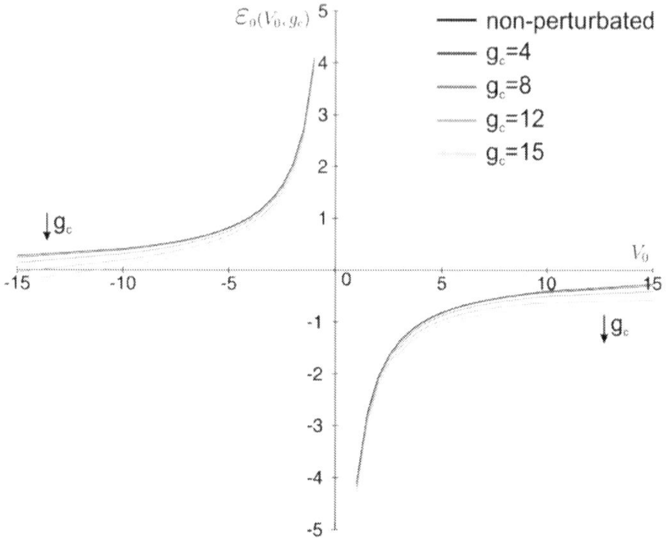

Figure 3. Energy level of the localized electron interacting with the induced magnetic moment for different values of the coupling constant g_c. The energy in the units of hopping t.

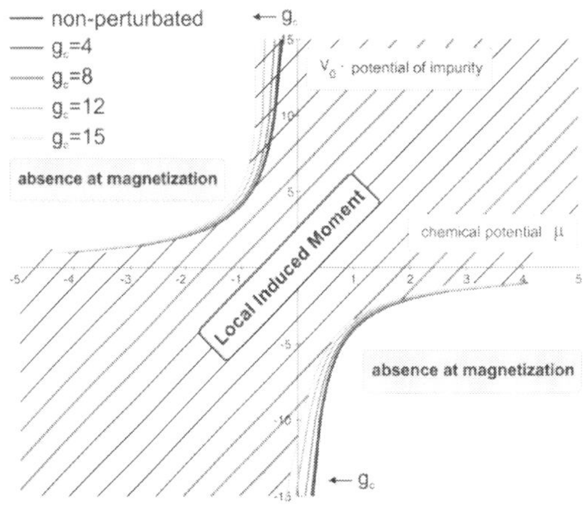

Figure 4. The phase diagram with the dashed area corresponding to the local magnetic moment induced by a nonmagnetic impurity.

6. Phase diagram

The obtained results can be presented in the form of a phase diagram, Fig. 4, which shows the regions of the parameters V_0 and μ, for which the impurity level is occupied with electron inducing the local magnetization around impurity. It should be pointed out that the level can be occupied even if the nonrenormalized position of the level is above the Fermi energy [13]. This result is related to the effect of onsite magnetic correlations.

7. Conclusion

We calculated the energy level and the wave function of a localized nonmagnetic impurity in graphene. We also calculated numerically the induced magnetization around this impurity and presented a phase diagram demonstrating the transition to local ferromagnetic state. The results of our calculations show that the nonmagnetic impurity creates an induced localized magnetic state due to the local electronic correlations. It should be emphasized that this effect is not related to free electrons because the density of free electrons in graphene can be vanishingly small. However, the magnetic polarizability of the electron system is large and long-range even without free electrons, which is the main reason for the local ferromagnetism.

Acknowledgments

This work is supported by the Polish Ministry of Science and Higher Education as a research project in years 2007 – 2010 and by FCT Grant PTDC/FIS/70843/2006 in Portugal.

References

[1] Novoselov K S, Geim A K, Morozov S V, Jiang D, Katsnelson M I, Grigorieva I V, Dubonos S V, and Firsov A A 2005 *Nature* **438** 197
[2] Geim A K and Novoselov K S 2007 *Nature Materials* **6** 183
[3] Castro Neto A H, Guinea A, Peres N M R, Novoselov K S and Geim A K 2009 *Rev Mod Phys* **81** 109
[4] Avouris P, Chen Z and Perebeinos V 2007 *Nature Nanotechnology* **2** 605
[5] Vozmediano M A H, López-Sancho M P, Stauber T and Guinea F 2005 *Phys Rev* B **72** 155121
[6] Peres N M P, Guinea F and Castro Neto A H 2005 *Phys Rev* B **72** 174406
[7] Dugaev V K, Litvinov V I and Barnaś J 2006 *Phys Rev* B **74** 224438
[8] Yazyev O V and Helm L 2007 *Phys Rev* B **75** 125408
[9] Uchoa B, Kotov V N, Peres N M R and Castro Neto A H 2008 *Phys Rev Lett* **101** 026805
[10] Yazyev O V 2008 *Phys Rev Lett* **101** 037203

[11] Abrikosov A A Gorkov L P and Dzyaloshinski I E 1963 *Methods of Quantum Field Theory in Statistical Physics* (New York: Dover)
[12] Abramowitz M and Stegun I A 1965 *Handbook of Mathematical Functions* (New York: Dover)
[13] Abrikosov A A 1973 *Sov Phys JETP* **65** 814
[14] Güçlü A D, Patosz P, Voznyy O, Karkusinski M and Hawrylak P 2009 *arXiv:0907.5431v2*

Photo-galvanic effect in asymmetric quantum wells

K Majchrowski, W Paśko and I Tralle

Institute of Physics, University of Rzeszów
Al. T. Rejtana 16A, 35-310 Rzeszów, Poland

E-mail: tralle@univ.rzeszow.pl

Abstract. In this work we studied the charge carriers behaviour in quantum structures where the symmetry with respect to space coordinates and time-reversal symmetry are broken simultaneously. As the models of such structures we considered *finite* triangular as well as *finite* semiparabolic quantum wells placed in external magnetic field. We have shown by numerical analysis that the energy spectra of charge carriers in such structures are anisotropic with respect to in-plane (transverse) motion $\epsilon_n(+k_x) \neq \epsilon_n(-k_x)$. This leads to the anisotropy of charge carriers in-plane momentum transfer which, in its turn leads to the anisotropy of photo-conductivity $\sigma(+k_x) \neq \sigma(-k_x)$ and as it follows from our calculations, the effect though not very great, could be measurable for the magnetic field of about few T.

1. Introduction

Since the time of Emmy Noether we know how important role plays the idea of symmetry in Physics. It is interesting enough that the symmtery breaking is almost equally important. Perhaps the first person who stated it clearly was Pierre Curie: *"C'est la dissymétrie qui crée le phénomène"*.The same idea was stressed by late Prof. B. Wybourne: *"Every symmetry can be considered as a statement that a certain experiment is impossible. If the experiment is possible then the symmetry must at least be broken"* [1].

Several propositions to observe a number of new effects in asymmetric quantum structures were recently discussed in scientific literature [2-4]. Thus, in Ref.[2] it was suggested that if the symmetry with respect to the space coordinates and time-reversal symmetry are broken simultaneously in the nanostructures, then some photovoltaic and magnetoelectric effects have to occur in them. As an example of such structure, in Ref. [2] the asymmetric double-quantum well (QW) in an external magnetic field was considered. In this model, the space asymmetry is introduced by the δ-barriers of different heights, while external magnetic field parallel to the layers provides time-invariance breaking. Later on in [5], the photogalvanic effect in an asymmetric undoped system of three GaAs/AlGaAs quantum wells was studied experimentally and the predictions by [1] concerning this effect were in general confirmed. In Ref. [6], the anisotropy of electron momentum transfer occurring in the *infinite* triangular quantum well in an external magnetic field due to electron-phonon interaction was considered theoretically. The models discussed in Refs [2,6], despite of their remarkable insights should be emended however, since the model of δ-barriers, as wells as the infinite triangular quantum well, are not very realistic. So, the aim of this work is twofold: first, to make some revision of the model proposed in [6] and to study more realistic models of finite triangular as well as semiparabolic QWs which could be produced by means of different modern techniques such as, for instance

MBE-technology [7]; second, to analyze the anisotropy of photo-conductivity which can occur in such QW in an external magnetic field.

2. Finite triangular quantum well in an external magnetic field

To treat the two-dimensional electron gas in an external magnetic field, we start with conventional approach based on effective mass equation of the form [8]:

$$\left[E_c + \frac{(i\hbar\nabla + e\mathbf{A})^2}{2m^*} + U(z)\right]\psi(\mathbf{r}, z) = \mathrm{E}\psi(\mathbf{r}, z), \tag{1}$$

where E_c stands for the bottom of the semiconductor conduction band, $\mathbf{A}$ -vector potential, e and m^* are the electron charge and effective mass, respectively, $\mathbf{r} = (\mathrm{x}, \mathrm{y})$ is in-plane two-dimensional vector. Let us take the gauge: $\mathbf{A} = (Bz, 0, 0)$ and assume the potential $U(z)$ to be:

$$U(z) = \begin{cases} U_0, & z \le 0 \\ eEz, & 0 < z \le d, \quad |eEd| = U_0, \end{cases} \tag{2}$$

where E is the electric field. This electric field arises, for instance, at the interface of a heterostructute,in this way making a QW.

Suppose also that the electron wave function is of the form

$$\phi_k(x) = C\varphi(k_x, z)\exp(ik_x x + ik_y y),$$

where C is the normalizing constant and in-plane wave vector $\mathbf{k} = (\mathrm{k_x}, \mathrm{k_y})$ Then, by means of (1)-(2) and the last formula, one obtains the following ordinary differential equation for the $\varphi(k_x, z)$-function:

$$-\frac{\hbar^2}{2m^*}\frac{d^2\varphi(k_x, z)}{dz^2} + \left[\frac{\hbar^2 k_x^2}{2m^*} + \frac{\hbar eBk_x z}{m^*} + \frac{(eBz)^2}{2m^*} + eEz - \epsilon\right]\varphi(k_x, z) = 0, \tag{3}$$

where $\epsilon = E_c + \epsilon(k_x) - \frac{\hbar^2 k_y^2}{2m^*}$.

Treating the infinite triangular QW and using the analytical solutions of corresponding Schrödinger equation, the author of Ref.[6] has come to the conclusion that the electron spectrum is anisotropic with respect to transverse motion, that is $\epsilon(v_x) \neq \epsilon(-v_x)$. According to [6], the physical reason for that is this: as an electron travels at the velocity v_x, the Lorentz force acts on it in the direction $\langle -z\rangle$, with the result that the maximum of the electron wave function is shifted in the direction $\langle -z\rangle$. If the electron velocity changes to $-v_x$, the Lorentz force reverses direction and the maximum of the electron wave function shifts to the direction $\langle z\rangle$. Hence in asymmetric potential $U(z) \neq U(-z)$ the electron energy $\epsilon(v_x) \neq \epsilon(-v_x)$. The claim is correct in principle and in case of *finite* triangular QW the physical explanation of such asymmetry is similar and even a bit simpler. It immediately follows from (3) that the energy spectrum of electrons in QW depends on the direction of electron movement along x-axis. Indeed, since the electrons moving along x-axis in opposite directions, are shifted by Lorentz force in opposite directions with respect to z-axis, they behave as if they were in QWs of different "effective" depths and hence, even without any calculations, one infers that $\epsilon_n(+k_x) \neq \epsilon_n(-k_x)$, where n is the corresponding quantum number.

Having in mind of what was said above concerning $\epsilon_n(+k_x) \neq \epsilon_n(-k_x)$, it is convenient, in what follows, to consider two equations of the form:

$$\frac{d^2\varphi(\zeta)}{d\zeta^2} = (\zeta - \tilde{\epsilon}_1)\varphi(\zeta), \quad \frac{d^2\varphi(\zeta)}{d\zeta^2} = (\zeta - \tilde{\epsilon}_2)\varphi(\zeta), \tag{4}$$

where ζ ,$\tilde{\epsilon}_1$ and $\tilde{\epsilon}_1$ are dimensionless quantities which are equal: $\zeta = z/z_{01}$ for the first of equations (4) and $\zeta = z/z_{02}$ for the second one; $z_{01} = \left[\hbar^2/2m^*(eE + eB\hbar k_x/m^*)\right]^{1/3}$, $z_{02} = \left[\hbar^2/2m^*(eE - eB\hbar k_x/m^*)\right]^{1/3}$; $\tilde{\epsilon}_1 = \epsilon/\epsilon_{01}, \tilde{\epsilon}_2 = \epsilon/\epsilon_{02}$, while

$$\epsilon_{01} = \left[((eE + eB\hbar k_x m^{*-1})\hbar)^2/2m^*\right]^{1/3} \quad \epsilon_{02} = \left[((eE - eB\hbar k_x m^{*-1})\hbar)^2/2m^*\right]^{1/3}.$$

In deriving Eqs. (4) we also took into account the condition $(e^2B^2d/2m^*)/((eB\hbar k_x/m^*) + eE) << 1$ which plays the role of small parameter and is valid for the reasonable values of d and B. For instance, at the electric field $E = 2.12 \times 10^7$ V/m, the width of QW $d = 15$ nm and magnetic field $B = 1$ T this parameter is 0.063 and for the values $E = 3.18 \times 10^7$ V/m, $d = 10$ nm, $B = 2$ T the above mentioned parameter is 0.085.

The existence of such parameter enables to neglect the term $\sim B^2$ in the Eqs.(4).

In Fig.1 an electron energy in the ground state of finite triangular QW is plotted versus $|k_x|$; it is clearly seen that definitely, starting from some values of $|k_x|$, $\epsilon(+k_x) \neq \epsilon(-k_x)$. In the same picture the second derivatives of energy $\epsilon(\pm k_x)$ with respect to k_x are also shown; notice, that they are represented by straight lines up to the $k_x = k_F$, where k_F is the Fermi wave vector.

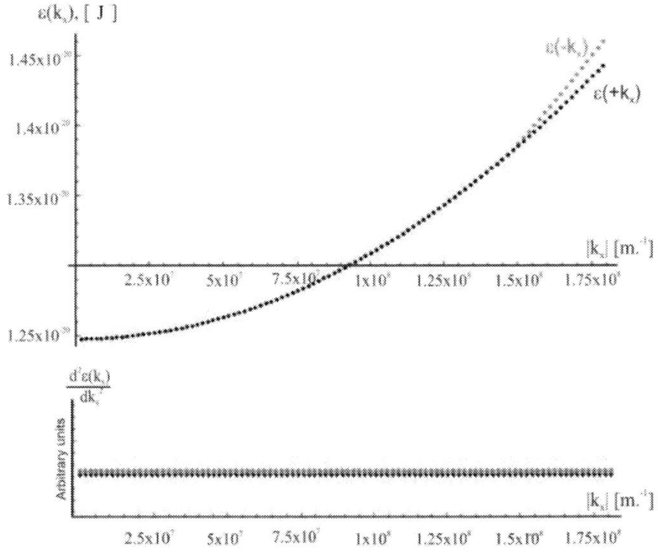

Figure 1. Electron energy in the ground state of finite triangular QW and its second derivative versus $|k_x|$

3. Anisotropy of the photoconductivity

The anisotropy of energy spectrum $\epsilon_n(+k_x) \neq \epsilon_n(-k_x)$ suggests that the anisotropy of electron momentum transfer could occur under light absorption within this QW and in an external magnetic field. Indeed, suppose that the light linearly polarized along x-axis with the wave vector $\mathbf{k}_{phot} = (0, 0, k_{z,phot})$ is incident on the semiconductor structure containing finite triangular QW (Fig.2). That the second derivatives of $\epsilon(\pm k_x)$ with respect to electron wave number k_x are the straight lines withing the wide range of k_x-values, enables to characterize the electron dynamics by means of 'renormalized' effective masses (see below), which are *different* for $\langle -x \rangle$ and $\langle +x \rangle$.

Then, despite the fact that absorbed photons supply the electrons with $+k_x$ and $-k_x$ the *same momentum*, the electrons' momenta corresponding to $\langle -x\rangle$ and $\langle +x\rangle$ are *different*, because the renormalized effective masses corresponding to $+k_x$ and $-k_x$ are different. Defining the renormalized effective masses as

$$\tilde{m}_n^*(\pm k_x) = \hbar^2 \left(\frac{d^2\epsilon_n(\pm k_x)}{dk_x^2}\right)^{-1}$$

and calculating them numerically, we can see that $\tilde{m}_n^*(+k_x) \neq \tilde{m}_n^*(-k_x)$. We term these effective masses 'renormalized' and denote them by tilde, because they depend on magnetic field B as well as an electric field E.

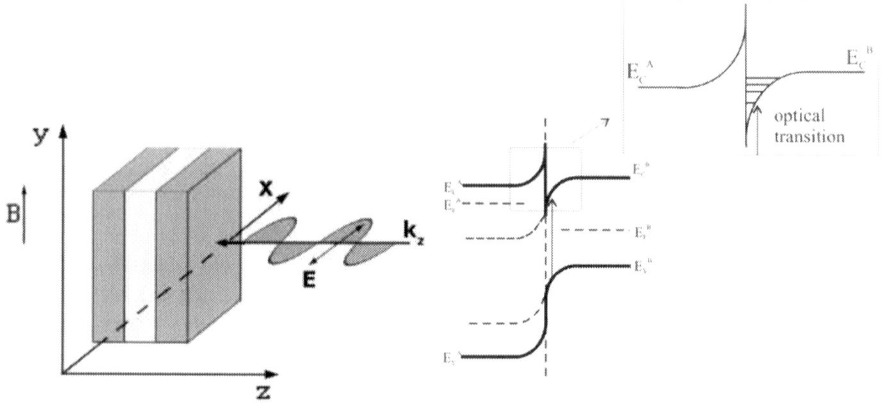

Figure 2. Schematic representation of the structure containing triangular QW and the optical transitions

Proceed now to the analysis of anisotropy of photoconductivity, induced by the polarized light beam incident on the structure, as it is shown in Fig. 2. Here two scenarios are possible. The first one corresponds to optical inter-subband transitions between the states of two triangular QWs, one for the electrons in conduction band and another one for the holes in valence band. The second scenario corresponds to optical transitions between valence band and the states of electrons in triangular QW in the conduction band (see Fig.2).

Using the standard approach (see, for instance [9]), one gets the general expression for the real part of the photoconductivity which we denote by σ:

$$\sigma = \frac{\pi e^2}{m_e^2\omega}\frac{2}{\Omega}\sum_{i,j} |\langle |j|\mathbf{e}\cdot\hat{\mathbf{p}}|i\rangle|^2 \left[f(E_i) - f(E_j)\right]\delta\left(E_j - E_i - \hbar\omega\right),$$

where, since the electromagnetic wave is polarized along z, $\mathbf{e} = (0,0,1)$ and $\mathbf{e}\cdot\hat{\mathbf{p}} = -i\hbar\partial/\partial z$. The factor of 2 in front of the summation is for spin, $f(E_i)$, $f(E_j)$ are the corresponding Fermi-factors and Ω stands for the volume of the system. Then doing in a similar way as in Ref.[9], after some manipulations which include the summation of Dirac-comb, one gets the next expression for the inter-subband transitions in triangular QWs):

$$\sigma^{\pm} = \frac{\pi e^2}{m_e^2 d\omega}\sum_{n,m} |\mathbf{e}\cdot\mathbf{p_{cn,vm}}|^2 |\langle cn|vm\rangle|^2 \left(m_e\tilde{m}^{*(\pm)}_{cn,vm}/\pi\hbar^2\right)\Theta\left[\hbar\omega - (E_g + \epsilon_{cn} - \epsilon_{vm})\right]. \quad (5)$$

Here

$$\mathbf{e}\cdot\mathbf{p_{cn,vm}}\langle cn|vm\rangle \equiv \mathbf{e}\cdot\mathbf{p_{cn,vm}}\int\varphi_{cn}^*(z)\varphi_{vm}dz \approx \langle cn\mathbf{k}|\mathbf{e}\cdot\hat{\mathbf{p}}|vm\mathbf{k}\rangle,$$

where 'c' and 'v' stand for the conduction and valence bands, n and m enumerate the bound states within the QWs and $\mathbf{k}$ stands for the in-plane (transverse) wave vector, while the matrix element $\mathbf{e}\cdot\mathbf{p_{cn,vm}}$ depends on the nature of Bloch functions and on the polarization $\mathbf{e}$; $\Theta(x)$ is the step function and the $\pm$-superscripts of σ and $\tilde{m}^*_{cn,vm}$ correspond to $+k_x$ and $-k_x$, respectively. The main difference between formula (6) and the analogous formula from Ref.[9], is that here, instead of reduced effective mass m^*_{cv} we have $\tilde{m}^{*\pm}_{cn,mv}$, which are defined as follows:

$$\left(\tilde{m}^{*+}_{cn,vm}\right)^{-1}=\left(\tilde{m}^{*+}_{cn}\right)^{-1}+\left(\tilde{m}^{*+}_{vm}\right)^{-1},\left(\tilde{m}^{*-}_{cn,vm}\right)^{-1}=\left(\tilde{m}^{*-}_{cn}\right)^{-1}+\left(\tilde{m}^{*-}_{vm}\right)^{-1},$$

for the first scenario and

$$\left(\tilde{m}^{*+}_{cn,vm}\right)^{-1}=\left(\tilde{m}^{*+}_{cn}\right)^{-1}+\left(m^*_h\right)^{-1},\left(\tilde{m}^{*-}_{cn,vm}\right)^{-1}=\left(\tilde{m}^{*-}_{cn}\right)^{-1}+\left(m^*_h\right)^{-1},$$

for the second one. Here $\tilde{m}^{\pm}_{cn}$ and $\tilde{m}^{\pm}_{vm}$ are the renormalized effective masses of charge carriers in the QWs of conduction and valence bands, respectively, while m^*_h is the usual hole effective mass.

The results of our calculations corresponding to the first scenario are presented in Fig.3 as the $\Delta\sigma_1(B)/\sigma_1(0)$-curves plotted versus B for different d, that is the different QW-widths.

Here $\Delta\sigma=\sigma^+-\sigma^-$ and $\sigma(0)$ is the usual photoconductivity at zero magnetic field. The results for optical transitions between valence band and the states of electrons in triangular QW in the conduction band are very similar to that ones shown in Fig.3.

4. Finite semiparabolic quantum well in an external magnetic field

Let us notice that the galvano-magnetic effect described above is universal in the sense that it could be observed in QW of different shapes, if only the QW-potential is asymmetric with respect to the space-coordinates inversion. Indeed, the fact that $\epsilon_n(+k_x)\neq\epsilon_n(-k_x)$ means the existence of charge current along x-axis, whose denisty in some quantum state n can be defined as:

$$j_x=-e\frac{1}{\hbar}\frac{d\epsilon_n(+k_x)}{dk_x}-e\frac{1}{\hbar}\frac{d\epsilon_n(-k_x)}{dk_x}.$$

Remember however, that after integration over all populated states with taking into account the equilibrium distribution function, the total current along x-axis has to become zero. The non-zero current is possible, only if there is some external energy source which gets the system out of equilibrium and in the example considered in Sec.3 it was the source of light.

In order to verify the conjecture concerning the universality of the effect, proceed to the semiparabolic QW. In case of semiparabolic QW, the Eq.(3) reads:

$$\left(\frac{\hat{p}_z^2}{2m^*}+\frac{(eBz+\hbar k_x)^2}{2m^*}+\alpha z^2-\varepsilon\right)\varphi(k_x,z)=0. \tag{6}$$

Here $0\leq z\leq d$, d is the QW-width while the coefficient α can be defined as follows: $\alpha=\frac{1}{2}m^*\omega_0^2=U_0/d^2$ and $\omega_0^2=2U_0/d^2m^*$, where U_0 is the QW-depth. Introduce also two additional auxiliary parameters, $\omega_c=|e|B/m^*$, $\omega_{c0}^2=\omega_c^2+\omega_0^2$; then instead of Eq.(6) we arrive at the following effective mass equation:

$$\left(\frac{\hat{p}_z^2}{2m^*}+\frac{1}{2}m^*\omega_{c0}^2\left(z+\frac{\omega_c^2}{\omega_{c0}^2}z_k\right)^2-\varepsilon_1\right)\varphi(k_x,z)=0,$$

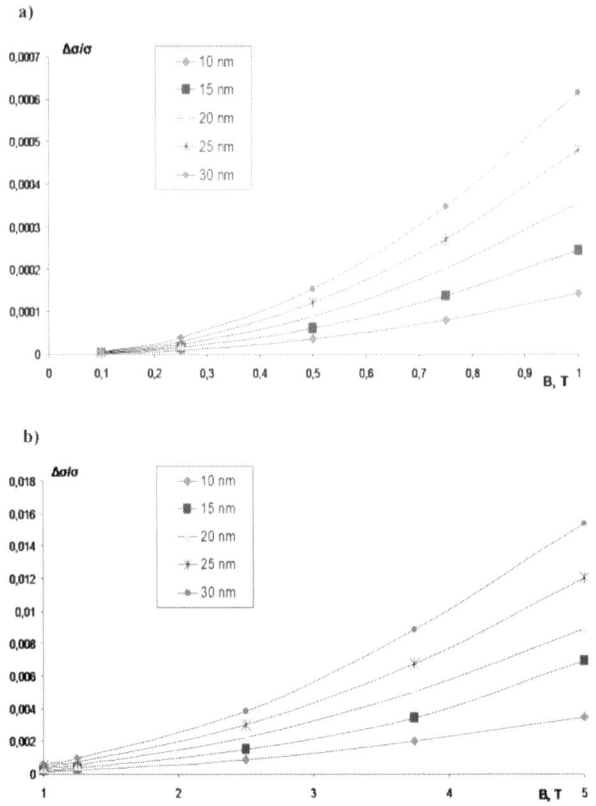

Figure 3. $\Delta\sigma/\sigma(0)$ versus magnetic field for different QW-widths

where $\varepsilon_1 = \varepsilon - \frac{1}{2}m^*(\omega_0^2\omega_c^2/\omega_{c0}^2)z_k^2$, $z_k = \hbar k_x/|e|B$. Introducing new variable $\zeta = \sqrt{m^*\omega_{c0}/\hbar}z$, one gets the next equation to be solved numerically:

$$\left(\frac{d^2}{d\zeta} - \left(\zeta + \frac{\omega_c^2}{\omega_{c0^2}}\zeta_k\right)^2 + \tilde{\varepsilon}\right)\tilde{\varphi}(\zeta, \zeta_k) = 0.$$

Here $\zeta_k = \sqrt{m^*\omega_{co}/\hbar}z_k$, $\tilde{\varepsilon} = 2\varepsilon_1/\hbar\omega_{c0}$, and $0 \le \zeta \le \zeta_0 = \sqrt{m^*\omega_{co}/\hbar}d$. The results of numerical solution of the last equation at the same boundary conditions as in case of triangular QW and some values of magnetic field B are shown in Table 2. It is worthy to mention that unlike the previous case of finite triangular QW, here we do not use any small parameter and the term $\sim B^2$ is included.

The dispersion curve for the subband $n = 1$ is shown in the Fig.4. while the results of our calculations for the anisotropy of the photoconductivity are presented in Fig.5.

It is clearly seen from Figs. 3, 5 that the anisotropy of the photoconductivity should occur in such asymmetric nanostructures in an external magnetic field and that the effect should be measurable, since $\Delta\sigma/\sigma(0)$-ratio is about 0.016 at the moderate magnetic field of about $\sim 5T$.

5. Conclusion

The results of the work can be summarized as follows. We studied the charge carriers behaviour in a quantum structure where the symmetry with respect to space coordinates and time-

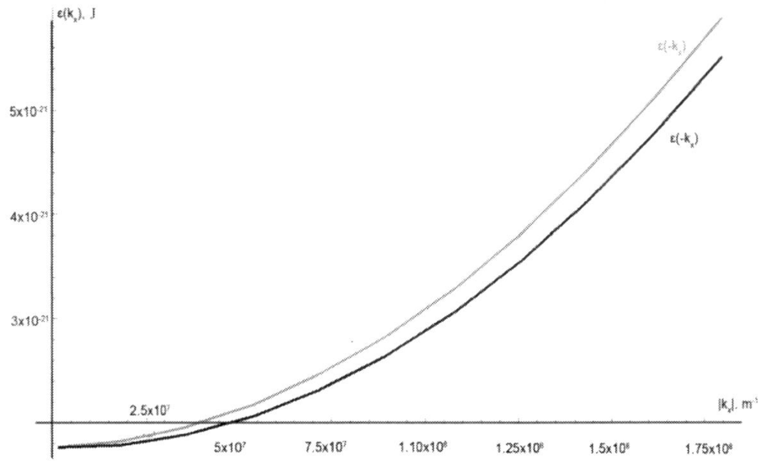

Figure 4. The ground-state electron energy versus $|k_x|$ in case of semiparabolic QW

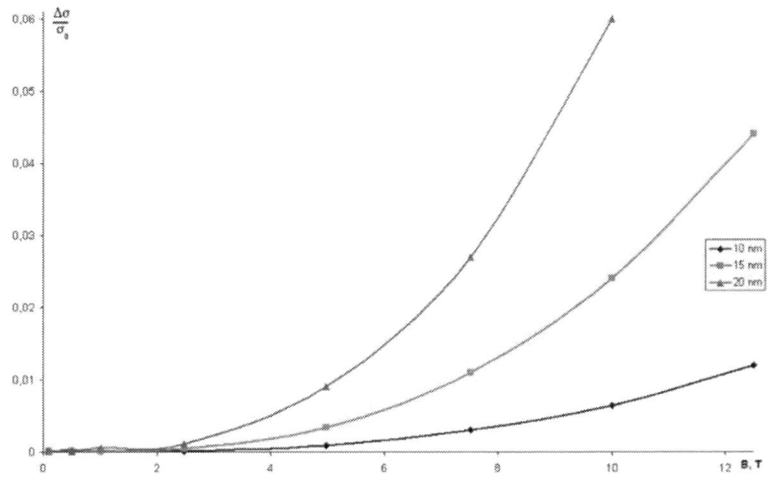

Figure 5. The anisotropy of the photoconductivity

reversal symmetry are broken simultaneously. As the models of such structure we considered the finite triangular, as well as semiparabolic quantum wells placed in external magnetic field. We have shown, by numerical analysis of corresponding Schrödinger equation, that the energy spectrum of charge carriers is anisotropic with respect to in-plane (transverse) motion $\epsilon_n(+k_x) \neq \epsilon_n(-k_x)$. This leads to the anisotropy of momentum transfer under the carriers interaction with polarized light, which can be very naturally explained by introducing the concept of charge carriers 'renormalized' effective mass. Indeed, despite the fact that absorbed photons supply to the charge carriers (electrons, for definiteness) with $+k_x$ and $-k_x$ the same momentum, their momenta corresponding to $\langle -x \rangle$ and $\langle +x \rangle$ are different. The anisotropy of momentum transfer leads to the interesting galvano-magnetic effect, anisotropy of photoconductivity $\sigma(+k_x) \neq \sigma(-k_x)$ and, as it follows from our calculations, the effect though not very great, could be measurable for the magnetic field of about few T.

Acknowledgments
The authors are gratefully acknowledged the support from polish government grant 'Advanced Technologies for the Infrared Optoelectronics' (10/G017/T02/2007).

References
[1] Wybourne B G 1998 *Physics as a Journey* (Toruń: Nicholas Copernicus University Press)
[2] Gorbatsevich A A, Kapaev V V, Kopaev Yu V 1993 *JETP Lett.* **57** 565
[3] Zhang Li, Xie Hong-Jing 2003 *Mod. Phys. Lett.* B **17** 347
[4] Zhang Li, Xie Hong-Jing 2003 *Phys. Rev.* B **68** 235315
[5] Omel'yanovskii O E, Tserbo V I and Kadushkin V I 1996 *JETP Lett.* **63** 209
[6] Kibis O 1999 *JETP* **88** 527
[7] Kowalik K, Kudelski A, Gaj J A, Wojtowicz T, Krebs O, Voisin P 2003 *Solid State Commun.* **126** 467
[8] Datta S 1997 *Electronic Transport in Mesoscopic Systems* (Cambr. Univ. Press)
[9] Davies J H 2005 *The Physics of Low-dimensional Semiconductors* (Cambr. Univ. Press)

Soliton theory of incommensurate phases in $LiCsSO_4$ crystals

A Lehmann-Szweykowska, R J Wojciechowski, M Kurzyński, M Wiesner and B Mróz

Faculty of Physics, Adam Mickiewicz University, ul. Umultowska 85
61-614 Poznan, Poland

E-mail: als@amu.edu.pl

Abstract. An extended version of the soliton theory is derived and used to explain the observed low-temperature phase transitions in $LiCsSO_4$ which are assumed to be driven by the planar orientation of SO_4 tetrahedrons. Two possible states of this orientation are expressed as S=1/2 spin states. Within the framework of the hcp Ising model, the phase transitions are described in terms of two order parameters representing spin-density wave along two mutually orthogonal directions in the basic hcp plane. The first order parameter describes the modulated phase below the temperature, T_1. The ground-state configuration of the incommmensurate phase is a soliton lattice of domain walls which get narrower as the temperature decreases till the lock-in point, T_2. The second order parameter describes the Curie second-order phase transition at T^* and represents a collinear coherence of the planar turns of the tetrahedrons SO_4 in each of the single magnetic planes, being the building units of the domain walls.

Introduction

The $LiCsSO_4$(LCS)crystal belongs to the family of the general formula A'A"BX_4. The structure and phase transitions in LCS and other members of A'A"BX_4 family have been extensively described in many review papers (see, for instance, [1] and references therein). All the compounds can be presented as a set of BX_4 tetrahedrons and A" cations which are regularly placed in the sites of a hcp structure. The A' cations form a regular simple hexagonal lattice. A sequence of structural phase transitions in $LiCsSO_4$ is determined by a competition between two types of orientations of the SO_4 tetrahedrons in the cationic environment : the first corresponds to the vertical up/down orientation of the tetrahedron apexes and the second planar one is associated with the turns of the tetrahedrons either to the right or to the left. Orientation ordering of the tetrahedrons breaks the original hexagonal symmetry $P6_3/mmc$ and drives a sequence of structural transformations when the temperature decreases [1]. Within the framework of the Kurzyński and Halawa model [2, 3],the vertical and planar orientations are described by two binary $S = 1/2$ spin variables with the exchange interactions between neighbouring tetrahedrons. The model is equivalent to a double-spin Ising Hamiltonian and provides a large variety of ground states, fitting beautifully the experimentally observed phases in $A'A''BX_4$ compounds [2, 3]. Since in our further discussion, the Ising model is explored,so we use terminology appropriate for the magnetic systems. There is a common fact, concerning all the compounds, that is that the vertical orientations are established at high temperatures (for instance, T=600K in LCS [4]). The structural transformations of the planar ordering typically

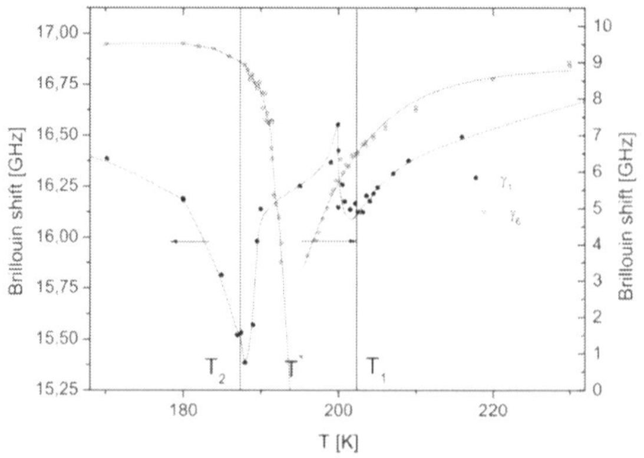

Figure 1. Temperature dependence of the Brillouin shifts.

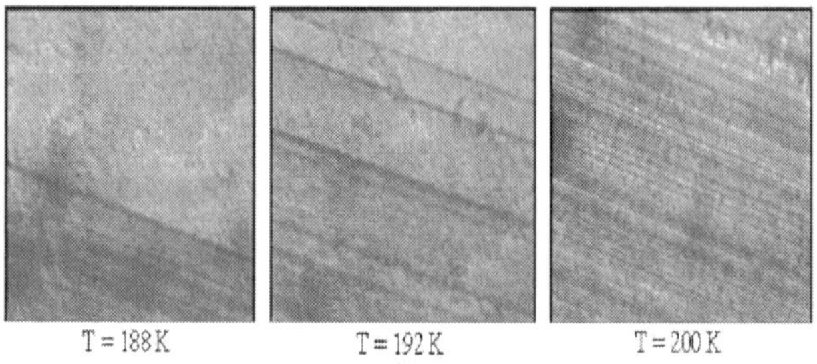

Figure 2. The temperature variation of the domain structure.

occur at much lower temperatures (near T=200 K in LCS [1]). The basic idea of a model, suggested by Luk'yanchuk et al [5], consists in a separation of the vertical ordering from the planar one, and consequently, we focus our attention on a sequence of the low-temperature transitions, which introduce and modify the planar ordering. Inelastic Brillouin light scattering reveal several low-temperature phase transitions in LCS, with a number of unusual features (for details see: [1] and references therein).

It appears that the crystal undergoes an order-disorder phase transition at $T_C = 201$ K. The other transition occurs at $T_2 = 189$ K. Apart from them, however, the Brillouin light scattering experiments, indicate also the third anomaly in the Brillouin shift at $T^* = 193$ K. It seems to be interesting that while the first two anomalies are observed for the longitudinal phonon γ_1, the third one is observed for the transverse phonon γ_6, and, moreover, the respective phonons travel in mutually perpendicular directions in the hcp basal $(\vec{a}, \vec{b})$ - plane. The temperature dependence of their respective Brillouin shifts is given in Fig.1 [1]. As a result of the ferroelastic phase transition, a ferroelastic domain structure can be observed in polarised light under the microscope [1]. The structure turns out to be strongly temperature dependent which remains consistent with a suggestion that there are incommensurate modulations associated with the planar ordering of the tetrahedrons in the temperature region $T_2 < T < T_1$. The temperature variation of the domain structure is shown in Fig.2. The main objective of the paper is to propose a two-order parameter soliton model of the low-temperature structural phase transformations in

LCS, providing an interpretation of the empirical anomalies characterising the transformations, with the special emphasis on an origin of the third anomaly at T^*.

Soliton theory of the low-temperature structural phase transitions in the $LiCsO_4$ crystals

The model of the structural phase transitions, proposed by Kurzyski and Halawa [2, 3] can be supplemented with a detailed description of the low-temperature phase transitions in $LiCsSO_4$. The two-spin (σ and τ) model can be effectively reduced to the three-dimensional one-spin Ising model with competing interactions, and spin $S = 1/2$ describing planar rotations of the BX_4 tetrahedrons. The reduction, suggested by Luk'yanchuk et al [5], is based on the idea, that, in A'A"BX_4 crystals, the vertical ordering of the tetrahedrons, connected with breaking of the initial symmetry, occurs at high temperatures. In the low-temperature region, the apexes of the tetrahedrons are "up" or "down" oriented in the alternating $(\vec{a}, \vec{c})$ planes and this antiferromagnetic orientation remains constant with decreasing temperature. In the high-temperature region, the planar orientation of the tetrahedrons, is disordered and, consequently, the planar spin variable is equal to zero. There is ample empirical evidence that, in the LCS crystals, the process of planar ordering begins at $T_1 = 201K$ and is completed at $T_2 = 189K$ [1]. Commensurate and the intermediate incommensurate modulated phases, which appear to be characteristic for the whole A$'$A$''$$BX_4$ family, are observed between T_1 and T_2. Below T_2, the planar ordering is ferromagnetic which means that the tetrahedrons in the whole system rotate in the same way. While in the paper [5], the spin-reduced Ising model is employed in an analysis of the high-temperature vertical order-disorder phase transitions, the following analysis is restricted to the sequence of the low-temperature planar phases. The empirical data clearly indicate that all the tetrahedrons display a tendency to rotate in the same way, thus the planar ordering should be ferromagnetic contrary to the antiferromagnetic vertical ordering. Considering the electrostatic nature of the interactions between the tetrahedrons, one can expect all the local correlations between their two respective orientations to be the same i.e. either positive or negative. In LCS, however, it does not seem to be true. We assume that this contradiction can mediate an occurrence of the modulated phases in the system. Therefore, although the vertical ordering is not explicitly introduced in our analysis, it plays a role of the catalyst in the low-temperature planar-oriented transformations. The three-dimensional Ising model (ANNI model), with competing interactions, has a very rich phase diagram. The Lifshitz point separates a uniformly ordered ferromagnetic phase, a modulated ordered phase and a paramagnetic disordered phase [6]. The empirical data, concerning the planar phase transitions in LCS, are interpreted in terms of the model. On the basis of the empirical data, we assume that, between T_1 and T_2, the system display "devil's staircase" behaviour i.e. the wave vector jumps between an infinite number of different rational fractions of the reciprocal lattice vector via intermediate incommensurate values. T_1 is located at the line of the phase diagram of the Elliott model, separating the paramagnetic (P) and modulated phases, so it can be identified with the mean-field Curie temperature (T_c). Consequently, $T_2 = T_{CI}$ is the lowest of the lock-in temperatures, located at the line of the phase diagram of the Elliott model, separating the modulated and ferromagnetic phases [6]. Between T_1 and T_2 the crystal displays a spatially-modulated phase with a temperature-dependent wave vector ("devil's staircase"). The wave vectors, which characterise subsequent commensurate phases, are rational fractions of the reciprocal lattice vectors. For the intermediate phases, it is not the case, however, which means that the intermediate periodicity is incommensurate. These additional phases can be stable in extremely narrow temperature intervals and probably most of them are not observable in experiments. In our analysis, the sequence of phases is presented as follows [7].

$$P - (T_c = T_1) - I - (T_{CI} = T_2)) - C \qquad (1)$$

The temperature region between T_1 and T_2 is described in terms of only one incommensurate phase (I) with temperature - dependent wave vector. A commensurate phase (C) below T_2 corresponds to the ferromagnetic ordering in our pseudo-spin model which means the full planar order of the tetrahedron turns. A soliton theory can explain stability of the commensurate phase. This theory in the form, developed by Bak and von Boehm in their study of magnetic modulated structures in CeSb [8], is extensively used in our analysis of the structural phase transformations in LCS. The solitons are the domain walls between commensurate domains with different planar rotations of the tetrahedrons. The soliton lattice, which is obtained near the incommensurate-to-commensurate phase transition, is expressed in terms of one order parameter S_{q_0}. The second one, $S_{q_0'}$ describes a different situation [7]

$$P' - (T_c' = T^*) - C' \tag{2}$$

where again P' is the paramagnetic and C' - commensurate (ferromagnetic) phase. However, it should be noted that the primed phases cannot be identified with those mentioned before. The phase transition at $T'_c = T^*$ results directly in a commensurate phase C', which, as we believe, corresponds to a ferromagnetic order of the turns of tetrahedrons in a single $(\vec{a}, \vec{c})$ plane inside the domain wall. As mentioned before, the Hamiltonian is the three-dimensional spin-1/2 Ising model

$$\hat{H} = -1/2 \sum_{i,j} J_{ij} \vec{S}_i \vec{S}_j \tag{3}$$

where $J_{ij} = J_1$ for the NN pairs in the hcp $(\vec{a}, \vec{b})$ plane, $J_{ij} = J_2$ for the NN out-plane pairs (see: Fig.1 in [5]). Because of the incompatibility of the local correlations between vertical and planar orientations of SO_4 tetrahedrons, it is assumed that J_1 can be either ferro- or anti-ferromagnetic, whereas the second exchange integral J_2 remains ferromagnetic. Let us consider, in a more detailed manner, the commensurate-to-incommensurate (CI) phase transition. We follow the procedure, worked out and described in a great detail, in the papers [8, 9, 10]] by Bak et al. The mean field transition temperature T_c between the disordered paramagnetic phase and the spatially modulated phase is determined as that which the paramagnetic susceptibility divergence condition yields

$$kT_c = J(\vec{q}_c) \tag{4}$$

where $J(\vec{q}_c)$ is a Fourier transform of the exchange integral with the wave vector $\vec{q}_C$, describing periodicity at T_c.

For the hcp structure, the Fourier transform $J(\vec{q})$ can be expressed as (cf.[5])

$$J(\vec{q}) = J_1(\cos q(n_1 - n_2) + \cos q(n_2 - n_3) + \cos q(n_3 - n_1))$$

$$+ J_2 \cos q_z (\cos qn_1 + \cos qn_2 + \cos qn_3) \tag{5}$$

where, following the notation in the paper [4], we have defined $\vec{n}_1 = (0, b/3, 0)$, $\vec{n}_2 = (a/2, -b/6, 0)$, $\vec{n}_3 = (-a/2, -b/6, 0)$ with the basic hexagonal orthogonal vectors $\vec{a}$ (or x), $\vec{b}$ (or y) and $\vec{c}$ (or z).

The point of departure for generating an effective Landau-Ginzburg functional for the Ising model is the formula for the mean-field free energy per spin

$$F = -\frac{1}{2N} \sum_{i,j} J_{ij} < S_i >< S_j > + \frac{T}{N} \sum_i \int_0^{<S_i>} d\sigma \tanh^{-1} \sigma \tag{6}$$

where N is the number of spins. In the next step of the procedure, which we follow, the free energy F is expanded in the Taylor series in terms of the Fourier transformed thermal average spin $S_{\vec{q}}$. The power series can be terminated at the fourth-order term. The expansion includes both translationally invariant terms and the umklapp terms, where the wave vectors add up to appropriate reciprocal lattice vectors. Near the commensurate-to-commensurate phase transition it is sufficient to consider only two mutually perpendicular wave vectors, $\vec{q_0}$ and $\vec{q'}_0$, with small deviations parallel ($q_{\|}$) and perpendicular ($q_{\perp}$) to their directions, respectively [8]. $\vec{q_0}$ and $\vec{q'}_0$ describe the two-dimensional (planar) commensurate ferromagnetic phase, while the deviations determine the incommensurate one. Considering the hexagonal symmetry of the system and the empirical data [1], which we are about to interpret, it is assumed that the vectors are $\vec{q_0} = (0, 2\pi, 0)$ and $\vec{q'}_0 = (2\pi, 0, 0)$, whereas the deviations from them are both along the hexagonal reciprocal axis $(0, q_b, 0)$. The Fourier transform $J(q)$ can be now expanded into a power series in terms of the fluctuations from the both commensurate wave vectors. Consequently, also the formula for the free energy can be rewritten in the form of the Taylor power series up to the terms proportional to the quadratic deviations

$$
\begin{aligned}
F = \sum_q (r + aq + cq^2) S_{q_0+q} S_{-q_0-q} + \sum_{q'} (r' + c'q'^2) S_{q_0'+q'} S_{-q_0'-q'} + \\
T \sum_{q_1 \cdots q_4} (S_{q_0+q_1} S_{q_0+q_2} S_{-q_0-q_3} S_{-q_0-q_4} + \\
S_{q_0+q_1} S_{q_0+q_2} S_{q_0+q_3} S_{q_0+q_4} + \\
S_{-q_0+q_1} S_{-q_0+q_2} S_{-q_0+q_3} S_{-q_0+q_4}) \\
\delta(q_1 + q_2 + q_3 + q_4) + \ldots + \\
T \sum_{q,q'} S_{q_0+q} S_{-q_0'-q'} + \ldots
\end{aligned}
\tag{7}
$$

where in the first term of the free energy the coefficients are $r = T - J(q_0), a = J_1 \sin(q_0/2) + J_2/3(\sin(q_0/3) + \sin(q_0/6)), c = J_2/2(-\cos(q_0/3) + \cos(q_0/6))$ in the second one we have $r' = T - J(q_0'), c' = -J_1/4 \cos(q_0'/2) - J_2/18(1 + \cos(q_0'/2))$. Our notation is deliberately analogous to that used by Bak and Boehm in Ref. [8]. The last term in the free energy expansion introduces an entropy-like coupling of two order parameters of the system. For the sake of simplicity, we restrict the expansion of the free energy to the terms which are absolutely necessary for our analysis. The first continuous order parameter can be now introduced

$$
S_{q_0}(y) = \int dq \exp(iqy) S_{q_0+q}(y) \tag{8}
$$

$S_{q_0}(y)$ describes a one-dimensional spin density wave along one of the two hexagonal axes in the plane $(\vec{a}, \vec{b})$

$$
S(y) = S_{q_0}(y) \exp(iq_0 y) \tag{9}
$$

The second order parameter $S_{q_0'}(x)$ can be defined in an analogous way albeit along the orthogonal axis i.e. along $\vec{a}$. There is an interesting conclusion resulting from our analysis that is that the choice of the directions of the wave vectors is not arbitrary. The incommensurate phase (I) can be obtained along $\vec{b}$ which means that the domain walls should perpendicular to this direction if the commensurate phase (C) is expected to be ferromagnetic i.e. described by the wave vector $(0, 2\pi, 0)$. This is not true, however, if the commensurate phase is "fractional" i.e. its wave vector is $(2\pi/p, 0, 0)$, where p is an integer. The expression for the free energy can be greatly simplified once the commensurate wave vectors are fixed. We arrive at the following effective Landau-Ginzburg density functional [8]

$$
\begin{aligned}
F(\vec{r}) = r/2 \mid S_{q_0} \mid^2 + T/8 \mid S_{q_0} \mid^4 + \\
T/48(S^4_{q_0} + S^4_{-q_0}) + \\
c/2 \mid (\partial/\partial y - ia/2c) S_{q_0}(y) \mid^2 + \\
r'/2 \mid S_{q'_0} \mid^2 + T \mid S_{q'_0} \mid^4 + \\
c'/2 \mid \partial S_{q'_0}/\partial y \mid^2 + \\
T/4(S_{q_0} S_{-q'_0} + S_{-q_0} S_{q'_0})
\end{aligned}
\qquad (10)
$$

Evidently two contributions to the free energy, associated with the two order parameters,respectively, are not equivalent to each other.

A Discussion of the Results

Following Refs 4 and 5, we assume $S_{q_0}(y)$ with its amplitude $A(T)$ essentially constant in space, while the phase, $\phi(y)$, allowed to be space dependent in order to accommodate to the underlying lattice

$$S_{q_0}(y) = A(T)\exp(\pm i\phi(y)) \qquad (11)$$

Neglecting the entropy coupling between the order parameters, we can discuss separately the two contributions to the effective Landau-Ginzburg free energy density. The amplitude $A(T)$ can be found by minimisation of the free energy in the commensurate phase

$$A^2(T) = 3(J(q_0) - T)/T \qquad (12)$$

The contribution to the free energy density per unit area perpendicular to y becomes

$$
\begin{aligned}
F = cA^2(T)(\partial\phi/\partial y - \delta_0)^2 + \\
v(1 + \cos(4\phi)) - 1/2\delta_0^2
\end{aligned}
\qquad (13)
$$

where we denote $v = -TA^2/24c$ after Ref.8. $\delta_0 = a/2c$ indicates the wave vector (relative to the commensurate wave vector q_0), which would be stable in the absence of the umklapp term. The phase function $\phi(y)$ for which the free energy F has an extremum, fulfills the sine-Gordon equation

$$\phi''(y) + 4v\sin(4\phi) = 0 \qquad (14)$$

Solutions of the sine-Gordon equation are static solitons describing domain walls which separate almost commensurate regions. Close to T_c, the incommensurate phase consists of thick overlapping solitons. The ground-state configurations, however, correspond to a soliton lattice of domain walls whose thickness gets narrower as the temperature decreases while the inter-soliton distance L grows [7, 8]. This is also confirmed by the experimental data (see: Fig.2). Very close to the lock-in commensurate phase i.e. as $T \to T_2$, the formula for the first contribution F to the free energy density can be rewritten in an asymptotic form as a linear function of either the inter-soliton distance L or the thermal average $\bar{q}$ of the incommensurate parallel deviation vector, since $\bar{q} = 2\pi/L$ [8]. The critical value of $\bar{q}$ turns out to be proportional to δ_0. From the asymptotic form of the free energy, it becomes relatively easy to determine the temperature of the commensurate-to-incommensurate phase transition

$$T_{CI} = J(q_0) + a^2/c \tag{15}$$

The mean-field value of the critical temperature, corresponding to an anomaly in the temperarure variation of amplitude A(T), is corrected by a factor resulting from the presence of the soliton lattice. The same analysis performed for the second contribution F' to the free energy density provides only one critical temperature $T'_c = T^*$, which is compatible both with the Brillouin scattering observations. Analogously to the first order parameter, the second one can be also defined as an exponential function of the position along the vector $\vec{a}$ i.e. the spin density wave takes the following form

$$S(x, y) = B(T) \exp(iq'y) \exp(iq'_0 x) \tag{16}$$

The amplitude B(T) is found by minimising the free energy. The anomaly in its temperature variation occurs at T'_c which is equal to T^*. The phase transition introduces a commensurate phase which means a long-range ferromagnetic order in each of the $(\vec{a}, \vec{c})$ single layers. The domain walls, which ,in our model, are found to be perpendicular to the axis $\vec{b}$, are constructed from such the layers. The perpendicular deviations (q') from the direction the commensurate wave vector q'_0 give rise to temperature mediated shifts of the magnetic layers inside the domain wall. Considering the entropy-like coupling of the two order parameters we can obtain an additional correction of the temperature of the lock-in phase transition.

References

[1] Wiesner M 2009 *Phase Transitions* **82** 699
[2] Kurzyński M and Halawa M 1986 *Phys. Rev.* B **34** 4846
[3] Kurzyński M 1994 *Phase Transitions* **52** 1-56
[4] Zamkova N G and Zienenko V I 1993 *J. Phys: Cond. Matter* **6** 9043
[5] Luk'yanchuk I, Jorio A and Pimenta M A 1998 *Phys. Rev* B **57** 5086
[6] Elliott R 1961 *Phys. Rev.* **124** 346
[7] Blinc R 1982 *Physica Scripta* T**1** 138
[8] Bak Per and von Boehm J 1980 *Phys. Rev.* B **21** 5297
[9] Bak Per and Emery V J 1976 *Phys. Rev. Lett.* **36** 978
[10] Bak Per and Timonen J 1980 *J. Phys. C: Solid State Phys.* **11** 5297

Journal of Physics: Conference Series **213** (2010) 012035 doi:10.1088/1742-6596/213/1/012035

Theoretical models and EPR study of Cr based diluted magnetic semiconductors

M Kuzma*, I Stefaniuk and M Bester

Institute of Physics, University of Rzeszow, 35-959 Rzeszow, Rejtana 16A, Poland

*kuzma@univ.rzeszow.pl

Abstract. The II-VI diluted magnetic semiconductors (DMS) with chromium are good candidates for room temperature ferromagnetic semiconductors. We review the theoretical models of sp-d and d-d exchange interactions applicable for semiconductors containing transition metal ions. The unique magnetic properties of Cr based DMS originate from the following properties of Cr ion in tetrahedral environment: strong Jahn-Teller distortion the position of the donor level and occupation of the t_{2g} orbital. The magnetic properties of samples $Cd_{1-x}Cr_xTe$, x=0.02 we investigate by electron paramagnetic resonance in the temperature range T=290-380 K. The temperature broadening of the epr line points on a ferromagnetic phase transition in the vicinity of the room temperature.

1. Introduction

The main goal of the material engineering for spintronics is to find a magnetic semiconductor which exhibits electrically tunable ferromagnetism at room temperature. It allows to join logic and storage operations on the same chip [1]. The studies on this subject have a long history as they were started in 1970's when the sp-d exchange interaction in solid solutions of Mn-based II-VI semiconductors was extensively studied. The wide range of experimental as well as theoretical results of such and related investigations were summarised in several reviews (see [2] and references therein). The other types of diluted magnetic semiconductors (DMS) with manganese were also designed, grown and investigated: IV–VI [2] and III-V [3] DMS. The main progress was made since the theoretical paper written by Dietl et al. [4, 5] at al on 5% Mn doped DMS's showing high Curie temperatures. From calculations presented in these papers the semiconductors Si, Ge, AlP, AlAs, GaP, GaAs, InP, InAs, ZnSe, ZnTe with hole concentration 3.5 x 10^{20} attain the Curie temperature in the range 30-100K. Moreover, the room Curie temperature was derived for C, GaN, ZnO. Unfortunately, the extended experimental investigations do not confirm this optimistic predictions until nowadays, and the task is still open.

A promising group of DMS's are Cr-based II-VI [6] IV-VI [2, 7-11] and other DMS's. In a few among them the room temperature ferromagnetism was detected e.g. in Cr:InN [12], Cr:GaN [13], CdCrTe [14], ZnCrTe [15]. This list of room temperature ferromagnetic semiconductors can be completed by Mn-doped InSb, $CdGeAs_2$ and $CdGeP_2$ [16].

In this paper we review the main properties of Cr - DMS's, focusing the study on microscopic models of the exchange interaction in II-VI diluted magnetic semiconductors with transition metal ions. We also present our experimental results on CdCrTe. The significant aim of the paper is to communicate the unusual magnetic property, room temperature ferromagnetism of our samples of

$Cd_{1-x}Cr_xTe$. The result was obtained from EPR measurements and it is in agreement with theoretical predictions outlined in Section 2.

2. Exchange interactions in DMS's

Magnetic properties of DMS's follow from the exchange interactions between spins of the band carriers and localised spins of magnetic ions (sp-d exchange) or between localised spins (d-d exchange). Several microscopic models were used to explain nature of these interactions. Dominated mechanisms in II-VI DMS's are:

1. Direct or potential exchange interaction based on the Coulomb interaction between spins [17]. This model very well describes the s-d interaction.
2. Kinetic exchange interaction involving hybridization effects [17, 18] (is used for p-d coupling).
3. Indirect exchange of the type of superexchange [19] (provides good results for the d-d interaction).

2.1 Model of the s-d interaction

The interaction between conduction s band electron spins and magnetic ion spins resulting from the Coulomb interaction is of the ferromagnetic type. Its value is close to $N_0\alpha \approx 0.2\ eV$ almost for all DMS's. This s-d interaction is described by Kondo-like Hamiltonian [20]

$$H_{ex} = -J(r - R_i)\vec{s} \cdot \vec{S}_i \tag{1}$$

where $\vec{s}$ is a spin operator of the band electron, $\vec{S}_i$ is a spin operator of the magnetic ion which is localized in the crystal lattice point $\vec{R}_i$, $J(r - R_i)$ is the exchange integral.

For electrons from conduction band (s electrons) and from valence band (p electrons) the matrix elements of the exchange integrals are denoted as

$$N_0\alpha = N_0\langle S|J|S\rangle, \quad N_0\beta = N_0\langle X|J|X\rangle \tag{2}$$

respectively.

2.2 Model of the p-d interaction

The experimentally measured interaction between valence charge carriers and localized spins (p-d interaction) is of antifferomagnetic type (negative value of β). The value $N_0\beta$ is six to eight times greater than s-d interaction $N_0\alpha$. Thus the resulting magnetic properties of DMS are antiferromagnetic ones. Such large value of the p-d interaction does not arise from the simple Kondo model. Dietl [18], Bhattacharjee et al. [17] and Larson [19] explained the experimentally observed values and the sign of the p-d interaction for Mn based DMS by the p-d hybridization effects.

The model Hamiltonian of the multisite Anderson Hamiltonian consists of Hamiltonian term H_0 of the s, p energy bands $\varepsilon_n(k)$ of the crystal, Hubbard Hamiltonian H_d of Mn d electrons on site i with linearized electron – electron interaction, the Mn 3d – Te 5p hybridization Hamiltonian H_{pd} (termed as kinetic part of H) and Mn-d – sp-band potential exchange Hamiltonian H_χ [19]:

$$H = H_0 + H_d + H_{pd} + H_\chi \tag{3}$$

In the second order perturbation approach the p-d exchange energy consists of two contributions represented by formulae [19]:

$$N\beta = -32V_{pd}^2\left\{(\varepsilon_d + U_{eff} - E_v)^{-1} + (E_v - \varepsilon_d)^{-1}\right\} \tag{4}$$

where E_v is the valence- band-edge energy, ε_d and U_{eff} are parameters of the Hubbard model: the energy of the d orbital on site i and the energy of an electron-electron interaction respectively (see Fig 1a).

Due to symmetry, the V_{pd} vanishes for s states; therefore the allowed transitions are those between t_{2g} states only. Such virtual transitions for p electrons from the vicinity of the valence band edge are illustrated by diagram in Fig.1a (without changing of k vector) and in Fig 1b (with change of k vector).

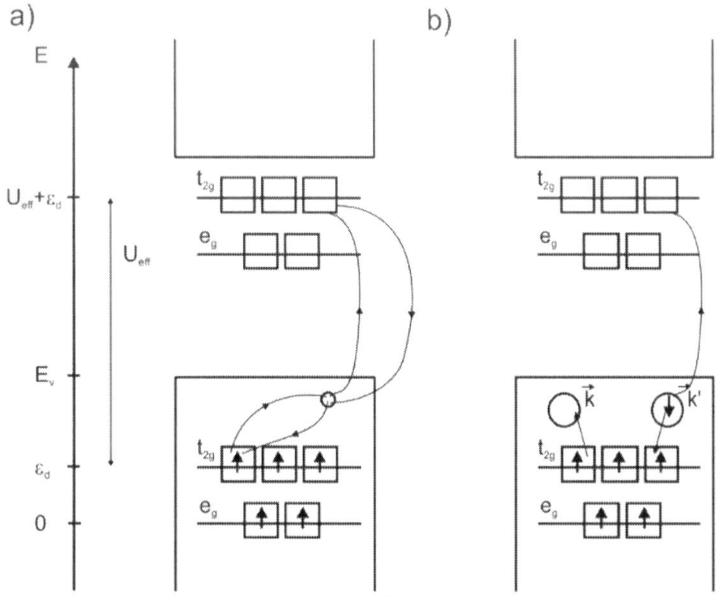

Fig.1. Virtual transitions in a second order p-d interaction between localized d and band p electrons a) transition at the top of the valence band ($\vec{k} = 0$) b) transition for $\vec{k} \neq 0$.

Blinowski and Kacman [21, 22] derived kinetic exchange Hamiltonians of Luttinger Γ8 states interacting with t_{2g} impurity states for whole set of transition metal ions using initial Hamiltonian in the form

$$H = H_i + H_{cr} + H_{hyb} \tag{5}$$

Here H_i is the Parmenter like Hamiltonian [23]. Such Hamiltonian commutes with both orbital $\vec{L}$ and spin $\vec{S}$ momentum operators, therefore the calculations are applicable also for states different from S-state.

$$H_i = \varepsilon_d N + \left(J - \frac{U'}{2}\right)\hat{N} + \left(\frac{1}{2}\right)\left(U' - \frac{1}{2}\right)\hat{N}^2 - J\vec{S}\vec{S} \tag{6}$$

where ε_d is here the energy of d state, $\hat{N}$ is an occupation number operator, U' is nondiagonal Coulomb integral and J is non-diagonal exchange integral, H_{cr} is the Hamiltonian of the band electrons restricted to electrons with energy ε_p in a vicinity of the center of the Brillouine zone.

$$H_{cr} = \varepsilon_p \sum_{v,k} n_{v,k} \tag{7}$$

The hybridization Hamiltonian

$$H_{hyb} = \sum_{nk} \sum_{m\sigma} \left(V_{m\sigma,nk} a^{*}_{m\sigma} a_{nk} + h.c. \right) \tag{8}$$

couples the p electrons with the t_{2g} d electrons only.
For Mn^{2+} the effective kinetic p-d exchange Hamiltonian is [22]

$$H_{ex} = -(1/\Omega) 2 B_5 \vec{S}\vec{J} \tag{9}$$

After thermodynamic averaging over randomly distributed Mn ions, the mean-field Hamiltonian takes the form

$$H_{ex} = -\frac{1}{3} x N_0 \beta_N \langle S_{\|} \rangle J_{\|} \tag{10}$$

where xN_0 is a concentration of the magnetic ions, $J_{\|}$ is the projection of the fictitious angular momentum j=3/2 of the valence band electron onto external magnetic field, β_N is exchange constant:

$$\beta_N = -\frac{V_{pd}^2}{S} \left[\frac{1}{E_{N-1}^{S-1/2} + \varepsilon_p - E_N^S} + \frac{1}{E_{N+1}^{S-1/2} - \varepsilon_p - E_N^S} \right] \tag{11}$$

where V_{pd} is hybridization constant

2.3 Model of the d-d interaction
The direct Heisenberg exchange interaction between localized spins (direct d-d interaction)

$$H_{ij} = -2 J_{ij} \vec{S}_i \vec{S}_j \tag{12}$$

is much smaller than the sp-d interaction.

Larson et al [19] studied the d-d mediated superexchange interaction between two Mn spins in II-VI DMS. It appears that this process is dominated in Mn-Mn interaction. The theory was developed in the frame of the band picture of MnTe and CdMnTe compounds. The Mn-Mn interaction has been described by Heisenberg Hamiltonian

$$H_N = -\sum_{i \neq j}{}' J^{dd}(R_{ij}) \vec{S}_i \vec{S}_j \tag{13}$$

To calculate the exchange integral J^{dd} Larson took into account the effect of the H_{pd} Hamiltonian on an intermediate states in which the occupation of Mn d states is different from five. In the forth-order perturbation formalism the Mn-Mn interaction constant J^{dd} is expressed as [19]:

$$-2\left(\frac{5}{2}\right)J^{dd}(R_{ij}) = \sum_{I_1,I_2,I_3} \frac{\langle f|H_{pd}|I_1\rangle\langle I_1|H_{pd}|I_2\rangle\langle I_2|H_{pd}|I_3\rangle\langle I_3|H_{pd}|i\rangle}{(E_0-E_1)(E_0-E_2)(E_0-E_3)} \quad (14)$$

where $|i>$ and $|f>$ are initial and final states and $|I_i>$ (where $i = 1,2,3$) are intermediate states. The sum of terms in formulae (16) can be decomposed into three terms $J^{dd}{}_{hh}$, $J^{dd}{}_{ee}$, $J^{dd}{}_{he}$ consisting of Mn intermediate states formed by transfer of two electrons respectively from the valence band, from conduction band, as well as one electron from conduction band and one from valence band.

$$J^{dd}(R_{ij}) = J^{dd}_{hh}(R_{ij}) + J^{dd}_{he}(R_{ij}) + J^{dd}_{ee}(R_{ij}) \quad (15)$$

These three mechanisms are presented schematically in Fig 2.

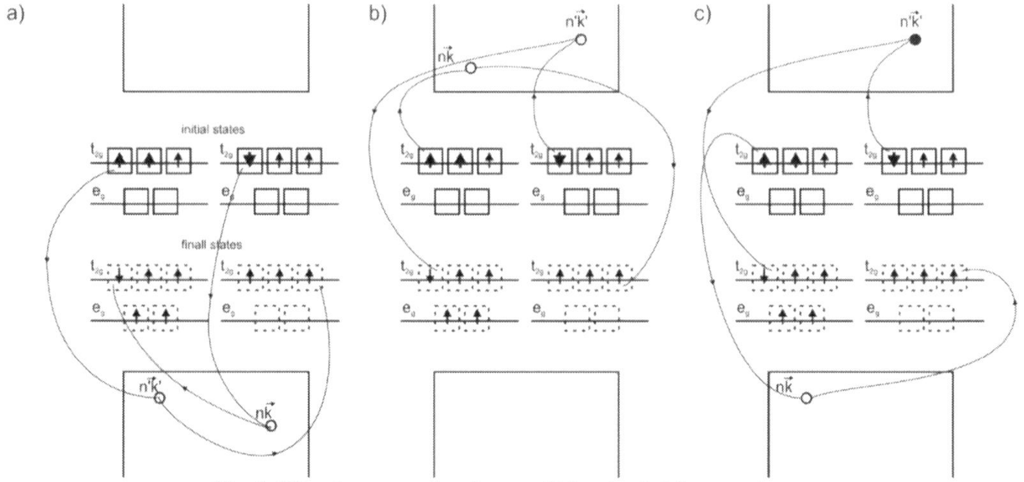

Fig.2. The three mechanisms of kinetic d-d interactions.

For II-VI DMS the dominant contribution follows from the $J^{dd}{}_{hh}$ terms (Fig.2a). They can be calculated as [19]:

$$\begin{aligned} J^{dd}_{hh}(R_{ij}) &= [A,B,C,D]+[C,D,A,B]+[A,C,D,B]+[C,A,D,B]+[C,A,B,D]= \\ &= -2\sum_{k,k'}\sum_{n,n'}\left|\tilde{V}_{pd}(n,k)\right|^2\left|\tilde{V}_{pd}(n',k')\right|^2 \cos\left[(\vec{k}-\vec{k}')\cdot\vec{R}_{ij}\right]\times \\ &\times\left\{U_{eff}^{-1}\left[\varepsilon_n(k)-\varepsilon_d-U_{eff}\right]^{-1}\left[\varepsilon_{n'}(k')-\varepsilon_d-U_{eff}\right]^{-1}-\left[\varepsilon_n(k)-\varepsilon_d-U_{eff}\right]^{-2}\left[\varepsilon_{n'}(k')-\varepsilon_d-U_{eff}\right]^{-1}\right\} \end{aligned} \quad (16)$$

where the symbols [A,B,C,D] and its permutations are sums of terms of Eq. (14) corresponding to the intermediate state sequences showed. Exemplary calculations performed by Larson for real CdMnTe crystals give results being in good agreement with experiments.

2.4 Three level model of the hole mediated superexchange.

Basing on calculations presented above Larson [19] proposed the simplified expression for the superexchange in materials related to CdMnTe as well as to many other DMS

$$J_{hh}^{dd}(R_{ij}) = -2V_{pd}^{4}\left[U_{eff}^{-1}\left(E_v - \varepsilon_d - U_{eff}\right)^{-2} - \left(E_v - \varepsilon_d - U_{eff}\right)^{-3}\right]f(r) \tag{17}$$

where ε_d is energy of occupied d level, $\varepsilon_d + U_{eff}$ is the energy of an unoccupied d level, E_v is energy of p electrons at the top of valence band, V_{pd} is hopping parameter which connects the p level to both d levels, $f(r)$ is the constant describing the influence of R_{ij}/a on J^{dd}_{hh}. This constant should be determined for each material within particular approximation separately.

3. Magnetic properties of Cr-DMS

The sp-d exchange interaction in magnetic semiconductors controls their unusual magnetic properties such as giant magneto – optical effects (e.g. Zeeman splitting, Faraday rotation). The nature of the exchange interaction depends on the magnetic ion and determine the type of long–ranged interactions in material which results in paramagnetism, antiferromagnetism, spin-glass or ferromagnetism depending on the magnetic ion and temperature. Almost all DMS's show antiferromagnetic properties. In the paramagnetic temperature regime the DMS's with Mn, Co are Brillouin paramagnets (these ions have multiplet ground states and orbital momentum L=0) whereas in Fe: DMS (ground state is singlet (L=2, S=2) the Van Vleck–type paramagnetism is observed. The separate group of magnetic semiconductors are those containing chromium. In these semiconductors the sp-d interaction is ferromagnetic and paramagnetic properties are intermediate between Brillouin and Van Vleck type (ground state is semidublet but excited states lie close enough to generate magnetic moment).

The unique magnetic properties of the II-VI DMS based on Cr follow from several features of Cr ion in a tetrahedral environment:

1. Cr ion is strongly affected by Jahn–Teller distortion.
2. The position of chromium donor level t_d is in close vicinity to the top of the valence band.
3. One t_{2g} orbital of Cr^{2+} is not occupied.

The model of p-d interaction proposed by Blinowski and Kacman [22] presented in preceding section (Eq. (9)) is applicable for all transition metal ions. Kacman [24] interpreted the dependence of one electron orbital filling on magnetic properties in that model. The Mn, Fe, Co ions have the same filling (singly with spin up) of t_2 orbitals (see Fig. 3). The exchange interaction between these ions is antiferromagnetic.

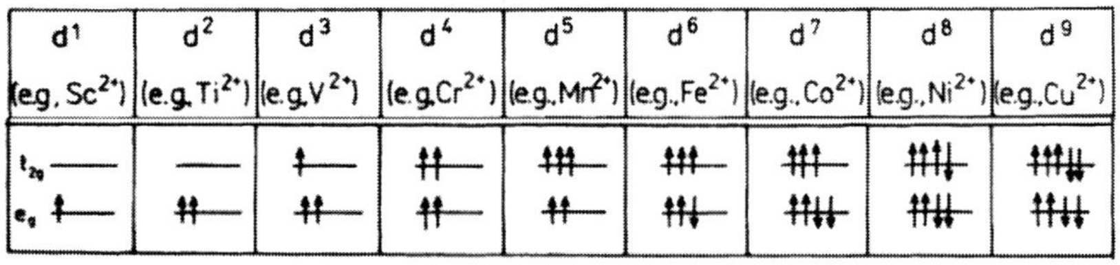

Fig.3 Filling of one-electron orbitals in transition metal ions.

Opposing, the ions Sc, Ti, have the t_{2g} orbitals empty. Therefore the virtual transition are in direction from the band to an ion only, independently on the spin direction. In this case the exchange constant was obtained in the form.

$$\gamma_N = \frac{V_{pd}^2}{\Omega\left(S+\frac{1}{2}\right)} \times \left[\frac{1}{E_{N+1}^{S+1/2} - \varepsilon_p - E_N^S} - \frac{1}{E_{N+1}^{S-1/2} - \varepsilon_p - E_N^S}\right] \tag{18}$$

The second term in (18) is smaller in comparison with the first one, thus the exchange constant is positive giving ferromagnetic properties of the system. The free Cr^{2+} ion has N = 4 electrons on d shell. The ground state is a triplet state T_2. One state among the one electron t_{2g} states is empty. In this case the kinetic exchange Hamiltonian between d electrons and p (Γ_8) band electrons is [22].

$$H_{ex} = \left(\frac{1}{\Omega}\right)\left\{\left[(C_N - B_N)\cdot\vec{J} \pm (C_N + B_N)\cdot\left(\vec{J} + \left[\hat{H}_Q - \vec{L}\cdot\vec{J}, \vec{L} + \vec{J}\right]_+\right)\right]\cdot\vec{S} + D_N\left(\hat{H}_Q - \vec{L}\cdot\vec{J}\right)\right\} \quad (19)$$

The constants B_N, C_N, D_N (corresponding to the constants β_N, γ_N, δ_N in notation of the paper [24]) and the operator H_Q are:

$$B_N = \beta_N = \left(\frac{V^2}{2S}\right)\left[\left(E_N^S - E_{N-1}^{S-1/2} - \varepsilon_p\right)^{-1} + \left(E_N^S + \varepsilon_p - E_{N+1}^{S-1/2}\right)^{-1}\right] \quad (20)$$

$$C_N = -\gamma_N = \frac{V^2}{2S+1}\left[\left(E_N^S + \varepsilon_p - E_{N\pm1}^{S+1/2}\right)^{-1} - \left(E_N^S + \varepsilon_p - E_{N\pm1}^{S-1/2}\right)^{-1}\right] \quad (21)$$

$$D_N = \delta_N = (S+1)C_N + SB_N \quad (22)$$

$$\hat{H}_Q = \left(\vec{L}\cdot\vec{J}\right)^2 + \frac{1}{2}\vec{L}\cdot\vec{J} - \frac{1}{3}l(l+1)j(j+1) \quad (23)$$

The parameters B_N, C_N, D_N depends on Parmenter parameters $\varepsilon_d - \varepsilon_p$, on the energy U' and J as well as on the hybridization constant V. The E_N^S denotes energy of the unperturbed d shell with N electrons and total spin S.

The exchange Hamiltonian H_{ex} (19) depends on the orbital momentum $\vec{L}$ of the ground state of the ion, contrary to the Hamiltonian (14)

$$H_{ex} = -\left(\frac{1}{\Omega}\right)2B_5\vec{S}\vec{J} \quad (24)$$

derived for Mn ions. Therefore the system is sensitive to a perturbation which can lift the degeneracy of the ground state such as Jahn-Teller distortion, external magnetic field, or spin-orbit coupling. In the II-VI DMS with Cr^{2+} ions the tetragonal Jahn-Teller distortion is static and very strong [25, 26]. In this case the exchange Hamiltonian has form [22]:

a) for a distortion in the [001] direction

$$H_{ex} = \frac{1}{6}xN_0\delta_4\left(\frac{5}{4} - J_z^2\right) + \frac{1}{3}xN_0\langle S_z\rangle J_z\left[\beta_4\left(\frac{5}{4} - J_z^2\right) - \gamma_4\left(\frac{9}{4} - J_z^2\right)\right] \quad (25)$$

b) for uniform distortion in three equivalent {001} directions

$$H_{ex} = -\frac{1}{9} x N_0 (2\beta_4 + \gamma_4) \langle S_{\|} \rangle J_{\|} \tag{26}$$

The second Hamiltonian is of the Heisenberg-like form with exchange constant proportional to $(2\beta_4 + \gamma_4)$. Thus the sign of this constant follows from the contributions of both values β_4 and γ_4. However, there are several experiments [27] pointed that the origin of ferromagnetic type of Cr based DMS follows from the position of the donor $Cr^{2+/3+}$ level which is near the top of the valence band as it is schematically drawn in Fig.4. We note in Fig.4 that the donor levels of Cr impurity is almost at the valence band maximum in two cases: of ZnTe and CdTe crystals. Blinowski and Kacman [21, 28] show that, in principle, the sign of β is determined by the energy e_1 of charge transfer from Cr^{2+} ion to the valence band. The β should be positive if this energy is negative. This conclusion is in good agreement with Kanamori model of the energy levels of Cr^{2+} in ZnS, ZnSe, ZnTe [29] as well as with experimental results of Mac et al [27].

Table 1. Values of the p-d exchange interaction constants in Cr based DMS.

	$N_0\alpha$ (eV)	$N_0\beta$ (eV)
$Cd_{1-x}Cr_xS$	-	-
$Cd_{1-x}Cr_xSe$	0.22 ± 0.02 [30]	0.48 ± 0.05 [31]
$Cd_{1-x}Cr_xTe$	-	-
$Zn_{1-x}Cr_xS$	-	0.62 [27]
$Zn_{1-x}Cr_xSe$	-	0.95 [27]
$Zn_{1-x}Cr_xTe$	-	4.25 [27]
$Zn_{1-x}Cr_xSe_{0.95}S_{0.05}$	-	0.93 [27]

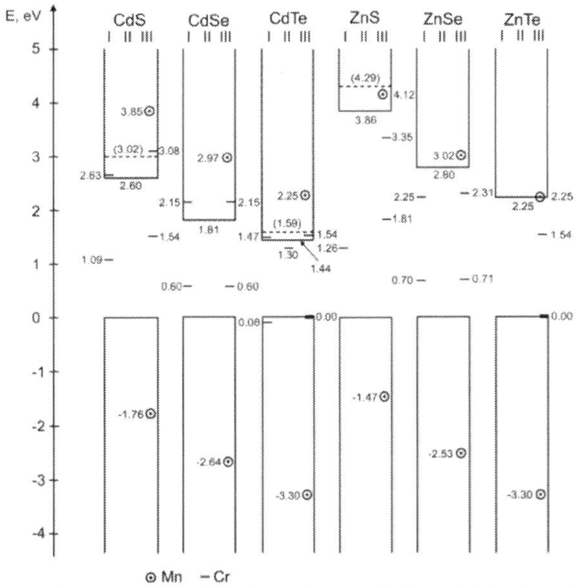

Fig. 4 Approximate energy levels position of Cr and Mn ions in $Cr^{2+/3+}$, $Mn^{2+/3+}$ (lower levels) and $Cr^{1+/2+}$, $Mn^{1+/2+}$ (upper levels) charge states. The diagram is constructed basing on the data given in [32] and [33].

The ferromagnetic p-d exchange interaction of Cr ions in Zn-chalcogenides has been experimentally studied in several papers of Mac, Twardowski at al [27, 31, 34, 35]. The values of p–d exchange constants in such materials are collected in Table 1.

4. EPR measurements in CdCrTe

The measurements were carried out on $Cd_{1-x}Cr_xTe$ sample. The sample was prepared by synthesis of the CdTe and Cr_2Te_3 components with the nominal Cr concentration x = 5%. The liquid high temperature synthesis was carried out in quartz ampoule at temperature 1300^{0}C. The concentration of Cr in the ingot obtained was measured by XRF method and was x = 0.02. The EPR studies were carried out on X-band spectrometer in high temperature regime of 290 -380K.The set of EPR spectra obtained at various temperatures are collected in Fig.5a.

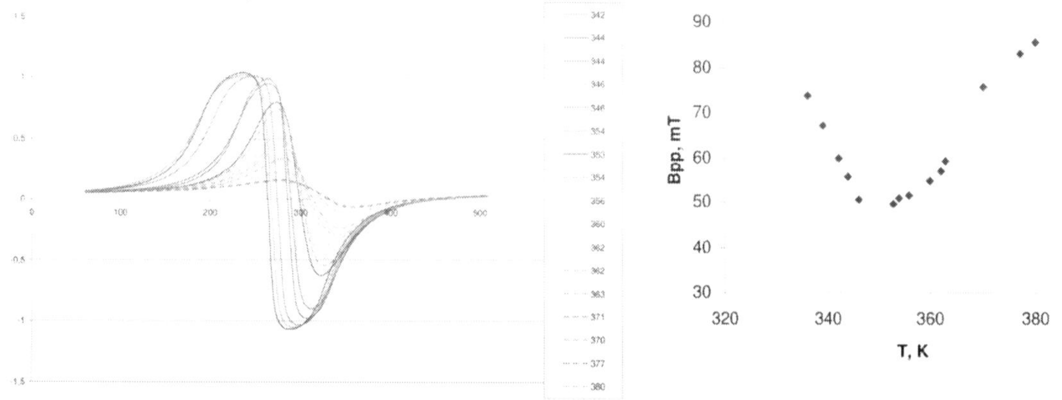

Fig.5. EPR spectra of CdCrTe DMS (a), Temperature dependence of the EPR lines width (b).

The spectrum consists of one broad line with resonance field at 260 – 320 mT depending on temperature. The shape of the line at two different temperatures are presented in Fig. 6a and b. The shape was adjusted to the Lorentz function. The good agreement with this function is visible. The width of the lines was derived from the theoretical adjustment.

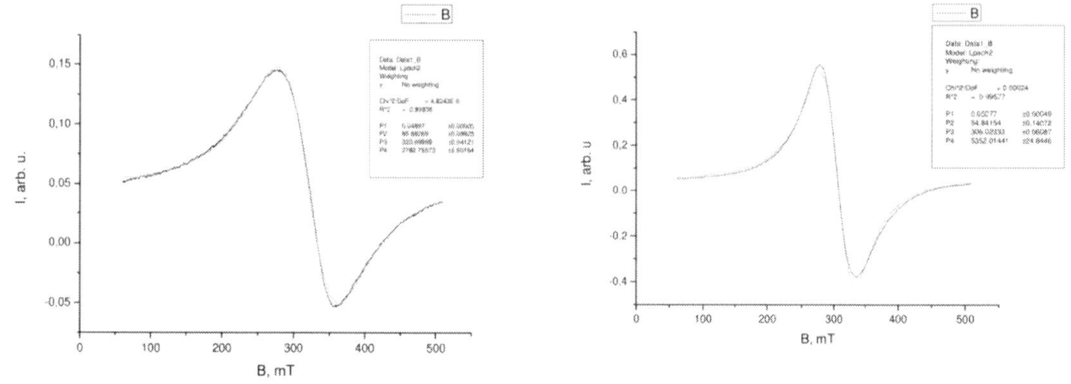

Fig 6 High temperature shape of the resonance lines of CdCrTe approximated with Lorentz function: T= 380K (a), T=360K (b).

The temperature variation of the line width is presented in Fig. 5b. This dependence does not relate to that observed for Cd or Zn based antiferromagnetic DMS with Mn (e.g. $Zn_{1-x}Mn_xS$, $Zn_{1-x}Mn_xSe$, $Zn_{1-x}Mn_xTe$, $Cd_{1-x}Mn_xS$, $Cd_{1-x}Mn_xSe$, $Cd_{1-x}Mn_xTe$, [36]). For samples with Mn there is a dramatic broadening of the resonance at low temperatures and then the resonance narrows continuously as the temperature increases. Such a behaviour is typical for antiferromagnet DMS. In our sample we note (in Fig.5b) that the linewidth decreases starting from high temperature until temperature T_{min}= 320K at which the linewidth takes minimal value. Below this temperature it increases with temperature decreasing. Such a behaviour was observed in studies of CMR manganites and is not fully understood. Huber et al [37] relates the EPR linewidth in $LaMnO_3$ to the interplay of crystal field anisotropy, Dzialozhinsky-Mariya (antisymmetric) exchange and anisotropic superexchange. Moreover, the same type of the temperature dependence of the EPR linewidth was observed by Story at al [38] in $Pb_{1-x}Sn_yMn_x$ Te in the regime of high carrier concentration exceeding the value of threshold (see Fig.8 in [38]). However, in this paper this effect was attributed to the Korringa relaxation mechanism, where linear temperature dependence is expected in the high temperature regime. This mechanism was also addressed to interpretation of results of HgFeSe DMS [39].

In our material investigated the resonance field decreases strongly at temperatures T< 300 K, the linewidth becomes large and the shape of lines is asymmetric considerably. Therefore at low temperatures the magnetic resonance of the material studied is addressed as ferromagnetic resonance (FMR). The properties of CdCrTe studies by FMR in low temperature regime 77-300K were presented in our previous paper [40].

5. Conclusions

Chromium based diluted magnetic semiconductors have been theoretically many years ago to reveal exceptional magnetic properties [28]. The theoretical models of the exchange interactions in DMS show that the unusual magnetic properties of $Cd_{1-x}Cr_xTe$ arises from the strong Jahn-Teller distortion, from the position of donor Cr^{2+} level, as well as from half occupation of t_{2g} orbital of Cr^{2+}. These properties should lead to room temperature ferromagnetism in such materials. In the present paper the EPR measurements were carried out for $Cr_xCd_{1-x}Te$ crystal with a high concentration of Cr ions (x=0.02). Results presented in the paper, seem to correlate with the theoretical expectation. The shape of resonance line is Lorentzian. The temperature dependence of linewidth in high temperature regime (paramagnetic regime) differ considerably from that observed for Mn DMS. We observe broadening of the line with temperature instead the constant behaviour. Obtained minimum in the temperature dependence of the linewidth relates to the Curie temperature as in other ferromagnetic semiconductors. This temperature in our material is near room temperature. But the results do not explain which interaction is responsible for the observed phenomena. Further supplementary studies are necessary.

References

[1] Ohno H, Chiba D, Matsukora F, Omiya T, Abe E, Dietl T, Ohno Y and Ohtani K 2000 *Nature* **408** 944
[2] Dobrowolski W, Kossut J and Story T 2003 *Handbook of Magnetic Materials* vol. 15, ed K H Bushow Amsterdam: Elsevier Science
[3] MacDonald A H, Schiffer P and Samarth N 2005 *Nature Materials* **4** 195
[4] Dietl T, Ohno H, Matsukura F, Cibert J and Ferrand D 2000 *Science* **30** 1019
[5] Dietl T, Ohno H and Matsukura F 2001 *Phys. Rev. B* **63** 195205
[6] Blinowski J, Kacman P and Majewski J A 1996 *Phys. Rev.* **53** 9524
[7] Wilamowski Z and Story T 1995 *Journal of Magnetism and Magnetic Materials* **140-144** 1731
[8] Dobrowolski W, Arciszewska M and Brodowska M 2006 *Sci. of Sintering* **38** 109
[9] Grodzicka E, Dobrowolski W, Story T, Wilamowski Z and Witkowska G 1996 *Cryst. Res. Technol.* **31** 651
[10] Story T, Wilamowski Z, Grodzicka E and Dobrowolski W 1995 *Acta Phys. Pol.* **87** 229
[11] Grodzicka E, Dobrowolski W, Kossut J, Story T and Witkowska B 1994 *J. of Crystal Growth* **138** 1034
[12] Ney A, Rajaram R, Kammermeier T, Ney V, Dhar S, Ploog K H and Parkin S S 2008 *J. Phys. Condens. Matter.* **20** 20

[13] Cui Y X, Medvedeva J E, Delley B and Freeman A J 2005 *Phys. Rev. Lett.* **95** 256404.1
[14] Ko K Y and Blamire M G 2006 *Appl. Phys. Lett.* **88** 1721101
[15] Saito H, Zayets V, Yamagata S and Ando K 2003 *Phys. Rev. Lett.* **116** 207202
[16] Story T, Galazka R R, Frankel R B and Wolf P A 1986 *Phys. Rev. Lett.* **56** 777
[17] Bhattacharjee A K, Fishman G and Coqblin B 1983 *Physica* **449** 117
[18] Dietl T 1981 *Physics in High Magnetic Field*, vol. 24 ed M Miura Berlin: Springer
[19] Larson B E, Haas K C, Ehrenreich H and Carlsson A E 1988 *Phys. Rev. B* **37** 4137
[20] Kossut J 1976 *Phys. Stat. Sol. (b)* **78** 537
[21] Blinowski J, Dietl T and Kacman P 1992 *Acta Phys. Pol.* **82** 641
[22] Blinowski J and Kacman P 1992 *Phys. Rev. B* **46** 12298
[23] Parmenter R H 1973 *Phys. Rev. B* **8** 1273
[24] Kacman P 2001 *Semicond. Sci. Technol.* **16** R25
[25] Vallin J T, Slack G A and Roberts S 1970 *Phys. Rev. B* **2** 4313
[26] Vallin J T and Watkins D G 1974 *Phys. Rev.* **B9** 2051
[27] Mac W, Twardowski A and Demianiuk M 1996 *Phys. Rev. B* **54** 5528
[28] Blinowski J, Kacman P and Majewski J A 1995 *Acta Phys. Pol.* **88** 683
[29] Kanamori J 1963 *Progr. Theor. Phys.* **30** 275
[30] Twardowski A, Heiman D, Liu M T, Shapira Y and Demianiuk M 1996 *Phys. Rev. B* **53** 10728
[31] Herbich M, Mac W, Twardowski A, Ando K, Shapira Y and Demianiuk M 1998 *Phys. Rev. B* **58** 1912
[32] Langer J M and Heinrich H 1985 *Pys. Rev. Lett.* **55** 1414
[33] Dietl T 1994 *Handbook on Semiconductors* vol. 3, ed T S Moss Amsterdam: Elsevier Science
[34] Mac W, Khoi Nguyen The, Twardowski A and Gaj J A 1993 *Phys. Rev. Lett.* **1a** 2327
[35] Mac W, Twardowski A, Eggenkamp P J, Swagten H J, Shapira Y and Demianiuk M 1994 *Phys. Rev. B* **50** 14144
[36] Kremer R E and Furdyna J K 1985 *Phys. Rev. B* **31** 1
[37] Huber D L, Alejandro G, Caneiro A, Causa M T, Prado F, Tovar M and Oseroff S B 1999 *Phys. Rev.* **60** 12155
[38] Story T, Eggenkamp P J, Swuste C H, Swagten H J and de Jonge W J 1993 *Phys. Rev. B* **47** 227
[39] Wilamowski Z, Mycielski A, Jantsch W and Hendorfer G 1988 *Phys. Rev. B* **38** 3621
[40] Stefaniuk I, Bester M and Kuzma M 2008 *J. of Physics: Conference Series* **104** 012010

AUTHOR INDEX

Institute of Physics
Dirac House, Temple Back
Bristol BS1 6BE UK